IUTAM SYMPOSIUM ON MICROSTRUCTURE-PROPERTY
INTERACTIONS IN COMPOSITE MATERIALS

SOLID MECHANICS AND ITS APPLICATIONS
Volume 37

Series Editor: **G.M.L. GLADWELL**
Solid Mechanics Division, Faculty of Engineering
University of Waterloo
Waterloo, Ontario, Canada N2L 3G1

Aims and Scope of the Series

The fundamental questions arising in mechanics are: *Why?, How?,* and *How much?* The aim of this series is to provide lucid accounts written by authoritative researchers giving vision and insight in answering these questions on the subject of mechanics as it relates to solids.

The scope of the series covers the entire spectrum of solid mechanics. Thus it includes the foundation of mechanics; variational formulations; computational mechanics; statics, kinematics and dynamics of rigid and elastic bodies; vibrations of solids and structures; dynamical systems and chaos; the theories of elasticity, plasticity and viscoelasticity; composite materials; rods, beams, shells and membranes; structural control and stability; soils, rocks and geomechanics; fracture; tribology; experimental mechanics; biomechanics and machine design.

The median level of presentation is the first year graduate student. Some texts are monographs defining the current state of the field; others are accessible to final year undergraduates; but essentially the emphasis is on readability and clarity.

For a list of related mechanics titles, see final pages.

IUTAM Symposium on

Microstructure-Property Interactions in Composite Materials

Proceedings of the IUTAM
Symposium held in Aalborg, Denmark,
22–25 August 1994

Edited by

R. PYRZ

Institute of Mechanical Engineering,
Aalborg University, Aalborg, Denmark

SPRINGER-SCIENCE+BUSINESS MEDIA, B.V.

Library of Congress Cataloging-in-Publication Data

IUTAM Symposium on Microstructure-Property Interactions in Composite
   Materials (1994 : Rebild Bakker Conference Centre, Denmark)
   IUTAM Symposium on Microstructure-Property Interactions in
   Composite Materials / edited by R. Pyrz.
         p.   cm. -- (Solid mechanics and its applications ; v. 37)
   Includes index.
   ISBN 978-94-010-4031-0   ISBN 978-94-011-0059-5 (eBook)
   DOI 10.1007/978-94-011-0059-5
   1. Composite materials--Mechanical properties--Congresses.
2. Microstructure--Congresses. 3. Micromechanics--Congresses.
I. Pyrz, R.   II. International Union of Theoretical and Applied
Mechanics.   III. Title.   IV. Series.
TA418.9.C6I943   1994
620.1'1892--dc20   95-7863

ISBN 978-94-010-4031-0

Printed on acid-free paper

CONTENTS

PREFACE

The IUTAM Symposium on Microstructure Property Interactions in Composite Materials was held during the dates 22nd to 25th August 1994 in Rebild Bakker Conference Centre, situated in the heart of one of Denmark's most beautiful natural areas.

Participation in the Symposium was reserved for invited participants, suggested by members of the Scientific Committee. The cooperation with the Scientific Committee is highly appreciated.

The Symposium brought together 76 researchers from 15 countries representing a broad range of backgrounds relevant to the topic of the meeting. The participants represented the disciplines of materials science and engineering, applied mechanics, applied mathematics and scientific computations. The Symposium comprehensively addressed the analytical, numerical and experimental methods that provide an estimation of the overall, effective properties from microstructural data. The 41 contributions emphasized the significance of the microstructure morphology in understanding the nature and origin of a multitude of properties such as viscoelasticity, plasticity, strength and fracture for a variety of polymer, metal and ceramic based composite materials. Specifically, the Symposium examined and reviewed the current state of the art of micromechanical modelling, experimental investigations and morphological quantification of composite materials' microstructure.

The volume contains 35 papers published in an alphabetic order after the name of the first author. Much to regret of the Scientific Committee some manuscripts were not submitted.

The financial support of the IUTAM, the Obels Family Foundation and the Institute of Mechanical Engineering, Aalborg University, is gratefully acknowledged.

Finally, I would like to express my appreciation to the members of the Local Organizing Committee for their help which cannot be overestimated.

Aalborg, November 1994

Ryszard Pyrz

IUTAM Symposium on
Microstructure-Property Interactions in Composite Materials
August 23 - 25, 1994

Institute of Mechanical Engineering, Aalborg University, Denmark

IUTAM Symposium on
Microstructure-Property Interactions in Composite Materials
August 23 - 25, 1994

Scientific Committee:

E.C. Aifantis,	Aristotle University of Thessaloniki, Greece and Michigan Technological University, U.S.A.
B.A. Boley,	Columbia University, U.S.A. (ex officio)
J.L. Chermant,	ISMRA - Université, France
G.J. Dvorak,	Rensselaer Polytechnic Institute, U.S.A.
K. Friedrich,	University of Kaiserslautern, Germany
Z. Hashin,	Tel Aviv University, Israel
K. P. Herrmann,	University of Paderborn, Germany
Z. Mróz,	Polish Academy of Sciences, Poland
A. Needleman,	Brown University, U.S.A.
S. Murakami,	Nagoya University, Japan
S. Nemat-Nasser,	University of California, San Diego, U.S.A.
M.R. Piggott,	University of Toronto, Canada
R. Pyrz,	Aalborg University, Denmark *(Chairman)*
K.L. Reifsnider,	Virginia Polytechnic Institute and State University, U.S.A.
V. Tamužs,	Latvian Academy of Science, Latvia
V. Tvergaard,	Technical University of Denmark, Denmark
J. R . Willis,	University of Cambridge, United Kingdom

Local Organizing Committee:
M. S. Axelsen
P. S. From
E. S. Knudsen
R. Pyrz
K. H. Winter
L. Kolmorgen *(Secretary)*

Sponsors:
International Union of Theoretical and Applied Mechanics
Det Obelske Familiefond
Institute of Mechanical Engineering, Aalborg University

Institute of Mechanical Engineering, Aalborg University, Denmark

COMMENTS ON A VARIATIONAL MICRO-MACRO MODEL FOR RANDOM COMPOSITES AND THE INTEGRATION OF MICROSTRUCTURAL DATA

M. ARMINJON, A. BOTTERO, B. GUESSAB, S. TURGEMAN
Laboratoire "Sols, Solides, Structures", Université de Grenoble & CNRS
B.P. 53X, 38041 GRENOBLE cedex, France

1. Introduction

Any model for homogenization-localization tries to establish, in a "macro-homogeneous" situation, a correspondence between macro-fields and micro-fields of stimulus and response (e.g. strain and stress, or pressure gradient and filtration velocity, etc.). More exactly, from the data of the asymptotically unique value of the macro-stimulus S, a micro-field s, depending on the micro-position x, is first deduced, using the local constitutive equation binding $s(x)$ to the response $r(x)$:

$$r(x) = f(s(x), x). \tag{1}$$

This essential localization step is generally envisaged as the solution of a boundary value problem for a partial differential equation (PDE), and a first difficulty is to specify which is the domain whose boundary is considered and which boundary values should be relevant. For one does not wish to schematize directly the real physical situation, in which some external input, such as a surface traction, acts at the boundary of the piece of material: It leads only occasionnally to macro-homogeneous fields, and this only in some (central) part of the whole piece. Whereas one seeks to study in detail what happens in a such macro-homogeneous part. In a such part, "equivalent macro-elements are constrained by one another, not by the apparatus" (Hill 1984), thus the real surface tractions (say) at the boundary of one such macro-element are inherently unknown. Due to the asymptotic nature of the notions of macro-homogeneity and statistical homogeneity, the relevant ideal situation is that of an infinite medium in which a macro-homogeneous situation prevails, although the micro-fields fluctuate randomly in the whole space. Considering a PDE in an infinite domain often leads, however, to mathematical and numerical problems, e.g. as to the definition of convolution integrals (this is well-known in the case of Poisson's equation). Moreover, due to the random fluctuation of the micro-fields, the boundary conditions remain undetermined, i.e. one does not have limit conditions at infinity. One may only hope that uniform conditions at infinity, as they are postulated e.g. in inclusions problems, are relevant to this situation. It was proved by Sab (1992) that, for an ergodic random elastic material, the uniform conditions for stress (at the boundary of a cube with side R), and the uniform conditions

1

R. Pyrz (ed.), IUTAM Symposium on Microstructure-Property Interactions in Composite Materials, 1–14.
© 1995 Kluwer Academic Publishers.

for strain, give as $R \to \infty$ the same *macroscopic* behaviour [this asymptotic equivalence was conjectured, though not in a closed mathematical form, by Suquet (1982)]. Yet this result does not seem to guarantee that uniform conditions at infinity are correct conditions to determine the micro-fields in the infinite medium. This is all the more so for non-linear behaviour, for which the sensitivity to the boundary conditions may be very strong, so that one might find a chaotic behaviour, like a turbulent flow.

A second difficulty is this: the microscopic constitutive equation [Eq. (1)] is not really known, certainly not as a determined function of the local position **x** (save in exceptional cases). Actually one has to assume a phenomenological form, in which the inhomogeneous local behaviour appears directly as a dependence, not on **x**, but instead on a set **X** of internal variables (here collectively designed under the short name "*state*"): crystal orientation, hardening parameters, etc. These variables *may include parameters of the local geometry*- e.g. those defining the size (cf. the Hall-Petch relation) and the shape of any grain containing a point **x** where the state is assumed to have a given value[1]. And what may be measured, or rather estimated, by microscopic observations, is generally not the dependence of **X** on **x**, but instead a set of statistical functions that characterize indirectly this dependence **X**=**F**(**x**). Thus if one seeks to enter an experimental characterization of the microscopic structure into whatever micro-macro model, so as to compare its predictions with experimental findings (regarding either the macroscopic behaviour or the micro-fields), the following occurs. One has to reformulate the model so that its algorithm for homogenization-localization can be expressed in terms of the state variable **X** instead of the local position. For example, if a polycrystal simulation is to be done by using a self-consistent model, it turns out to be possible to pass from a general integral formulation, based on the relevant Green tensor, to an interaction formula relating the average values of stress and strain in a given crystal orientation to the macro-averages (Molinari et al. 1987, Lipinski & Berveiller 1989). Of course, any such reformulation involves some closure assumptions; but only in that way can one take into account the existing microstructural information (e.g. the texture function). Note, however, that even for the case where the reformulation is possible, it does not solve the first problem, that of the appropriate domain and boundary conditions.

It seems better not to stake all on formulations of the homogenization-localization problem as a boundary value problem for a PDE, when (i) the boundary conditions and the local behaviour are not known in the desired form, and (ii) the solution may depend sensitively on the boundary conditions. Thus it has been determined which general statistical conditions must be fulfilled by the medium itself (i.e. by the spatial distribution of the states) and by the micro-fields, in order that it just *make sense* to formulate, as explained hereabove, a micro-macro algorithm in terms of the state variable (Arminjon 1991a). It has also been examined the extent to which the solution of the localization problem in terms of the state variable **X** may provide a physically acceptable solution in terms of the position **x** in physical space. This was obtained (Arminjon 1991a) in combining the solution of the "compatibility problem" for deformed aggregates (Arminjon 1991b) with the statistical theory of the distribution of the states. We

---

[1] Of course, if one includes geometrical parameters into the definition of the state **X**, then the latter will be a piecewise constant function of the position **x**, i.e. one adopts the scheme of an aggregate.

emphasize that the latter theory considers a deterministic medium, i.e. a non-random function $\mathbf{X}=\mathbf{F}(\mathbf{x})$.

In this paper, a more detailed comparison of this new statistical theory with the classical approach of random media is given. The necessity to supplement Hill's (1967,1984) macro-homogeneity conditions by additional statistical assumptions, in order to justify the reformulation of micro-macro problems in terms of the state $\mathbf{X}$, is illustrated. The algorithm of the proposed variational model is briefly recalled. Its basic assumptions: (i) assumptions on the dependence of the inhomogeneity parameter r_0 on the macro-stimulus $\mathbf{S}$, (ii) the principle of minimal inhomogeneity (Arminjon et al. 1994), are examined in more detail than before. Lastly, the integration of microstructural data into micro-macro models, as proposed by Arminjon et al. (1993,1994), is recalled, justified theoretically and experimentally checked for the case of fiber-reinforced mortars.

2. Statistical Homogeneity: Deterministic vs. Ergodic-Random Definition

2.1 COMMENTS ON THE CLASSICAL (ERGODIC-RANDOM) DEFINITION

The notion of statistical homogeneity is related to invariance by translation in some statistical sense. There already exists a general frame for discussing this and other statistical aspects of micro-macro models: this is the theory of *ergodic random media* [e.g. Beran (1968), Kröner (1986); see also Sab (1992)]. In this theory, all micro-fields, including the internal variables (thus the state $\mathbf{X}$) depend on the micro-position $\mathbf{x}$ *and* on the realization $\omega \in \Omega$, where the set of possible realizations, Ω, is assumed to be equipped with a probability measure P. Physically, one may think of the realizations ω as of different samples of the inhomogeneous medium (Kröner 1986). The law P *is not specified physically*, only the so-called "spatial laws" may be physically defined for any random field $Z(\mathbf{x},\omega)$, using the notion of the ensemble average. The ensemble average $<<Z>>$ of some random variable Z, i.e. of some function defined on the set Ω (e.g. $Z(\omega)$ might be the maximum stress in the realization ω), is the limit

$$<<Z>> = \lim_{N \to \infty} \frac{1}{N} \sum_{i=1}^{N} Z(\omega_i). \tag{2}$$

The existence of this limit and the fact that it does not depend on the sequence (ω_i) [for "P-almost every sequence (ω_i)"] are a consequence, within the assumed existence of the law P, of the "strong law of large numbers" [e.g. Guichardet (1969)]. From the physical point of view, it is rather the reverse: we do not know the law P, but we might check whether the arithmetic average in Eq. (2) does not fluctuate too much, provided we take enough samples. Then one defines the spatial laws, e.g. for the random field $\mathbf{X}$. A law of first-order (for $\mathbf{X}$) is, at given $\mathbf{x}$, the probability law

$$p_{\mathbf{x}}(A) = << \phi_{\mathbf{x}A} >> , \quad \phi_{\mathbf{x}A}(\omega) = \begin{cases} 1 \text{ if } \mathbf{X}(\mathbf{x},\omega) \in A , \\ 0 \text{ otherwise} \end{cases} \tag{3}$$

where A is any [measurable] subset in the set of the values of the considered field, here the space of states E. A law of order 2, $p_{x_1 x_2}$, is defined in a similar way [take A in E^2 and check whether $(\mathbf{X}(\mathbf{x}_1,\omega), \mathbf{X}(\mathbf{x}_2,\omega))$ is in A], and so on. The *statistical homogeneity* of a random field, e.g. the state $\mathbf{X}$, is defined as its stationarity, i.e. all the spatial laws of $\mathbf{X}$ are assumed to be translation invariant $(p_{\mathbf{x+h}}(A) = p_{\mathbf{x}}(A)$, etc.). For any stationary field $Z(\mathbf{x},\omega)$, the ensemble average $<<Z(\mathbf{x})>>$ (an average over a sequence of different samples ω, taken at the same place $\mathbf{x}$ in any sample), is independent of $\mathbf{x}$. Then a stationary field Z is said to be *ergodic* if the *asymptotic* volume average[2], $\overline{Z}$, (i) is well-defined and independent of the realization ω (for almost any ω) *and* (ii) is equal to the ensemble average $<<Z>>$.

This is an interesting theory. The reasons we find to formulate and to use a different theory are the following: (i) It is rather difficult and complex, appealing to quite advanced domains of probability theory; this may be the reason why definite statistical notions are not often used in the literature on mechanical micro-macro models. (ii) It may hardly be said that the notion of statistical homogeneity that emerges from this theory is an operational notion: the reason, developed by Matheron (1989), is that in many relevant physical situations, not only do we not have an infinity of realizations of the inhomogeneous material, but actually we only have one [3]. Hence, we cannot really check whether our material is stationary, and we even less can check whether it is ergodic; even if we would have enough realizations, we could not assess the *extent* to which our material is ergodic, because the definitions are too involved. (iii) Actually, all what we need for our purpose is the volume average. We thus have no reason to introduce an ensemble average and try to equate this to the asymptotic volume average $\overline{Z}$.

2.2 COMPARISON WITH THE PROPOSED DETERMINISTIC DEFINITION

The deterministic approach has been presented in some detail and in a mathematically rigorous form by Arminjon (1991a), so here we give only a sketch and some new comments. For any domain D of finite volume in the medium, we define the spatial distributions probabilities of the state $\mathbf{X}$ in D, of any order, *by using only the volume measure V*. Thus the law of order 1 is given by

$$P_D(A) = P_D\{\mathbf{X} \in A\} = \frac{1}{V(D)}V\{\mathbf{x} \in D; \mathbf{F}(\mathbf{x}) = \mathbf{X} \in A\}, \quad A \subset E. \tag{4}$$

The laws of order 2 are defined in a similar way:

---

[2] I.e. the limit volume average in larger and larger domains (cubes, say), reached independently of their position [Eq. (7) below]; thus the theory needs, strictly speaking, an infinite number of infinite media.
[3] We may consider different samples (subdomains) in a unique material, but they are different only in so far as they are ostensibly finite; thus if we take larger and larger samples so as to examine the "asymptotic" average, we really take the whole material so that there is indeed only one realization.

$$P_{\mathbf{h}D}(A_2) = \frac{1}{V(D)} V\{\mathbf{x} \in D; (\mathbf{F}(\mathbf{x}), \mathbf{F}(\mathbf{x}+\mathbf{h})) \in A_2\}, \quad A_2 \subset E^2, \tag{5}$$

and the like for n>2. A domain (sample) D is said to be ε-representative (for the laws of order 1, say) if, for any sufficiently large cubic sample D' in the material, some definite measure of the difference between the laws P_D and $P_{D'}$ [4] is not more than ε. And the *material*, that is, the distribution of the states $\mathbf{X}$, is said to be *statistically homogeneous* (S.H.) if ε-representative samples may be found for any ε, however small it is. This implies that the laws $P_{D'}$ tend towards a limit probability law P as the size of the cubic domain D' increases, independently of the position of D'. Note that the existence of ε-representative samples is something that really *can be checked experimentally*: e.g. in a polycrystal we can measure the orientation density function for different samples and we can check whether the difference between the densities, averaged over the orientation space, is negligible for couples of large enough samples. Also note that, due to the definition, large samples must be representative but that, conversely, representative samples are not assumed large: in a *periodic* medium, the elementary cell is ε-representative for *any* $\varepsilon >0$. The existence of a unique limit law P allows to define a notion of average for any (P-integrable) function ϕ of the state:

$$<\phi> = \int_E \phi(\mathbf{X}) dP(\mathbf{X}) = \int_E \phi(\mathbf{X}) f(\mathbf{X}) d\mathbf{X} \tag{6}$$

(the last equality assumes that the law P has a density f, as will usually be the case).

In order to express the relevant fields (stress, etc.) as function of the local state instead of the local position, one must give a precise form to the vague idea that "on an average, the local value of the field depends only on $\mathbf{X}$ ". Essentially, one defines, for a given field (of stimulus, say) $\mathbf{s}(\mathbf{x})$ and for any given sample D, a function σ^D of the state $\mathbf{X}$, by taking the volume average of the values $\mathbf{s}(\mathbf{x})$ at those positions $\mathbf{x}$ *in D* where the state is $\mathbf{X}$ (e.g. *one defines the average strain in those grains of the sample that have a given crystal orientation* $\mathbf{g}$). One computes the average δ of $\|\sigma^D(\mathbf{X}) - \sigma^{D'}(\mathbf{X})\|$ for a couple of samples D and D' (with $\| \ \|$ the norm defined for stimulus and response tensors). If for any couple of large enough samples, δ is small enough, then there exists a limit function σ, so one can speak of the average value $\sigma(\mathbf{X})$ of the field $\mathbf{s}$ at those points of the microstructure where the state is $\mathbf{X}$, without specifying the sample which was considered to compute these average values. The *field* $\mathbf{s}$ is then said to be S.H.. Note that in our definitions, not only the notion of an S.H. *material*, but also the notion of an S.H. *field* depend on what has been defined as the local state $\mathbf{X}$. This is important, because in practice one will consider different definitions of $\mathbf{X}$ for the same physical material, allowing to take into account more and more information on the microscopic behaviour and micro-geometry. Now we have the result that for any bounded S.H. field $\mathbf{s}$, the

---

[4] This measure is simply the average difference between the densities f_D and $f_{D'}$ of the laws P_D and $P_{D'}$, in the case of a "continuum" i.e. when the density exists for any sample, but it is a bit more technical for the case of an aggregate (Arminjon 1991a).

asymptotic average is well-defined and is equal to the average (6) of the corresponding function of the state, σ:

$$< \sigma > = \bar{s}, \quad \bar{s} \equiv \lim_{R(D)\to\infty} \frac{1}{V(D)} \int_D s \, dV \tag{7}$$

(the limit is reached e.g. for cubes D with side $R(D)$, independently of their position). We emphasize that Eq. (7) is totally different from the equality between ensemble average $<<s>>$ and (asymptotic) volume average $\bar{s}$ in the theory of random media: $\bar{s}$ has indeed the same definition in both theories (although in the theory of random media $\bar{s}$ a priori depends on the realization ω), but $< \sigma >$ is very different from an ensemble average (the latter has no meaning in the proposed theory since we have only one realization[5]). Moreover, we obtain here Eq. (7) for any S.H. field whereas, in the classical theory, $<<s>> = \bar{s}$ is true only for an S.H. (i.e. stationary) *and ergodic* field.

Thus, *if* the stimulus and response micro-fields $s(x)$ (e.g. strain) and $r(x)$ (e.g. stress) are S.H. in the proposed sense, we *may ask whether* the corresponding state-averaged values $\sigma(X)$ and $\rho(X)$ can be related together by a constitutive equation:

$$\rho(X) = \phi(\sigma(X), X). \tag{8}$$

The existence of a such equation is tacitly assumed in the operation of numerous micro-macro models for materials with randomly distributed inhomogeneity, such as the self-consistent models (e.g. Molinari et al. 1987, Lipinski & Berveiller 1989), the simple models of uniform strain (Voigt model, referred to as "Taylor model" in plasticity) or uniform stress (Reuss model), the "relaxed" Taylor model (e.g. Van Houtte 1984) and the variational model proposed at first for polycrystals by Arminjon (1987). But since $\sigma(X)$ and $\rho(X)$ are averaged values of inhomogeneous micro-fields $s(x)$ and $r(x)$, we know that these averages can be bound together via a constitutive equation only if these micro-fields fluctuate reasonably around their respective averages, in the sense precised by Hill (1967) for the case where s and r are indeed a strain and a stress field. Thus, at least for the particular case where the local domains D_X with given state X are well-identified constituents, like grains in a polycrystal, Hill's analysis applies here really as well as for the macroscopic average envisaged by Hill. In particular, the fields s and r should have no "correlation" in the domain D_X. Since this should apply simultaneously to any domain D_X for all values of X (e.g. in every grain, for the case that each can be characterized by its orientation), this would be a rather severe condition (Arminjon 1991b). However, one may content oneself with the weaker condition that the volume average, for the different states X present in a cubic sample D, of the deviation to the no-correlation of s and r in D_X, tends towards zero as $R(D)\to\infty$. This condition, if it is fulfilled by two fields s and r that are S.H. and also macro-homogeneous in the sense of Hill, implies that the corresponding functions of X verify the "transported no-correlation condition" (in which : means the scalar product),

---

[5] Strictly speaking, one may define here a trivial probability space with one unique element ω_0, equipped with the Dirac measure. Thus the ensemble average of a "random" field $Z(x,\omega)$ would be $Z(x)$!

$$< \sigma : \rho > = < \sigma > : < \rho >. \tag{9}$$

When trying to introduce precise definitions in a formal situation, assumed to represent a correct idealization of some physical situation, one must show that the formal situation is mathematically feasible, since otherwise one might come to contradictions. Thus we have the example of the space-filling periodic medium, the state $\mathbf{X}=\mathbf{F}(\mathbf{x})$ being defined as the position of the equivalent point to $\mathbf{x}$, in the elementary cell C (thus $E=C$). It is a simple exercise to verify that this is an S.H. continuum. It is now classical that periodic fields (of compatible strain and self-equilibrated stress) are macro-homogeneous in the sense of Hill, i.e. verify Hill's no-correlation condition in physical space (Suquet 1982, 1987). For any bounded periodic field $\mathbf{s}$, the function σ^D tends uniformly towards the restriction of $\mathbf{s}$ to C as $R(D) \rightarrow \infty$, hence any periodic field is S.H. and any two periodic fields satisfy the above-recalled asymptotic condition for the average deviation to the no-correlation (Arminjon 1991a). Hence, for admissible strain and stress fields in a periodic medium, condition (9) indeed applies to the associated functions of the so-defined state (and brings nothing more than Hill's condition on the elementary cell).

3. The inhomogeneous variational model: basics, comments and proposals

3.1 DETERMINATION OF THE MACROSCOPIC ACTUAL POTENTIAL

From now on, we consider the special case where the microscopic constitutive law [Eq. (1) or (8)] derives from a potential u. It has been established, for several relevant situations in the mechanics of materials, that the average of the micro-potential u is a potential U for the macroscopic constitutive law, and that Voigt's uniform strain assumption and Reuss' uniform stress assumption give an upper bound and a lower bound to U, respectively (Hill 1952, 1967). We have previously emphasized (Arminjon 1991a, Arminjon et al. 1994) that these three results depend *only* on the assumption of a convex potential and on the no-correlation condition between stimulus and response micro-fields, and thus can be extended to a number of situations (also in other fields of physics). The general proof of these results (Arminjon 1991a) has been given for the case where the "state" variable $\mathbf{X}$ is substituted for the micro-position $\mathbf{x}$, using the transported no-correlation (9). It has also been shown that they can be expressed, using a potential depending on a parameter r,

$$U_r(\mathbf{S}) \equiv \mathrm{Min}\{< u(\sigma) >; \ < \sigma > = \mathbf{S}, \ h(\sigma) \le r\}, \tag{10}$$

[where $< u(\sigma) > \equiv \int_E u(\sigma(\mathbf{X}), \mathbf{X}) f(\mathbf{X}) \, d\mathbf{X}$ and $h(\sigma) \equiv \left(< \|\sigma - < \sigma > \|^p > \right)^{1/p} \ (p>1)$ [6]], as follows:

---

[6] Note that $h(\sigma)$ is a homogeneous function [$h(\lambda\sigma) = |\lambda| \, h(\sigma)$] and that, if $p=2$, $h(\sigma)$ *is simply the standard deviation*. The number p is uniquely determined from the requirements that (i) $\|\partial u/\partial \mathbf{s}\| \le A \|\mathbf{s}\|^{p-1}$ for all $\mathbf{s}$ and $\mathbf{X}$, and (ii) $u(\mathbf{s},\mathbf{X}) \ge B \|\mathbf{s}\|^p$ for all $\mathbf{X}$ and all $\mathbf{s}$ with $\|\mathbf{s}\| \ge a$. However, (ii) and the condition $p>1$ are needed only to ensure that the minimum $U_r(\mathbf{S})$ is indeed reached by some function $\sigma_{r,\mathbf{S}}(\mathbf{X})$, whereas numerically an infimum can hardly be distinguished from a true minimum. In standard

$$U_R(\mathbf{S}) = U_\infty(\mathbf{S}) \le U(\mathbf{S}) \le U_0(\mathbf{S}). \tag{11}$$

The equality in (11) means that the minimum U_∞, which is obtained in dropping the inequality constraint in (10), is reached by a function σ_{Reuss} such that $h(\sigma_{\text{Reuss}}) \equiv R < \infty$; this indeed corresponds to the Reuss-Hill bound, because it turns out that the response function ρ_{Reuss}, associated with σ_{Reuss} by Eq. (8), is uniform. Clearly, U_0 (i.e. $r = 0$ in (10)) is the average potential corresponding to a uniform stimulus function $\sigma(\mathbf{X}) \equiv \mathbf{S}$, hence corresponds to the Voigt-Hill bound. From (11), it follows that *there exists a unique value* r_0 (with $0 \le r_0 \le R$), depending a priori on the macro-stimulus $\mathbf{S}$, *such that the actual macro-potential* $U(\mathbf{S})$ *is equal to* $U_{r_0}(\mathbf{S})$.

So the data r_0 determines the actual potential U as the minimum value corresponding to the minimum problem (10), but r_0 in turn is not determined by this theory. As long as we merely wish to determine U, we have thus replaced the scalar unknown U by the other one r_0. The point is that r_0 is an average inhomogeneity of the micro-stimulus, as expressed as a function of the state $\mathbf{X}$, and is likely to depend slowly on the macro-stimulus $\mathbf{S}$. More precisely, the value $r_0 = r_0(\mathbf{S})$ for a given $\mathbf{S}$ sets the exact potential $U(\mathbf{S})$ as $U_{r_0}(\mathbf{S})$ between the Reuss and Voigt bounds (and, conversely, $r_0(\mathbf{S})$ is hence determined by the data $U(\mathbf{S})$). For another stimulus $\mathbf{S}'$ (with $\|\mathbf{S}'\| = \|\mathbf{S}\|$), $U_{r_0}(\mathbf{S}')$, with the same value r_0, is a good approximation of $U(\mathbf{S}')$. This is at least what has been found for two very strongly inhomogeneous fiber-reinforced mortars, schematized as rigid-plastic (with $\mathbf{S} \equiv \mathbf{D}$, the strain-rate, and $\mathbf{R} \equiv \mathbf{T}$, the stress), for which the *ratios* of the Voigt bound to the Reuss bound in tension were 6 and 4.6 (Arminjon et al. 1993, 1994). Since the approximation $r_0(\mathbf{D}) = \text{Const.}$ (for $\|\mathbf{D}\| = 1$) works well for such strongly inhomogeneous materials, we may expect that it will do so for materials with more usual (smaller) size of the Reuss-Voigt band. But we recall that the existence of $r_0(\mathbf{S})$ involves only very general statistical assumptions (plus convexity), so actually the *model* for calculating the macro-potential from the micro one begins when one tries to guess the dependence $r_0 = r_0(\mathbf{S})$. This is unshamedly a *phenomenological* model for micro-macro transition, since the microscopic data are not enough. Once this phenomenological nature has been accepted and it has been recognized that it leads to powerful predictions, one must yet realize that the assumption $r_0(\mathbf{S}) = a \|\mathbf{S}\|$ is quite simplistic. Thus if we have more data, we can assume a more complex dependence of r_0 on $\mathbf{S}$, using e.g. the theory of invariants: as for a phenomenological yield criterion, the interest will be to predict U for values of $\mathbf{S}$ that were *not* incorporated in the input data. E.g. if $\mathbf{S}$ is a general symmetric tensor of order 2 (as for compressible plasticity, where $\mathbf{S} \equiv \mathbf{D}$ is not traceless), there is little doubt that an assumption like

$$r_0(\mathbf{S}) = [\, a^2 \|\mathbf{S}\|^2 + b\,(\mathrm{tr}\,\mathbf{S})^2\,]^{1/2} + c\,\mathrm{tr}\,\mathbf{S}, \qquad \|\mathbf{S}\| \equiv (\mathrm{tr}\,\mathbf{S}^2)^{1/2}, \tag{12}$$

---

plasticity, **s** is the *strain-rate*, **r** is the *stress*, u is the *rate of work* and $p=1$. In the applications to plasticity, we have nevertheless used the standard deviation for $h(\sigma)$. This is probably harmless, though the correct value $p=1$ might be more appropriate after all.

will be more accurate than the assumption $r_0(\mathbf{S}) = a \, \|\mathbf{S}\|$. Again, the decisive point is to predict: (i) correctly, and (ii) more than is entered into the model. *Summary*: the variational definition of the actual potential as $U(\mathbf{S})=U_{r_0(\mathbf{S})}(\mathbf{S})$ is predictive because it *turns out to be easy* to get the dependence $r_0=r_0(\mathbf{S})$, in so far as it influences the values $U_{r_0(\mathbf{S})}(\mathbf{S})$. This means that the average inhomogeneity h is a relevant, "heavy" parameter in micro-macro problems.

3.2 LOCALIZATION: THE PRINCIPLE OF MINIMAL INHOMOGENEITY

Consistently with the followed approach, the localization problem is seen as the search for the actual distribution of the local stimulus as a function $\sigma_{\mathbf{S}} = \sigma_{\mathbf{S}}(\mathbf{X})$, from the data of the macro-stimulus $\mathbf{S}$. Clearly, we have a good candidate, namely the solution $\sigma_{r_0}\,\mathbf{s}$ to the minimum problem (10), for the correct value $r = r_0(\mathbf{S})$ of the inhomogeneity parameter (*any* solution, if there are several; the uniqueness is guaranteed if the local potential u is a strictly convex function of $\mathbf{s}$ at fixed $\mathbf{X}$). A first point is that the (any) solution σ to (10) has exactly the inhomogeneity $h=r$, save in the exceptional case where the value of the minimum is the lower bound (in which case a solution to (10) with some value r is also a solution to (10) with any value $r' > r$ and so has an inhomogeneity $h<r'$). It has also been proved that assuming that $\sigma_{\mathbf{S}} = \sigma_{r_0}\,\mathbf{s}$ is equivalent to a *principle of minimal inhomogeneity* (PMI) according to which the actual distribution $\sigma_{\mathbf{S}}$ is a solution to:

$$h(\sigma^*) = \text{Min, under the constraints}: < \sigma^* > = \mathbf{S} \text{ and } <u(\sigma^*)> = U(\mathbf{S}) \qquad (13)$$

(Arminjon et al. 1994). Thus it is equivalent to assuming that, among the distributions of local stimulus whose average is the macro-stimulus $\mathbf{S}$ and whose average potential is the corresponding actual potential, the actual distribution has the least inhomogeneity h. Since, for $h=0$, the average potential is Voigt's upper bound, it still means that *the inhomogeneity occurs only in so far as it allows to lower the average potential*. We find this principle plausible, in any case it is this principle that underlies the success, as regards the prediction of deformation textures in polycrystals (Van Houtte 1984, Arminjon 1987), of the relaxed Taylor theory and the proposed variational model.

However, we do not have plausible assumptions allowing to deduce this principle from deterministic micro-macro arguments. Another problem is the apparent ambiguity in the statement of this principle. First, we may exchange stimulus and response; e.g. for elastic or viscoplastic behaviour, a potential for strain or strain-rate as the response, may be deduced from the potential for stress, by Legendre transformation (Hill 1956). If the corresponding statements are not equivalent (as seems likely), which is the correct form? It seems that the PMI is plausible *in so far as the potential may have a direct interpretation as an energy* (since in that case the PMI amounts to a principle of minimal energy consumption). Now consider the general class of inelastic materials with Green-elastic domain depending on internal variables H, the latter evolving only for inelastic strain increments, as envisaged by Hill & Rice (1973). For such materials, the elastic potential (ϕ in their notation) for the *stress* $\mathbf{t}$ i.e. that one which is a function of the *strain*

e, is a true energy, since the variation of ϕ during an elastic strain increment is indeed the work done, $\delta\phi = $ **t**:δ**e**. Since $\delta\psi = $ **e**:δ**t**, the same cannot be said of the complementary potential $\psi = $ **t**:**e** $- \phi$, although it has the dimension of an energy. In standard plasticity, the rate of work is a potential for the stress (Hill 1986) and a true complementary potential cannot be defined (the yield function, which is surely not an energy, is a potential for the *direction* of strain-rate only), hence there is no ambiguity. Thus in the mechanics of materials at least, the physical meaning of the PMI seems to imply that *the primary ("stimulus" **s**) variable should be taken as the strain (or strain-rate) and the response as the stress-* not the reverse. Second apparent ambiguity in the statement of the principle: that of the *norm* $\|$**s**$\|$. One norm appears naturally in the theory, e.g. in the proof of the general form of the Reuss-Voigt bounds (Arminjon 1991a): it is the Euclidean norm, derived from the scalar product, $\|$**s**$\| = ($**s** : **s**$)^{1/2}$; for every relevant tensor space, we have one and only one natural scalar product, e.g. **s** : **r** $\equiv s_{ij}\, r_{ij}$ for second-order tensors. Thus, although all theoretical results remain valid if one takes any other norm (since all norms are equivalent in finite dimension), there seems to be little reason to do so. Lastly, we have seen that the real exponent $p \geq 1$ is imposed[6] and thus the average inhomogeneity h is uniquely defined. We conclude that the statement of the PMI is unambiguous, unless one artificially defines a different norm for stimulus tensors[7].

4. Microstructure as an internal, state variable

4.1 THE METHOD AND ITS THEORETICAL JUSTIFICATION

The way we propose to account for micro-geometry and additional parameters such as interface behaviour (Arminjon et al. 1994), is not particularly bound to the variational model recalled in §3. Indeed, it depends only on the statistical theory of the distribution of the states in S.H. media, summarized in Sect. 2.2. Thus this way could be used with different micro-macro models, e.g. it may be used with the self-consistent models proposed by Molinari et al. (1987) and Lipinski & Berveiller (1989). Our starting point is similar to the basic idea of "cluster models" (Kocks & Canova 1986) and to that of "morphological representative patterns" (Stolz & Zaoui 1991): it is the idea (Arminjon 1991a) that the near environment of a material (micro-) volume has a greater influence on the inhomogeneity of the local fields than the long-range one. This idea, however, translates to the statistical theory in a rather original form. The crucial point is that, in this theory, the inhomogeneity of the local fields (**s** and **r**) within the constituents is recognized from the beginning, i.e. from the definition of the distributions of the local

---

[7] The possibility of selecting an anisotropic norm on morphological grounds was evoked by Arminjon et al. (1994). One may in that way recover the relaxed Taylor model (e.g. Van Houtte 1984) as the limiting case of a degenerate (semi-) norm. The relaxed Taylor model has not a very sound theoretical basis. Morphological effects are consistently described using the proposed methodology for integration of microgeometry (§4). In addition to the fact that it derives from the scalar product, a distinctive feature of the Euclidean norm is this: if one introduces an anisotropic norm N on physical grounds, then its coefficients must have a physical dimension, hence N will not make sense for stimulus *and* response.

stimulus and response as functions σ and ρ, not of the position $\mathbf{x}$, but of the state $\mathbf{X}$. Thus the "constituents", i.e. *the domains D_X with given state* $\mathbf{X}$, are subjected to fields which are homogeneous only in a strongly statistical sense, since (i) those "constituents" are not necessarily contiguous (e.g. a group of grains with the same orientation) and (ii) only an average, for the different states i.e. for *all* constituents, of the *deviation to the no-correlation condition* between the local fields $\mathbf{s}$ and $\mathbf{r}$, must cancel. The concrete consequence is that one may take as "constituents" a *group of* n *neighbouring constituents* $C_1,..., C_n$, which is formally done in replacing the initial "simple" state $\mathbf{X}^0$ (identified as the set of parameters indexing the local constitutive law) by a "complex" state $\mathbf{X}$ involving the states $\mathbf{X}^0_1,..., \mathbf{X}^0_n$, and the parameters describing the geometry of this cluster (sizes and shapes of the C_i's and their relative positions)(Arminjon 1991a).

In order to run a micro-macro model one has to know the constitutive law, so the question arises: how can we get the "constitutive law" of a such cluster? The answer will be obtained if we remember the statistical meaning of the local constitutive law (8) in the theory: One assumes that, in any statistically homogeneous situation, the average values $\sigma(X)$ and $\rho(X)$ of the fields $\mathbf{s}$ and $\mathbf{r}$, over the domain D_X, are related by Eq. (8). By hypothese, this relation depends *only* on the particular state $\mathbf{X}$, thus it does not depend on the *distribution* of (all the other) states. Hence, if we consider an inhomogeneous medium for which the state is (the same) $\mathbf{X}$ everywhere, i.e. for which the domain D_X is the *whole* (space-filling) material, the same law ϕ will relate the *asymptotic volume averages* $\bar{\mathbf{s}}=\mathbf{S}$ and $\bar{\mathbf{r}}=\mathbf{R}$ of the fields $\mathbf{s}$ and $\mathbf{r}$ in S.H. situations. Now consider the case where the state $\mathbf{X}$ involves geometrical parameters, more precisely describes the geometry and microscopic behaviour of some cell C[8]. Then a space-filling medium with state $\mathbf{X}$ everywhere is nothing else than *a periodic medium with elementary cell* C. And the relation between asymptotic averages $\bar{\mathbf{s}}$ and $\bar{\mathbf{r}}$ is none other than *the homogenized law of the periodic medium* (Suquet 1987). Thus the constitutive law of our micro-cluster, in precisely the meaning it has in the proposed statistical theory, will be most rigorously obtained as the homogenized law of a periodic medium consisting of the endless repetition of this same cluster- provided, of course, that the space can be filled with this cluster, i.e. provided it has the form of e.g. a parallelepiped.

In order to take into account short-range effects of micro-geometry, Arminjon et al. (1993,1994) envisaged any S.H. medium as consisting of a periodic array of cells with identical shape, but with a "random" variation of $\mathbf{X}$ from one cell to another (this may always be envisaged, since the array is only in our thought). They defined and obtained the behaviour of a cell C, depending on the "state" parameters $\mathbf{X}$ describing this cell, as that delivered by the homogenization theory for a periodic medium with cell C [this is hence done successively for each different cell. Of course, this will be feasible in practice only if the medium is schematized so that there is only a (small) finite number of different cells (states); however, symmetries allow to reduce this number]. We have just proved that this method is rigorously correct in the frame of the statistical theory considered.

---

[8] Note that this situation does not imply that the state $\mathbf{X}$ contains an infinite number of parameters, since in practice the geometry will be simple: a few elliptic inclusions, one or two fibers,...However, the whole of the statistical theory remains valid for cases with an infinity of parameters (Arminjon 1991a).

4.2. APPLICATION TO RIGID-PLASTIC REINFORCED MORTARS

The variational model (§3) was applied to predict the failure criterion of two mortars reinforced by steel fibers by Arminjon et al. (1993,1994), using the rigid-plastic scheme. The model was first applied directly as a "*volume-fraction model*", in defining the state **X** as the phase identifyier i (in view of the isotropy of each phase and its perfectly plastic behaviour, no other parameter intervenes in that case), thus **X**=1 for mortar and **X**=2 for steel. The inhomogeneity parameter r_0 of the model was assumed independent of the macro-stimulus **D** (with $\|\mathbf{D}\| = 1$), and was adjusted so that the predicted and measured loads coincide for the tension test of a plate structure. It was found that the predicted loads were also close to the experimental band for bending test and compression test. Another application of the model ("*refined model*") was obtained in combination with the proposed method for taking the short-range effects of micro-geometry into account: the material was schematized as an array of rectangular cells, each of which containing one whole fiber and two half-fibers (cut by the walls of the cell), all three with same orientation. The contact at the fiber-mortar interface was schematized by a Coulomb friction with the same friction coefficient $f = \tan \phi$ for all cells. The cells differed only by the fiber in-plane orientation, so the state was this orientation angle **X**=α. Consistently with the real distribution, a uniform angle distribution was assumed. The "refined model" (**X**=α) is much heavier to run, due to the preliminary steps of periodic homogenization of the different cells; its interest for this material was the dramatic reduction of the Voigt-to-Reuss load ratio ξ (with the retained value of the friction coefficient f, see below): in tension, ξ passed from 6. to 1.15 for one material (Arminjon et al. 1993) and from 4.6 to 1.04 for the second one (Arminjon et al. 1994). However, it was found that the experimental band was not exactly within the Reuss-Voigt (RV) bounds of the refined model (*Fig. 1*), which did not give a better overall agreement with experiment. We work out this question hereafter.

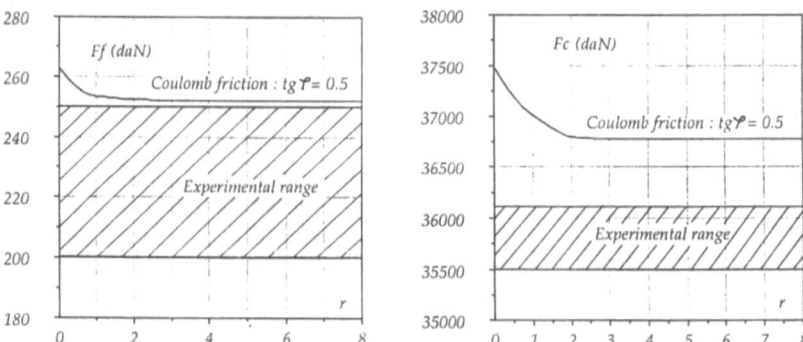

Figure 1. Comparison between loads computed with the refined model (double-scale homogenization), depending on the inhomogeneity parameter r, and experimentally observed load range, for a fiber-reinforced mortar with 0.6% volume fraction of steel fibers: compression test of a plate (*left*) and bending test on a beam (*right*). The macroscopic inhomogeneity of these tests is accounted for via the use of the finite element method in addition to the homogenization model.

The experimental band is of the same order as the RV band of the refined model. So any micro-macro model is good enough to predict the behaviour of the *aggregate of cells* from the behaviour of the cells and their volume fractions, and there is less need to adjust the inhomogeneity parameter in the refined model. Assume that the (carefully checked!) experimental and numerical results are correct. To understand why there is some (small) *distance* between the experimental and RV bands, we must identify in which respect the aggregate of cells may not be a good enough schematization of the *real material*. Thus: (i) Did we assume an adequate constitutive scheme (here rigid-plasticity with isotropic quadratic yield criteria, plus Coulomb friction)? (ii) Are the values or rather, in view of the inhomogeneity, the *bands* of the corresponding constitutive parameters correct? (iii) Is the (volume) *distribution* of the geometrical and constitutive parameters of the cells correct? As long as we merely look at the maximum loads, we may argue that the answer to (i) is "yes", cf. Chambard (1993). (ii) As it should be, we have tried to get the correct experimental values of the material parameters for the phases (mortar and steel), but the values are not yet very accurate and may account for a good part of the distance. Furthermore, for the friction coefficient, no direct measurement was possible. We checked 4 values: "f=0" (in fact, sliding contact), f=0.2, 0.5, "f=∞" (in fact, adhesive contact) and retained that one (f=0.2 or 0.5, depending on the material) which gave the best overall agreement, but noted that the value adopted in the technical norm for reinforced concretes (a similar material), f=0.4, would have given close results. It still remains worth to examine point (iii), since it is specific to micro-macro models and since the main point is perhaps that *the RV bounds of the refined model were simply too close*. Thus we are currently investigating the effect of allowing f to vary from one cell to another, as the fibres may be more or less closely bound to the matrix (in future work, the effect of varying the geometry of the cell will be also investigated). To this end, we now define the state as the *pair* (α, f). Preliminary results have been obtained for the case where the four values of f are uniformly distributed and independent of the value of the angle α. As expected, they show a wider RV band than for the case where the friction coefficient f is assumed uniform. Of course, the interest of these computations with the refined model is mainly illustrative, since for this material we are not able to measure f and even less its distribution.

5. References

Arminjon, M. (1987) Théorie d'une classe de modèles de Taylor "hétérogènes". Application aux textures de déformation des aciers, *Acta Metall.* **35**, 615-630.

Arminjon, M. (1991a) Limit distributions of the states and homogenization in random media, *Acta Mechanica* **88**, 27-59.

Arminjon, M. (1991b) Macro-homogeneous strain fields with arbitrary local inhomogeneity, *Arch. Mech.* **43**, 191-214.

Arminjon, M., Chambard, T., and Turgeman, S. (1993) Homogénéisation d'un mortier renforcé par une distribution aléatoire de fibres, *C.R. Acad. Sci. Paris* **316**, *Série II*, 1505-1510.

Arminjon, M., Chambard, T., and Turgeman, S. (1994) Variational micro-macro transition, with application to reinforced mortars, *Int. J. Solids Structures* **31**, 683-704.

14

Beran, M. J. (1968) *Statistical Continuum Theories*, Interscience, New York.

Chambard, T. (1993) *Contribution à l'Homogénéisation en Plasticité pour une Répartition Aléatoire des Hétérogénéités*, thèse de Doctorat, Université de Grenoble.

Guichardet, A. (1969) *Calcul Intégral (Maîtrise de Mathématiques)*, Armand Colin, Paris.

Hill, R. (1952) The elastic behaviour of a crystalline aggregate, *Proc. Phys. Soc. Lond.* **A65**, 349-354.

Hill, R. (1956) New horizons in the mechanics of solids, *J. Mech. Phys. Solids* **5**, 66-74.

Hill, R. (1967) The essential structure of constitutive laws for metal composites and polycrystals, *J. Mech. Phys. Solids* **15**, 79-95.

Hill, R. (1984) On macroscopic effects of heterogeneity in elastoplastic media at finite strain, *Math. Proc. Camb. Phil. Soc.* **95**, 481-494.

Hill, R. (1986) Extremal paths of plastic work and deformation, *J. Mech. Phys. Solids* **34**, 511-523.

Hill, R. and Rice, J.R. (1973) Elastic potentials and the structure of inelastic constitutive laws, *SIAM J. Appl. Math.* **25**, 448-461.

Kocks, U.F. and Canova, G.R. (1986) Effective-cluster simulation of polycrystal plasticity, in J. Zarka and S. Nemat-Nasser (eds.), *Large Deformations of Solids*, Elsevier, London- New York.

Kröner, E. (1986) Statistical modelling, in J. Gittus and J. Zarka (eds.), *Modelling Small Deformations of Polycrystals*, Elsevier, London- New York, pp. 229-291.

Lipinski, P. and Berveiller, M. (1989) Elastoplasticity of micro-inhomogeneous metals at large strains, *Int. J. Plasticity.* **5**, 149-172.

Matheron, G. (1989) *Estimating and Choosing*, Springer, Berlin- Heidelberg- New York.

Molinari, A., Canova, G.R., and Ahzi, S. (1987) A self-consistent approach of the large deformation polycrystal plasticity, *Acta Metall.* **35**, 2983-2994.

Sab, K. (1992) On the homogenization and the simulation of random materials, *Eur. J. Mech. A/Solids* **11**, 585-607.

Stolz, C. and Zaoui, A. (1991) Analyse morphologique et approches variationnelles du comportement d'un milieu élastique hétérogène, *C.R. Acad. Sci. Paris* **312**, *Série II*, 143-150.

Suquet, P. (1982) *Plasticité et Homogénéisation*, thèse de Doctorat d'Etat, Université Paris VI.

Suquet, P. (1987) Elements of homogenization for inelastic solid mechanics, in E. Sanchez-Palencia and A. Zaoui (eds.), *Homogenization Techniques for Composite Media*, Springer, Berlin- Heidelberg-New York, pp. 193-278.

Van Houtte, P. (1984) Some recent developments in the theories for deformation texture prediction, in C.F. Brakman, P. Jongenburger and E.J. Mittemeijer (eds.), *Textures of Materials*, Netherlands Soc. Materials Sci., Zwijndrecht, pp. 7-23.

CORRELATION BETWEEN FRACTURE TOUGHNESS AND THE MICROSTRUCTURE MORPHOLOGY IN TRANSVERSELY LOADED UNIDIRECTIONAL COMPOSITES

M.S. AXELSEN and R. PYRZ
Institute of Mechanical Engineering
Aalborg University
Aalborg, Denmark

1. Introduction

The dispersion of fibers and cracks in the transverse direction of a unidirectional composite material has very strong influence on local stress field and therefore, it may affect the durability of the material. The local stress field influences the fracture toughness of cracks situated among the fibers and new cracks initiates at various positions depending on the surrounding fibers and existing cracks. Thus it is necessary to investigate the correlation between the microstructure variability and different mechanisms of crack nucleation.

Usually, the dispersion of fibers is assumed to have some form of regularity or fibers are assumed to be sparsely distributed. In these cases each fiber is either exposed to the same amount of interaction in regular distributions or it is exposed to no interaction in the dilute distributions. Therefore it is possible to establish a repetitious unit cell containing only one fiber and it may be analyzed thoroughly within reasonable limits. In order to investigate a non–regular distribution of fibers the unit cell concept is not sufficient. Each fiber is in this case exposed to different amount of interaction and the local stress field varies throughout the whole microstructure. Therefore it is necessary to re–define the unit cell concept so that it contains enough fibers and cracks to be representative for the non-regular microstructure. In relation to this re–defined concept a method for calculating the stress field in a material with randomly dispersed fibers as well as determining

15

R. Pyrz (ed.), IUTAM Symposium on Microstructure-Property Interactions in Composite Materials, 15–26.
© *1995 Kluwer Academic Publishers.*

the stress intensity factors for cracks situated among the fibers must be established.

Pijaudier-Cabot and Bažant(1991) presented a method to calculate the stress field in a solid containing multiple fibers and to determine the stress intensity factors for a single crack situated among the fibers. A method for stress analysis in an elastic solid with randomly distributed cracks was presented by Kachanov(1987). Both methods are based upon a superposition scheme and take into account the interaction between fibers and cracks. In the present work a new calculation procedure is developed that allows to treat multiple fibers and cracks in a unified way.

The local stress field is dependent upon the exact position of fibers and cracks and this also affects the initiation of new cracks. Consider the case of cracks initiating at the interface around the fibers. Two scenarios are possible in this case; matrix and interface cracks. The matrix cracks initiate at the interface and extend radially to the fiber into the matrix material. Interface cracks initiate tangentially to the fiber and also extend in this direction. At which angle around the fibers the cracks appear depends on the local stress field. It is reasonable to assume that matrix cracks initiate at positions where maximum tangential stress occurs and interface crack initiate at positions where maximum radial stress occurs. The magnitude of the maximum stress components and the angle where they occur are strongly affected by the dispersion of fibers and existing cracks.

Having the criterion of the crack initiation it possible to determine how these microcracks affect the fracture toughness of a composite material. Such an investigation is performed by situating a macrocrack in the vicinity of distributed fibers. Interface cracks are then allowed to initiate during a load increase. As a result, the fracture toughness of a material is affected depending on the position of fibers and interface cracks.

2. Stress and Fracture Analysis

In order to determine various mechanical properties of materials with randomly dispersed fibers and cracks it is necessary to introduce a method for calculating the stress field in an infinite solid containing multiple fibers and exposed to uniform tractions at the remote boundaries. Also it is necessary to determine the stress intensity factors for cracks located in the matrix material.

2.1. STRESS ANALYSIS METHOD

In the following only a short introduction to the stress analysis method is given. The stress field solution for the single fiber configuration, Fig. 1a, may be obtained analytically from the complex potential theory or the

Eshelby solution, see e.g: Muskhelishvili(1962), Mura(1987)). Since the method must be extended to include multiple fibers another iterative procedure is applied. A heterogeneous solid is replaced by an equivalent

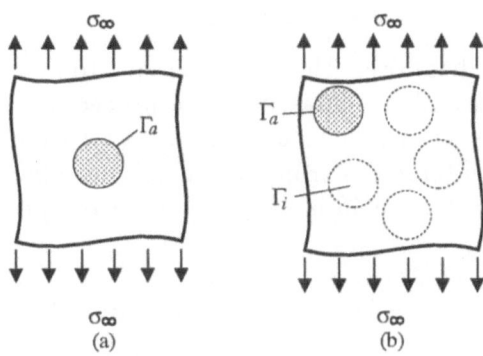

Figure 1. (a) Single fiber configuration, (b) multiple fiber configuration.

homogeneous solid where tractions are applied on the imaginary contour of the circular fiber and an unbalanced stress field inside this contour is added. Then the iterative procedure is applied and as a starting point the stress field in the whole solid is $\sigma = \sigma_\infty$. Using the theory of eigenstresses, Mura(1987), the unbalanced stress field becomes

$$\Delta\sigma = (\mathbf{D}_m - \mathbf{D}_a)\mathbf{D}_a^{-1}\,\sigma \tag{1}$$

where $\mathbf{D}_a$ and $\mathbf{D}_m$ are the stiffness matrix for the fiber and matrix material, respectively. Tractions are applied at the imaginary contour of the fiber in order to account for the unbalanced stresses inside the fiber

$$\mathbf{p}_a = -\Delta\sigma\mathbf{n}_a \tag{2}$$

where $\mathbf{n}_a$ is a unit outward normal to the circular contour Γ_a. The tractions are substituted by concentrated forces for which the stress field can be obtained from the complex potential theory. The forces are integrated along the circular contour and then the stress field is calculated at an arbitrary point within the contour

$$\sigma_a = \oint_{\Gamma_a} \sigma(\mathbf{p}_a)\,ds \tag{3}$$

The new stress field inside the fiber is obtained by

$$\sigma = \sigma_\infty + \sigma_a - \Delta\sigma \tag{4}$$

From this expression the unbalanced stress is re-calculated according to Equation 1 and the iterations are repeated until $\mathbf{p}_a$ does not change significantly. The method converges quite rapidly to the analytical solution and the stress field outside the fiber may be determined as follows

$$\sigma = \sigma_\infty + \sigma_a \tag{5}$$

For the multiple fibers, Fig. 1b, it is necessary to account for the interaction between fibers and a similar iterative procedure is applied. In this case the stress field in each fiber is determined as it were alone in the matrix except that the interaction from the remaining fibers is added in the calculation. Thus the stress field is based upon a superposition scheme and Equation 4 yields

$$\sigma = \sigma_\infty + \sigma_a + \sigma_i - \Delta\sigma \tag{6}$$

where σ_i is the interacting stress field from the remaining fibers. When the stress field inside the fibers is determined, the stress field in the matrix material is calculated similarly to the single fiber solution.

2.2. DETERMINATION OF STRESS INTENSITY FACTORS

The determination of the stress intensity factors is based on a superposition scheme in which the interaction between fibers and cracks is taken into account. First the original problem is decomposed into an initial and subsidiary problem, Fig. 2. The initial problem consists of calculating the

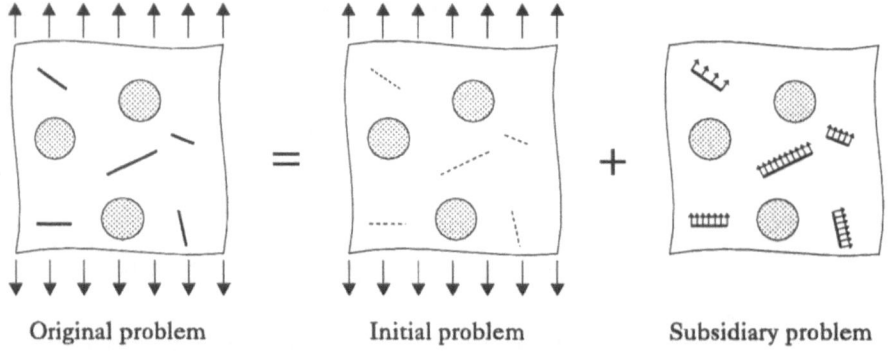

| Original problem | Initial problem | Subsidiary problem |

Figure 2. Superposition scheme for the multiple fiber-crack problem.

stress field at the imaginary contours of the cracks by applying the method described in the previous section. These stress fields are added in the subsidiary problem so that the configuration is in equilibrium. In order to solve the subsidiary problem it is divided into a number of subproblems corresponding to the number of cracks. Each subproblem consists of one crack with applied tractions, a number of imaginary cracks and the surrounding fibers. The tractions may be written as

$$\mathbf{p} = \mathbf{p}_c + \mathbf{p}_{int} \tag{7}$$

where $\mathbf{p}$ is the applied traction, $\mathbf{p}_c$ is the real yet unknown traction including the interaction from the fibers and cracks and $\mathbf{p}_{int}$ is the interacting stress. The interacting stresses arise because the real tractions interact with fibers and other cracks and they are subsequently reflected back. In order to determine the real tractions Equation 7 is averaged and the interacting stresses are written as a function of real stresses

$$\langle \mathbf{p} \rangle = (\mathbf{I} + \Lambda)\langle \mathbf{p}_c \rangle \tag{8}$$

where Λ is a transmission factor which takes into account the whole interaction between fibers and cracks. The averaged real tractions are found by rearranging Equation 8

$$\langle \mathbf{p}_c \rangle = (\mathbf{I} + \Lambda)^{-1}\langle \mathbf{p} \rangle \tag{9}$$

Having calculated these uniform tractions for all cracks the non-uniform tractions may now be determined as

$$\mathbf{p}_c = \mathbf{p} - \Lambda \langle \mathbf{p}_c \rangle \tag{10}$$

These tractions may now be calculated at any point of the crack lines and by numerical integration the stress intensity factors are determined as

$$K_I(\pm c) = \frac{1}{\sqrt{\pi c}} \int_{-c}^{c} \sqrt{\frac{c \pm x}{c \mp x}}\, \mathbf{p}_c \cdot \mathbf{n}_c \, dx \tag{11}$$

$$K_{II}(\pm c) = \frac{1}{\sqrt{\pi c}} \int_{-c}^{c} \sqrt{\frac{c \pm x}{c \mp x}}\, \mathbf{p}_c \cdot \hat{\mathbf{n}}_c \, dx \tag{12}$$

where c is the half crack length. The method is more thoroughly described in Axelsen(1994).

3. Effect of Fiber Distribution on Matrix and Interface Cracks

The stress field for randomly dispersed fibers is determined in two steps: first the stress field inside the fibers is determined and then the stress field in the matrix material may be calculated. In order to investigate how the dispersion of fibers affects both the stress field inside the fibers and the local stress field in the matrix two fiber distributions are analyzed, Fig. 3. The clustered distribution exists in composites made of bundles of fibers

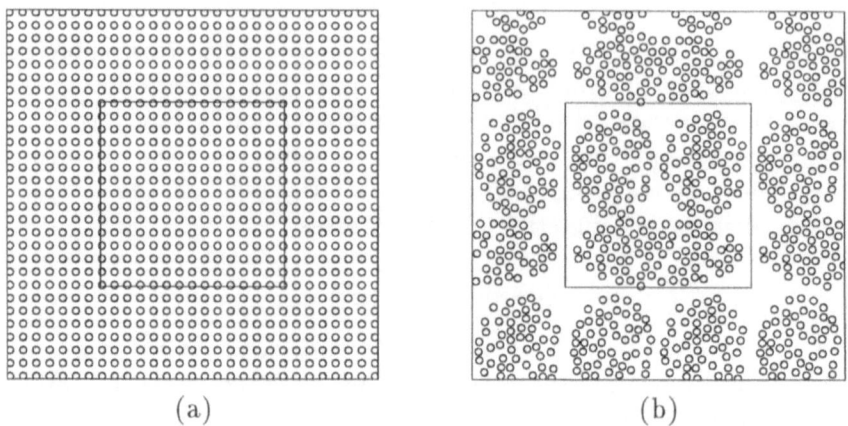

(a)   (b)

Figure 3.   Distribution of fibers in the re–defined unit cell; (a) regular distribution, (b) clustered distribution.

containing a very dense dispersion of fibers and matrix rich areas. The area within the box represents the sample area whereas the area outside the box represents the boundary area. With this type of unit cell the boundary conditions may be altered by changing the distribution of fibers in the boundary area. The fibers in the boundary area constitute periodic boundary conditions for both distributions. The number of fibers included in the sample and boundary areas is dependent on what must be analyzed and adjusted, accordingly. In this case 841 and 798 fibers are dispersed in the regular and clustered distributions, respectively. The distributions are exposed to a unidirectional load applied at the remote boundaries in the vertical direction. The ratio between the Young's moduli for the matrix and fiber material is $E_a/E_m = 23$, and the Poisson ratios are $\nu_a = 0.3$ and $\nu_m = 0.35$. The stress field inside the fibers is represented by the von Mises stresses calculated for fibers within the sample area, Fig. 4.   The fiber stresses within the sample area are almost uniform for the regular distribution, which has been also expected as each fiber is exposed to the same amount of interaction. For the clustered distribution the fiber stresses are non-uniform due to the non-regularity of the dispersion of fibers.

The fiber stresses are used to calculate the stress field in the matrix material. Particularly, two important damage modes, matrix and interface cracking, are affected by the local stress field around the fibers. Figure 5a shows the tangential and radial stress components around the fibers which

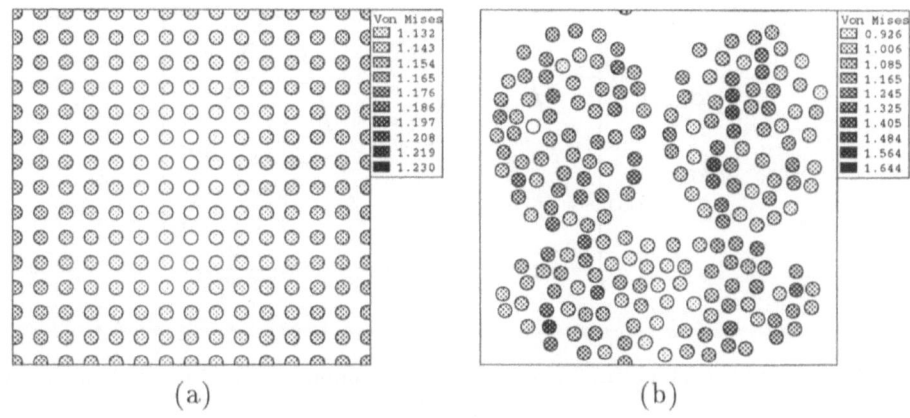

Figure 4. Stress field inside the fibers represented by Von Mises stress; (a) regular distribution, (b) clustered distribution.

are responsible for the matrix and interface cracking, respectively. It is

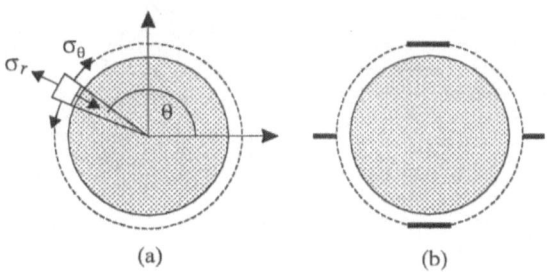

Figure 5. (a) Tangential and radial stress components around the fibers, (b) estimated positions of matrix and interface cracks.

reasonable to assume that they are nucleated at positions where the tangential and radial stress components reach their maximum. Figure 6 shows the normalized tangential and radial stress components for one fiber alone in the matrix. The maximum values of the tangential stress component are located at $\theta = 0°, 180°$ resulting in a possible crack initiation at these positions, Fig. 5b. For the radial stress component the maximum values are located at $\theta = 90°, 270°$.

In order to show how the dispersion of fibers affects the magnitude of the maximum stress components and at which angles they appear, the regular and clustered distributions are analyzed. The maximum tangential stress component is calculated for all fibers and depicted in Figure 7. It appears that the dispersion of fibers is very influencial on both the magnitude of

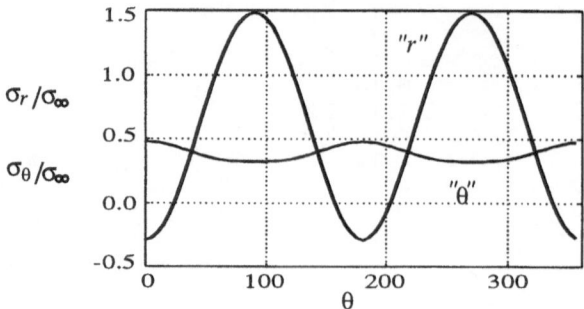

Figure 6. Normalized tangential and radial stress components for the single fiber configuration.

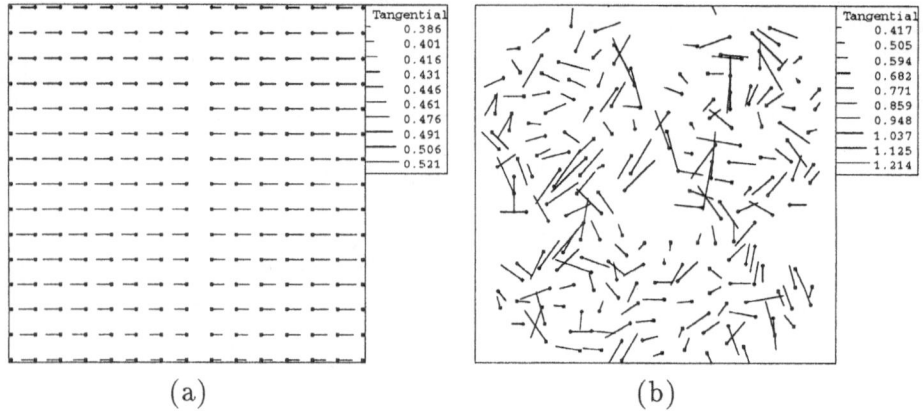

(a)   (b)

Figure 7. Distribution of the maximum tangential stress component for; (a) regular distribution, (b) clustered distribution.

tangential stresses and the angle at which they are detected. For regular distribution the maximum values are equally distributed and they appear at $\theta = 0°, 180°$ due to the symmetrical dispersion of fibers. The maximum radial stress component is influenced as well, Fig. 8. In this case the angles at which the maximum values occur are less affected by the dispersion of fibers.

The analyses show that the dispersion of fibers is a very important parameter in the investigation of damage modes.

4. Influence of Interfacial Cracks

In order to investigate the effect of microcracks on the stress field at the tip of preexisting macrocracks two selected configurations are analyzed. Only

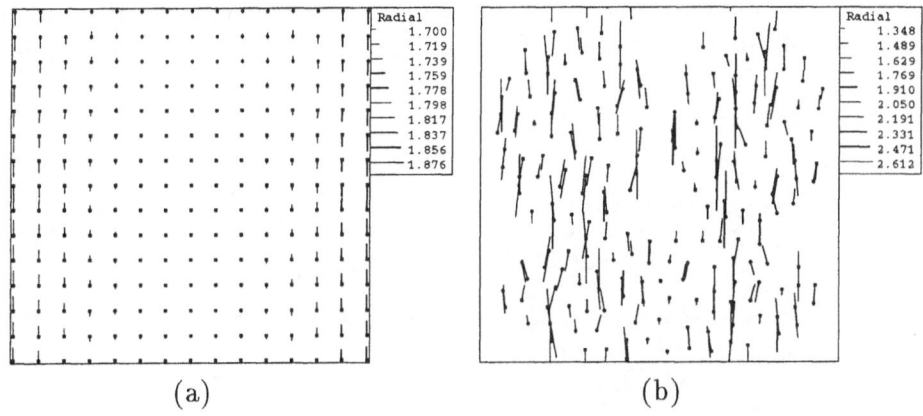

Figure 8. Distribution of the maximum radial stress component for; (a) regular distribution, (b) clustered distribution.

the initiation of interface cracks is taken into consideration as the matrix cracks show much less influence on the stress field at the tip of macrocracks.

The interface cracks do not initiate simultaneously because the magnitude of the radial stress component depends on the dispersion of fibers. Therefore an iterative procedure is applied while the load is increased until interface cracks appear in all fibers. During this load increase the stress intensity factor is calculated. The procedure is summarized in Table 1. The

TABLE 1. The iteration procedure for initiation of interface cracks.

---

initial state   no interface cracks are initiated

    $\sigma_{r,max}$ is calculated (incl. interaction from the macrocrack)

If $\sigma_{r,max} > \sigma_{r,critical} \Rightarrow$ crack initiation

iterations

    – A crack is introduced at θ_{max} with length $a = a(\sigma_{r,max})$

    – The load is increased

    – $\sigma_{r,max}$ is calculated for the remaining fibers (incl.interaction from the macrocrack and the initiated interfacial cracks)

    – If $\sigma_{r,max} > \sigma_{r,critical} \Rightarrow$ crack initiation

stop iteration when   interface cracks appear in all fibers

end

---

interface cracks are initiated at small distance from the fibers for numerical purposes.

The first configuration consists of a macrocrack situated in front of the cluster of 50 fibers, Fig. 9a. This configuration is subjected to far field

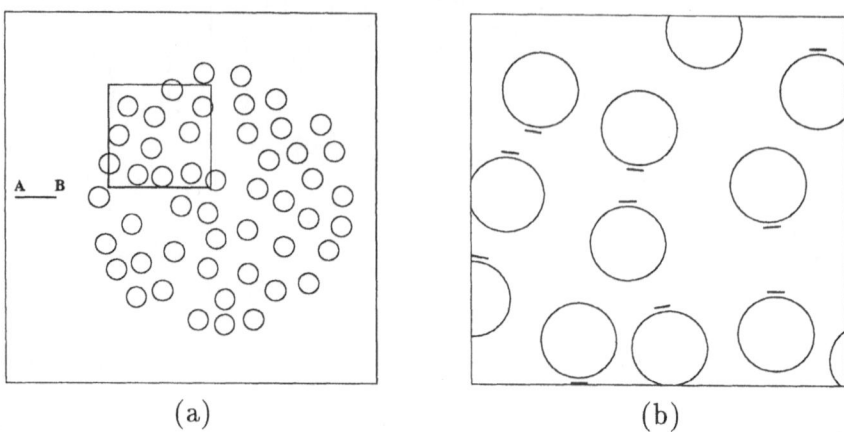

(a)   (b)

Figure 9. Configuration with a macrocrack situated in front of the cluster of 50 fibers; (a) initial state, (b) enlarged view of the enclosed area in the final state.

loading in vertical direction. The ratio between the macrocrack length and the fiber radius is 2. In the initial state no interface cracks have been initiated and this particular configuration will increase the fracture toughness as compared with pure matrix by lowering the stress intensity factors for both crack tips A and B. The iterative procedure is repeated until interface cracks appear in all fibers. An enlarged picture of the distribution in the final state is shown in Figure 9b. The position of interface cracks varies due to the influence of neighbouring fibers. Cumulative distribution of interface cracks is shown in Figure 10a. Appearence of interface cracks results in increased values of the normalized stress intensity factor for both tips of the macrocrack, Fig. 10b. It is interesting to notice that embedding fibers into the matrix material improves the fracture toughness by itself as the stress intensity factor K_I is less than the corresponding stress intensity factor K_{I0} for pure matrix. Thus the nucleation of interface cracks deteriorates the reinforcing effect of fibers.

The second configuration is shown in Figure 11a where two clusters, each consisting of 25 fibers, are situated below and above the macrocrack. The ratio between the macrocrack length and the fiber radius is in this configuration 3. In the initial state this configuration will decrease the fracture toughness by increasing the stress intensity factor. An enlarged picture of the final state is shown in Figure 11b where interface cracks appear in all fibers. As the interface cracks are initiated, Fig. 12, the stress intensity factor decreases and thus the fracture toughness is increased while in the

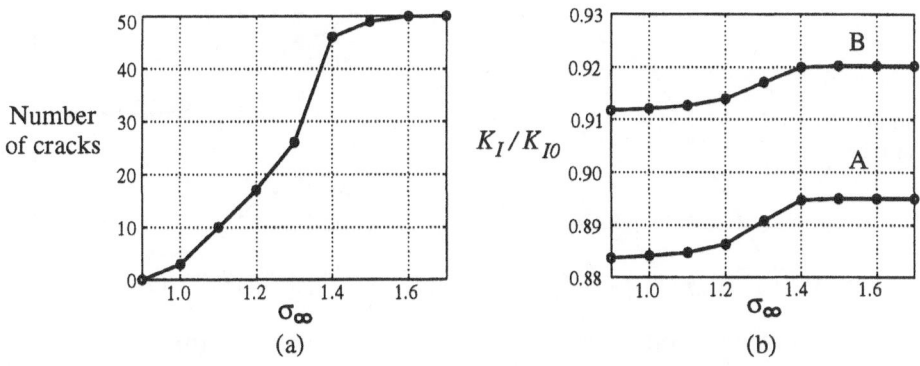

Figure 10. (a) Number of initiated cracks as a function of the applied load, (b) normalized stress intensity factors as a function of the applied load.

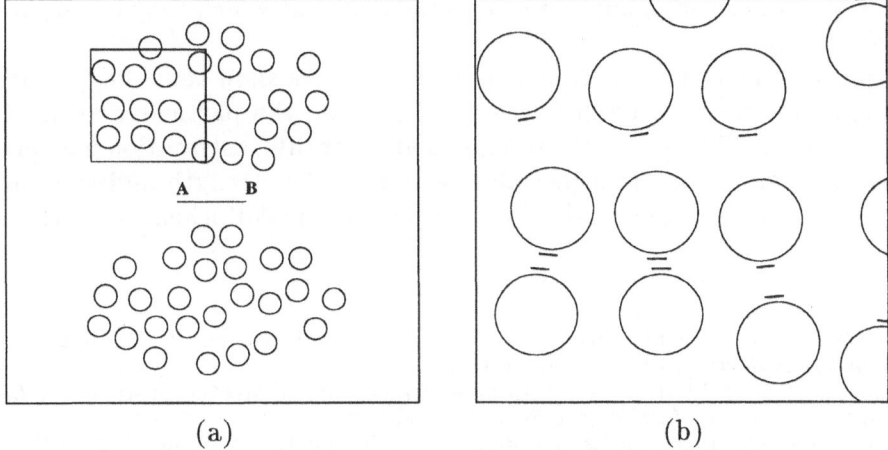

Figure 11. Configuration with two clusters of 25 fibers situated below and above a macrocrack; (a) initial state, (b) enlarged view of the enclosed area in the final state.

initial state the existence of fibers decreases the fracture toughness, i.e. $K_I/K_{I0} > 1$.

5. Conclusion

A method for calculating the stress field in a solid containing randomly dispersed fibers as well as determination of the stress intensity factors for cracks situated among the fibers has been established.

The dispersion pattern of fibers seems to be very influencial on the initiation of matrix and interface cracks around the fibers. Both the magnitude

26

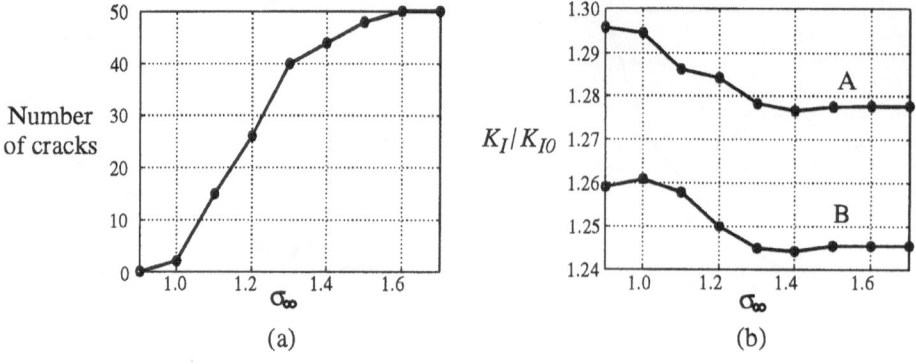

Figure 12. (a) Number of initiated cracks as a function of the applied load, (b) normalized stress intensity factor as a function of the applied load.

of the maximum stress components and the angle, at which they occur, are affected.

Typical configurations of fibers and a macrocrack show that the fracture toughness may increase or decrease depending on the particular arrangement of fibers. Therefore, descriptors that quantify distribution pattern of fibers should become indispensable factors in the strength and fracture analysis of composite materials, Pyrz(1994), Pyrz and Bochenek(1994).

References

Axelsen, M.S. (1994). *Quantitative Description of the Morphology and Microdamages of Composite Materials.*, Ph.D. thesis, in preperation

Kachanov, M. (1987). Elastic Solids with Many Cracks-Simple Method of Analysis., *International Journal of Solids and Structures*, **Vol. 23**, pp. 23-43

Mura, T. (1987). *Micromechanics of Defects in Solids.* 2nd Ed., Martinus Nijhoff Publishers, the Hauge, the Netherlands.

Muskhelishvili, N.I. (1962). *Some Basic Problems of the Mathematical Theory of Elasticity.* Noordhoff International Publishing, Leyden.

Pijaudier-Cabot, G. and Bažant, Z. P. (1991). Cracks Interacting with Particles or Fibers in Composite Materials., *Journal of Engineering Mechanics*, **Vol. 117, No. 7**, pp. 1611-1630.

Pyrz, R. (1994). Correlation of Microstructure Variability and Local Stress Field in Two-phase Materials., *Materials Science and Engineering*, **A177**, pp. 253-259.

Pyrz, R. and Bochenek, B. (1994). Statistical Model of Fracture in Materials with Disordered Microstructure., *Science and Engineering of Composite Materials*, **Vol. 3, No. 2**, pp. 95-110.

THE QUANTITATIVE MICROSTRUCTURE-PROPERTY CORRELATIONS OF COMPOSITE AND POROUS MATERIALS: AN ENGINEERING TOOL FOR DESIGNING NEW MATERIALS

A. R. BOCCACCINI, G.ONDRACEK
Institut für Gesteinshüttenkunde
-Glas, Bio- und Verbundwerkstoffe-
Mauerstr. 5, D-52056 Aachen, Germany

ABSTRACT. This paper reports theoretical and experimental work carried out in the field of microstructure-property correlations of porous and composite materials. It deals with the aim, to get a better scientific insight into the effects of microstructure on the properties of multiphase materials and to use the results technologically for designing purposes. In this context porous materials are considered to be the limiting case of multiphase materials, when one phase becomes gaseous. Equations for the mechanical properties of two phase materials are presented and the theory of the microstructure-property correlation via microstructural modelling is described. To satisfy the demand of maximun reliability from a theoretical as well as practical point of view, no fitting factors have been introduced into the equations and the properties of a matrix type composite (porous) material remain only dependent on the microstructural features and the concentration of the included phase. Finally, the case of the thermal shock resistance of porous ceramics is presented as an example of application of the microstructure-property correlations to design new materials with improved properties.

1. Introduction

In materials science the interrelationship between microstructure and properties became important not only to get a better scientific insight into the behaviour of materials but also because of the necessity to develop methods for designing new materials, showing the required properties and being economically advantageous and ecologically not suspicious. These new materials have to substitute less available, scarce or ecologically polluting conventional engineering materials. The essential problem that has to be considered is that of how the properties of composite materials depend on the properties of the separate phases, their volume fractions and their geometrical configuration or microstructure.

The selection of a suitable plan for the discussion of the physical properties of composites presents some difficulties. Therefore, treating properties in their relation to microstructure, it is instructive to define

27

R. Pyrz (ed.), IUTAM Symposium on Microstructure-Property Interactions in Composite Materials, 27–38.
© 1995 *Kluwer Academic Publishers.*

property groups for which similar considerations and treatments are valid. These groups are:

a) Thermochemical properties, which describe the behaviour of the material during -non mechanical- energy transfer and which are directly correlated with the atomistic bonding conditions as chemical bonding and thermal vibration of the atoms. Heat capacities, transformation heats or thermal expansion by heat absorption are examples for this group.

b) Field properties, which characterize the behaviour of materials under electrical, magnetic or thermal fields, as for example electrical and thermal conductivity.

c) Mechanical properties, referred to the behaviour of materials under stress-strain conditions, as for example modulus of elasticity, Poisson's ratio and fracture strength.

d) Technological properties, which are of especial practical interest and consist of a mathematical combination of primary properties. An example is the thermal shock resistance of brittle materials, which is influenced by the following material properties: Young's modulus, thermal conductivity, mechanical strength, coefficient of thermal expansion and Poisson's ratio.

While the thermochemical and field properties of two phase materials have been extensively treated in previous works [1-5] in this article the mechanical properties are considered. Theoretical approaches leading to the microstructure dependence of the mechanical properties of composite materials are reviewed and discussed in section 2 while section 3 presents an example of application of the microstructure-property correlations in designing a composite (porous) material with improved thermal shock resistance. Therefore the article discusses representatively the microstructure-property-correlation of two-phase materials and its use as a tool of "materials engineering" to design composite materials with predetermined "tailor-made" properties.

2. Mechanical properties of two phase materials

2.1. GENERAL CONSIDERATIONS

Since the aim of the present paper is to demonstrate, how a technological property like the thermal schock resistance of a composite or porous material may be tailored by proper microstructural design, only those mechanical properties which influence the thermal shock behavior are considered. These are: Young's modulus of elasticity (E), Poisson's ratio (v) and fracture strength (σ).

Although bound equations exist for both Young's modulus of elasticity [5] and Poisson's ratio [6] as well as for the fracture strength [7] of composite materials, the treatment here will be restricted to microstructure-property equations based on the spheroidal modelling of the microstructure as described in the next paragraph.

2.2. SPHEROIDAL MICROSTRUCTURAL MODEL

The chosen model proceeds from a real two-phased material, whose microstructure consists of a continuous matrix phase in which the particles of the inclusion phase are embedded discontinuously but macroscopically quasi-homogeneously. These particles, which are normally irregularly shaped in real materials, are replaced by spheroids, i.e. particles with a regular mathematically definable geometry and geometrical arrangement within the material. The mean shape is given by the ratio of the rotational axis (z) to the minor axis (x) of the spheroid (z/x), see figure 1-a. To obtain this, each real particle is considered to be replaced by an spheroid having the same surface-to-volume ratio as the real particle and therefore a specific axial ratio. For a given axial ratio there are two alternatives for substituting the real particles of the inclusion phase, namely, either by an oblate (z/x <1) or a prolate one (z/x >1). The mean orientation of the substituting spheroid is determined by the orientation of the rotational axis to the stress direction, see figure 1, and is given by $\cos^2\alpha_D$. As shown in previous studies [3,4], if one assumes an statistically homogeneous distribuion of the second phase in the matrix only these parameters, shape and orientation, are required for the complete characterization of the microstructure in addition to the volume fraction of the included phase, or phase concentration factor. Spheroidal characterization of the inclusion phase particles offers the advantage of high adaptability to real irregular geometries by changing the axial ratio. The extreme cases include disc-shaped (z/x -> 0, platelets) and cylindrical inclusions (z/x ->∞ , fibres), whilst spherical inclusions are realized as a special case (z/x =1). How to determine the substituting spheroid best-suited to a real structure by quantitative microstructural analysis and stereological functions has been shown in previous works [4,5].

2.3. YOUNG´S MODULUS OF ELASTICITY

Contrarily to the bound concept, where variational methods are used [5,8], the model concept to be discussed here uses direct methods in which the averaged stresses and strains are calculated with the aid of an effective Hooke's tensor. The derivation starts assuming a two-phase material with matrix phase microstructure, where the two phases behave isotropically. The two-phase material then is subdivided into elementary cells (finite elements), where the elementary cell consists of a cube of given elastic materials in which the spheroidal inclusions in any orientation are discontinuosly embedded in the matrix phase, as shown in figure 1-a. The mean stresses and strains are calculated for this elementary cell by dividing it into small, disjunct prisms (see figure 1-b). An effective modulus of elasticity is approximately calculated for each prism. The final effective modulus of elasticity is determined on the basis of a new averaging over all prisms. The exact derivation of the equation has been published recently [9]. The effective Young's modulus of elasticity can be given in terms of the elastic moduli of the matrix and inclusion phases and the microstructural parameters as:

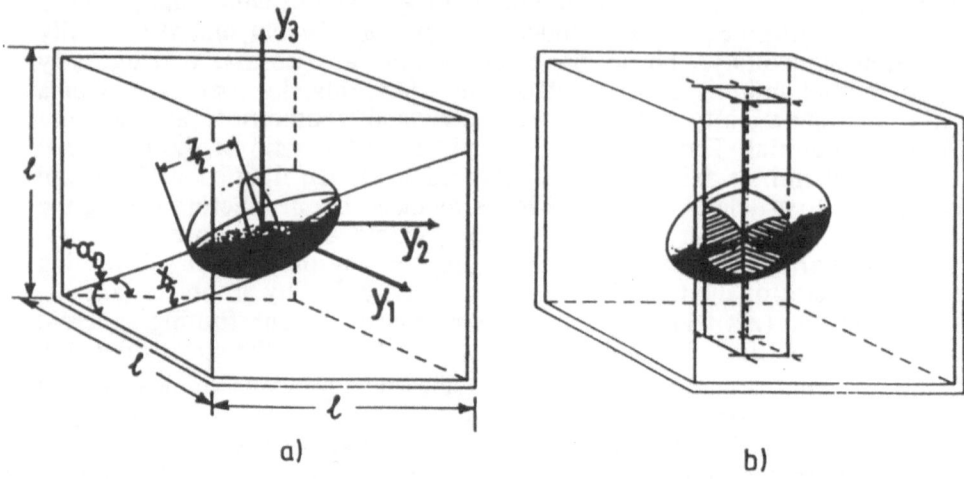

Figure 1. Spheroidal microstructural modelling: a) definition of shape and orientation, b) elementary cell and prism for derivation of eq. 1

$$E_C = E_M \left\{ 1 - \frac{\pi}{A} \left[1 - \frac{1}{9 \left(1 + \frac{1,99}{B} \left(\frac{E_M}{E_D} - 1 \right) \right)} - \frac{1}{3 \left(1 + \frac{1,68}{B} \left(\frac{E_M}{E_D} - 1 \right) \right)} - \frac{1}{\frac{9}{5} \left(1 + \frac{1,04}{B} \left(\frac{E_M}{E_D} - 1 \right) \right)} \right] \right\} \quad (1)$$

with

$$A = \frac{\left(\frac{4\pi}{3\,c_D} \right)^{2/3} \left(\frac{z}{x} \right)^{-1/3}}{\sqrt{1 + \left(\left[\frac{z}{x} \right]^{-2} - 1 \right) \cos^2 \alpha_D}} \qquad \text{and}$$

$$B = \left(\frac{4\pi}{3\,c_D} \right)^{1/3} \left(\frac{z}{x} \right)^{1/3} \sqrt{1 + \left(\left[\frac{z}{x} \right]^{-2} - 1 \right) \cos^2 \alpha_D}$$

E_D and E_M are the Young's moduli of the inclusion and the matrix phase respectively.

If in eq. 1 the Young's modulus of the inclusion is assummed to be zero, then the effective modulus of elasticity of porous materials is obtained as a function of volume fraction porosity (P) and pore structure as follows, with a correction made for considering the boundary condition $E_P = 0$ at $P = 1$ [10]:

$$E_P = E_M \left(1 - P^{2/3}\right)^S$$

$$S = 1.21 \left[\frac{z}{x}\right]^{1/3} \sqrt{1 + \left(\left[\frac{z}{x}\right]^{-2} - 1\right) \cos^2\alpha_D} \tag{2}$$

E_M represents the Young's modulus of the fully dense matrix. The comparison between calculated and experimental data for particulate composites with glass, ceramic and polymer matrix [7] and for porous metals and ceramics [7,10] has been carried out and the suitability of the equations has been demonstrated.

2.4. POISSON'S RATIO

Being a dimensionless parameter Poisson's ratio is a very useful elastic property because it enters in a number of equations describing the fracure and deformation behavior of materials. Therefore it is theoretically interesting and practically useful to obtain its dependence on microstructure and second phase content accurately.

The derivation of the equations has been carried out recently [7] so that only details will be given here.

The way to derive the dependence has been to consider the relationship between the Poisson's ratio and the elastic constants Young's and bulk modulus in isotropic materials and the known microstructural depencence of these elastic constants. While the microstructural dependence of the Young's modulus is accurately known (eqs. 1 and 2), for the bulk modulus only equations for spherical geometry are available [11,12]. Therefore the derived equations for the Poisson's ratio are strictly valid for spherical inclusion phase. For the case of porous materials, which is the relevant case for this study, the final equation has been derived to be valid on the whole porosity range [13] as:

$$v_P = 0.5 - \frac{\left(1 - P^{2/3}\right)^{1.21}}{4\left[(1-u)\dfrac{(3-5P)(1-P)}{2(3-5P)(1-2v_M)+3P(1+v_M)} + u\dfrac{(1-P)}{3(1-v_M)}\right]} \tag{3}$$

$$u = \frac{1}{1 + e^{-100(P-0.4)}}$$

where v_M represents the Poisson's ratio of the porous-free matrix. For the high porosity range the theoretical variation of the Poisson's ratio exhibits a trend of converging to a value $v_P = 0.5$, when the porosity increases to P=1. A similar convergence trend has been found in other theoretical studies [14] but a rigorous experimental verification of such variations is still to be done. For the low porosity range eq. 3 has been tested succesfully by comparison with extensive experimental data on porous ceramics [15].

2.5. FRACTURE STRENGTH

The starting point to study the microstructural dependence of the fracture strength is to consider the case of porous materials not only because it is more simple but also because of its practical relevance.

A rigorous derivation, in which the shape of the included phase is a variable of the system is formidable even with nowadays computers. This is why again the spheroidal modelling of the microstructure is applied, permitting the quantification of the shape and orientation effect of the pores. In order to assess the porosity dependence of the rupture strength two effects have to be considered:

i) the reduction of the load-bearing area due to the presence of pores and

ii) the stress concentration originated at the pores.

While the load-bearing area reduction can be calculated knowing the volume fraction of porosity [16], the model substitution allows the calculation of the stress concentration by using the equations of the theory of elasticity in three dimensions [17]. This path provides the equations in order to calculate the effective rupture strength of porous materials σ_P in dependence of the volume fraction of porosity P and the pore structure as follows [7]:

$$\sigma_P = \sigma_M (1 - P)^K$$

$$K = f\left(\left[\frac{z}{x}\right], \cos^2\alpha_D\right) \tag{4}$$

where σ_M is the strength of the fully dense matrix and K is the stress concentration factor, which results as a function of the shape and orientation of the pores and of the Poisson's ratio v_M of the matrix. Due to the complicated mathematical nature of the equations involved in the calculation [17] a computational programm was developed to calculate the stress concentration factor K as a function of shape and orientation of the pores for variing Poisson's ratio of the matrix phase [7]. The results for a matrix phase with Poisson's ratio $v_M = 0.25$ are plotted in figure 2, where the stress concentration factor K varies with the orientation angle for different axial ratios of oblate (figure 2-a) and prolate (figure 2-b) spheroids. The calculations [7] also showed, that there is no significant effect of the Poisson's ratio of the matrix on the stress concentration factor. The results in figure 2 can therefore be used with accuracy for all materials with v_M between 0.1 and 0.4. Calculated values using eq. 4 have been satisfactorily compared with experimental data on porous glasses, ceramics and metals with different porous structures [7,18].

The determination of the fracture strength of a composite material is *by an order of magnitude more difficult* than the problem of property prediction of other properties as pointed out by Hashin [8]. A recent survey [7] has shown that there is considerable work dedicated to this topic in the literature and that the different theoretical approaches can not predict all experimental results. The reason is the great number of variables influencing the problem, which can not be considered in a unique formula at the present state-of-the-art. Besides the strength of the phases involved and their microstructural arrangement, the mismatch between the elastic moduli and thermal expansion coefficients of the phases and the strength of the bonding and further mechanisms at the interfaces are of great importance in determining the fracture strength of the composite.

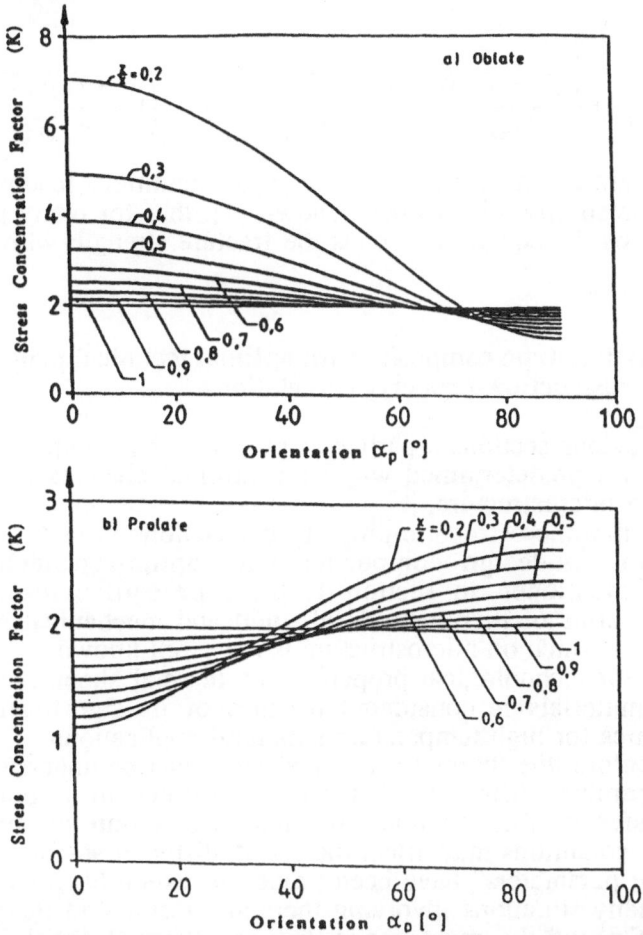

Figure 2. Stress concentration factor (K) for a) oblate and b) prolate spheroids of different axial ratios as a function of their orientation to stress direction

For composites without thermal expansion mismatch and perfect bonding between matrix and inclusions a load-sharing mechanism has been proposed [19] to determine the final strength of the composite. If the components in the system share the applied load in proportion to their elastic moduli, the strain in all components in unidirectional tension will be the same, i.e. both the second phase and the matrix must deform equally. It follows that the load to failure, and consequently σ_C, varies proportionally to the Young's modulus of the composite. Using the known microstructural dependence for the Young's modulus (eq. 1) the following equation has been proposed [7] for the variation of the fracture strength of composite materials with matrix-type microstructure:

$$\sigma_C = \sigma_M \left\{ 1 - \frac{\pi}{A} \left[1 - \frac{1}{9\left(1 + \frac{1,99}{B}\left(\frac{E_M}{E_D} - 1\right)\right)} - \frac{1}{3\left(1 + \frac{1,68}{B}\left(\frac{E_M}{E_D} - 1\right)\right)} - \frac{1}{9\left(1 + \frac{1,04}{B}\left(\frac{E_M}{E_D} - 1\right)\right)} \right] \right\} \quad (5)$$

where A and B are given by eq. 1. Although the conditions leading to eq. 5 seem to be too restrictive, it has been shown [7], that for many particulate composite systems the equation predicts the fracture strength with sufficient accuracy.

3. Design of a matrix-type composite with optimal technical properties using the microstructure-property correlations

As shown in previous sections a particular property of a composite material can be varied in a predetermined way by controlled changes in the phase composition and microstructure.

In this context, the microstructure-property correlations provide a powerful engineering tool to design composites with optimized technological properties. As mentioned in section 1, these properties result from a combination of terms of thermochemical, field and mechanical properties, the dependence of which on microstructure is now well known.

As an example for technological properties the thermal shock resistance of porous brittle materials is considered because of its significance in the choice of ceramics for high temperature structural applications.

As mentioned before, the thermal shock resistance can be understood as the maximum temperature difference that can be tolerated in a ceramic body under heat transfer conditions without thermal stress failure occuring. Since different testing conditions may affect the result, different so-called "thermal shock resistance parameters" have been proposed, which have been already compiled for many situations involving thermal stresses and thermal stress fracture [20]. One of the most generally used thermal shock resistance parameters, named R_{TS}, and originally derived in the last century [21] is defined as:

$$R_{TS} \equiv \frac{\sigma \ \phi}{\alpha \ E} (1 - \nu) \quad (6)$$

where ϕ is the thermal conductivity and α is the thermal expansion coefficient of the material. The R_{TS}-parameter characterizes the resistance to fracture initiation under steady state heat conduction. The materials properties involved in eq. 6 can now be substituted by porosity functions from the microstructure-property correlations. As shown in previous studies [1] the thermal expansion coefficient of porous materials does not depend on porosity. Moreover the variation of the Poisson's ratio with porosity has only a minor effect on the thermal shock behavior of the composite, as a recent theoretical and experimental study has demonstrated [22]. Therefore only the porosity functions of the thermal conductivity, the Young's modulus and the fracture strength have to be considered and substituted in eq. 6. Eqs. 2 and 4 give the porosity dependence of the Young's modulus and the fracture strength respectively, while the following relation has been derived [4] for the thermal conductivity of porous materials:

$$\phi_P = \phi_M (1 - P)^R$$
$$R = \frac{1 - \cos^2\alpha_D}{1 - F_D} + \frac{\cos^2\alpha_D}{2 F_D}$$

(7)

The thermal conductivity, of the porous material (ϕ_P) appears as a function of the conductivity of the matrix phase (ϕ_M), the porosity, the shape factor (F_D), which is a function of the axial ratio of the pores [4], and the orientation factor ($\cos^2\alpha_D$).
Substituting eqs. 2, 4 and 7 in eq. 6 the thermal shock resistance $R_{TS}(P)$ of a porous ceramic material normalized to the property of the fully dense material $R_{TS}(0)$ results as:

$$\frac{R_{TS}(P)}{R_{TS}(0)} = \frac{(1 - P)^{K+R}}{\left(1 - P^{2/3}\right)^S}$$

(8)

where R, S and K depend on the porosity structure, i.e. shape and orientation of the pores. Thus three variables remain influencing the material's property. By changing these variables properly it is possible to optimize the final value of the property and hence to design a porous ceramic with improved thermal shock resistance.
Figure 3 shows, for example, the variation of the relative thermal shock resistance with the pores axial ratio (z/x) for parallel oriented pores ($\cos^2 \alpha_D$ =1). Values for three different porosities (P) are shown. While for porosities above P=0.1 the porous ceramic behaves invariably worse than the fully dense body ($R_{TS}(P)/R_{TS}(0)<1$), for much lower porosities (P=0.01), the porous body has a better thermal shock resistance than the fully dense one for a wide range of pores axial ratios. It is significant to note that the same behaviour of the fully dense body can be reached by a porous body containing about 5% of residual porosity providing the pores have an axial ratio z/x $\approx$ 0.7.
    A possibility of improving the thermal shock behavior of porous ceramics at higher porosities is shown in figure 4, where the variation of the relative thermal shock resistance is represented as a function of the porosity

for pores having an axial ratio $z/x = 10$ (cylindrical porosity). The curves shown represent the values for statistical ($\cos^2 \alpha_D = 0.33$), perpendicular ($\cos^2 \alpha_D = 0$) and parallel ($\cos^2 \alpha_D = 1$) orientation of the pores. For this kind of pore structure, the perpendicular orientation provides the best result for improving the thermal shock behavior, reaching a maximum at a volume fraction porosity of about 7%. Moreover, the property of the porous body remains above that of the fully dense one up to a porosity of aprox. 20%.

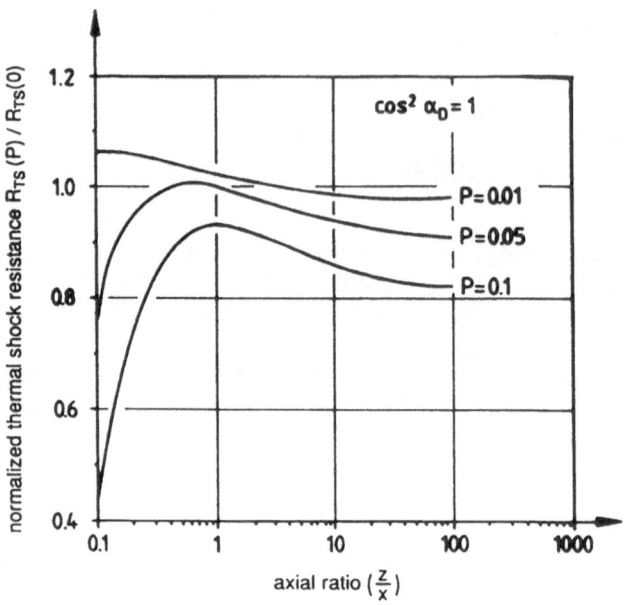

Figure 3. Relative thermal shock resistance of a porous brittle material as a function of the axial ratio of the pores for different porosities and orientation $\cos^2 \alpha_D = 1$.

These theoretical results are alltogether in qualitative agreement with many statements found in the literature [23] concerning the initial increase of thermal shock resistance with the volume fraction of pores of ceramic materials. Experimental verification of the predictions of eq. 8 have been made for sintered glass and $CaTiO_3$-TiO_2 ceramics containing spherical pores [22,24]. Work is in progress to test the theoretical predictions with experiments for porosity structures other than spherical [7], since the experimental verification of Eqs. 2, 4 and 7 is an indirect confirmation for eq. 8 too. Thus, microstructure-property correlations together with appropiated processing parameters form a useful basis to obtain desired porosity structures in order to design porous ceramics with optimized thermal shock resistance. A similar treatment as the one presented here for porous materials is being investigated for the improvement of the thermal shock resistance of dense brittle matrices by addition of second phase particles [22]. These considerations may obviously be extended for other technological properties or group of properties making the microstructure-property correlation an essential tool for composite design.

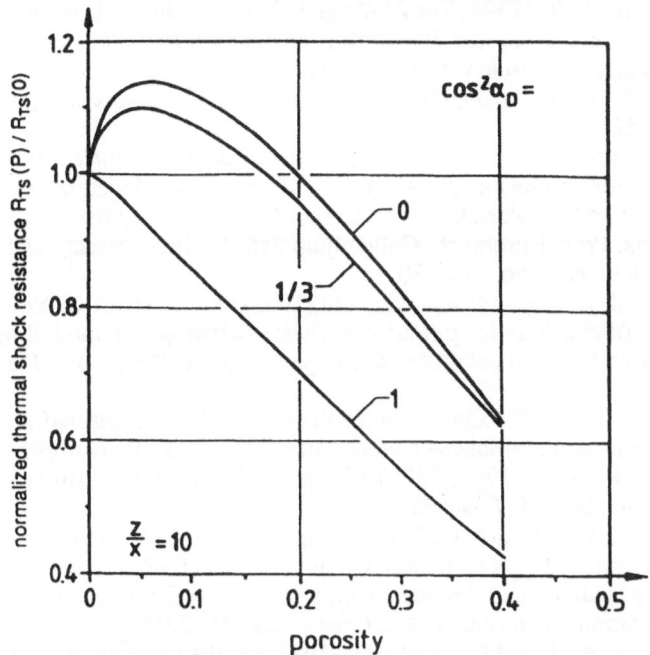

Figure 4. Relative thermal shock resistance of a porous brittle material as a function of porosity for different orientations of pores and axial ratio z/x=10.

4. References

[1]   Nazare, S. and Ondracek, G. (1978) Zum Zusammenhang zwischen Eigenschaften und Gefügestruktur mehrphasiger Werkstoffe. *Z. Werkstofftech.* 9, 140-147.

[3]   Bossert, J. (1993) Zur Haftung von Verbundwerkstoffen im festen Aggregatzustand. PhD Thesis Rheinisch-Westfälische Technische Hochschule RWTH Aachen.

[3]   Ondracek, G. (1982) Zur quantitativen Gefüge-Feldeigenschafts-Korrelation mehrphasiger Werkstoffe. *Metall* 36, 523.

[4]   Ondracek, G. (1987) The quantitative microstructure field property correlation of multiphase and porous materials. *Reviews on Powder Metallurgy and Physical Ceramics* 3, 205-322.

[5]   Ondracek, G. (1986) Microstructure-thermomechanical-property correlations of two phase and porous materials. *Mat. Chem. Phys.* 15 281-312.

38

[6]   Kreuzberger, S. (1993) Zum Grenzwertkonzept für mechanische und thermochemische Eigenschaften mehrphasiger Werkstoffe. PhD Thesis RWTH Aachen.

[7]   Boccaccini, A. R. (1994) Zur Abhängigkeit der mechanischen Eigenschaften zweiphasiger und poröser Werkstoffe von der Gefüge- bzw. Porositätsstruktur. PhD Thesis, Rheinisch-Westfälische Technische Hochschule RWTH Aachen.

[8]   Hashin, Z. (1983) Analysis of composite materials. A survey. *J. Appl. Mech.* 50, 481-505.

[9]   Mazilu, P and Ondracek, G. (1989) On the effective Young's modulus of elasticity for porous materials. Part I: The general model equation, in K. Herrmann and Z. Olesiak (eds.).*Thermal Effects in Fracture of Multiphase Materials*. Proc. Euromech. Colloquium 255. Springer Verlag Heidelberg Tokyo New York pp. 214-230.

[10]   Boccaccini, A. R., Ondracek, G., et al.(1993) On the effective Young's modulus of elasticity for porous materials: microstructure modelling and comparison between calculated and experimental values. J. *Mech. Behav. Mat.* 4, 119-126.

[11]   Ondracek, G. (1978) Zum Zusammenhang zwischen Eigenschaften und Gefügestruktur mehrphasiger Werkstoffe. Teil III. *Z. Werkstofftech.* 9 31-36.

[12]   Walsh, J. B., et al. (1965) Effect of porosity on compressibility of glass. *J. Am. Ceram. Soc.* 48, 605-608.

[13]   Arnold, M., Boccaccini, A. R. and Ondracek, G. (1994) Prediction of the Poisson's ratio of porous materials. Submitted to *J. Mat. Sci.*

[14]   Ramakrishnan, N. and Arunachalam, V. S. (1993) Effective elastic moduli of porous ceramic materials. *J. Am. Ceram. Soc.* 76, 2745-52.

[15]   Boccaccini, A. R. and Ondracek, G. (1993) On the porosity dependence of the Poisson's ratio in ceramics. *Ceramica Acta* 5, 61-66.

[16]   Griffiths, T. J., Davies, R. and Basset, M. B. (1979) Analytical study of effects of pore geometry on tensile strength of porous materials. *Powd. Metall.* 22, 119-123.

[17]   Sadowsky, M. A. Sternberg, E. (1949) Stress concentration around a triaxial ellipsoidal cavity. *J. Appl. Mech.* 16 149-155.

[18]   Boccaccini, A. R. and Ondracek, G. (1993) On the porosity dependence of the fracture strength of ceramics, in P. Durán and J. F. Fernandez (eds.). *Third Euro-Ceramics* 3, 895-900.

[19]   Borom, M. P. (1977) Dispersion-strengthened glass matrices - glass-ceramics, A case in point. *J. Am. Ceram. Soc.* 60, 17-21.

[20]   Hasselman, D. P. H. (1970) Thermal stress resistance parameters for brittle refractory ceramics: a Compendium. *Am. Ceram. Soc. Bull.* 49, 1033-37.

[21]   Winkelmann, A., Schott, O. (1894) Über thermischen Widerstandskoeffizienten verschiedener Gläser in ihrer Abhängigkeit von der chemischen Zusammensetzung. *Ann. Phys. und Chemie* 51, 730.

[22]   Boccaccini, A. R. and Ondracek, G. (1993) Erhöhung der Thermoschockbeständigkeit von Sinterglas und Keramik über das Verbundwerkstoffkonzept. *Mat.-wiss. u. Werkstofftech.* 24, 450-456

[23]   Salmang, H. and Scholze, H. (1982) *Keramik*. Springer Verlag, p. 253.

[24]   Jauch, U. (1988) Zur Thermoschockfestigkeit mehrphasiger Werkstoffe. PhD Thesis, Rheinisch-Westfälische Technische Hochschule RWTH Aachen.

CRACK GROWTH IN A COMPOSITE WITH WELL ALIGNED LONG FIBERS

J. BOTSIS, D. ZHAO and C. BELDICA
Department of Civil and Materials Engineering
University of Illinois at Chicago
842 W. Taylor St. Chicago, IL 60607, U.S.A.

ABSTRACT. Longitudinal strength σ_c of an epoxy reinforced with one layer of long aligned and equally spaced glass fibers has shown that for a certain fiber spacing λ, $\sigma_c\sqrt{\lambda} = \kappa$, where κ is a constant. A similar expression was suggested for the strength of a borosilicate glass - SiC fibers composite under bending and the maximum fiber spacing. Steady crack growth under fatigue was observed in the glass - epoxy system. Using experimental data and a simple analysis, the forces on the fibers in the bridging zone were found equal. Analysis indicated that the total stress intensity factor was constant at the steady growth mode. Power relations were used to correlate steady speed with the stress intensity factor and the rate of debonding with the applied stress.

1. Introduction

When long aligned fibers are used as a reinforcement in a brittle matrix the result is a composite material with improved mechanical properties and enhanced resistance to crack growth. Resistance to crack growth in this class of materials comes from two important sources. The first one results from bridging of the crack faces by fibers. The second one is from crack bowing and trapping. Depending on the material types, interfacial characteristics, geometry and loading conditions these two mechanisms may lead to crack deceleration or crack arrest.

Analytical research on the effects of bridging and crack bowing on stress intensity factors and crack growth behavior of composite with long aligned fibers has been reported by several researchers [1-12]. Although important progress has been achieved in understanding the role of reinforcement in the composites' behavior, experimental studies have received much less attention in the literature. In particular, issues related to the effects of fiber spacing and fiber types on strength and crack growth have not been addressed experimentally.

An experimental research program was initiated to address issues related to strength and crack bridging in model composite systems with long aligned fibers [13-15]. In this paper, a summary of results on strength and crack propagation on specimens with one layer of well aligned and equally spaced glass fibers (mono-layer) in an epoxy matrix are reported. Work on multi-layer composites will be reported elsewhere [16].

39

R. Pyrz (ed.), IUTAM Symposium on Microstructure-Property Interactions in Composite Materials, 39–50.
© 1995 *Kluwer Academic Publishers.*

2. Experimental Approach

The specimens used in the experimental studies were, unidirectional, single lamina composites. The matrix material was an epoxy and the reinforcement consisted of glass fibers in a bundle form with an overall diameter of 0.4 mm. The Young's moduli of the matrix and the fibers were E_m = 3.5 GPa, and E_f = 72.5 GPa. The strain-to-failure of the matrix is less than that of the fibers. Therefore the fibers in the crack wake do not fail and thus all of them contribute to the bridging of the crack faces. This is considered to be a typical case of large-scale bridging. Moreover, the system is sufficiently simple to allow for an in situ observation of crack growth, debonding, crack opening displacement (COD), any dissipative mechanisms in the matrix material as well as crack front changes due to the reinforcement. Details of the specimen preparation procedures and experimental methods can be found in [13].

So far in this work we have dealt with strength and fatigue crack propagation. Longitudinal strengths of the composite material were determined using smooth specimens with different fiber spacing that were pulled to fracture. For the fatigue testing, a 60° angle notch of 1 mm depth was milled at the middle of the specimen edge. Various experiments each having different fiber spacing and load levels were performed [13,14]. All experiments were load controlled with a sinusoidal waveform function and various levels of stress. It should be noted that because we were interested in the steady state, one value of crack speed was obtained from each experiment.

3. Results and Discussion

3.1. STRENGTH

The strength characteristics of the composite material were determined using smooth specimens that were pulled to fracture. To identify the effects of fiber spacing on strength ramp tests were performed on specimens with different fiber spacing and the same overall specimen dimensions. In all cases, fracture occurred first in the matrix while the fibers remained intact until the crack run across the specimen width.

Conventionally, longitudinal strength of composite materials is described in terms of volume fractions and the strength of the fiber or the matrix material. This approximation is based on the rule of mixture and that the fibers possess uniform strength [17]. The volume fraction used in the rule of mixture does not always represent the local morphology (especially when there is no regularity in fiber spacing) because it is a volume average parameter. Therefore, the effects of scale on strength can not be formulated. To directly relate the composite's strength with a characteristic length scale i.e., fiber spacing, we turn to another treatment of the experimental data.

A useful tool in analysis of experimental data as well as in the study of size effects is dimensional analysis. Through dimensional analysis the important parameters of the problem at hand can be identified and the relationship between the depended and independent variables can be illustrated.

In the present work, we are interested in the longitudinal strength of a composite specimen σ_c. It is then assumed that the governing parameters of the material system are: fracture toughness of the matrix material K_{mc}, the Young's moduli of the matrix

E_m, and of the reinforcing fibers E_f; the respective Poisson's ratios v_m, and v_f; the fiber spacing λ, and fiber diameter D; the specimen width w, and thickness h. Choosing $\mathbf{K_{mc}}$, and λ as fundamental parameters σ_c can be expressed as

$$\sigma_c = \frac{\mathbf{K_{mc}}}{\sqrt{\lambda}}\varphi_1(\frac{E_m\sqrt{\lambda}}{\mathbf{K_{mc}}}, \frac{E_f\sqrt{\lambda}}{\mathbf{K_{mc}}}, \frac{D}{\lambda}, \frac{w}{\lambda}, \frac{D}{h}, v_m, v_f) \tag{1}$$

Where, the dimensionless parameters in the function φ_1 are Π parameters, $\Pi_1 = E_m\sqrt{\lambda} / \mathbf{K_{mc}}$, $\Pi_2 = E_f\sqrt{\lambda} / \mathbf{K_{mc}}$, $\Pi_3 = D/\lambda$, $\Pi_4 = w/\lambda$, $\Pi_5 = D/h$, $\Pi_6 = v_m$, $\Pi_7 = v_f$. The first four Π parameters depend on the fiber spacing λ. Furthermore, because the explicit form of the function is not known, it is difficult to examine its dependence on λ. For the glass - epoxy system [13] $\Pi_1 \gg 1$ ($\approx 10^2$), $\Pi_2 \gg 1$ ($\approx 10^3$), $\Pi_3 \gg 1$ (≈ 60), $\Pi_4 \gg 1$ (≈ 10). According to dimensional analysis, a complete self - similarity implies that large or small parameters (compared to unity) can be eliminated from the function φ_1 as long as there exists a limit of φ_1 for very large or very small variables [18]. With respect to the problem at hand, Π_1, Π_2 and Π_3 are very large and can be eliminated from φ_1. The fourth parameter is not very large in comparison to unity especially in the case of the largest fiber spacing used in this work (for $\lambda \approx 3$ mm, w/$\lambda \sim 8$). However, it assumed that it is sufficiently large and can also be eliminated. Note that the comparison with the experimental results will attest this assumption. Moreover, the ratio D/h remains the same in all specimens. Thus, the following limiting similarity law is obtained for the longitudinal strength of the composite specimen with well aligned fibers

$$\sigma_c = \frac{\mathbf{K_{mc}}}{\sqrt{\lambda}}\varphi_2(v_m, v_f, \frac{D}{h}) \tag{2}$$

Strength values plotted against $1/\sqrt{\lambda}$ are shown in Fig. 1a. Note that with the exception of one datum point, a straight line through the origin represents the data very well. Thus, the strength of the composite specimen scales with the square root of the fiber spacing because the dimensionless function, $\varphi_2(v_m, v_f, D/h)$, and $\mathbf{K_{mc}}$ can be taken as constants. It is worth noticing that for those specimens where $\sigma_c\sqrt{\lambda} =$ constant, fatigue crack growth and the rate of debonding in single edge notched specimens reached a steady state mode, i.e., independent of the crack length. In specimens with larger fiber spacing, crack speed versus crack length increased with large fluctuations [14].

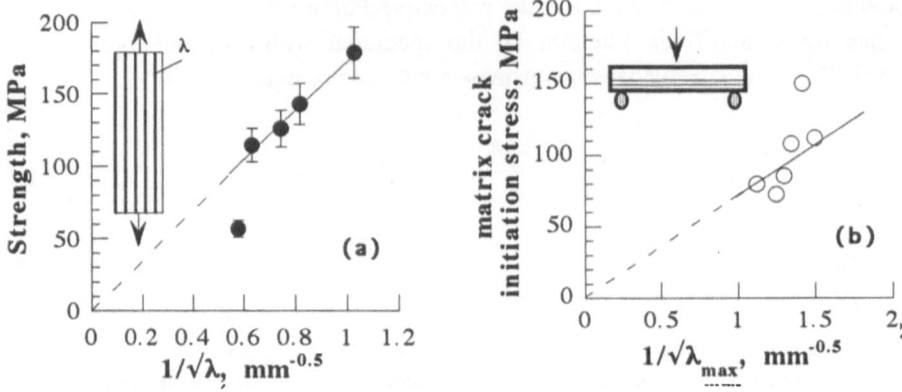

Fig. 1: strength with spacing for (a) epoxy - glass, (b) borosilicate glass - SiC

The similarity expression (2) is an important result for the model composite system with the range of fiber spacing examined in the present studies. However, in a real composite system fiber spacing is not regular. Thus, it is important to contemplate a similarity law akin to (2) in composite systems without a regular fiber spacing. Experimental data on strength versus a characteristic length scale are not available in the literature because strength usually is expressed in terms of fiber volume fraction. A set of such data on matrix initiation strength versus fiber spacing, recorded under three point bending on borosilicate glass reinforced with SiC fibers, has been reported [19]. Although, it is stated by the authors that the data were preliminary, they are examined here in an attempt to deduce a scaling expression similar to (2). Since there was no regularity in the fiber spacing in the borosilicate glass - SiC system, the parameters that reflect the specimen size and a set of length scales are introduced Λ_i , i = 1,..., n that represent fiber distances.

Recognizing that strength is a strong function of extreme local material heterogeneity, the fundamental parameters are taken as $\lambda_{max} = \Lambda_k = \max(\Lambda_1,..., \Lambda_{k-1}, \Lambda_k, \Lambda_{k+1},..., \Lambda_n)$ and K_{mc} . The choice of λ_{max} is also justified by the experimental observations that matrix cracking originated between the largest fiber spacing. Thus, the strength of the composite can be written as

$$\sigma_c = \frac{K_{mc}}{\sqrt{\lambda_{max}}}\Phi_1(\frac{E_m\sqrt{\lambda_{max}}}{K_{mc}}, \frac{E_f\sqrt{\lambda_{max}}}{K_{mc}}, v_m, v_f, \frac{t}{D}, \frac{w}{\lambda_{max}}, \frac{w}{D}, \frac{L}{\lambda_{max}}, \frac{\Lambda_1}{\lambda_{max}},..., \frac{\Lambda_{k-1}}{\lambda_{max}}, \frac{\Lambda_{k+1}}{\lambda_{max}},...\frac{\Lambda_n}{\lambda_{max}})$$

Here $K_{mc}, E_m, E_f, v_m, v_f$ and d have the same meaning as before; L, w, and t are the support span, width, and height of the beam, respectively. The parameters containing the Young's moduli and the ratios L / λ_{max} , t/D, and w/D are very large in comparison to unity. Moreover, because the distribution of Λ_i is not known it is assumed that their ratios with λ_{max} are small. Thus the function Φ_1 can be substituted by its limit when these parameters approach very large or very small values

$$\sigma_c = \frac{K_{mc}}{\sqrt{\lambda_{max}}} \Phi_2(v_m, v_f, \frac{w}{\lambda_{max}}) \tag{3}$$

The ratio w / λ_{max} is between 6 and 12. Such a quantity is not very large with respect to 1 when λ_{max} is largest. Nevertheless, considering its value to be large, one obtains $\sigma_c \sqrt{\lambda_{max}} = K_{mc} \Phi_3(v_m, v_f) = $ constant. Note that the data will determine the validity of this approximation.

The data in Fig. 1b show that, except for one datum point with strength of about 50 MPa, the matrix initiation stress σ_c, is well correlated with $1 / \sqrt{\lambda_{max}}$. Although this is a preliminary result, its implications are important because fiber spacing reflects the microstructure better that volume fraction, an average quantity, commonly used to correlate strength data in composite materials. This may be one of the reasons for the large scatter in strength versus volume fractions observed in brittle composites.

3.2. CRACK PROPAGATION BEHAVIOR

A typical behavior of crack speed with the crack length is shown in Fig. 2 [16]. Interestingly, three distinct regimes were observed: two transient phases separated by a steady phase. In all cases, the time to crack initiation largely depended upon the fiber spacing, the applied load and the distance of the notch tip to the first fiber. After initiation, a significant decrease in the crack speed was observed (Fig. 2). The extent of this behavior was dependent upon the fiber spacing, applied load and the distance of the notch tip to the first fiber. In some cases the crack speed increased upon initiation followed by a decrease [14].

The decrease in crack speed may be explained with the effects of the reinforcement on the stress field at the crack tip. It has been reported [20] that an inclusion, in front of a crack, with higher stiffness than the surrounding material lowers the stress intensity factor at the crack tip. Therefore when the crack approaches a fiber that is stiffer than the matrix, the local stress intensity factor is reduced leading to a deceleration of the crack speed.

After crack initiation and a transient phase, crack growth behavior depended upon the fiber spacing. In specimens with $\lambda \approx 3$ to 3.5 mm a tendency of increasing crack speed was seen albeit with large fluctuations [14]. In test pieces with smaller fiber spacing, fatigued under various loads, the crack speed reached a steady phase. That is, the crack speed and the rate of energy dissipation were independent of the crack length and cycle number.

After the steady phase of fracture, a decrease in the crack speed was observed. This decrease was accompanied by an increase in the energy dissipation (Fig. 2). To identify the sources of this behavior, attention was focused on the bridging zone and the bulk of the specimen. In particular two photoelastic sheets were used during testing to observe the specimen upon loading. Two photographs of the specimen are shown in Fig. 3. The morphology shown in Fig. 3A is typical of the steady phase of fracture. The photograph 3B was taken at the phase when a decrease in the crack speed was recorded.

The behavior shown in Fig. 3 may be explained in terms of time and/or temperature effects primarily on the matrix material [16]. At relatively short times, the material behaves as an elastic one with most of energy dissipated within the bridging zone and in creating crack surfaces.

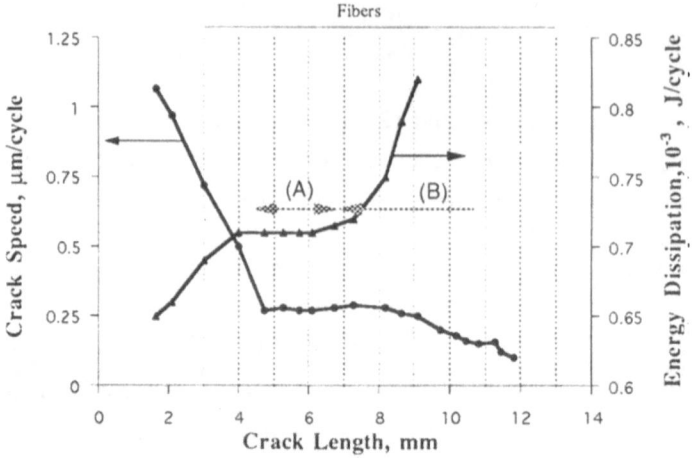

Fig. 2: crack speed plotted against the crack length

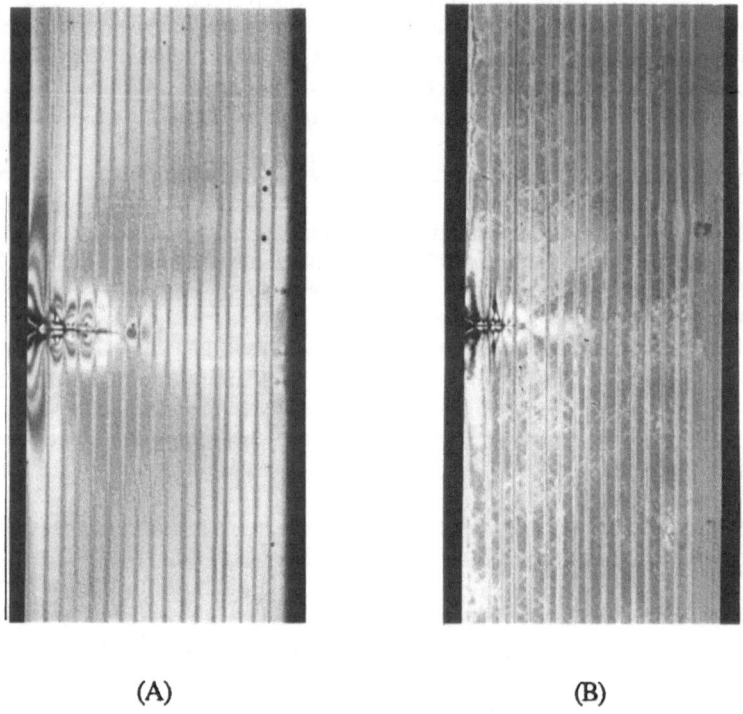

(A)   (B)

Fig. 3: photoelastic patterns at the steady phase (A) and transient phases (B)
(specimen width 25 mm)

As time progress, however, the rheological behavior of the matrix material changes due to temperature increase resulting from the cyclic load and/or due to creep. These changes in the matrix material result in an increase of energy absorbed by the matrix material. Consequently, less energy is spent within the bridging zone and crack propagation. Additional experimental and theoretical work is required to elucidate these important findings.

Most of the work so far has been aimed at understanding the steady phase of crack growth. Crack propagation rates $\Delta l/\Delta N$, plotted against the crack length l, are shown in Fig. 4. For the same fiber spacing the steady crack speed was an increasing function of the applied stress. For the same load level the steady speed depended upon the fiber spacing, i.e. the smaller the fiber spacing, the smaller the crack speed. For the same fiber spacing, the level of the average steady state speed depended upon the applied load. It is worth pointing out that steady crack speed was observed only in the specimens where $\sigma_c\sqrt{\lambda} = \kappa$. Moreover, for $\lambda \approx 3.5$ mm the average crack growth rate did not reach a constant value suggesting that for a given load, the steady state can be obtained for fiber spacing below a critical value. Thus both strength measurements and fatigue crack growth suggest a transition in the behavior of the composite material in terms of the fiber spacing.

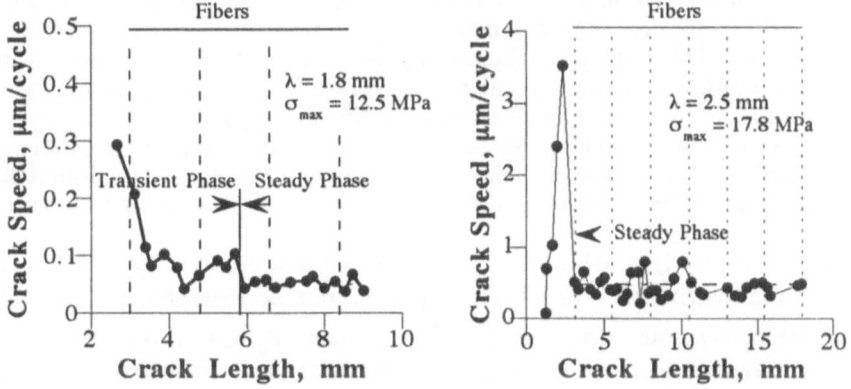

Fig. 4: crack speed for two typical fiber spacing

Crack opening displacements, measured at the maximum load of the fatigue cycle, are shown in Fig. 5a. Data points were obtained along the crack where fibers were located as a function of crack length. Note that the COD at a point where a fiber in the bridging zone is located is linearly related to debonding (Fig. 5a) and that the linear relationship is the same for the fibers in the bridging zone. Moreover, CODs at the points where the fibers are located vary linearly with crack length. The linearity with crack length indicates a constant rate of growth in COD in the steady state because crack length and cycle number are linearly related [13]. Within the resolution of the observations, fiber debonding, friction and some filament fracture were assumed to have contributed to energy dissipation.

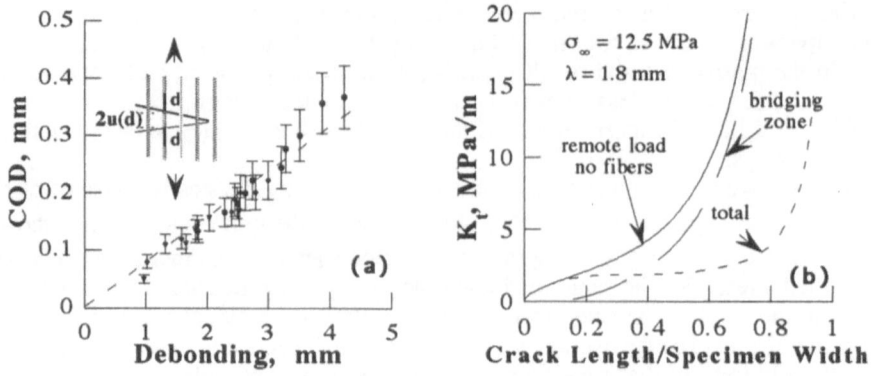

Fig. 5: (a) COD versus debonding, (b) evolution of stress intensity factors

4. ANALYSIS

The experimental results reported in this paper have demonstrated that crack propagation in uniaxially reinforced composite specimens exhibit a steady state behavior. Theoretical works on the dependence of matrix cracking stress on material properties in the steady state have been reported. These analyses use a shear lag approximation for the stresses on the fibers and energy balance [1-3] or a stress intensity factor based approach with a uniform distribution of tractions in the bridging zone [4-5]. Although these analyses provide an important understanding of the composite's fracture, they are not easily applicable here because the reinforcement cannot be substituted by continuous tractions on the crack planes and the effects of the fiber spacing cannot be investigated.

Because of the steady growth mode, it is assumed that the total stress intensity factor K_t, at the matrix crack tip, arising from the remote load and the fibers, is constant at the steady state. The next step is to calculate K_t and its dependence on fiber spacing. Towards this end, it was assumed that for a crack bridged by fibers the principle of superposition applies and that the level of residual stresses due to specimen preparation was negligible. Thus the stress intensity factor K_t, is expressed as

$$K_t = K(\sigma_\infty, l) - \sum_{i=1}^{n} K_i^f(P_i, c_i) + \Delta K \qquad (4)$$

Where $K(\sigma_\infty, l)$ is the stress intensity factor due to the applied stress σ_∞, on a homogeneous specimen with a crack of length l, originating from the middle of the edge. $K_i^f(P_i, c_i)$, is the contribution to K_t of the i-th fiber, in the bridging zone, expressed as the effect of a closure force P_i acting at a distance c_i from the specimen edge. The sum is over the number of fibers bridging the crack. The correction ΔK is due to an effect arising from the fibers ahead of the crack front. Both $K(\sigma_\infty, l)$ and $K_i^f(P_i, c_i)$ can be evaluated using standard procedures. For the fiber spacing employed in the present studies and considering in the calculations that the crack tip was located in the middle of two consecutive fibers, ΔK was presumed negligible [13].

Evaluation of $K(\sigma_\infty, l)$ does not possess any particular difficulty. The contribution of the reinforcements to the total stress intensity factor, however, can be evaluated only if the forces carried the fibers are known. These quantities are difficult to determine experimentally. To obtain a better insight as well as relationships between the forces on the fibers bridging the crack, the debond length and the COD, the stress - displacement relations of a fiber in the bridging zone were analyzed using a frictional model similar to that in [4]. According to this analysis [13] the COD $u(d)$ at a typical fiber in the bridging zone is

$$u(d) = \frac{2}{E_f r}\left(kd \int_0^d \tau(d,x)dx + (1-k)\int_0^d dx \int_0^x \tau(d,\eta)d\eta \right) \tag{5}$$

Where k is a constant, d is the debonding along the fiber and τ is the shear stress.

The experimental data have shown that, at steady state, the crack opening displacement and debonding are linearly related (Fig. 5a). Therefore the integral in (5) $\int_0^d \tau(d,x)dx$, should not depend on d. A simple expression for the shear stress can be derived, considering that for fixed d it does not vary along the interface, i. e. $\tau(d,x) = \tau(d)$. However, when the debonding increases the shear stress decreases such that the product $\tau(d)d$ remains constant, say to p_0. If $\tau(d,x) = p_0/d$ one obtains $\sigma(d) = 2(1+k)p_0/r$ and $u(d) = (1+k)dp_0/E_f r$.

To continue, the constant p_0 should be identified. Assuming that no fiber failure occurs and that the applied stress σ_∞ is equally taken by all fibers, the maximum stress on fiber is $(\sigma_f)_{max} = \sigma_\infty/V_f$, where V_f denotes the fiber volume fraction. Considering that the stress carried by the fibers is distributed across the thickness, the maximum value of the closing traction due to a fiber in the bridging zone is $P_{max} = \sigma_\infty B/N$ where N is the total number of fibers in the composite specimen. Assuming that the fibers in the bridging zone carry the same load and equal to $P_{max} = \sigma_\infty B/N$ simulations were carried out to evaluate K_t. Typical results of simulations are shown in Fig. 5b. Note that K_t is constant for $l/B \sim 0.2 - 0.4$. Afterwards, the interaction of the crack front and the specimen edge leads to an increase of K_t that results in specimen fracture.

Considering that K_t is constant at the steady state, the next step would be to correlate the steady speed for various fiber spacing and applied loads with K_t. Towards this end, an empirical power relation can be used. It would be useful, however, to explore the importance of the fiber spacing on crack speed and ascertain any similarity parameters. Due to lack of analytical work at present to help in such understanding an analysis of the steady state fracture process using dimensional arguments has been attempted to obtain guidance towards a better experimental design as well as analytical research [13]. From this analysis it was shown that for the case when the constituent materials, specimen geometry and fiber spacing are kept the same, the crack speed can be expressed as

$$\frac{\Delta l}{\Delta N} = A \, K_t^{n_1} \tag{6}$$

Where both A and $n_1 = 2+\alpha$ are parameters that depend upon the specimen geometry, material and interfacial properties. The applied load controls the crack speed only through K_t. Dimensional analysis suggested certain procedures to be followed when investigating the effects of fiber spacing and applied load on the steady crack speed. Namely, when studying the effects of load level on steady speed, the fiber spacing should be kept the same from experiment to experiment otherwise the exponent may vary in a way that is difficult to explain because the explicit form of A and the exponent $n_1 = 2+\alpha$ are unknown. This is also the case when the fiber spacing is changed from experiment to experiment while the stress level remains the same. This procedure was followed in the experimental part of this research [13].

Plots of $\Delta l / \Delta N$ these data as a function of the total stress intensity factor on a Log - Log plane are shown in Fig. 6a. For the three sets of data, the parameter n_1 (Eq. 6) was found equal to 4.26, 4.77, and 4.04, respectively. Using an average value of 4.35 and calculating the steady speeds for each set, a difference of about 20% was observed between the experimental and calculated steady speeds.

A power relation was also used for the rate of debonding as a function of stress level on specimens with $\lambda=1.8$ mm. It has been reported that for an interfacial crack the energy release rate at steady state is proportional to $t^2 r$, where t is the stress carried by a fiber of radius r [21]. Drawing on this results and considering that the forces carried by the fibers are proportional to the applied load, the experimental data on the steady evolution of debonding were correlated with the following expression

$$\frac{\Delta D}{\Delta N} = B \left(\sigma \sqrt{r} \right)^{n_2} \tag{7}$$

Where B and n_2 are constants and σ is the maximum stress of the fatigue cycle. The data shown in Fig. 6b are rates of debonding and the straight line represents Eq. (7). Note that the exponent $n_2 = 4.5$ is practically equal to the exponent n_1 (= 4.77) obtained from the correlation of the steady speed on specimens with the same fiber spacing. This is consistent with the steady state of the fracture phenomenon.

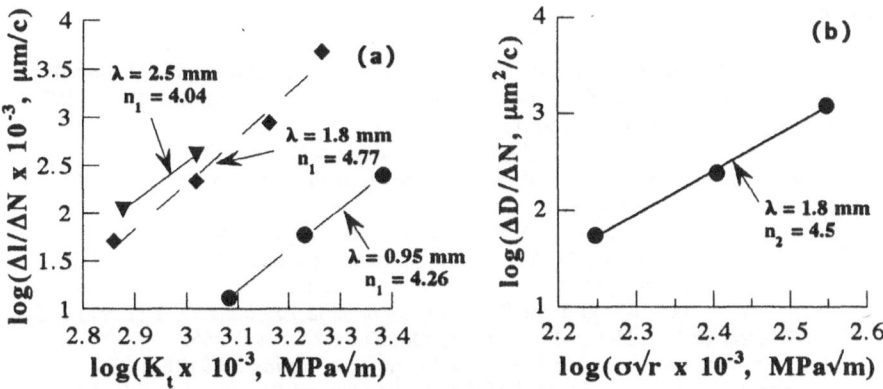

Fig. 6: (a) steady crack speed plotted against stress intensity factor,
(b) steady rate of debonding plotted against $\sigma\sqrt{r}$

5.   Summary

The results outlined in the work have shown that experimental and analytical research on systems with well controlled reinforcement can be useful in our attempts to understand the phenomena related to strength and fracture of composites with long aligned fibers. Despite the lack of analytical tools for an in depth modeling, some important results that emerged from this work are: (a) The scaling expression $\sigma_c\sqrt{\lambda} = \kappa$ relates the strength of the composite with a characteristic structural size, namely the fiber spacing. A similar expression is suggested for a ceramic composite without regular fiber spacing. (b) During the steady phase of propagation, crack speed, the rates of debonding, crack opening displacement and energy dissipation are constant. A decrease in crack speed is recorded after the steady phase. It is attributed to changes of the matrix material and is manifested in an increase in energy dissipation. (c) Strength measurements and crack growth suggest a transition in the behavior of the composite material in terms of fiber spacing. (d) Using the COD measurements and a simple analysis, the product of the shear stress, along the debonded interface and the debond length was found constant. Using this finding and simulations for the total stress intensity factor K_t, it was shown that K_t was approximately constant in the steady phase of crack growth. The steady crack speed and the steady rate of debonding seem to have a similar power dependence on stress level.

Acknowledgment

The authors wish to acknowledge the financial support from the AFOSR under grant 92-J-0493.

50

References

1. J. Aveston, G. Cooper and A. Kelly, Single and Multiple Fracture, in *The Properties of Fiber Composites*, Proceedings National Physical Laboratory, Guilford, UK. IPC Science and Technology Press Ltd. (1971) 15.
2. J. Aveston and A. Kelly, *Journal of Materials Science* 8 (1973) 352.
3. B. Budiansky, J. W. Hutchinson and A. G. Evans, *Journal of the Mechanics and Physics of Solids* 34 (1986) 167.
4. D. B. Marshall, B. N. Cox and A. G. Evans, *Acta Metallurgica* 33 (1985) 2013.
5. L. N. McCartney, *Proceeding of the Royal Society of London* A-409 (1987) 329.
6. T. Mori and T. Mura, *Mechanics of Materials* 3 (1984) 193.
7. C. C. Yang, W. B. Tsai, S. Quin and T. Mura, *Composite Engineering* 1 (1991) 113.
8. S. Nemat-Nasser and M. Hori, *Mechanics of Materials* 6 (1987) 245.
9. L. R. F. Rose, *Journal of the Mechanics and Physics of Solids* 35 91987) 383.
10. A. A. Rubinstein and K. Xu, *Journal of the Mechanics and Physics of Solids* 40 (1992) 105.
11. G. A. Kardomateas and R. L. Carlson, private communication
12. A. F. Bower and M. Ortiz, *Journal of the Mechanics and Physics of Solids* 39 (1991) 815.
13. J. Botsis and C. Beldica, *International Journal of Fracture*, submitted.
14. J. Botsis and A. B. Shafiq, *International Journal of Fracture* 58 (1992) R3.
15. J. Botsis, C. Beldica and D. Zhao, *Intentional Journal of Fracture*, Submitted.
16. J. Botsis and D. Zhao, in preparation.
17. R. F. Gibson, *Principles of Composite Materials Mechanics*, McGraw Hill, New York, N. Y. (1994).
18. G. I. Barenblatt, *Dimensional Analysis*, Gordon and Breach Science Publishers, New York, N. Y. (1987).
19. M. W. Barsum, P. Kangutkar and A. S. D. Wang, *Journal of Composite Science and Technology* 43 (1992) 257.
20. A. A. Rubinstein, *Journal of Applied Mechanics* 53 (1986) 505.
21. P. G. Charalambides and A. G. Evans, *Journal of the American Ceramics Society* 72 (1989) 746.

FIBER ARRANGEMENT EFFECTS ON THE MICROSCALE STRESSES OF CONTINUOUSLY REINFORCED MMCS

H.J. BÖHM AND F.G. RAMMERSTORFER

Institute for Light Weight Structures and Aerospace Engineering
Vienna Technical University
Gußhausstr. 27-29, A-1040 Vienna, Austria

Abstract. A unit cell based numerical approach is used for investigating fiber arrangement effects on the microscale stress and strain fields and on the overall thermomechanical response of a continuously reinforced unidirectional B/Al MMC. Simple periodic fiber arrays, clustered hexagonal configurations as well as modified and clustered square geometries are considered. The mean values and standard deviations of microstress and microstrain parameters in the matrix are computed and discussed for axial and transverse mechanical loading as well as for thermal loading.

1. Introduction

A number of methods have been developed for theoretical investigations of phase arrangement effects on the microscopic and macroscopic behavior of composites. One group of strategies employs statistical concepts for characterizing the phase arrangements of multiphase materials and for investigating relationships between microstructure and material properties. Such work includes, among others, studies based on correlation functions, e.g. (Pyrz, 1994), and on metallographic parameters, e.g. (Fan et al., 1994).

A different approach, which is followed in the present study, consists of analyzing the predicted responses of selected model microgeometries, which typically take the form of periodic phase arrangements of various levels of complexity, see e.g. (Bigelow, 1992; Nakamura and Suresh, 1993; Sørensen and Talreja, 1993; Böhm et al., 1994). By interpreting such results within a statistical framework, it was found possible in some cases to link the above strategies, see e.g. (Siegmund et al., 1993).

R. Pyrz (ed.), IUTAM Symposium on Microstructure-Property Interactions in Composite Materials, 51-62.
© 1995 *Kluwer Academic Publishers.*

2. Micromechanical Modelling

2.1. FIBER ARRANGEMENTS

Eight periodic microgeometries, each corresponding to a fiber volume fraction of ξ=0.475, are considered, see Fig.1. Configurations PH0 and PS0 are periodic hexagonal and square arrays, respectively. The clustered hexagonal arrangements CH1 and CH3, the modified square configuration MS5 as well as the clustered square geometries CS7 and CS8 were selected such that the minimum nearest-neighbour distance between fiber centers, a, takes the same value as that of the "honeycomb" arrangement RH2. For ξ=0.475 this corresponds to a=1.282d, where d stands for the fiber diameter.

Even though their nearest-neighbour distances are equal, the clustered microgeometries differ considerably in the average number of nearest neighbours per fiber and in the distribution of the thickness of matrix material around the fibers. This can be clearly seen from Fig.2, which shows the widths of the "matrix bridges" (in terms of the fiber diameter) as functions of the circumferential angle for the eight arrangements. Two curves each are given for the clustered arrangements CH1, CH3 and CS7 to account for the "inner" (dotted lines) and "outer" fibers of the clusters. As expected, the distributions of the widths of the matrix bridges are smoothest for the hexagonal and square arrays, and "matrix islands" are evident for the other configurations, especially arrangements CH1, RH2 and CH3.

It is worth noting that, whereas the overall elastic behavior of the four hexagonal microgeometries is transverse isotropic, arrangements PS0, CS7 and CS8 show tetragonal elastic symmetry, and MS5 is monoclinic. The thermal expansion behavior of all configurations except MS5 is transverse isotropic (Nye, 1957). Once yielding has taken place under non-axial mechanical loading, the above symmetry properties are typically degraded.

2.2. FINITE ELEMENT MODELS

The microstress and microstrain distributions for the eight periodic fiber arrangements were evaluated numerically via suitable unit cells, and the overall responses were obtained by homogenization. The unit cells and the associated boundary conditions were designed to be capable of handling axial and transverse normal mechanical loading as well as thermal loading (which in the present context is understood to involve no spatial temperature gradients), for a detailed discussion see (Böhm, 1993; Böhm et al., 1993; Böhm et al., 1994). Generalized plane strain models were used, the computations being performed with the FE-code ABAQUS (HKS, 1992).

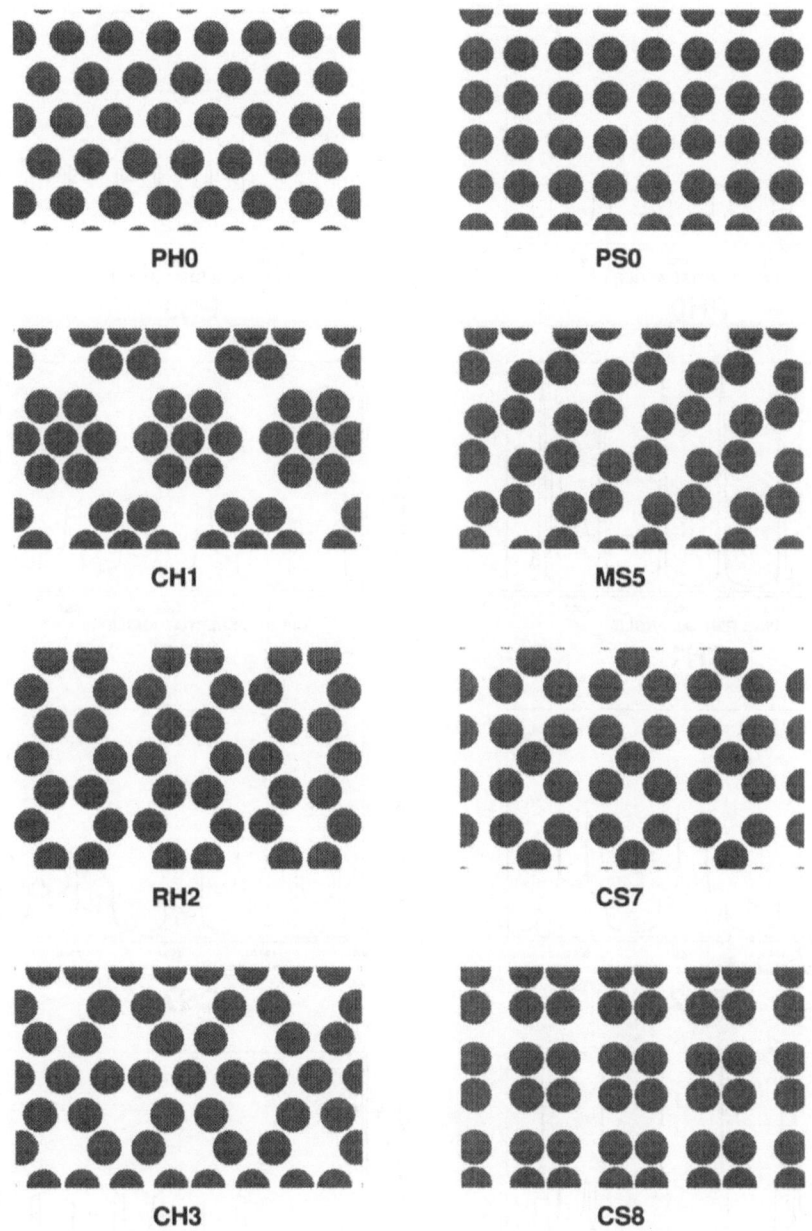

Figure 1. The eight periodic fiber arrangements considered (ξ=0.475).

54

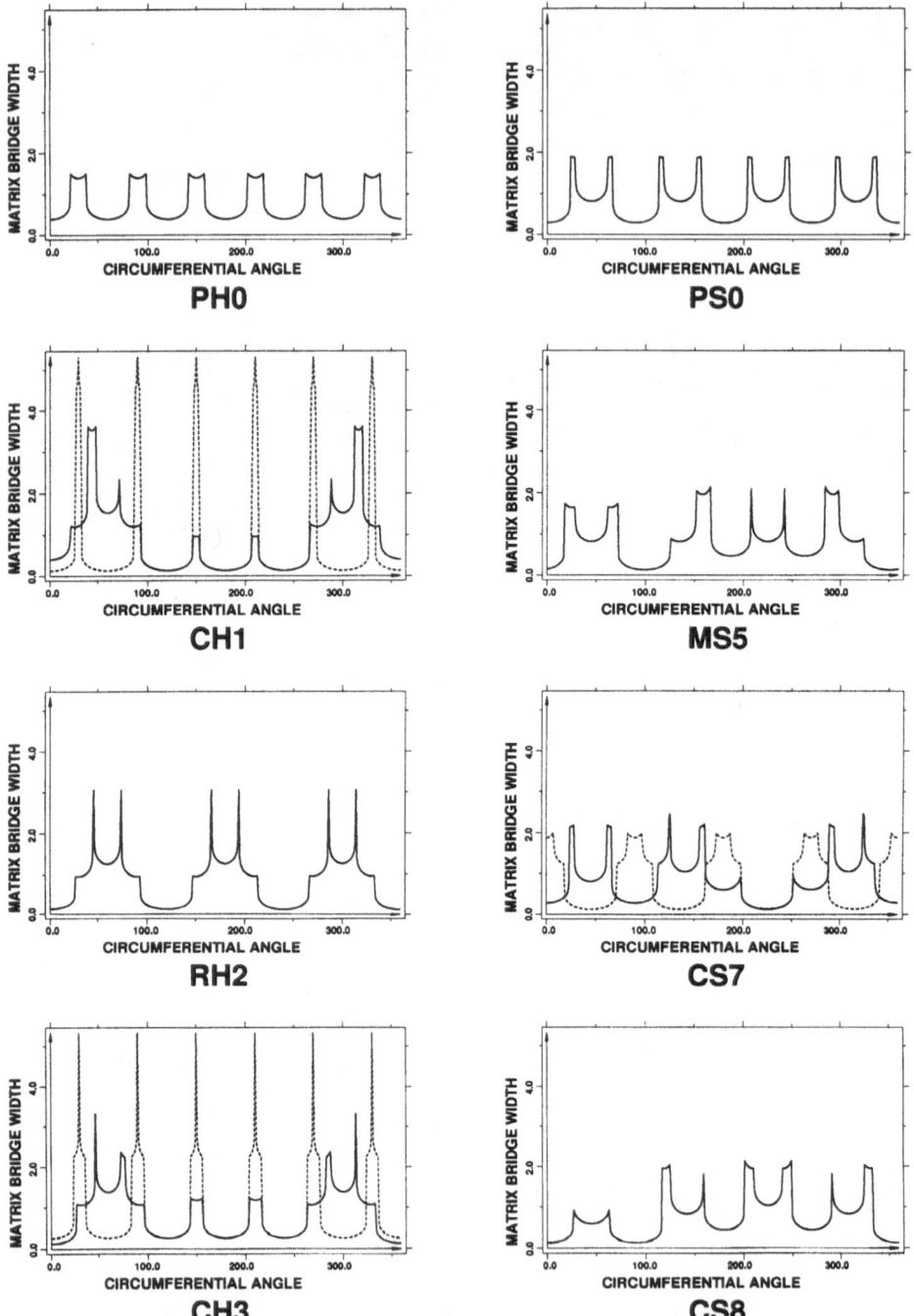

Figure 2. Angular distributions of the widths of the matrix bridges (in terms of fiber diameters) for the eight periodic arrangements considered.

TABLE 1. Material parameters used for the boron monofilaments and the Al6061-0 matrix

	T [°C]	E [GPa]	ν	σ_y [MPa]	E_t [MPa]	α [K^{-1}]
fibers	0–400	400.0	0.230	—	—	5.00×10^{-6}
matrix	20	68.9	0.350	68.9	1710	22.85×10^{-6}
	50	67.8	0.350	67.9	1683	23.50×10^{-6}
	100	66.2	0.350	65.7	1643	24.50×10^{-6}
	150	64.5	0.345	62.2	1601	25.50×10^{-6}
	200	63.0	0.335	55.0	1564	26.40×10^{-6}
	250	61.6	0.330	34.5	1529	27.30×10^{-6}
	300	60.3	0.330	25.1	1497	28.20×10^{-6}
	350	58.7	0.330	18.2	1457	29.20×10^{-6}
	400	56.3	0.335	11.3	1397	30.20×10^{-6}

The boron monofilaments were treated as isotropic elastic continua. For describing the behavior of the Al6061–0 matrix, a simple thermoelastoplastic material model with linear kinematic hardening was employed, the Young's modulus E, the Poisson's ratio ν, the yield stress in uniaxial tension σ_y, the hardening modulus E_t, and the total coefficient of thermal expansion (CTE) α being piecewise linear functions of the temperature, see Table 1. The interface between fibers and matrix was assumed to be perfect. Damage effects and the relaxation of microstresses were not considered for the present study.

3. Discussion of Results

The investigated load cases comprise axial normal loading to 400 MPa, transverse normal loading to 100 MPa and cooling down from 400°C to 20°C, the MMC being assumed to be initially stress free. In order to account for the anisotropic elastic and/or elastoplastic transverse behavior of the models, transverse loads acting in two directions were applied to all models except MS5. The directions of these loads are referred to the horizontal in Fig.1.

3.1. EVALUATION PROCEDURES

For comparing the microscale stress and strain variables predicted for the different fiber arrangements, histograms of their frequency distributions (weighted by volume) within each phase as well as the corresponding mean values and standard deviations were computed. Following standard Finite

Element practice, these evaluations used the "nodal averaged" microstress and microstrain fields. The phase averages of the stress components were checked with respect to the overall equilibrium conditions, good compliance being found.

3.2. OVERALL RESPONSE

The predicted overall axial and transverse Young's moduli, E_A^* and E_T^*, and CTEs, α_A^* and α_T^*, respectively, of the B/Al MMC at room temperature are listed in Table 2 together with analytical results obtained with the Mori–Tanaka method (Benveniste, 1987). The elastic transverse anisotropy of arrangements PS0, CS8 and, to a lesser degree, CS7 is evident from the difference in the transverse Young's moduli corresponding to loading in the 0° and 45° directions. Asterisks have been placed with the transverse data for arrangement MS5 to indicate their incompleteness.

TABLE 2. Numerical and analytical predictions for thermoelastic properties of a unidirectional B/Al MMC (room temperature, $\xi=0.475$)

arrangement	E_A^* [GPa]	E_T^* [GPa]	α_A^* [K^{-1}]	α_T^* [K^{-1}]
PH0	226.2	139.2	8.13×10^{-6}	15.7×10^{-6}
CH1	226.3	142.2	8.18×10^{-6}	15.5×10^{-6}
RH2	226.3	142.9	8.24×10^{-6}	15.2×10^{-6}
CH3	226.3	142.4	8.20×10^{-6}	15.4×10^{-6}
PS0	226.2	152.9/128.3	8.14×10^{-6}	15.6×10^{-6}
MS5	226.3	149.0/***	8.18×10^{-6}	15.5×10^{-6}/***
CS7	226.3	142.4/140.9	8.19×10^{-6}	15.5×10^{-6}
CS8	226.2	154.3/128.7	8.16×10^{-6}	15.6×10^{-6}
Mori–Tanaka	226.5	137.4	8.15×10^{-6}	15.6×10^{-6}

The dependence of the overall elastoplastic responses on the fiber arrangements was found to be very small for axial loading, of limited importance for thermal loading, and strong for transverse mechanical loading. For the latter case, the yielding and hardening behavior of the models PH0, CH1, RH2 and CH3 shows some anisotropy, and there are marked differences between the nonlinear responses to transverse loads in the 0° and 45° directions for the square-type arrangements CS0 and CS8. Such behavior is well known from the literature, compare e.g. (Nakamura and Suresh, 1993; Böhm et al., 1993).

3.3. AXIAL LOADING

In Table 3, the microscale axial stresses, $\sigma_A^{(m)}$, von Mises effective stresses, $\sigma_{eff}^{(m)}$, hydrostatic stresses $\sigma_H^{(m)}$, and effective plastic strains $\varepsilon_{eff,p}^{(m)}$ evaluated for the matrix of a B/Al MMC subjected to an axial load of 400MPa are listed in terms of their mean values and standard deviations.

TABLE 3. Microscale parameters (mean•standard deviation) predicted for the matrix of an initially stress free B/Al MMC (ξ=0.475) subjected to an axial load of 400MPa

arrangement	$\sigma_A^{(m)}$ [MPa]	$\sigma_{eff}^{(m)}$ [MPa]	$\sigma_H^{(m)}$ [MPa]	$\varepsilon_{eff,p}^{(m)}$ [$\times 10^{-4}$]
PH0	73.9•0.6	70.4•0.0	27.6•0.3	8.91•0.12
CH1	74.8•4.5	70.4•0.1	28.5•3.8	8.86•0.42
RH2	76.5•8.0	70.4•0.1	30.1•7.0	8.80•0.66
CH3	75.5•6.4	70.4•0.1	29.2•5.8	8.84•0.59
PS0	74.1•2.3	70.4•0.0	27.8•1.8	8.90•0.26
MS5	74.8•4.1	70.4•0.1	28.5•3.5	8.87•0.38
CS7	75.3•5.7	70.4•0.1	28.9•5.0	8.83•0.49
CS8	74.4•3.4	70.4•0.1	28.0•2.9	8.89•0.31

The dependence of the mean values of the above microfield parameters on the fiber arrangements is very small, the effective stresses being practically equal. There are, however, noticeable differences in the standard deviations, which are smallest for the periodic hexagonal array and reach the highest values for the clustered arrangements and the honeycomb geometry RH2. The larger standard deviations of the microstresses are due to regions of reduced axial and hydrostatic stresses at the points of closest approach between neighbouring clusters (arrangements CH1 and CH3) and at positions of minimum width of the matrix bridges (configurations RH2, CH3, MS5, CS7 and CS8). The effective plastic strains also reach their maxima in these zones. Interestingly, the inner fibers of the hexagonal clusters are strongly shielded (i.e. the clusters in geometries CH1 and CH3 behave like fiber-rich regions), but no such behavior is shown by arrangement CS7.

3.4. TRANSVERSE LOADING

Under transverse normal loading, both the mean values and the standard deviations of the matrix microstresses and microstrains depend noticeably on the fiber arrangement. This can be seen from Table 4, which lists matrix microscale parameters corresponding to a transverse load of 100MPa

$(\sigma_T^{(m)}$ stands for the transverse matrix stresses in the loading direction). There is also a clear correlation between the microfields and the direction of the transverse loads, which is most marked in the case of the tetragonal microgeometries PS0 and CS8.

TABLE 4. Microscale parameters (mean•standard deviation) predicted for the matrix of an initially stress free B/Al MMC (ξ=0.475) subjected to a transverse load of 100MPa

arrangement	$\sigma_A^{(m)}$ [MPa]	$\sigma_T^{(m)}$ [MPa]	$\sigma_{eff}^{(m)}$ [MPa]	$\sigma_H^{(m)}$ [MPa]	$\varepsilon_{eff,p}^{(m)}$ [$\times 10^{-3}$]
PH0(0°)	48.8•25.3	94.1•25.7	73.9•7.4	52.7•23.1	3.81•2.73
PH0(90°)	37.7•17.8	83.7•20.3	74.7•4.7	42.0•20.4	3.69•2.50
CH1(0°)	40.8•29.4	88.4•30.0	69.8•8.8	48.1•27.8	2.09•2.15
CH1(90°)	35.0•23.4	83.6•26.9	70.6•6.5	42.8•25.6	2.03•1.99
RH2(0°)	36.5•24.3	81.8•26.1	72.7•5.9	41.9•24.6	2.86•2.24
RH2(90°)	45.0•24.8	91.4•24.8	73.1•5.4	50.7•23.7	2.90•2.47
CH3(0°)	40.1•28.0	87.0•29.2	70.8•7.7	46.7•27.1	2.30•2.29
CH3(90°)	37.2•23.9	85.2•25.8	71.7•5.3	44.2•25.0	2.26•2.26
PS0(0°)	32.4•30.1	79.0•34.6	57.4•16.4	43.9•31.3	1.04•1.46
PS0(45°)	47.1•8.1	97.3•9.0	83.5•6.6	48.8•7.2	8.76•3.50
MS5(0°)	34.1•25.5	82.6•29.5	63.7•16.9	43.8•5.6	2.78•4.11
CS7(0°)	37.4•31.8	84.7•35.3	66.5•14.1	45.2•32.1	2.62•2.57
CS7(45°)	37.8•20.6	88.3•22.9	72.1•14.1	45.3•19.4	4.57•4.83
CS8(0°)	30.2•27.4	79.5•32.7	58.0•13.6	43.0•29.6	0.56•0.99
CS8(45°)	47.2•7.9	97.3•8.9	83.8•6.2	48.8•7.2	8.72•3.54

This behavior can be explained in terms of the spatial distributions of the effective plastic strains and the effective stresses in the matrix, which tend to become concentrated in "bands" where allowed by the geometry. The periodic arrangements discussed here show straight "channels" of infinite free path length in the matrix at angles of 0° and ±60° (PH0, CH1, CH3), of ±30° and 90° (RH2), of 0° and 90° (PS0, CS8), of 0°, 90° and −45° (MS5) and of 0°, ±45° and 90° (CS7). The highest mean values of the effective plastic strains and the effective stresses as well as the softest overall responses are predicted when wide bands can form in straight matrix channels at ±45° to the loading direction. The opposite behavior is found when the only channels are parallel or perpendicular to the loading direction, and shear strains activated at ±30° or ±60° give rise to intermediate results.

Random arrangements of fibers (and thus real composites) do not show straight matrix channels of infinite path length, which are a typical feature of periodic arrays and clusters. The latter are, however, well suited for describing the microfields in subregions with approximately periodic or clustered geometries, which are typically found in MMCs.

3.5. COOLING DOWN FROM PROCESSING TEMPERATURE

Among the load cases considered here, the strongest dependence of the matrix microstresses on the fiber arrangement was predicted for cooling down the B/Al MMC from a processing temperature of 400°C to room temperature, see Table 5. The effective plastic strains, however, showed only a limited sensitivity to the microgeometries.

TABLE 5. Microscale parameters (mean•standard deviation) predicted for the matrix of a B/Al MMC (ξ=0.475) cooled down from 400°C to 20°C

arrangement	$\sigma_A^{(m)}$ [MPa]	$\sigma_{eff}^{(m)}$ [MPa]	$\sigma_H^{(m)}$ [MPa]	$\varepsilon_{eff,p}^{(m)}$ [$\times 10^{-2}$]
PH0	84.0•9.6	87.2•3.9	45.2•4.4	1.10•0.21
CH1	93.6•49.7	87.7•10.8	50.5•40.5	1.11•0.63
RH2	113.2•95.8	88.8•19.4	68.7•86.4	1.19•1.14
CH3	105.1•96.0	88.4•15.0	63.3•88.6	1.16•0.92
PS0	87.0•22.8	87.3•7.1	45.3•13.4	1.08•0.41
MS5	91.1•43.7	88.3•10.2	51.0•37.3	1.15•0.59
CS7	100.1•79.4	88.2•13.6	57.7•71.5	1.15•0.79
CS8	89.2•41.3	87.7•8.7	48.1•34.8	1.11•0.51

A noteworthy feature of the above results are the very high standard deviations computed for the axial and hydrostatic microstresses of some arrangements, especially RH2, CH3 and CS7. This behavior is explained by Fig.3, which shows the histograms of the relative frequencies of the hydrostatic stresses corresponding to Table 5 (in order to improve the legibility of the figure, the negative tails of the distributions, which cannot be resolved at the scale of Fig.3, were clipped for RH2, CH3, MS5 and CS7). The distributions for configurations CH3 and CS7 are clearly bimodal with widely separated peaks, and RH2 has a flat second peak around −225MPa (clipped in Fig.3). These bimodal distributions are due to a tendency for the hydrostatic stresses to be markedly tensile in large matrix islands and to be strongly compressive around positions of narrow matrix bridges between fibers and/or fiber clusters (reduced axial and hydrostatic microstresses at similar positions were found for axial loading). This behavior was discussed in the context of the interfacial stresses in (Böhm et al., 1994).

60

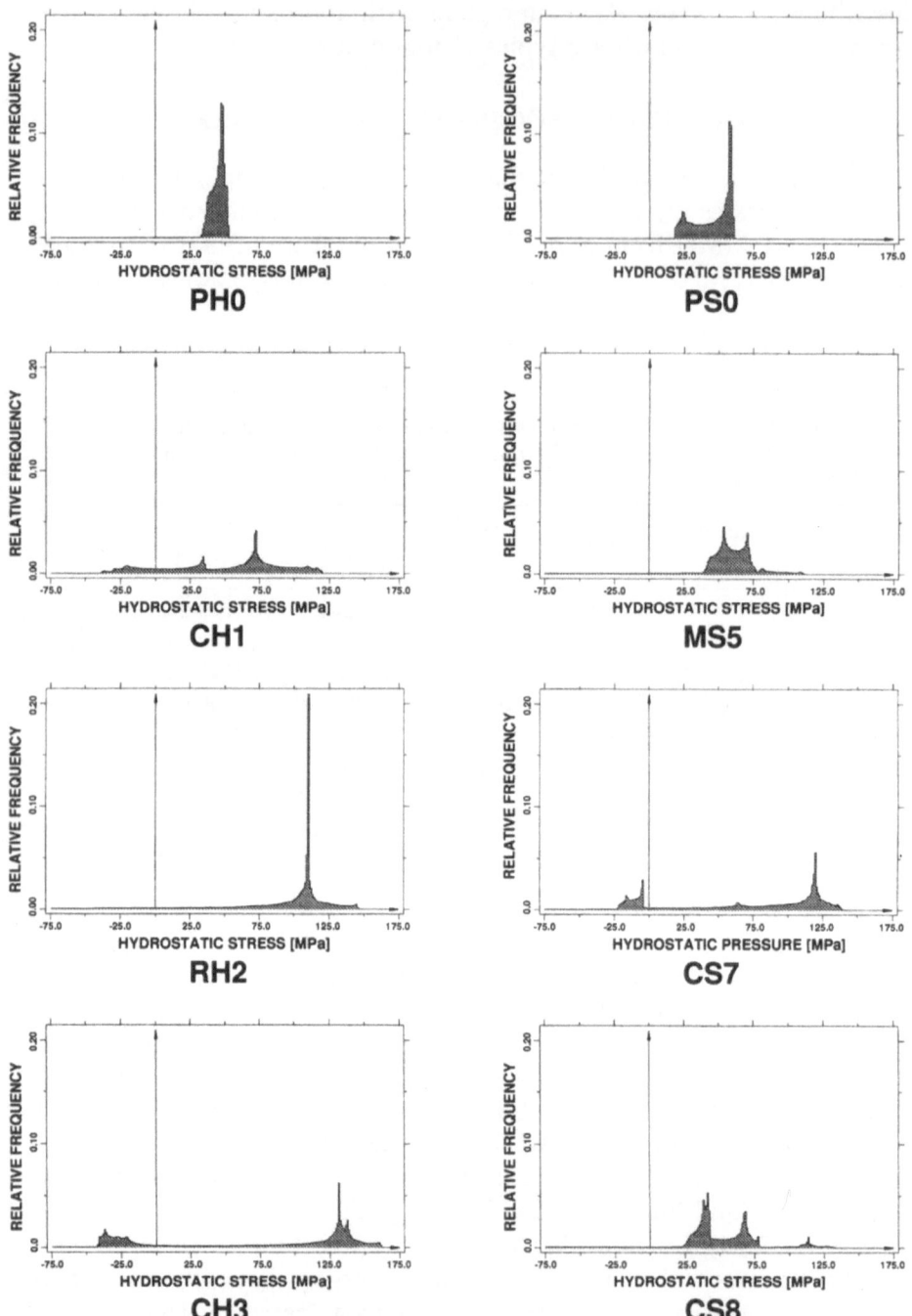

Figure 3. Relative frequencies of the hydrostatic stresses predicted for periodic B/Al MMCs (ξ=0.475) cooled down from 400°C to 20°C.

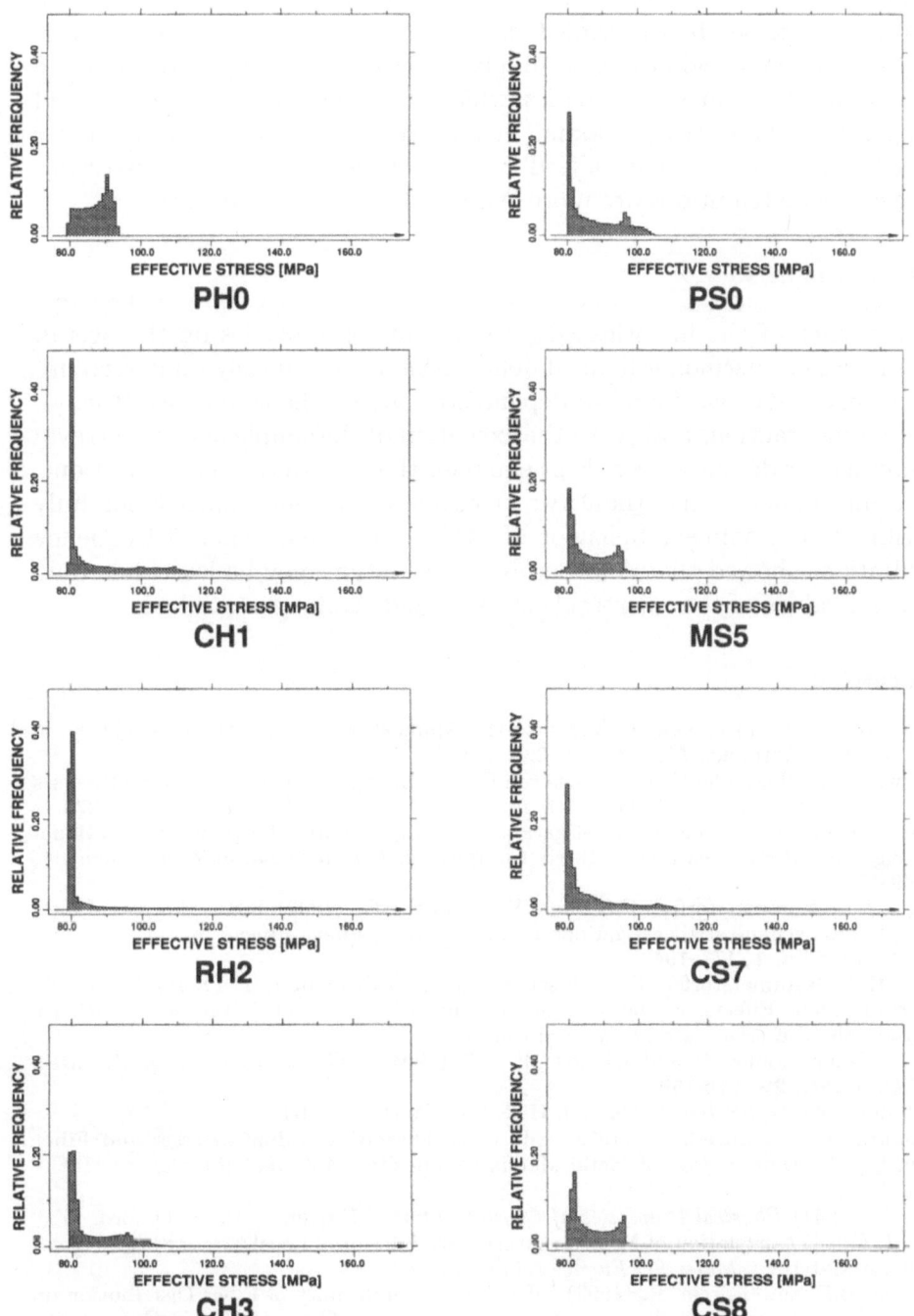

Figure 4. Relative frequencies of the von Mises effective stresses predicted for periodic B/Al MMCs (ξ=0.475) cooled down from 400°C to 20°C.

Whereas matrix islands are typically associated with marked tensile hydrostatic stresses in the cooled down state, the effective stresses tend to be low in such regions. Accordingly, in the histograms of the relative frequencies of the von Mises stresses, which are shown in Fig.4, the clustered arrangements show strong maxima at low stresses and long tails at higher values (corresponding to localized regions of elevated effective stresses), whereas the distributions are more "compact" for the simple arrays.

4. Conclusions

The influence of the investigated periodic microgeometries on the overall and microscale thermomechanical behavior of a continuously reinforced unidirectional MMC was found to depend strongly on the load cases. Because fiber volume fractions and (with the exception of the simple periodic arrays) minimum fiber distances were kept equal for the investigated configurations, the results indicate that these two parameters are not sufficient for fully describing the nonlinear behavior of MMCs. The discussion of frequency distributions showed that considerable information may be lost when only the mean values of the microscale stresses and strains are studied.

References

Benveniste, Y. (1987) A New Approach to the Application of Mori–Tanaka's Theory in Composite Materials, *Mech.Mater.* **6**, 147–157.

Bigelow, C.A. (1992) The Effects of Uneven Fiber Spacing on Thermal Residual Stresses in a Unidirectional SCS-6/Ti-15-3 Laminate, *J.Compos.Technol.Res.* **14**, 211–220.

Böhm, H.J. (1993) Numerical Investigation of Microplasticity Effects in Unidirectional Long-Fiber Reinforced Metal Matrix Composites, *Modell.Simul.Mater.Sci.Engng.* **1**, 649–671.

Böhm, H.J., Rammerstorfer, F.G. and Weissenbek, E. (1993) Some Simple Models for Micromechanical Investigations of Fiber Arrangement Effects in MMCs, *Comput.Mater.Sci.* **1**, 177–194.

Böhm, H.J., Rammerstorfer, F.G., Fischer, F.D. and Siegmund, T. (1994) Microscale Arrangement Effects on the Thermomechanical Behavior of Advanced Two-Phase Materials, *J.Engng.Mater.Technol.*, in print.

Fan, Z., Tsakiropoulos, P. and Miodownik, A.P. (1994) A Generalized Law of Mixtures, *J.Mater.Sci.* **29**, 141–150.

HKS (1992) *ABAQUS User's Manual.* HKS Inc., Pawtucket, RI.

Nakamura, T. and Suresh, S. (1993) Effects of Thermal Residual Stresses and Fiber Packing on Deformation of Metal-Matrix Composites, *Acta metall.mater.* **41**, 1665–1681.

Nye, J.F. (1957) *Physical Properties of Crystals.* Oxford University Press, Oxford.

Pyrz, R. (1994) Correlation of Microstructure Variability and Local Stress Field in Two-Phase Materials; *Mater.Sci.Engng.* **A177**, 253–259.

Sørensen, B.F. and Talreja, R. (1993) Effects of Nonuniformity of Fiber Distribution on Thermally-Induced Residual Stresses and Cracking in Ceramic Matrix Composites; *Mech.Mater.* **16**, 351–363.

Siegmund, T., Werner, E. and Fischer, F.D. (1993) Structure-Property Relations in Duplex Materials, *Comput.Mater.Sci.* **1**, 234–241.

ANALYTICAL/EXPERIMENTAL STUDY OF INTERPHASE EFFECTS ON COMPOSITE COMPRESSION STRENGTH

GREG P. CARMAN AND SAEED ESKANDARI
Mechanical, Aerospace and Nuclear Engineering Department
University of California, Los Angeles
Los Angeles, CA 90024-1597

Abstract

In this work we present an analytical model of a composite containing coated cylindrical fibers subjected to a compressive loading. The modeling approach used in this paper is a blend of two conventional analytical techniques to predict the composites compression strength as a function of microparameters. The model presented here explicitly includes interphase influences and the circular geometry of the fiber which causes substantial stress concentrations at key locations. Analytical results suggest that a specific coating may increase the compression strength of a composite. Experimental studies are conducted on several composite systems fabricated with coated glass fibers as the reinforcing fiber. Results indicate that coating effects, adhesion, and fiber geometry play a critical role in compression strength. Comparison of the experimental results with the analytical model yield reasonable correlation between the two, thus supporting the theoretical claims of the model.

1. Introduction

The use of composite materials has grown considerably over the past two decades in applications ranging from sporting goods to aerospace vehicles. While these materials are widely used in today's society, the specific physical mechanisms which govern their performance are not completely understood. To address this concern, scientists and engineers have begun to re-focus their attention on constituent interaction occurring in the composite between the fiber, coating, and matrix regions. These micromechanical studies are being performed with the hopes of gaining a better understanding of the physical mechanisms governing a composite's strength and performance characteristics. In fact, recent publications indicate that an opportunity exists to improve a composites performance by tailoring the fiber coatings (interphase region) or fibers adhesion characteristics (interface).

R. Pyrz (ed.), IUTAM Symposium on Microstructure-Property Interactions in Composite Materials, 63–75.
© *1995 Kluwer Academic Publishers.*

Papers which review the effect of fiber coatings on the macro-response of a composite systems include Swain et al. (1990) and Jayaraman et al. (1993). Optimum fiber coatings for metal-matrix and polymeric-matrix composites subjected to thermal loading were studied by Ghosn and Lerch (1989) using an energy based criteria. Carman et al. (1993) presented an analysis to determine the optimum fiber coating for a composite subjected to transverse loading to minimize cross-ply cracking. Pak (1992) published an analysis of a composite material subjected to shear tractions for maximizing the shear load supported by the fiber. Schwartz and Hartness (1985) investigated the effect of fiber coatings on interlaminar fracture toughness and transverse strength of a composite. There also exists a number of theoretical and experimental papers discussing the effect of interphases on global properties, for example Pagano and Tandon (1988), Lesko et al. (1991), and Chang et al. (1992). While interphase/interface effects have been studied in a variety of contexts, probably their largest impact is reported in experimental literature discussing a composites compression strength. Greszczuk (1975) demonstrated with a pseudo-composite system that fiber-matrix interface strength and fiber size strongly influenced compression strength. Madhukar and Drzal (1991) experimentally evaluated several graphite epoxy composite systems containing varying degrees of fiber/matrix interface adhesion and concluded that the compression strength changed substantially. Swain (1992) experimentally showed that fiber surface treatments altered the fatigue life of a composite. These experimental studies, as well as a host of others, have yielded a great deal of insight, as well as a fair degree of confusion, into understanding the effect of microparameters on compression strength.

Analytical modeling efforts of compression strength Rosen (1965) are based on compression failure that occur in either an "extension" or "shear" mode, with the latter being more typical in standard composites. Shear mode failure initiates at or near the interphase region in the form of matrix cracking or fiber/matrix interface decohesion. These failures are due to the local stress concentrations which arise in the matrix/interphase region near/adjacent to the fiber. While compression strength appears to be a shear dominated phenomena, these models apparently overestimate compression strength by a factor of 3. Attempting to account for this overestimation Kulkarni (1975) introduced the concept of fiber/matrix adhesion, Greszczuk (1974) included end fixity conditions, Waas et al. (1990) identified the absence of free-edge traction, Lessard and Chang (1991) suggested that fiber-fiber interactions play a role. Still other researchers, such as Hahn and Tsai (1980) introduced a fiber waviness parameter and Steif (1990) looked at finite deformations. In more recent work, Wass (1992) introduced interphase effects in a compressive strength model to investigate failure strain and buckling wave lengths. A review of compression strength models can be found in Shuart paper (1985) summarizing various techniques for assessing the compressive strength of a composite.

It is the purpose of this paper to analytically model the influence of the interphase and the circular geometry of the fibers on compression strength. While researchers have indicated that the interphase region plays a role in compression strength, it is the supposition of this paper that the cylindrical geometry of the fiber is an additional parameter which needs to be considered. This geometry gives rise to specific variations in stress which vary azimuthally in the composite and generate substantial stress concentrations at key locations. Theoretical results generated suggest that a specific coating on a circular fiber might improve compression strength. Experimental results are also presented for a model composite system containing fibers coated with different materials. The agreement between experimental and theoretical results support certain analytical claims.

2. Analytical model

Consider an unidirectional composite subjected to compression in the direction of the structural fibers as presented in Figure 1. To analytically model the effect of fiber coatings on compression strength, we begin with the approach presented in Tsai and Hahn (1980). Assumptions used in this development include each constituent in the composite (fiber, interphase, and matrix) exhibits linear elastic behavior, the materials are transversely isotropic, the strains are small, and the bond between each constituent is "ideal", such that the traction and displacements are continuous across each interface. The fibers are also assumed to be much

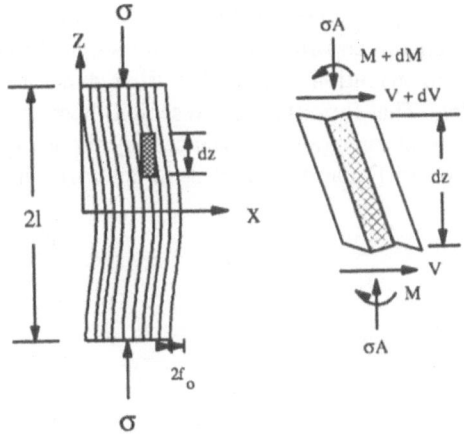

Figure 1 Illustration of compression strength model.

stiffer than the matrix. An initial deflection, v_0, of the plies is assumed to take the form:

$$v_0 = f_0\left[1 + \cos\left(\frac{\pi z}{l}\right)\right] \qquad (1)$$

where l is the half-length of the undulation and f_0 is the amplitude of the undulation. When a compressive load is applied to the composite, the deflection of the plies is assumed to be of a similar form to that presented in Equation 1 or:

$$v = f\left[1 + \cos\left(\frac{\pi z}{l}\right)\right] \qquad (2)$$

where f is the final amplitude of the deflection for a given applied load. While other deformation profiles are possible it is more plausible that the end deformation will resemble the initial one. Therefore, Equation 2 is a reasonable estimate of this deformation profile. Using a representative volume element of length dz and cross-sectional area A as presented in Figure 1, a mechanics of materials approach can be used to develop an equation for a composite's compressive strength X_c

$$X_c = C_{55}^c \left(1 - \frac{f_0}{f_c} \right) \tag{3}$$

where f_c is the amplitude of the undulation when the composite fails. To evaluate this amplitude, we assume that the composite fails by local shear failure in the matrix or interphase region when a critical shear strength has been reached. This critical value can be represented by the shear strength of the matrix S^m, or the interface strength between the constituents. It is our supposition, that this value depends upon the local stress distribution around the fiber and must be evaluated with a micromechanical model. The formulation presented here requires a quantitative assessment of the average shear strain experienced by the local element. Using the deformation profiles defined in Equation 1 and 2 the average shear strain can be determined

$$\gamma_{zx} = \gamma_{zx}^c \sin(\frac{\pi z}{l}) \tag{4}$$

where

$$\gamma_{xz}^c \frac{\pi}{l} (f_c - f_0)$$

where γ_{zx}^c is the local composites shear strain when the matrix region or interphase region fails. Using the strain information in a concentric cylinders model containing fiber/interphase/matrix constituents (Hashin & Rosen 1964), the local stress/strain state can be determined and an appropriate failure criteria based on the constituents, fiber geometry, and local stress concentrations developed. Since we are only interested in failure of the composite, we choose to investigate the stress distribution at a specific axial location, i.e. $z=0$. By assuming that the representative volume element experiences the same volume average strains as the bulk composite, the following boundary condition is applied to the element.

$$w^m(r_m) = \gamma_{xz}^c r_m \cos\theta \tag{5}$$

Where r_m is the outer radii of the element and w is the axial displacement. The pertinent governing differential equation for this problem written in terms of displacement variables is

$$w_{,rr} + \frac{1}{r} w_{,r} + \frac{1}{r^2} w_{,\theta\theta} = 0 \qquad (6)$$

The solution to this partial differential equation subjected to the boundary condition in Equation 5 is

$$w^n(r,\theta,z) = [\frac{F_1^n}{r} + F_2^n r] \cos\theta \qquad (7)$$

Employing the stain-displacement and stress-strain relations for a transversely isotropic material, the non-zero stresses in each constituent are found to be

$$\sigma_{rz}^n = C_{55}^n[-\frac{F_1^n}{r^2} + F_2^n] \cos\theta \qquad \sigma_{\theta z}^n = C_{55}^n[-\frac{F_1^n}{r^2} - F_2^n] \sin\theta \qquad (8)$$

where C_{55}^n is the axial shear modulus of the n'th phase (f=fiber, i=interphase, m=matrix, and c=composite) in the composite, and F_i^n are undetermined constants. These are evaluated by
  A. applying the boundary conditions stated in Equation 5,
  B. demanding continuity of the traction and displacements at each interface (i.e. σ_{rz}^n and u_z^n), and
  C. demanding that the stresses be bounded ($F_1^f = 0$).
Using this information the stress/strain state in each of the constituents of the composite can be determined. The matrix stress becomes

$$\sigma_{xz}^m = (\sigma_{rz}^m \cos\theta - \sigma_{\theta z}^m \sin\theta) \sin(\frac{\pi z}{l}) \qquad (9)$$

To investigate the magnitude of the stress concentrations in the constituents, we recognize that their are two possible locations for failure. Stress concentrations arise in composites at bi-material interfaces at either $\theta = 0°$ or $\theta = 90°$ (Carman & Case 1992) subjected to shearing loads. These locations are associated with classical problems similar to a hole or a circular rigid inclusion in a plate. Based on this argument, the magnitude of the stress concentrations in the constituents is

$$K = \frac{[-\frac{F_1^n}{r_n^2} \pm F_2^n]}{\gamma_{xz}^c} \qquad (10)$$

The presence of the plus or minus term in Equation 10 reflects the nature of the stress concentration occurring at $\theta = 0°$ or $\theta = 90°$. Assuming that the shear strength of the matrix has been reached Equation 9 can be recast in terms or the shear strength

$$S^m = C_{55}^m K \frac{\pi}{l} (f_c - f_0) \tag{11}$$

where K represents the stress concentration arising due to geometry and constituent properties defined in Equation 10. Solving for the critical value f_c in Equation 11 and substituting into Equation 3, we obtain an expression for the compression strength of the composite.

$$X_c = C_{55}^c \left(\frac{1}{1 + \dfrac{f_0 \pi}{l} \dfrac{C_{55}^m K}{S^m}} \right) \tag{12}$$

To predict the compression strength of a composite from constituent information the shear modulus of the composite must also be determined. This can be done by relating the volume averaged stresses calculated from Equation 8 to the applied strain in Equation 5. This is depicted by the following equation

$$C_{55}^c = \frac{1}{V^c \gamma_{xz}^c} \sum_n \int_{V_n} (\sigma_{rz}^n \cos\theta - \sigma_{\theta z}^n \sin\theta) dV_n \tag{13}$$

3. Manufacturing methods

The composite test specimens are fabricated by one of three manufacturing processes, autoclave (A), hot-press (HP), and resin transfer mold (RT). All of the composites were constructed with optical fiber approximately 200 microns in diameter as the reinforcing fiber. Optical fibers provide a convenient off the shelf commodity to construct a composite system containing fibers with well characterized coatings and geometries. The optical fibers used in our experiments are coated with either a silicon rubber (S), polyimide (P), acrylate (A), or a nylon (N) coating. Relative dimensions between the fiber (core-cladding) and the coating are silicon rubber 200/230, polyimide 225/245, nylon 210/230, and acrylate 125/215. The matrix used in these studies is an epoxy M-10E resin.

Autoclave and hot press manufacturing methods are based on a lamination approach. This common technique is presently utilized throughout the aerospace industry for manufacturing typical graphite epoxy composite specimens. Only acrylate fibers were employed in the hot press operations while acrylate, polyimide, and nylon coated fibers were incorporated into autoclave processes. The latter method provided highly repeatable results while the previous one yielded composites containing considerable voids. In either method, the fibers are first laid onto sheets of resin material using a filament winder, a hand layup technique, or an alignment fixture to provide a sheet of prepreg tape. Following this process, the prepreg is laid into a desired lamination scheme, for our purposes a unidirectional layup. Once constructed, the laminate is cured with either a hot-press or an autoclave.

The mold injection method used silicon rubber and acrylate coated fibers. To manufacture these specimens, the coated fibers are suspended between two perforated plates located at the ends of the die. These perforated plates are fabricated to accommodate various fiber size, fiber volume, fiber arrays, and can be used to control fiber spacing. The fibers are held in their position with the use of RTV (silicon rubber) or a mechanical gripper while liquid resin is injected into the mold. The RTV silicon permits pretension to be applied to the fibers during the resin transfer process to ensure fiber straightness. All of the manufacturing methods, hot press, autoclave, and resin transfer mold, provide a material system containing physically measurable micro-parameters. However, the resin transfer mold and the autoclave process yielded higher quality specimens than did the hot press.

4. Analytical results

The fiber volume fraction used in the analytical study is $v_f = 0.59$, the interphase volume fraction is $v_i = 0.13$, the pertinent fiber property is $C_{55}^f = 28$ GPa, and the pertinent matrix property is $C_{55}^m = 1.1$ GPa. Using these physical quantities (Carman & Case 1992) an optimum interphase shear modulus of $C_{55}^i = 0.11$ GPa is calculated. In discussing the results, a normalized shear modulus is defined as $C''_{55} = C_{55}^i/C_{55}^m$. The four interphase values studied here range from an extremely compliant interphase value (i.e. actually a hole) to the geometric mean of the fiber and matrix shear moduli. In the present study, a similar shear strain is applied to each composite element containing the different interphase materials. The stress quantities are normalized to a homogeneous matrix material subjected to the same strain.

The shear stress variations in the composite as a function of the normalized radial coordinate (r/r_m) for $\theta = 0°$ are presented in Figure 2. As the shear modulus of the interphase increases, the stress concentration in each of the constituents increases. This is expected, since the shear stress concentration at $\theta = 0°$ is related to a rigid inclusion effect. Another interesting feature of this graph is that as the interphase modulus decreases the stress supported by the fiber decreases. This suggests that for compliant coatings, the composite may buckle after the ultimate load is reached rather than resulting in fiber failure.

To understand the influence of fiber coatings on stress state at the bimateraial interfaces, a plot of shearing stresses at the matrix/interphase interface as a function of θ is presented in Figure 3. For stiff interphase values large shearing stresses occur at $\theta = 0°$ while for compliant interphases they occur at $\theta = 90°$. On the other hand, the composite containing the optimum interphase does not display a dependence on angular position. This is typical of a homogeneous material system subjected to axial shear. For compliant systems the larger shearing stresses exist along the neutral axis of the local element and may lead to premature buckling of the composite specimen. While the stress state in the composite is minimized with the application of the compliant coating there is detrimental effect to the composite. That is, the shear stiffness of the composite decreases with the application of a compliant coating. As presented in analytical results by Carman et al. (1992) the stiffness of the composite, including the transverse modulus (E_y) and the transverse shear modulus (C^c_{55}) decrease significantly. In regards to compression strength, the shear modulus has a significant impact on this property as evidenced by Equation 12. However, results published by Madhukar and Drzal (1991) suggests that this influence is not as significant as analytically predicted.

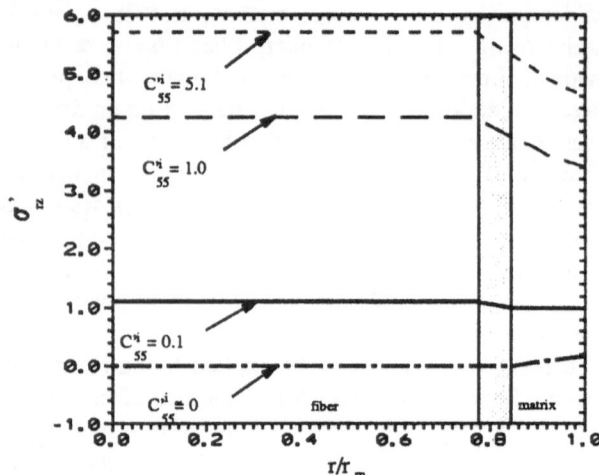

Figure 2: Shearing Stresses in Composite.

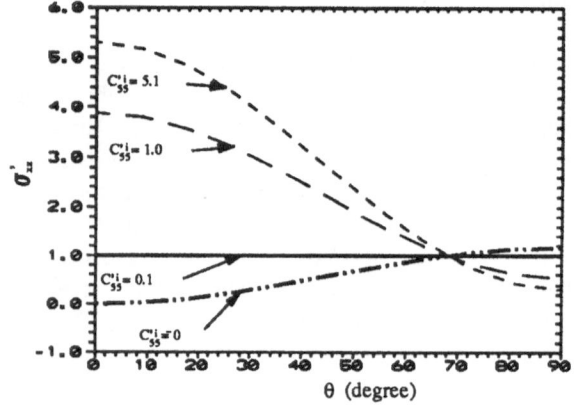

Figure 3: Shearing Stresses at Matrix Coating Interface.

In Figure 4, a parametric study of the effect that undulation size, matrix shear strength, and interphase properties has on a composite material subjected to a similar traction profile is presented. In presenting the results we have normalized the compression strength of the composite to the shear modulus of the composite. This normalization process helps remove any discrepancies which might be associated with inconsistencies in shear stiffness predictions. The curves discontinuity is caused by the maximum stress state shifting from $\theta = 0$ to 90. The sensitivity parameter, i.e. $(f_0 C_{55}{}^m)/(S^m l)$ represents changes in matrix shear stiffness, matrix shear strength and undulation amplitude of the fiber in the composite. With decreasing values of this parameter, the local undulation becomes less severe for constant matrix values. In the limit as the undulation approaches 0 or a straight line, Equation 16 indicates that compression strength is not a function of interphase stiffness values. Figure 4 indicates that an optimum interphase stiffness value exists to maximize the normalized compression strength of a composite, that is if the shear modulus of the composites does not change appreciably. Furthermore, the curves suggests that compliant interphase values reduce the compression strength to a larger extent than do stiff interphase values, as one might expect. In this light, it seems possible that an interphase might exist to maximize the compression strength/strain to failure of a composite.

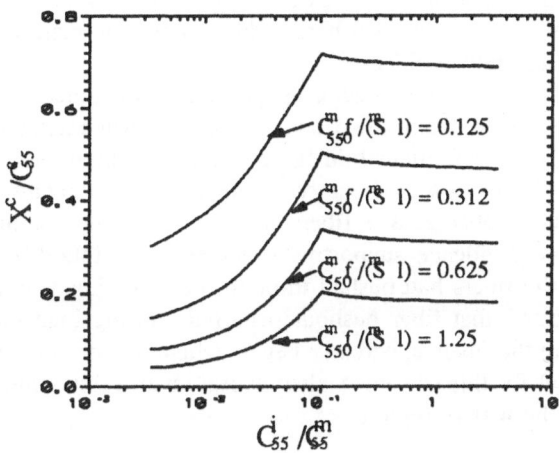

Figure 4: Normalized Compression Strength as a Function of Coating.

5. Experimental results

All compression tests were conducted on an Illinois Institute of Technology Research Institute (IITRI) test fixture in a 10 kip electromechanical driven load frame. Tests were conducted at a rate of 0.127 cm/min on specimens conforming to the ASTM D3410 standard. The gauge length of the specimens was 1.27 cm, the width of the specimens was approximately 0.635 cm, and the thicknesses of the specimens varied from 0.3 to 0.4 cm. Specimens were strain gauged to determine failure strains and evaluate buckling characteristics. The shear modulus values used for theoretical calculations presented in this section are core-cladding 28 GPa, polyimide 1.2 GPa, nylon 1.2 GPa, acrylate 0.13 GPa, silicon rubber 0.13 GPa, and epoxy matrix 1.6 GPa. The length of the undulation used in the theoretical calculations was the gauge length of the specimen with an amplitude equal to a nominal fiber diameter (i.e. f= 200 microns) and S^m= 3.5 MPa.

Experimental and theoretical results are presented in Table 1. Panel number 5, which was manufactured by the hot press method, contained excessive amount of voids in the epoxy resin. In this panel, all of the voids were spherical in geometry and fully encapsulated in epoxy resin. While the precise volume fraction of the voids is not known at this time, an estimate of 30% was used in a spherical inclusion model to calculate a 40% stiffness reduction for the matrix. Panels 3, 7, 8, 9, and 10 were all essentially void free.

During the manufacturing process it was noted that the nylon coating could easily be removed from the fiber. Therefore, when analytically modeling this panel an extremely compliant thin layer was placed between the fiber and the nylon material. In addition to this anomaly, compression tests conducted on composites containing silicon rubber coated fibers indicated that the coating does not adhere well to the matrix. Evidence supporting this claim was found by post-test inspections revealing that the fibers had pushed through the ends. Load time plots for these specimens also indicated that fiber pushout occurred. As the load increased, a plateau was reached where the fibers apparently began debonding from the matrix. Therefore, the analytical model for this specimen also incorporated a thin compliant layer between the coating and the matrix region.

Panel #	Coating	'Process	Fiber Volume Fraction %	Exp. Strength MPa	Theory MPa	Failure Strain %	Theory Strain %	σ/G %
5*	A	HP	7.5	122.7	140	2.3	2.2	19.8
7	A	RM	7.5	168.6	204	2.0	2.5	17.7
3	A	AC	14.9	148.8	125	1.8	1.1	15.0
10	S	RM	16.3	140.0	148	---	1.1	14.6
9	N	AC	33.0	121.3	100	---	0.5	13.1
8	P	AC	42.0	917.0	807	2.6	2.8	24.5

Analytical failure predictions for the specimens indicate that all of the composites would fail along the neutral axis ($\theta = 90$) with the exception of the polyimide coated fibers where theoretical predictions indicated that failure should occur ($\theta = 0$). Experimental results support this contention. That is, the polyimide specimens microbuckled followed by fast fracture severing the specimen into two pieces. All other specimen failures were typified by long wavelength fiber buckling leading to specimen buckling. However, several specimens from panel 5 failed by microbuckling leading to composite fast fracture. These latter specimens had the lowest ratio of stress concentration at $\theta = 90$ to $\theta = 0$, indicating a higher propensity for microbuckling than other ones.

Theoretical predictions for strength are comparable to the experimental results obtained on all specimens shown in Table 1. Clearly, the polyimide coated fiber was the strongest of the samples tested. However, for a composite with a comparable fiber volume fraction, i.e. nylon, the strength decreased by almost an order of magnitude when compared to the polyimide. This degradation is attributable to the lack of adhesion between the fiber and the coating and is explicitly depicted in the model through the presence of an extremely compliant layer. On the other hand, comparing the results obtained on a lower fiber volume fraction composite than the nylon, i.e.

silicon, which also exhibits poor adhesion between the coating and the matrix, the silicon exhibited a larger strength than did the nylon coated fiber, a result which is not intuitively obvious. The reason for the relative increase in strength for the silicon fiber system is because the nylon coated fiber has a larger stress concentration and a lower shear modulus. These two reasons are associated with the relatively larger area of decohered region within the nylon composite compared to the lower fiber volume fraction silicon composites.

Turning our attention to the acrylate coated specimens, some additional observations can be made. For panel 3, which contained a relatively larger fiber volume fraction than panel 7, the strength is lower. The reason that the strength is lower for a composite containing a larger fiber volume fraction is attributable to the compliant coating decreasing the shear modulus and increasing the stress concentrations in the matrix. The reason panel 5 exhibited a relatively lower strength when compared to panel 3 and 7 is attributable to the voids present in the resin material that were discussed previously. By decreasing the matrix modulus the theoretical model is able to predict this decrease. In all of the experimental results, the theoretical model provides an accurate representation of strength and a means to explain specific phenomena.

Theoretical failure strains presented in Table 1 were calculated by dividing the theoretical failure strengths by the analytically determined composites longitudinal Young's modulus. The theoretical failure strains compared to experimentally measured ones appear to be less accurate than the failure strength predictions. On the other hand by comparing the measured failure strains to the normalized values obtained from dividing the strength of the composite by the shear modulus, similar trends are observed. The normalized value was presented in Figure 4 and discussed in section 4.0. The correlation in the trends between the normalized value and the strain to failure in Table 1 suggests that the normalized parameter could be used to predict the strain to failure for the composite. One reason that this may be a plausible approach is due to the inaccuracies associated with predicting the shear modulus of the composite.

6. Conclusions

An analytical model was presented to investigate the influence of coated cylindrical fibers on a composites compression strength. Results indicate that compliant coatings cause final failures along the neutral axis leading to composite buckling, while stiff coatings fail by microbuckling. The compliant coatings also cause significant decreases in a composite shear modulus which suggests large reductions in compression strength. However, if we normalize the compression strength of the composite to the shear modulus of the composite a specific coating appears to maximize the strength.

An experimental methodology employing coated glass fibers was presented to study the influence of coatings on composite properties. The manufacturing techniques are typical of actual composites in aerospace applications. The model composite system provides the opportunity to systematically change physical micro-parameters within the composite with a high degree of confidence. Compression tests results for the composites containing varied coatings properties indicate that coatings properties, adhesion, and void content significantly influence compression strength. Experimental results closely parallel the theoretical predictions, which support the predictions capabilities of the model.

7. Acknowledgements

The authors gratefully acknowledge the support provided by the National Science Foundation under research grant MSS-9222515. The authors would also like to thank Ciba-Geigy and Matt Lowery for all the assistance they provided during the manufacturing of these specimens.

8. References

Carman G.P., Averill R.C., Reifsnider K.L., and Reddy J.N., "Optimization of Fiber Coatings to Minimize Micromechanical Stress Concentrations in Composites," *J. of Comp. Mat.*, accepted, Jan. 1993.

Carman G.P. & Case S.W. , "Minimizing Stress Concentrations in Material Systems with Appropriate Fiber Coatings," American Society of Composites 7th tech conf. 1992, pp. 889-899.

Chang, Y.S., Lesko, J.J., Case, S.W., Dillard, D.A., Reifsnider, K.L., "Mechanical Properties of Thermoplastic Composites: The Interphase Effect," Proceedings of the 7th Tech. Conf. of American Society of Composites, Oct. 1992, pp. 817-826.

Ghosn, L.J. and Lerch, B., "Optimum Interface Properties for Metal Matrix Composites," NASA Technical Memorandum, no. 102295, August 1989, pp. 1-19.

Greszczuk, L.B., "Microbuckling of Lamina-Reinforced Composites," *Composite Materials: Testing and Design (Third Conference), ASTM STP 546*, American Society for Testing and Materials, 1974, pp. 5-29.

Greszczuk L.B., "Mircobuckling Failure of Circular Fiber-Reinforced Composites," AIAA Journal, V. 13, no 10, Oct. 1975 pp. 1311-1318.

Hashin, Z. and Rosen, B.W., "The Elastic Moduli of Fiber-Reinforced Materials," *J. of Appl. Mech.*, pp. 223-232, 1964 Vol. 31.

Jayraman K., et al., "Elastic and Thermal Effects in the Interphase: Part I. Comments on Characterization Methods," *J. of Comp. Tech. and Research*, 1993, pp. 3-22.

Kulkarni, S.V., Rice, J.R., and Rosen, B.W., "An Investigation of the Compressive Strength of Kevlar 49/Epoxy Composites," *Composites*, Vol. 6, 1975, pp. 217-225.

Lesko, J.J., Carman, G.P., Dillard, D.A., and Reifsnider, K.L., "Penetration Testing of Composite Materials as a Tool for Measuring Interfacial Quality, *"Composite Materials: Fatigue and Fracture" (4th Symposium ASTM)*, accepted for publication Nov. 1991, ASTM-STP 1156.

Lessard, L.B. and Chang, F.K., "Effect of Load Distribution on the Fiber Buckling Strength of Unidirectional Composites," *Journal of Composite Materials*, Vol. 25, Jan. 1991, pp 65-87.

Madhukar, M.S. and Drzal, L.T., "Fiber-Matrix Adhesion and its Effect on Composite Mechanical Properties: I. Inplane and Interlaminar Shear behavior of Graphite/Epoxy Composites", *J. of Comp. Mat.*, Vol. 25, Aug. 1991, pp. 932-958.

Pagano N. J. and Tandon G.P., "Elastic Response of a Multi-directional Coated-fiber Composites," *Comp. Sci. and Tech.*, Vol. 31, pp. 273-293 (1988).

Pak Y., "Longitudinal Shear Transfer in Fiber Optic Sensor," *Smart Mat. Struct.*, Vol. 1, 1992, pp. 57-62.

Rosen, B.W., *Fiber Composite Materials*, American Society for Metals, Metals Park, Ohio, 1965, Chap. 3.

Schwartz, H.S., and Hartness, J.T., "Effect of Fiber Coatings on Interlaminar Fracture Toughness of Composites," Symposium on Toughened Composites, ASTM STP 937, pp. 150-165, 1985.

Shuart, M.J., "Short Wavelength Buckling and Shear Failures of Compression-Loaded Composite Laminates," NASA TN87640, Nov. 1985.

Steif, P.S., "A Model for Kinking in Fiber Composites - 1. Fiber Breakage via Micro-Buckling," *International Journal of Solids and Structures*, Vol. 26, No. 5/6, 1990, pp. 549-561.

Swain, R., Reifsnider, K.L., Jayraman, K., and El-Zein M., "Interface/Interphase Concepts in Composite Material Systems," *J. of Thermoplastic Comp.*, Vol. 3, 1990, pp. 13-23.

Swain, R., "The role of the Fiber/Matrix Interphase in the Static and Fatigue Behavior of Polymeric Composite Laminates," Dissertation, Va. Tech, Eng. Sci. and Mech. Dept., Feb. 1992.

Tsai S.W. and Hahn T.H., *Introduction to Composite Materials*, Technomic, 1980, pp. 414-416.

Wass, A.M., Babcock, C.D., Jr., Knauss, W.G., "A Mechanical Model for Elastic Fiber Microbuckling," *Journal of Applied Mechanics*, Vol. 57, March 1990, pp. 138-149.

Wass, A.M., "Effect of Interphase on Compressive Strength of Unidirectional Composites," J. of Applied Mech., June 1992, V. 59, pp. s183-s188.

The Measurement and Modelling of Fibre Directions in Composites

A. R. CLARKE, N.C. DAVIDSON and G. ARCHENHOLD
Molecular Physics and Instrumentation Group
Department of Physics
University of Leeds
Leeds LS2 9JT
UK

ABSTRACT. The nearest neighbour angle frequency distributions, $f(\Phi_{NN})$ of fibre images on a 2D section plane through both glass fibre and carbon fibre reinforced, polymer composites are compared to the predictions of a 3D, 'Monte Carlo' computer simulation. Actual $f(\Phi_{NN})$ distributions can be simulated by a '3D random hard core' spatial distribution of fibres. The shapes of the $f(\Phi_{NN})$ distributions within all of our composite samples appear to be well approximated by $f(\Phi_{NN}) = A + B.\cos^2\Phi_{NN} + C.\cos^4\Phi_{NN}$ where A, B and C are functions of the fibre packing fraction and range of directions ($\Delta\theta$, $\Delta\Phi$).

1. Introduction

For the past five years, a research project has been undertaken into the quality measurement of glass and carbon fibre directions in a variety of polymer matrices. Both unidirectional, continuous fibres at high packing fractions ($40\% \leq V_f \leq 55\%$) as well as short fibre reinforcements at a number of different packing fractions ($25\% \leq V_f \leq 50\%$) have been studied. An automated, 2D image analyser has been designed to not only derive the best fit elliptical parameters of each fibre image on a section plane, but also to identify the absolute centre coordinates of each fibre image within a 2 mm x 2 mm area of the sample. The elliptical parameters of each fibre's image indicate the direction of the fibre in space, denoted by the angles (θ,Φ) where Φ is the in-plane angle, given by the orientation of the major axis of the elliptical image and θ is the colatitude angle or out-of-plane angle given by the ratio of the major axis 'a' and the minor axis 'b':

$$\theta = \cos^{-1}(b/a) \qquad (1)$$

see figure 1(a).

A few years ago, Davy and Guild (1988) published a paper where they modelled a composite as though the reinforcements followed a Gibbs 'hard core' distribution of random positions in a 2D section plane. However, the Gibbs hard core hypothesis is only

77

R. Pyrz (ed.), IUTAM Symposium on Microstructure-Property Interactions in Composite Materials, 77–88.
© *1995 Kluwer Academic Publishers.*

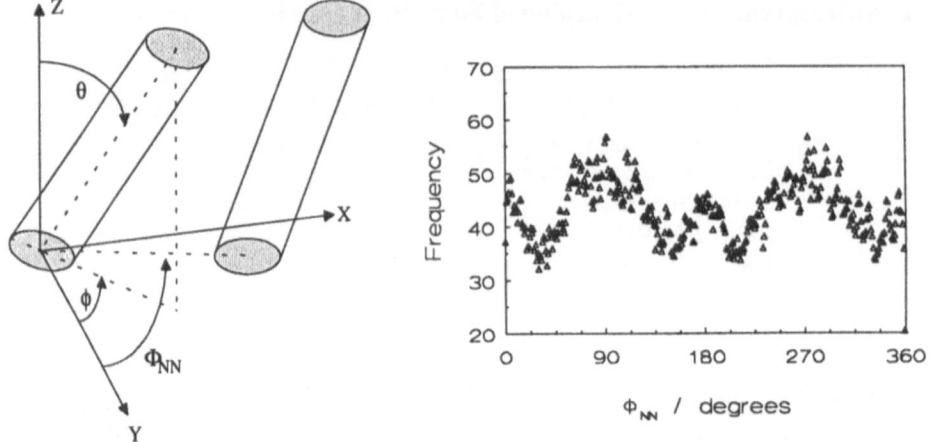

Figure 1a. Coordinate system adopted.
Φ *NN is the nearest neighbour angle.*

Figure 1b. Typical frequency distribution of nearest neighbour angle.

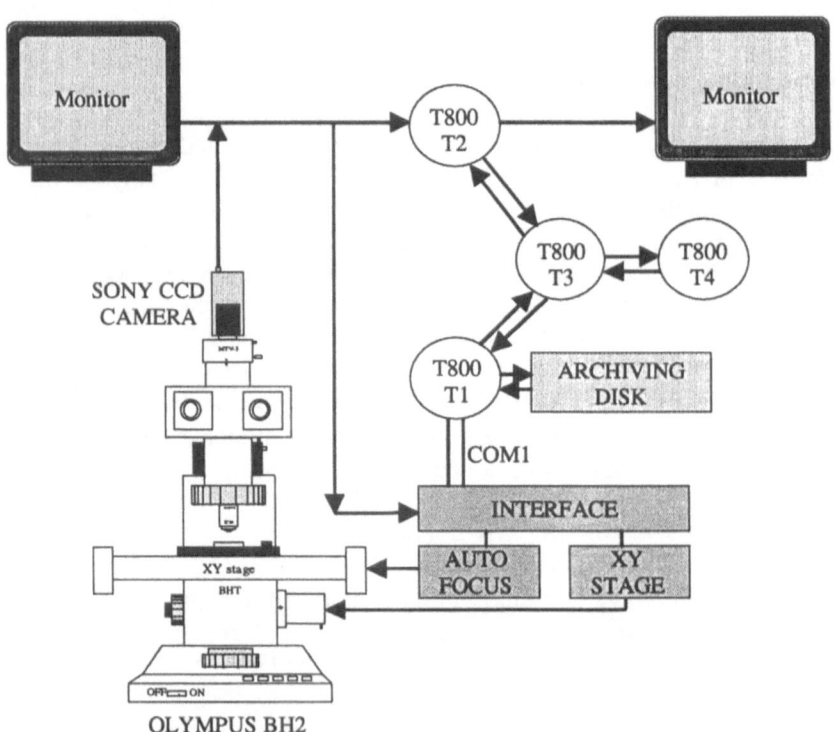

Figure 2. Schematic of Leeds Image Analyser, incorporating a network of transputers.

realistic if **either** the fillers are flat, 2D objects (rather than 3D rods) **or** if the packing fraction of short fibres in the composite is so low that there is essentially no interaction between the fibres!

It was decided to test Davy & Guild's hypothesis by checking whether the distribution of nearest neighbour angles, $f(\Phi_{NN})$ for the thousands of fibres in a typical 2 mm x 2 mm dataset was consistent with the predicted *isotropic* $f(\Phi_{NN})$ distribution. As reported in Clarke and Davidson(1991), no actual composite analysed at Leeds gave an isotropic $f(\Phi_{NN})$ distribution. Every composite sample studied has exhibited a remarkably similar functional form for the $f(\Phi_{NN})$ distribution, as shown in figure 1(b).

However, like Davy & Guild we acknowledged that the fibres in all real composites (with $V_f \leq 55\%$) have spatial distributions which are far from the ideal, regular 'square' or 'hexagonal' arrays assumed by many finite element analysis papers, e.g. Dubois *et al*(1993). Therefore, we embarked upon a 3D modelling exercise to generate the $f(\Phi_{NN})$ distributions at different levels within a simulated 3D space. Some of the model results have been puzzling us for the past three years, but with our improved 2D and 3D techniques, Archenhold *et al*(1992), Mattfeldt *et al*(1994), Clarke *et al*(1994) to analyse composite microstructures, we are now in a better position to interpret the original model data. Also, a recent paper on 2D microstructural issues by Pyrz(1993) has rekindled our interest in this work.

2. Measurement of Fibre Directions in Composites

A large area, high spatial resolution, 2D image analyser system has been developed to automate fully the collection of (θ, Φ) fibre data, see Clarke *et al*(1991). For speed of operation, the design uses a small network of transputer chips which form the basis of a parallel processing system, as shown in figure 2.

The image analyser design has a number of unique features. It scans automatically in X and Y, finding the best position of focus at each new XY location and merges overlapping image frames to create a data table of *absolute x, y* fibre coordinates over a 2 mm x 2 mm area. The system automatically performs image splitting, determines a quality factor for the elliptical fit to each fibre image, performs an 'autocalibration' in XY by following specific fibre images during the XY scanning, merges partial fibre images between XY frames and plots the (θ, Φ) angular distributions within seconds of the end of each large area scan.

When a sample containing well-aligned fibres is sectioned perpendicular to the main fibre direction and analysed by **any** 2D system, the *apparent* θ and Φ distributions will resemble those shown in figures 3(a) and 3(b). The θ distribution will show a spurious peak at around $\theta = 10 - 20^o$ and the Φ distribution will most probably show a very broad distribution of angles between 0° and 180° (note that every 2D system has a Φ ambiguity of 180°). These angular distributions occur because of errors in deriving θ from near

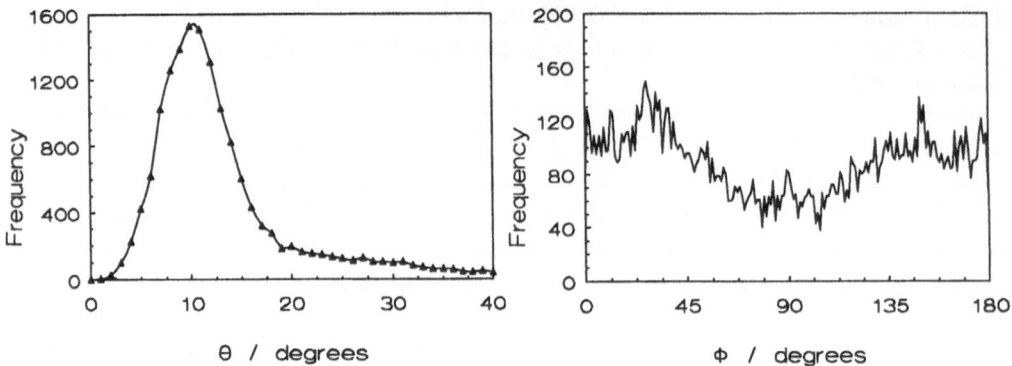

Figure 3a Apparent θ *distribution, glass fibres in epoxy (cf. Figure 10).*

Figure 3b. Apparent φ *distribution, glass fibres in epoxy (cf. Figure 11).*

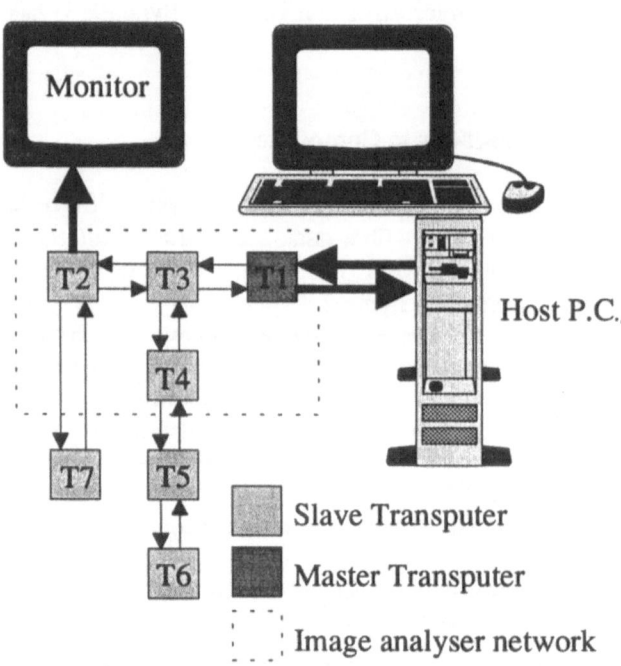

Figure 4. Transputer configuration for 3D modelling.

circular, digitised images, see Clarke *et al*(1993). Our original motivation for exploring the $f(\Phi_{NN})$ distributions was to improve upon these apparent (θ, Φ) distributions.

However we now know that, if the well-aligned sample is cut at an angle of 45° to the main fibre direction, the fibre ellipticities and hence the θ', Φ' values in this 45° section plane can be derived very accurately. Each fibre's direction may be mathematically transformed by 45° to derive more accurate (θ, Φ) distributions, see Hine *et al*(1993).

3. Modelling of Fibres in 3D

3.1 COMPUTER HARDWARE

The basic network of transputers used for the image analyser design was reorganised for the 3D modelling by adding another pipeline of three transputers as shown in Figure 4. In this configuration, the master transputer, T_1 was responsible for collating information from the other 'slave' transputers and archiving to disk. The master transputer together with each of the other six transputers was running identical code which enabled them to perform 1500 simulations in an 8 hour overnight run. Hence, a speedup of x7 over a single transputer was achieved and 10,500 points on the $f(\Phi_{NN})$ distributions every 20 μm in section depth were generated, as typified by the distributions in figure 5.

3.2 THE MONTE CARLO ALGORITHM

The algorithm allows us to vary the initial spatial locations of each fibre (e.g. on a basic square array or hexagonal array or placed randomly at the top surface) and also to choose the range of fibre directions.

In order to mimic the 'unidirectional', glass and carbon reinforced composites at our disposal, the fibres were constrained to have a range of angles, θ between 0° and $+ \Delta\theta$ and Φ between $(90^o \pm \Delta\Phi)$ and $(270^o \pm \Delta\Phi)$. The spatial positions of the fibres *at the surface of the simulated volume* were chosen randomly in x and y or were placed in a square array. For each candidate fibre, four random numbers were chosen. The first random number determined the $\Delta\theta$ value, the second determined the $\Delta\Phi$ value, the third and the fourth determined the x, y of the fibre centre on the section plane. The model assumed a fixed fibre length, $L = 1$ mm and a fixed diameter, $D = 10$ μm for ease of computation. Each fibre was followed in 3D space and if it was found to hit a fibre which was already in that 3D space, the fibre would be discarded and another fibre with another set of random numbers and hence a different set of $\{x, y, \theta$ and $\Phi\}$ was produced and checked. The process was repeated until the required packing fraction of fibres had been achieved at the surface of the simulated volume. Hence the model, in effect, simulates a 3D version of the Gibbs hard core process. Note that, as the nearest neighbour information is obtained from the *central fibre* and it's nearest neighbours, rather than to all of the fibres within the simulated volume, there are no 'edge effects' which might lead to an incorrect $f(\Phi_{NN})$ distribution.

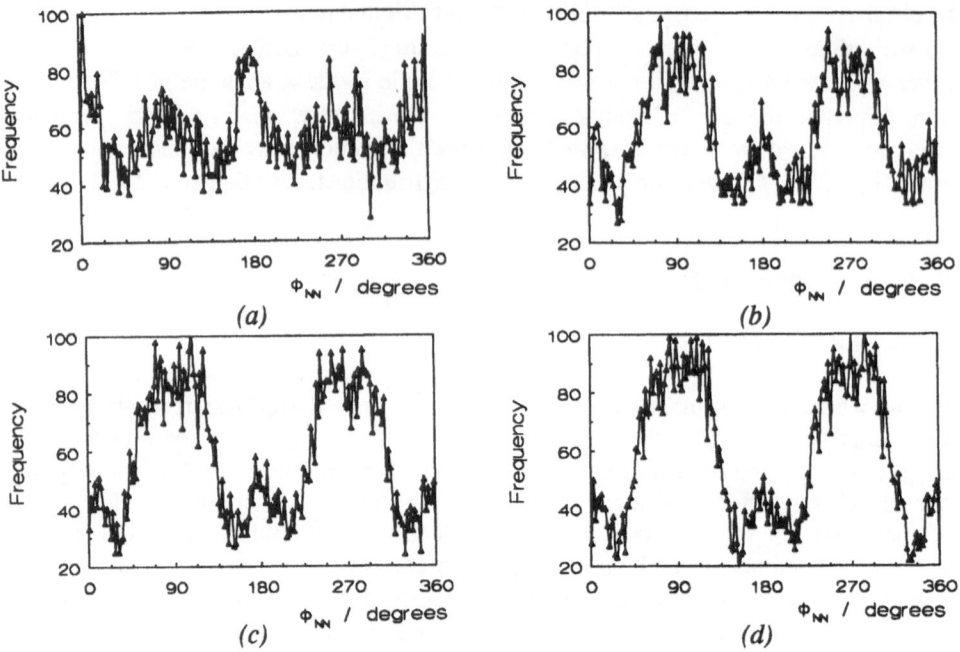

Figure 5. Φ $_{NN}$ simulations at different depths within a 3D volume (40% packing fraction at the surface), (a) top level, (b) -20μ m, (c) -80μ m, (d) -100μ m.

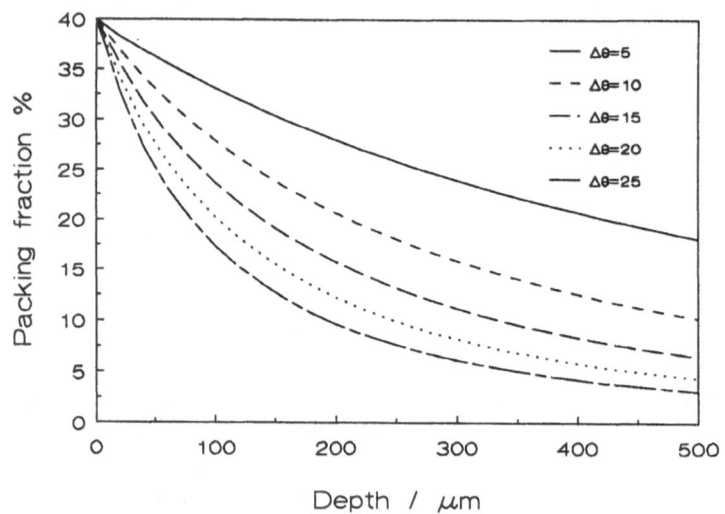

Figure 6. Maximum variation of packing fraction with depth for different Δ θ   ranges.

3.3 LIMITATIONS OF THE CURRENT 3D MODEL

The gradual evolution of the shape of the $f(\Phi_{NN})$ distribution as a function of depth puzzled us until we realised that our 3D model currently has two limitations:

a) Although a particular packing fraction is established at the 'surface level', there is a systematic change of effective packing fraction with depth, which depends upon the permissible range of angles, $\Delta\theta$ during the simulation. The worst case change of packing fraction as a function of $\Delta\theta$ is shown in figure 6.

b) Because the model only checked for non-intersecting fibres *within the 3D volume*, the $f(\Phi_{NN})$ distributions at and near the surface of the 3D space are physically unreasonable. However, the angular distributions simulated at lower levels within the 3D volume should be representative of those distributions within a real composite.

4. Characterisation of the $f(\Phi_{NN})$ Distribution

Instead of the expected 'isotropic distribution' of nearest neighbour angles for a '2D Gibbs hard core' process, the influence of the third dimension i.e. fibre length gave rise to characteristic, $f(\Phi_{NN})$ distributions in our 3D simulations.

The main point of our analysis was to explore the 'anisotropy' of the $f(\Phi_{NN})$ distributions and to seek correlations with the (θ, Φ) fibre directional distributions. All of the real and modelled nearest neighbour angular distributions seem to have the same functional form, shown idealised in figure 7(a). Recently we have found that a reasonably good fit to these frequency distributions, $f(\Phi_{NN})$ is given by the function

$$f(\Phi_{NN}) = a + b.\cos 4\Phi_{NN} - c.\cos 2\Phi_{NN} = A + B.\cos^2\Phi_{NN} + C.\cos^4\Phi_{NN} \qquad (2)$$

as shown in figure 7(b). However, when the model data was reduced originally, it was decided to characterise the distribution in terms of two, easily determined, 'probability amplitudes', A_{in} and A_{out}.

Referring to figure 7(a), the 'in-phase' amplitude of the $f(\Phi_{NN})$ distribution is defined as

$$A_{in} = (f_2 - f_1)/(f_2 + f_1) \qquad (3)$$

and the 'out-of-phase' amplitude of the $f(\Phi_{NN})$ distribution is defined as

$$A_{out} = (f_3 - f_1)/(f_3 + f_1) \qquad (4)$$

In this way, the systematic modification to the shape of the $f(\Phi_{NN})$ distributions can be followed as different constraints are placed on the range of individual fibre (θ, Φ) values. (Note that the definition of A_{in} and A_{out} is arbitrary when analysing a *real* composite and hence each value in the top part of the A_{in}-A_{out} plot has an equivalent value in the lower

84

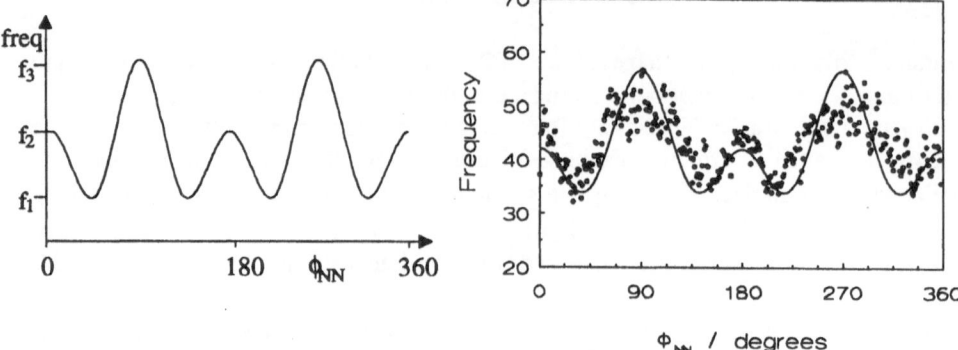

Figure 7a. Idealised functional form of the nearest neighbour angular distribution $f(\Phi_{NN})$.

Figure 7b. Real $f(\Phi_{NN})$ data (glass in epoxy) fitted to:-
$$f = 42 + 7.3(cos(4\Phi_{NN}) - 7.3(cos(2\Phi_{NN}))$$

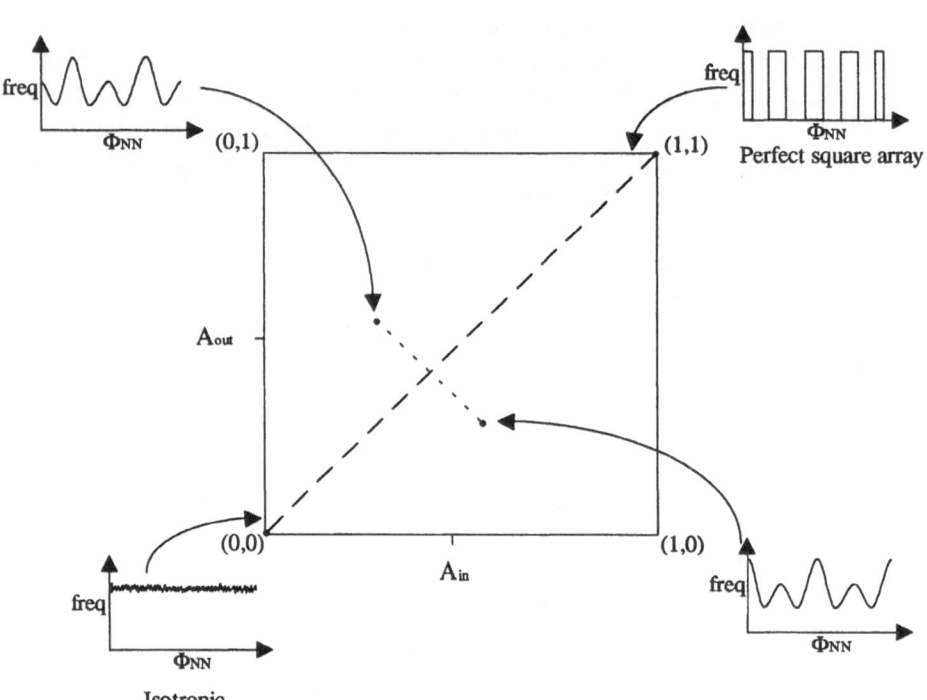

Figure 8. Mapping of probability amplitudes.

part of the A_{in}-A_{out} plot). Figure 8 illustrates the mapping of the probability amplitudes onto the A_{in}-A_{out} plot and figure 9 shows the result of a number of model simulations.

5. Comparison with real composites

Most of the composites that have been studied were produced by processes which involve the layering of plies and therefore it is not surprising if the fibres' exhibit a preferred orientation. When the samples were positioned on the XY stage, the operator lined up the edge of the sample (which was perpendicular to the plies) to correspond with the X or Y axis of the video system. Hence, the most probable nearest neighbour orientations should be at $\Phi_{NN} = 0^o$ or 90°.

Great care was exercised to ensure that the 'peaks' in the distributions were not artefacts of the image analyser (e.g. they could be produced spuriously if an incorrect 'aspect ratio' was chosen for the X-Y image frame analysis). The aspect ratio was checked by rotating the sample on the X-Y stage and confirming that the peaks appeared at the correct angles in the image analyser's coordinate system.

The probability amplitudes of various composites are plotted in figure 9.

5.1 CONTINUOUS GLASS FIBRES IN EPOXY

Recently, considerable research effort has been put into the analysis of a continuous, 'unidirectional', glass fibre reinforced composite with a view to characterise the 'waviness' of the glass fibres in 3D. The sample has been analysed with our 2D image analyser by sectioning at 45° to the main fibre orientation axis (in order to produce the most accurate θ distributions) and also with our new 3D confocal laser scanning microscope technique, Clarke et al(1994) which is capable of accurate 3D positional information. Both of these techniques confirm that the θ angular distribution of 100 μm segments of fibres exhibits a narrow range, $\Delta\theta_{FWHM} = \pm 2.5^o$, see figure 10, equivalent to $\Delta\theta = 2.5^o$ in our model. The Φ distribution has a range, $\Delta\Phi_{FWHM} = 40^o$ - 60° (i.e. $\Delta\Phi = \pm 25^o$), as shown in figure 11.

The 2D image analyser has also scanned three sections, each section being perpendicular to the main fibre direction and the $f(\Phi_{NN})$ distributions have been obtained for all three sections. The first section was a 2 mm x 2 mm scanned area in XY. The second section was prepared by removing approximately 10 μm of material and repolishing, thereby creating a section plane parallel to the first. Another 50 μm of material was removed to create a plane parallel to the first two sections. Care was taken to orient the sample for each scan so that there would be minimal $\Delta\Phi$ error between scans. The results are shown in figures 12(a), (b) and (c). The scans are statistically identical and have been added together and normalised in figure 12(d). These distributions give 'probability amplitudes' of $A_{in} = 0.19$ and $A_{out} = 0.24$. Note that, although the (θ, Φ) angular distributions are due to intrinsic 'waviness' of $continuous$ fibres (typical wavelengths in the range 0.5 mm through to 2 mm and typical amplitudes 25 to 50 μm), the $f(\Phi_{NN})$ distribution is characterised by a

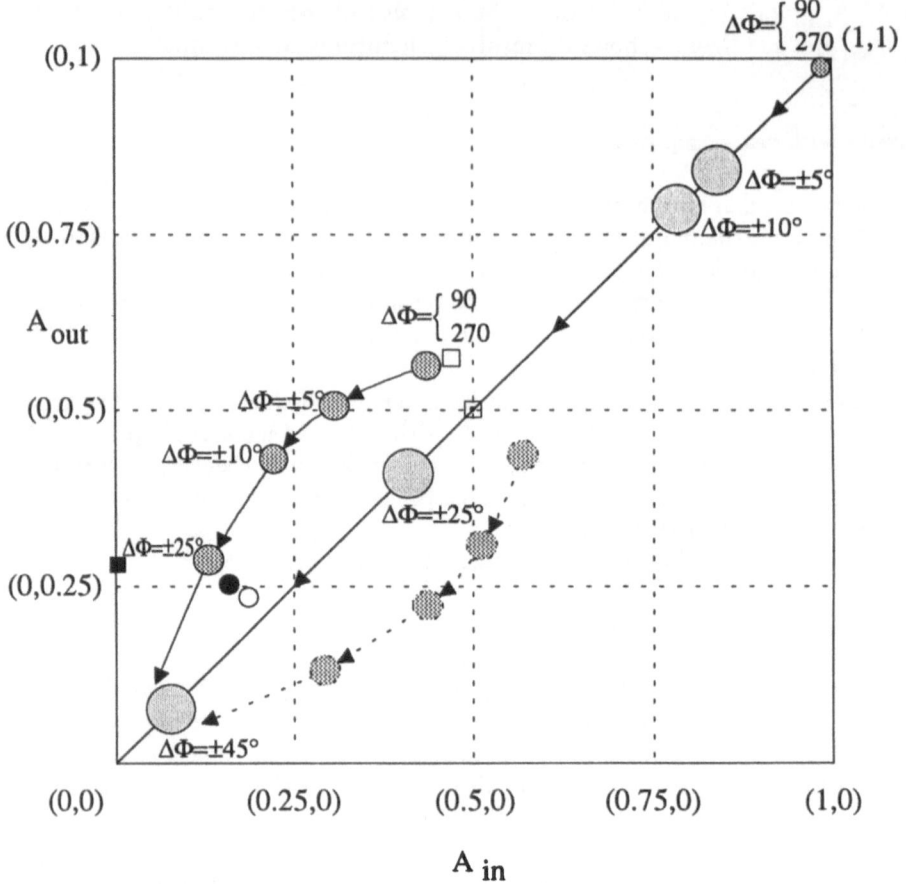

$\Delta\Phi=\begin{cases} 90 \\ 270 \end{cases}$ (1,1)

(0,1)

$\Delta\Phi=\pm5°$

(0,0.75) $\Delta\Phi=\pm10°$

A_{out} $\Delta\Phi=\begin{cases} 90 \\ 270 \end{cases}$

(0,0.5) $\Delta\Phi=\pm5°$

$\Delta\Phi=\pm10°$

$\Delta\Phi=\pm25°$

$\Delta\Phi=\pm25°$

(0,0.25)

$\Delta\Phi=\pm45°$

(0,0)   (0.25,0)   (0.5,0)   (0.75,0)   (1,0)

A_{in}

○ Continuous glass fibres in epoxy, packing fraction 52%, $\Delta\Phi_{FWHM}=50°, \Delta\theta=5°$.
● Cosmos, carbon samples, packing fraction 25%, $\bar{l}=120\mu m$.
□ FT14, Carbon reinforced, packing fraction 42%, $\bar{l}=1mm$.
⊡ FT15, Carbon reinforced, packing fraction 42%, $\bar{l}=1mm$.
■ UD913, continuous carbon reinforced, packing fraction 52%.
3D hard core model, $\bar{l}=1mm$, packing fraction 33%, $\Delta\theta=5°$.
3D hard core model (with change of Φ_{NN} origin).
Square array model, $\bar{l}=1mm$, packing fraction 45%, $\Delta\theta=5°$.

Figure 9. Probability Amplitudes for both real composites and model simulations.

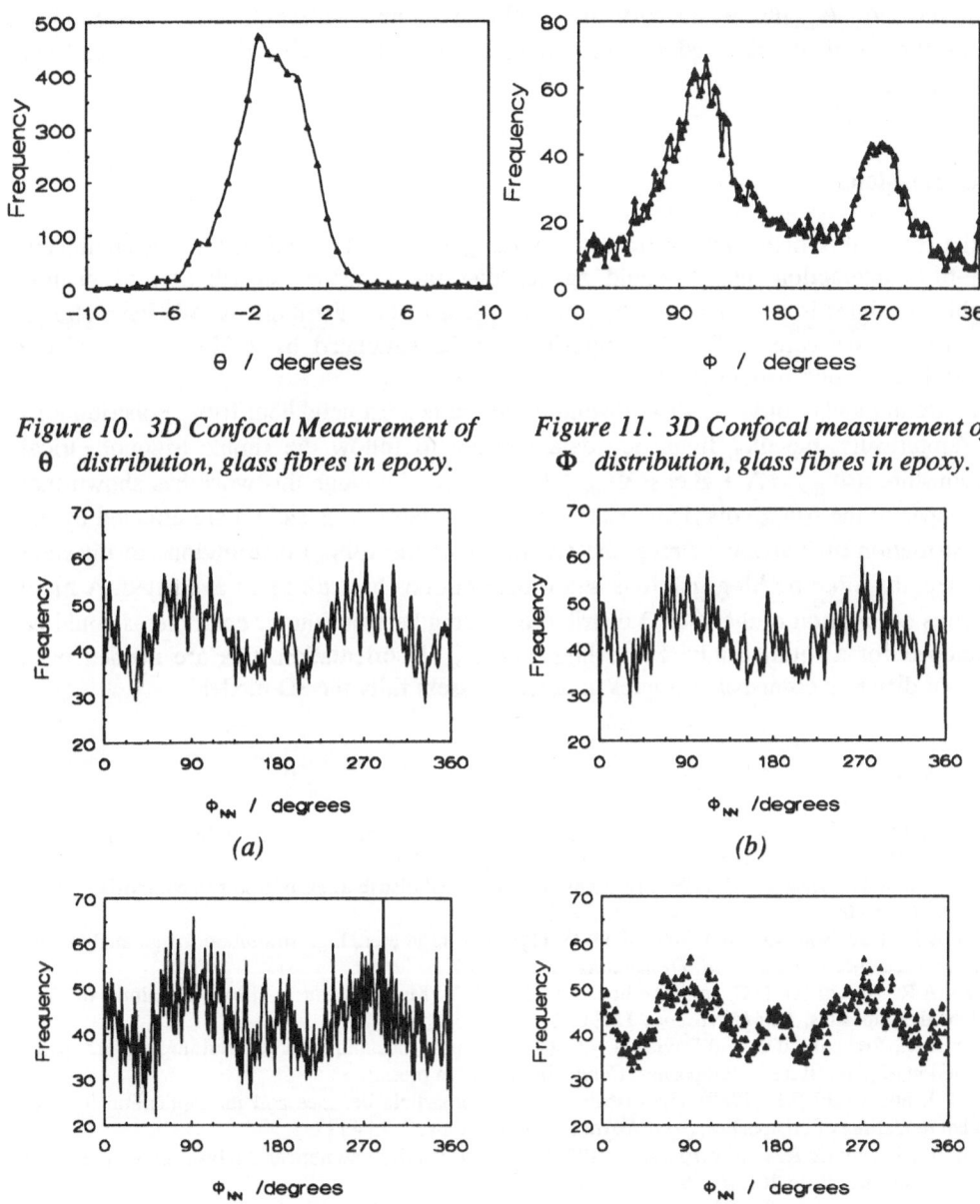

Figure 10. 3D Confocal Measurement of θ distribution, glass fibres in epoxy.

Figure 11. 3D Confocal measurement of Φ distribution, glass fibres in epoxy.

(a)

(b)

(c)

(d)

Figure 12. Glass fibres in epoxy, f(Φ $_{NN}$) frequency distributions, (a) 1st scan, (b) 2nd scan, (c) 3rd scan, (d) average of all three scans showing probability amplitudes A_{in}=0.19, A_{out}=0.24

88

point in our A_{in}-A_{out} plot which corresponds closely to the simulated $f(\Phi_{NN})$ distribution of 1 mm fibres with the allowed angular ranges set at $\Delta\theta = +5^o$ and $\Delta\Phi = \pm 25^o$ i.e. $\Delta\Phi_{FWHM} = 50^o$.

6. Conclusions

At the reasonably high packing fractions ($55\% \geq V_f \geq 40\%$) studied, there is significant fibre-fibre interaction and it would appear that the statistical distribution of nearest neighbour angles is correlated to the intrinsic (θ, Φ) fibre distributions. We have shown that the microstructure of real composites can be simulated by a 3D version of the classical 2D Gibbs 'hard core' process.

We are not aware of an *analytical* solution to the nearest neighbour $f(\Phi_{NN})$ distributions but, empirically, the Φ_{NN} frequency data appears to follow the simple trigonometrical relationship, $f(\Phi_{NN}) = A + B.\cos^2 \Phi_{NN} + C.\cos^4\Phi_{NN}$. Although this work has shown that the shapes of the $f(\Phi_{NN})$ distributions (i.e. the coefficients A, B and C) are affected by the Φ distribution of individual fibres, the *sensitivity* of the $f(\Phi_{NN})$ distributions to different fibre lengths, fibre packing fractions and θ distributions has still to be evaluated. A more rigorous description of the $f(\Phi_{NN})$ distributions using best fit Fourier coefficients could be developed for a future study. More high quality, 3D orientation data are needed on a range of different composite samples in order to assess fully the 3D model.

References

Archenhold G., Clarke A.R. and Davidson N.C. (1992) 3D microstructure of fibre reinforced composites, *SPIE Proc. Biomed Image Proc.& 3D Microscopy 1660*, 199 - 210.
Clarke A.R. and Davidson N.C. (1991) Determining the spatial distributions of fibres in composites, *Proc. of ICAM91*, 55 - 60.
Clarke A.R., Davidson N.C. and Archenhold G. (1991) A large area, high resolution image analyser for polymer research, *Proc. Int. Conf. Transputing 91 1*, 31 - 47.
Clarke A.R., Davidson N.C. and Archenhold G. (1993) Measurement of fibre directions in fibre reinforced composites, *J. of Microscopy 171*, Pt 1, 69 - 80.
Clarke A.R., Archenhold G. and Davidson N.C. (1994) A novel technique for determining the 3D spatial distribution of glass fibres in composites, *Comp. Sci. Tech.*, in press.
Davy P.J. and Guild F.J. (1988) The distribution of interparticle distance and its application in finite element modelling of composites, *Proc. Royal Soc. London A418*, 95 - 112.
Dubois F., Keunings R. and Verpoest I. (1993) Micromechanical numerical analysis of toughness in model composites, *Proc. ICCM9 III*, 375 - 382.
Hine P.J., Duckett R.A., Davidson N.C. and Clarke A.R. (1993) Modelling of the elastic properties of fibre reinforced composites I: Orientation Measurement, *Comp. Sci. Tech 47*, 65 - 73.
Mattfeldt T., Clarke A.R. and Archenhold G. (1994) Estimation of the directional distribution of spatial fibre processes using stereology and confocal laser microscopy, *J. of Microscopy 173*, 87-101.
Pyrz R. (1993) Morphological description of microstructure for composite materials, *Proc. ICCM9 III*, 383 - 390.

TRANSFORMATION ANALYSIS OF INELASTIC LAMINATES

GEORGE J. DVORAK and YEHIA A. BAHEI–EL–DIN
Rensselaer Polytechnic Institute, Troy, NY 12180–3590, USA
Structural Engineering Department, Cairo University, Giza, Egypt

Abstract

The transformation field analysis (TFA) of inelastic composite materials (Dvorak 1992) is extended here to fibrous composite laminates. Loading is limited to uniform in–plane stresses and out–of–plane normal stress, and to uniform changes in temperature. The solution for local stresses or strains in the plies is found in terms of elastic transformation influence functions and concentration factors which reflect a selected microgeometry representation of a unidirectional composite, and the constraints imposed on the in–plane strains of the perfectly bonded plies. This methodology is applied in simulations of hot isostatic pressing and subsequent loading of a $(0/90)_s$ Sigma/Timetal 21S laminate under axial tension/tension stress cycles applied at constant temperature.

1. Introduction

The transformation field analysis (TFA) is a method for incremental solution of thermomechanical loading problems in inelastic heterogeneous media and composite materials, described in recent papers by Dvorak (1991, 1992). In its application to composite materials reinforced by aligned continuous fibers, the local strain and stress fields in a representative volume of the material are modeled by piecewise uniform approximations using a selected micromechanical model such as the self–consistent (Hill 1965) and Mori–Tanak (1973) models, or the Periodic Hexagonal Array (PHA) model (Dvorak and Teply 1985, Teply and Dvorak 1988). Only elastic solutions under certain overall uniform loads and local transformation strains are required from these models to recover the transformation influence functions and concentration factors used in the TFA method to evaluate the local fields and the overall response. Implementation of this procedure for several constitutive laws of the matrix material is described by Dvorak and co–workers (1994a,b).

    The purpose of the present paper is to apply the TFA approach to inelastic fibrous composite laminates consisting of unidirectional layers bonded together with fibers oriented at different directions. Only symmetric layups under overall uniform stresses and temperature variations which produce membrane stresses in the individual plies are considered. These loading conditions are found in fabrication processes and in service under static and cyclic in–plane loads of symmetric composite laminates.

R. Pyrz (ed.), IUTAM Symposium on Microstructure-Property Interactions in Composite Materials, 89–100.
© *1995 Kluwer Academic Publishers.*

The TFA analysis for laminates is described in sections 2 and 3 of the paper. Application of the method to a $(0/90)_s$ titanium matrix laminate under hot isostatic pressing conditions and subsequent load cycles is presented in section 4.

2. Transformation Field Analysis of Laminated Plates

Consider a laminate consisting of 2N thin elastic plies arranged in a symmetric layup with respect to the midplane x_1x_2 of a cartesian coordinate system, Fig. 1. The ratio $c_i = t_i/t$, i=1,2,..N, of the ply thickness, t_i, and half the laminate thickness, t, denotes the ply volume fraction. In–plane membrane forces and the corresponding uniform stresses are applied, together with uniform change in temperature. In addition, we admit loading by uniform normal stresses in the thickness direction x_3; this is useful in applications to processing by hot pressing, and also in analysis of eigenstress states under in–plane constraint. Moreover, we also account for inelastic deformation of the phases in each ply, the resulting inelastic response of some or all of the plies, and of the laminate itself. Our goal is to find the ply and phase stresses as well as the overall strains under these loading conditions.

The local and overall inelastic strains, and the thermal strains, will be regarded as eigenstrains or transformation strains in an otherwise elastic laminate. Hence, we write the overall constitutive relations of the laminate in the overall coordinate system x_k, k = 1,2,3, as,

$$\sigma = \mathbf{L}\,\epsilon + \lambda, \qquad \epsilon = \mathbf{M}\,\sigma + \mu, \tag{1}$$

where, μ is eigenstrain, λ is eigenstress, $\mathbf{L}$ and $\mathbf{M}$ are the elastic stiffness and compliance matrices, respectively, and $\mathbf{L} = \mathbf{M}^{-1}$, $\lambda = -\mathbf{L}\,\mu$. Using contracted notation, the stress and strain vectors are:

$$\sigma = [\sigma_1,\, \sigma_2,\, \sigma_6,\, \sigma_3]^{\mathrm{T}} = [\sigma',\, \sigma_3]^{\mathrm{T}}, \tag{2}$$

$$\epsilon = [\epsilon_1,\, \epsilon_2,\, 2\epsilon_6,\, \epsilon_3]^{\mathrm{T}} = [\epsilon',\, \epsilon_3]^{\mathrm{T}}, \tag{3}$$

where the vectors σ' and ϵ' list the in–plane stress and strain components. In the sequel, we outline the solution for the elastic properties of the laminate, and develop a transformation field analysis for evaluation of the overall eigenstrains μ, or eigenstresses λ, and the corresponding local fields when inelastic or thermal strains are present in the phases.

2.1. LAMINA STRESSES

In analogy with (1), the ply constitutive relations of a ply (i) in the local coordinate system $\bar{x}_k$, k = 1,2,3, can be written as,

$$\bar{\sigma}_i = \bar{\mathbf{L}}_i\,\bar{\epsilon}_i + \bar{\lambda}_i, \qquad \bar{\epsilon}_i = \bar{\mathbf{M}}_i\,\bar{\sigma}_i + \bar{\mu}_i, \tag{4}$$

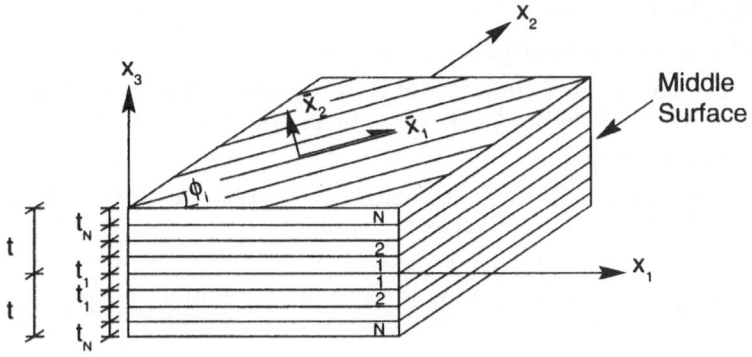

Figure 1. Geometry of a fibrous composite laminate.

where, $\bar{\mu}_i$ is the ply eigenstrain, e.g. thermal or inelastic strain, $\bar{\lambda}_i$ is eigenstress, $\bar{L}_i$ and $\bar{M}_i$ are stiffness and compliance matrices. The stress and strain vectors are:

$$\bar{\sigma}_i = [\bar{\sigma}_1^{i}, \bar{\sigma}_2^{i}, \bar{\sigma}_6^{i}, \bar{\sigma}_3^{i}, \bar{\sigma}_4^{i}, \bar{\sigma}_5^{i}]^T = [\bar{\sigma}_i', \bar{\sigma}_3^{i}, \bar{\sigma}_4^{i}, \bar{\sigma}_5^{i}]^T , \tag{5}$$

$$\bar{\epsilon}_i = [\bar{\epsilon}_1^{i}, \bar{\epsilon}_2^{i}, 2\bar{\epsilon}_6^{i}, \bar{\epsilon}_3^{i}, 2\bar{\epsilon}_4^{i}, 2\bar{\epsilon}_5^{i}]^T = [\bar{\epsilon}_i', \bar{\epsilon}_3^{i}, 2\bar{\epsilon}_4^{i}, 2\bar{\epsilon}_5^{i}]^T . \tag{6}$$

To simplify the subsequent analysis, we re–write eqs. (4) for the in–plane stress and strain components and account for the nonvanishing $\bar{\sigma}_3^{i}$, $\bar{\epsilon}_3^{i}$ components (Bahei–El–Din 1992, Dvorak et al. 1992);

$$\bar{\sigma}_i' = \bar{L}_i' \bar{\epsilon}_i' + \bar{k}_i \bar{\sigma}_3^{i} + \bar{\lambda}_i', \quad \bar{\epsilon}_i' = \bar{M}_i' \bar{\sigma}_i' + \bar{n}_i \bar{\sigma}_3^{i} + \bar{\mu}_i' , \tag{7}$$

where

$$\bar{L}_i' = \frac{1}{k+m} \begin{bmatrix} E_L k+mn & 2m\ell & 0 \\ & 4km & 0 \\ SYM. & & p(k+m) \end{bmatrix} = [\bar{M}_i']^{-1}, \quad \bar{k}_i = \frac{1}{k+m} \begin{bmatrix} \ell \\ k-m \\ 0 \end{bmatrix} , \tag{8}$$

$$\bar{M}_i' = \begin{bmatrix} 1/E_L & -\nu_L/E_L & 0 \\ & 1/E_T & 0 \\ SYM. & & 1/G_L \end{bmatrix} , \quad \bar{n}_i = \begin{bmatrix} -\nu_L/E_L \\ -\nu_T/E_T \\ 0 \end{bmatrix} = -\bar{M}_i' \bar{k}_i . \tag{9}$$

The overall longitudinal, and transverse Young's moduli, E_L and E_T, and Poisson's ratios, ν_L and ν_T, and the longitudinal shear modulus, G_L, are found from a selected micromechanical model of a unidirectional composite. The corresponding Hill's moduli (Hill 1964) are denoted by k, ℓ, n, m, p, where $E_L = n - \ell^2/k$, $\nu_L = \ell/2k$, $m = E_T/2(1+\nu_T)$, $G_L = p$.

When expressed in the overall coordinate system x_k, $k = 1,2,3$, eq. (7) is written as (Bahei–El–Din 1992, Dvorak et al. 1992)

$$\sigma'_i = L'_i\,\epsilon'_i + k_i\,\overset{i}{\sigma_3} + \lambda'_i\,, \quad \epsilon'_i = M'_i\,\sigma'_i + n_i\,\overset{i}{\sigma_3} + \mu'_i\,, \tag{10}$$

where

$$\bar{\sigma}'_i = R'_i\,\sigma'_i\,, \quad \bar{\epsilon}'_i = N'_i\,\epsilon'_i\,, \tag{11}$$

$$L'_i = N'^T_i\,\bar{L}'_i\,N'_i = [M'_i]^{-1}\,, \quad k_i = N'^T_i\,\bar{k}_i = -L'_i\,n_i\,, \tag{12}$$

$$R'^T_i = (N'_i)^{-1} = \begin{bmatrix} \cos^2\varphi_i & \sin^2\varphi_i & -\frac{1}{2}\sin 2\varphi_i \\ \sin^2\varphi_i & \cos^2\varphi_i & \frac{1}{2}\sin 2\varphi_i \\ \sin 2\varphi_i & -\sin 2\varphi_i & \cos 2\varphi_i \end{bmatrix}\,, \tag{13}$$

and φ_i is the angle between the local $\bar{x}_1$–axis and the overall x_1–axis, Fig. 1.

We now can address the problem of finding the ply stresses in a laminate loaded by overall in–plane stresses σ', out–of–plane normal stress σ_3, and ply eigenstresses λ'_i, introduced by certain prescribed in–plane eigenstrains $\mu'_i = -M'_i\,\lambda'_i$. Since the laminate is elastic, we write the ply stresses as the sum of the overall stress and local eigenstress contributions,

$$\sigma'_i = H'_i\,\sigma' + \kappa_i\,\sigma_3 + \sum_{j=1}^{N} F'_{ij}\,\lambda'_j\,, \quad i = 1, 2, ..N\,, \tag{14}$$

$$\overset{i}{\sigma_3} = \sigma_3\,, \quad \overset{i}{\sigma_4} = \overset{i}{\sigma_5} = 0\,, \quad i = 1, 2, ..N\,. \tag{15}$$

The lamina out–of–plane eigenstresses λ_3, λ_4, λ_5, are not introduced in eqs. (14), (15) since the in–plane equi–strian condition imposed on the perfectly bonded plies can be maintained under these eigenstresses without introducing additional ply stresses. The H'_i, κ_i are stress distribution factors for in–plane overall stresses, and out–of–plane normal stress, respectively, and F'_{ij} is transformation influence coefficient. In the absence of the overall stresses σ', σ_3, the ply in–plane stresses

σ_1, σ_2, σ_6 caused in lamina (i) by a unit eigenstress λ_k, $k = 1,2,3$, applied to lamina (j) are given by the kth column of matrix F'_{ij}. Evaluation of the distribution factors H'_i, κ_i, and the transformation coefficients F'_{ij} for a symmetric laminate is given in section 3.

Considering the full (6x1) eigenstress vector $\lambda_j = -L_j \mu_j$, we augment eqs. (14), (15) and write the (6x1) local stress vector as

$$\sigma_i = H_i \, \sigma - \sum_{j=1}^{N} G_{ij} \, \mu_j \,, \quad i = 1, 2, ..N \,, \tag{16}$$

where

$$H_i = \begin{bmatrix} H'_i & \kappa_i \\ 0 & 1 \end{bmatrix}, \quad G_{ij} = \begin{bmatrix} F'_{ij} L'_j & 0 \\ 0 & 0 \end{bmatrix}, \tag{17}$$

σ is given by (2), 0 is a (3x3) null matrix, and $1 = [1, 0, 0]^T$. Note that the order of H_i is (6x4), and the order of G_{ij} is (6x6).

The eigenstrains originate in the phases of each unidirectional ply and must be evaluated from a micromechanical model. This is usually achieved in the local coordinate system $\bar{x}_k$, $k = 1,2,3$, of the ply. Considering (6x1) strain vectors, the inverse transformation of $(11)_2$ can be written for the eigenstrain as

$$\mu_j = R_j^T \bar{\mu}_j \,, \quad R_j = \begin{bmatrix} R'_j & 0 \\ 0 & I \end{bmatrix}, \quad j = 1, 2, ..N \,, \tag{18}$$

where I is a (3x3) identity matrix. Substituting (18) into (16), and applying the transformation $(11)_1$ to the (6x1) stress vectors, the ply stress in the local coordinate system is found as

$$\bar{\sigma}_i = R_i \, \sigma_i = R_i \, H_i \, \sigma - R_i \sum_{j=1}^{N} G_{ij} \, R_j^T \bar{\mu}_j \,, \quad i=1,2,..N \,. \tag{19}$$

2.2 PHASE STRESSES

Consider a representative volume V_j of a unidirectional composite ply (j), $j = 1,2,..N$, which is divided into M subvolumes, V_η^j, $\eta = 1,2,..M$. We recall the modified Levin's formula (Dvorak and Benveniste 1992) and write the lamina eigenstrain $\bar{\mu}_j$ in terms of the local eigenstrains found in each of the subvolumes as

$$\bar{\mu}_j = \sum_{\eta=1}^{M} \bar{c}_\eta^j \, [B_\eta^j]^T \, \mu_\eta^j \, , \quad \bar{c}_\eta^j = V_\eta^j / V_j \, , \quad j=1,2,..N \, , \tag{20}$$

where B_η^j is elastic stress concentration factor. The columns of B_η^j are given by the average stress vectors σ_η caused in subvolume η by a ply overall stress $\bar{\sigma}_k = 1$, $k = 1,2,..6$. These factors can be obtained from a selected micromechanical model of a unidirectional composite, e.g. the self–consistent and Mori–Tanaka models, or the PHA model.

The local stresses in the subvolumes of a lamina (i) can now be written as (Dvorak 1992)

$$\sigma_\rho^i = B_\rho^i \, \bar{\sigma}_i - \sum_{\eta=1}^{M} F_{\rho\eta}^i \, L_\eta^i \, \mu_\eta^i \, , \quad \rho =1,2,..M, \quad i=1,2,..N \, . \tag{21}$$

The columns of the transformation influence coefficient $F_{\rho\eta}^i$ provide the stress vectors in subvolume ρ caused by an eigenstress $\lambda_k = 1$, $k = 1,2,..6$, introduced in subvolume η . These coefficients are derived from analysis of the selected representative volume of a unidirectional composite as described by Dvorak et al. (1994a,b). From eqs. (19)–(21), the local stress σ_ρ^i can be written as

$$\sigma_\rho^i = B_\rho^i \, R_i \, H_i \, \sigma - \sum_{\eta=1}^{M} F_{\rho\eta}^i \, L_\eta^i \, \mu_\eta^i - B_\rho^i \, R_i \sum_{j=1}^{N} G_{ij} \, R_j^T \left[\sum_{\eta=1}^{M} \bar{c}_\eta^j \, [B_\eta^j]^T \, \mu_\eta^j \right] ,$$

$$\rho = 1, 2, .. M, \quad i = 1, 2, .. N \, . \tag{22}$$

The first term in (22) is the local stress caused by the overall stress applied to the laminate, while the second and third terms are the contributions of the subvolume eigenstrains in all the plies to subvolume ρ of lamina (i). The second term provides the local stresses due to local eigenstrains in lamina (i). The in–plane constraint $\epsilon' = \epsilon_i'$ imposed on the lamina causes additional stresses in the subvolumes of the plies when eigenstrains μ_η^j are present in other layers (j). This effect is given by the third term in (22).

3. Distribution factors and Influence Coefficients

Here we find expressions for the distribution factors H_i', κ_i, and the influence coefficients F_{ij}' which appear in eq. (14). From in–plane strain compatibility of the perfectly bonded plies, $\epsilon' = \epsilon'_i$, and the force equilibrium condition, $\Sigma\, c_i\, \sigma'_i = \sigma'$, $i = 1,2,..N$, one can establish that (Bahei–El–Din 1992, Dvorak et al. 1992)

$$\mathbf{H}'_i = \mathbf{L}'_i\, \mathbf{M}'\,, \qquad \kappa_i = \mathbf{L}'_i\, (\mathbf{n} - \mathbf{n}_i)\,, \tag{23}$$

$$\mathbf{M}' = [\mathbf{L}']^{-1}\,, \qquad \mathbf{L} = \sum_{i=1}^{N} c_i\, \mathbf{L}_i\,, \tag{24}$$

$$\mathbf{n} = -\,\mathbf{M}'\,\kappa\,, \qquad \kappa = \sum_{i=1}^{N} c_i\, \kappa_i\,, \tag{25}$$

$$\sum_{i=1}^{N} c_i\, \mathbf{H}'_i = \mathbf{I}\,, \qquad \sum_{i=1}^{N} c_i\, \kappa_i = 0\,. \tag{26}$$

The transformation influence factors $\mathbf{F}'_{ij}$ in (14) are found from the solution of an elastic symmetric laminate in which an in–plane eigenstress vector λ'_j is applied to lamina j as the only load, i. e., $\sigma' = 0$, $\sigma_3 = 0$. The lamina and laminate are first constrained from in–plane deformation. Under this constraint, the eigenstress is equilibrated by an overall in–plane stress $\overset{*}{\sigma}{}'$ such that

$$\sigma'_i = \delta_{ij}\, \lambda'_j\,, \qquad \overset{*}{\sigma}{}' = \sum_{i=1}^{N} c_i\, \sigma'_i = c_i\, \lambda'_j\,, \tag{27}$$

where δ_{ij} is Kronecker's symbol. For the laminate to return to the unconstrained state under $\sigma' = 0$, $\sigma_3 = 0$, the overall stress $\overset{*}{\sigma}{}'$ must be removed. This is achieved by applying the stress $-\overset{*}{\sigma}{}'$ to the laminate. Under this stress, the lamina stresses are given by the first term in (14), and the net stress found in lamina (i) at the end of this loading/unloading sequence is given by the sum

$$\sigma'_i = \delta_{ij}\, \lambda'_j - c_i\, \mathbf{H}'_i\, \lambda'_j\,. \tag{28}$$

The (3x3) influence factors $\mathbf{F}'_{ij}$ follow from a comparison of (28) and the last term of (14). The result is

$$\mathbf{F}'_{ij} = \delta_{ij}\, \mathbf{I} - c_j\, \mathbf{H}'_i\,. \tag{29}$$

4. Application to Thermo–Viscoplastic Laminates

As an application of the TFA method, we consider viscoplastic deformation of the phases of a fibrous composite under thermomechanical loads applied to a symmetric laminate. In this case, the eigenstrain rates $\dot{\mu}_\eta^j$ in subvolume V_η^j, $\eta = 1,2,..M$, of lamina (j), j = 1,2,..N, can be written as the sum of thermal strain and inelastic strain rates. If the latter is specified by a power law of an internal stress variable Q_η^j, then

$$\dot{\mu}_\eta^j(t) = \left[\alpha\,(\theta) + \frac{\partial\,M_\eta^j(\theta)}{\partial\,\theta}\,\sigma_\eta^j(t) \right] \dot{\theta} + \chi\,(\theta)\,(Q_\eta^j)^{p(\theta)}\,q_\eta^j , \qquad (30)$$

where t is time, θ is temperature, $\chi\,(\theta)$ and p (θ) are material parameters for the element volume V_η^j, and q_η^j specifies the direction of the inelastic strain rate in the local stress space. The first term in eq. (30) is the thermal strain rate where α is the coefficient of thermal expansion, M is the elastic compliance, and σ is the current stress in the subvolume.

Substituting (30) into (22) yields a system of rate equations for the local stresses in the subvolumes of all plies of the laminate which can be integrated along a specified loading path $\sigma'(t)$, $\sigma_3(t)$, $\theta\,(t)$ applied to the laminate as described by Dvorak et al. (1994a,b). The ply stresses follow from eq. (16), or (19), with (18) and (20), and the ply as well as the laminate in–plane strain, $\epsilon' = \epsilon_i'$, from (10)$_2$.

Alternately, the lamina stresses may be computed using an elastic finite element routine which utilizes the modified Levin's formula (20) and the ply transformation field equation (21). This approach, encoded by Bahei–El–Din (1994) in the VISCOPAC routine, was used to compute the local stresses in the phases of a (0/90)$_s$, Sigma/Timetal 21S laminate under hot isostatic conditions and subsequent axial tension/tension load cycles at 650°C as shown in Fig. 2. The laminate was first pressed by a hydrostatic pressure of 103.5 MPa at 899°C for 2 hours, and then aged at 621°C for 8 hours (condensed in Figs. 2–6 to 2 hours). The 14 axial stress cycles applied at 650°C correspond to the number of cycles sustained by the laminate up to failure in actual experiment under the same loading conditions. In the analysis, the Mori–Tanak model was used to estimate the concentration factors B_η for the matrix and fiber, and the constitutive equations described by Bahei–El–Din et al. (1991) were used to compute the inelastic strains for the matrix, eq. (30). The fiber was assumed to be elastic. The elastic moduli and coefficients of thermal expansion of both phases vary with temperature.

Figures 3 and 4 show the evolution of the local stress averages in the fiber and matrix of the 0° ply during the applied loading history. The results show that the residual stresses caused by processing do contribute to the subsequent cyclic stress magnitudes. This contribution need not be detrimental, indeed, in the present case, the residual axial fiber stress σ_{11}^f is compressive and thus helps to reduce the tensile stress magnitude under the cyclic mechanical load. However, the final peak magnitude of the fiber stress is 976 MPa, and suggests that failure may be caused by overloading of the fiber in the 0° plies.

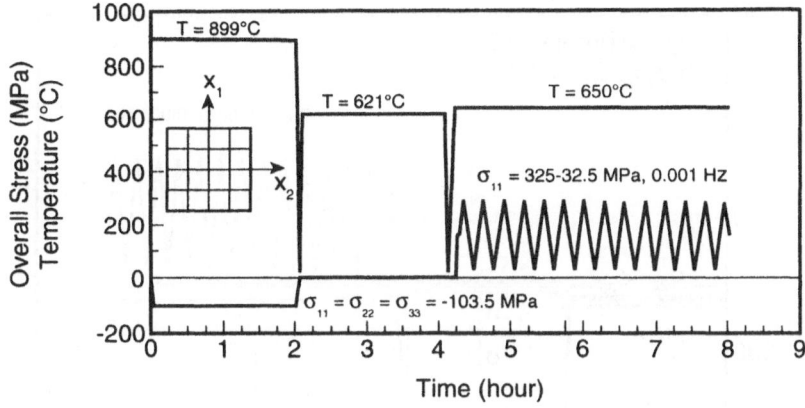

Figure 2. Thermomechanical loading history applied to a $(0/90)_s$ laminate

Figures 5 and 6 show the local stress averages in the 90° ply. The magnitudes of the transverse tension σ_{22} is rather large and can contribute to debonding of the fiber–matrix interface. In fact, the in–plane transverse stress σ_{22}^f in the fiber is as high as 227 MPa, probably suggestive of partial fiber debonding. However, the σ_{33}^f remains in the range of -107 to -133 MPa, offering support for the interface bond. The axial stress in the 90° fiber is compressive in the range 143–282 MPa.

These results suggest the existence of rather high internal stresses in the 90° plies that are not likely to be supported by the interfaces. The axial fiber stress is also very high, and would increase substantially after debonding of the 90° fibers.

5. Closure

The transformation field analysis (TFA) is a general method for solving inelastic deformation problems in heterogeneous media and can accommodate any uniform loading path, inelastic constitutive equation, and micromechanical model. The method can be also used in structural applications of heterogeneous materials such as fibrous composite laminates. The structural as well as model geometries are incorporated in the TFA method through mechanical transformation influence functions or concentration factor tensors derived from elastic solutions for the specified geometry and the elastic moduli. Thus, there is no need to solve inelastic boundary value problems either for heterogeneous materials or for their structural applications. As an example of the TFA application to fibrous composite laminates, the method was used to analyze the local stresses in the fiber and matrix phases of a titanium matrix laminate under fabrication conditions and fatigue loads.

Acknowledgments

This work was supported by the Air Force Office of Scientific Research and the Office of Naval Research.

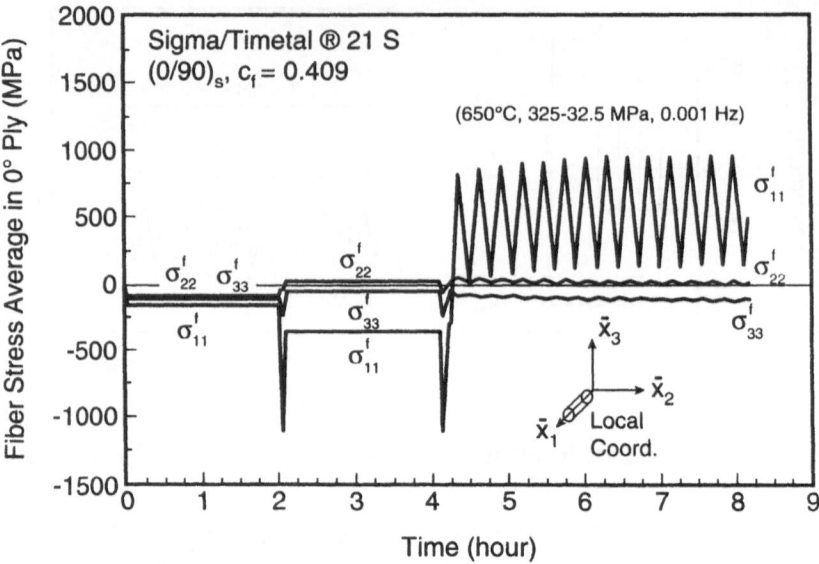

Figure 3. Fiber stress averages computed in the 0° ply of a (0/90)ₛ laminate subjected to the thermomechanical load history of Fig. 2.

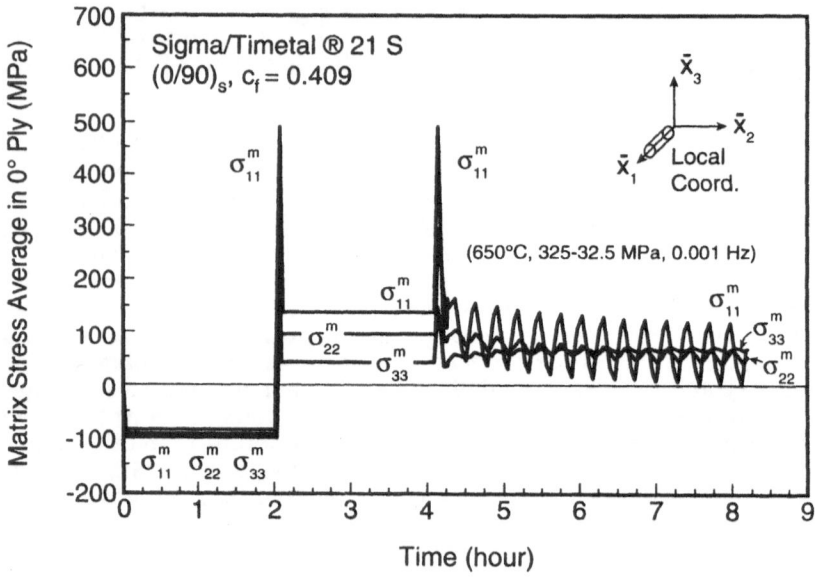

Figure 4. Matrix stress averages computed in the 0° ply of a (0/90)ₛ laminate subjected to the thermomechanical load history of Fig. 2.

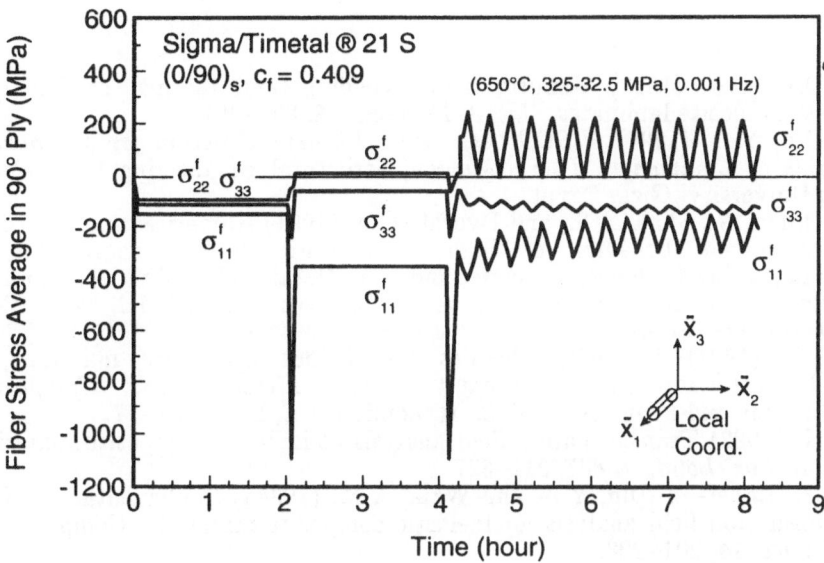

Figure 5. Fiber stress averages computed in the 90° ply of a (0/90)$_s$ laminate subjected to the thermomechanical load history of Fig. 2.

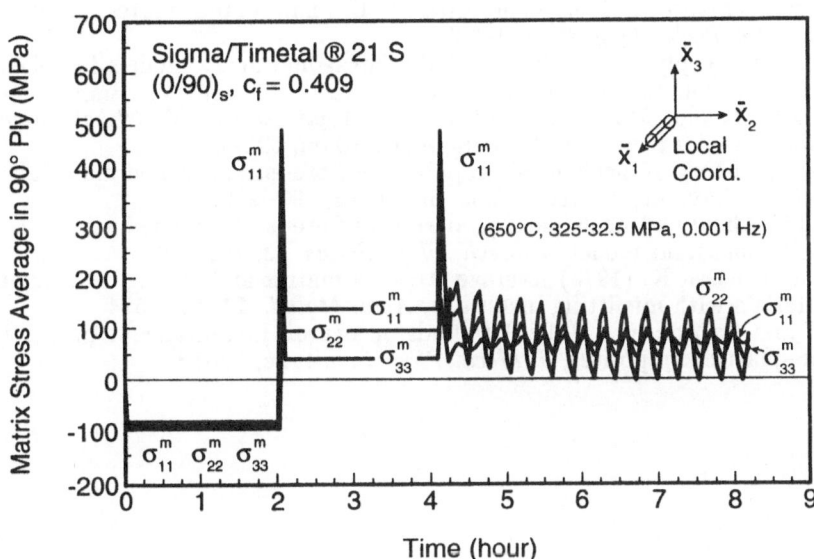

Figure 6. Matrix stress averages computed in the 90° ply of a (0/90)$_s$ laminate subjected to the thermomechanical load history of Fig. 2.

References

Bahei–El–Din, Y.A. (1992) Uniform fields, yielding, and thermal hardening in fibrous composite laminates, *Int. J. Plasticity* **8**, 867–892.

Bahei–El–Din, Y.A. (1994) *VISCOPAC Finite Element Program for Viscoplastic Analysis of Composites, User's Manual*, Structural Engineering Department, Cairo University, Giza, Egypt.

Bahei–El–Din, Y.A., Shah, R.S., and Dvorak, G.J. (1991) Numerical analysis of the rate–dependent behavior of high temperature fibrous composites, in S.N. Singhal, W.F. Jones, T. Cruse, and C.T. Herakovich (eds.), *Mechanics of Composites at Elevated and Cryogenic Temperatures*, ASME, New York, AMD–vol. 118, 67–78.

Dvorak, G.J. (1991) Plasticity theories for fibrous composite materials, in R.K. Everett and R.J. Arsenault (eds.), *Metal Matrix Composites, Mechanisms and Properties, vol. 2*, Academic Press, Boston, 1–77.

Dvorak, G.J. (1992) Transformation field analysis of inelastic composite materials, *Proc. R. Soc. Lond.* **A 437**, 311–327.

Dvorak, G.J., Bahei–El–Din, Y.A. and Wafa, A.M. (1994a) Implementation of the transformation field analysis for inelastic composite materials, *Computational Mechanics* **14**, 201–228.

Dvorak, G.J., Bahei–El–Din, Y.A. and Wafa, A.M. (1994b) The modeling of inelastic composite materials with the transformation field analysis, *Modelling Simul. Mater. Sci. Eng.* **2**, 571–586.

Dvorak, G.J. and Benveniste, Y. (1992) On transformation strains and uniform fields in multiphase elastic media, *Proc. R. Soc. Lond.* **A 437**, 291–310.

Dvorak, G.J., Chen, T. and Teply, J. (1992) Thermomechanical stress fields in high–temperature fibrous composites: II. Laminated plates, *Composites Science and Technology* **43**, 359–368.

Dvorak, G.J. and Teply, J.L. (1985) Periodic hexagonal array models for plasticity analysis of composite materials, in A. Sawczuk and V. Bianchi (eds.), *Plasticity Today: Modeling, Methods and Applications, W. Olszak Memorial Volume*, Elsevier Science Publishers, Amsterdam, 623–642.

Hill, R. (1964) Theory of mechanical properties of fibre–strengthened materials: I. Elastic behaviour, *J. Mech. Phys. Solids* **12**, 199–212.

Hill, R. (1965) Theory of mechanical properties of fibre–strengthened materials: III. Self–consistent model, *J. Mech. Phys. Solids* **13**, 189–198.

Mori, T. and Tanaka, K. (1973) Average stress in matrix and average elastic energy of materials with misfitting inclusions, *Acta. Metall.* **21**, 571–574.

Teply, J.L. and Dvorak, G.J. (1988) Bounds on overall instantaneous properties of elastic–plastic composites, *J. Mech. Phys. Solids* **36**, 29–58.

DEVELOPMENT OF ANISOTROPY IN POWDER COMPACTION

N. A. FLECK,
Cambridge University Engineering Dept.,
Trumpington St., Cambridge,
CB2 1PZ, England

Abstract

The cold compaction of an aggregate of powder is treated from the viewpoint of crystal plasticity theory. The contacts between particles are treated as compaction planes which yield under both normal and shear straining. The hardening of each plane represents both geometric and material hardening at the contacts between particles; the macroscopic tangent stiffness can be written down in terms of the hardening rate for active compaction planes. During the early stages of compaction the contacts yield in an independent manner, which can be interpreted within the crystal context as independent hardening. The macroscopic yield surfaces for isostatic and closed die compaction are estimated for a uniform distribution of an orthogonal pair of compaction planes. A vertex forms at the loading point and significant anisotropy develops for closed die compaction.

1. Introduction

The powder metallurgy industry is based upon the process of cold compaction of powders (usually, but not exclusively metallic) followed by sintering. This production route allows for the net shape forming of exotic alloys which are difficult to cast or shape by other methods. Cold compaction occurs within a closed die or in a cold isostatic press, and densification is by low temperature plasticity. At low relative densities (relative density $D < 0.9$) plastic deformation occurs local to the contacts between particles: this is 'Stage I ' compaction. As full density is approached 'Stage II' compaction takes over and plastic flow spreads throughout each particle; then, the powder aggregate is best viewed as a non-dilute concentration of cusp-shaped voids within a metallic matrix.

In this paper we consider stage I compaction within the framework of crystal plasticity theory. The central idea is to mimic the response at a contact between particles by a 'compaction plane', that is by a plane which can suffer both normal and shear straining, see Fig. 1. We consider the compaction plane to be smeared out through the neighbouring particles on each side of the contact, but sharing the same normal n as that of the contact plane, as shown in Fig. 1a. In this manner the compaction plane is analogous to a slip plane in crystal plasticity theory. Each contact is represented by a compaction plane, and the overall response of the aggregate is the sum of the responses of each compaction plane. The approach builds upon Calladine's micromechanical model of the yielding of clays, Calladine (1971). He assumed that compaction planes exist physically as rough surfaces of contact within the aggregate; here, we consider them to represent discrete contacts between particles.

R. Pyrz (ed.), IUTAM Symposium on Microstructure-Property Interactions in Composite Materials, 101–112.
© 1995 *Kluwer Academic Publishers.*

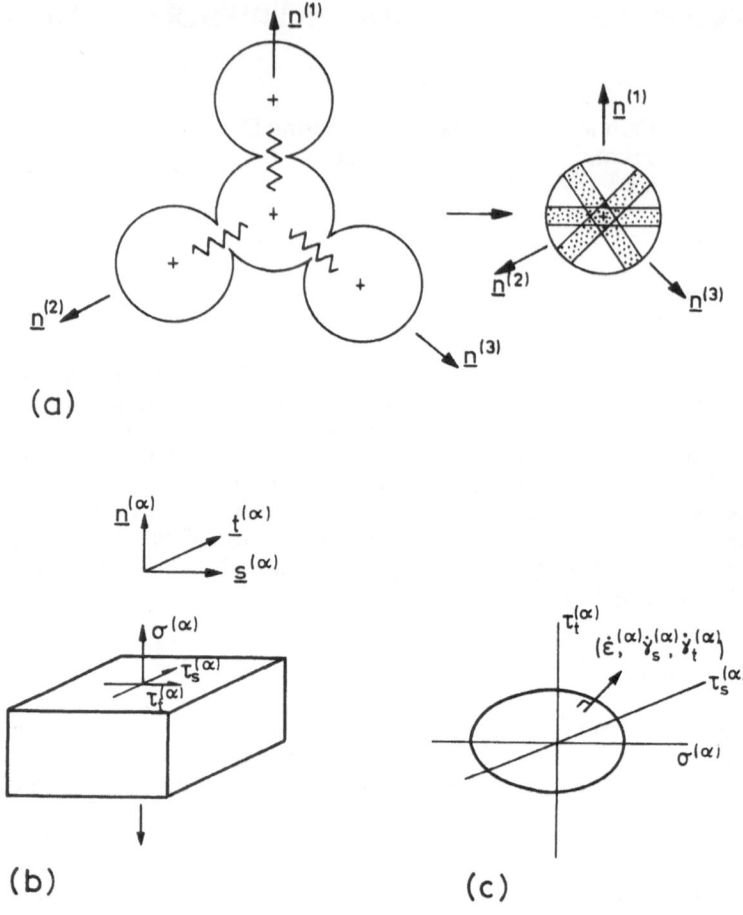

Figure 1. (a) Representation for the isolated contacts between particles by 'smeared-out' compaction planes. (b) Loading on a compaction plane (α). (c) Yield surface and normal plastic flow rule for a compaction plane (α).

The outline of the paper is as follows. First, a single crystal plasticity framework is summarised for a finite number of compaction planes. The Bishop and Hill (1951) method is then used to determine the macroscopic yield surface for a 'polycrystal' wherein each crystal comprises a pair of orthogonal compaction planes. The normal to the compaction planes is averaged over all orientations **n** within a plane, and the polycrystal is subjected to macroscopic in-plane biaxial loading. Using the crystal plasticity framework the effect of strain path upon yield surface evolution is explored by comparing the yield surfaces for isostatic compaction and closed die compaction. The marked development of anisotropy is evident for closed die compaction.

2. Crystal plasticity framework

We replace each contact between particles by a compaction plane within the particle. In this way the finite number of contacts around a representative particle are represented by a finite number of compaction planes within the particle: the particle is analogous to a single crystal with a set of slip planes.

Consider a representative contact with normal $n^{(1)}$ as shown in Fig. 1a. The neighbouring particles on each side of this contact can suffer a relative displacement in a direction parallel to $n^{(1)}$, and a shear displacement orthogonal to $n^{(1)}$. The resulting contact force between the particles depends upon the deformation mechanism at the contact; here we shall assume a non-linear response due to plastic dissipation. The non-linear contact law between the neighbouring particles can be treated as a non-linear spring at the contact, or as a 'smeared-out' compaction plane within each particle.

A representative compaction plane α is defined in Fig. 1b. The smeared-out plane is allowed to suffer a normal strain $\varepsilon^{(\alpha)}$ in the direction $n^{(\alpha)}$ and two shear strains $\gamma_s^{(\alpha)}$ and $\gamma_t^{(\alpha)}$ in the directions $s^{(\alpha)}$ and $t^{(\alpha)}$, respectively. We shall assume that strains are small, and shall neglect the effects of finite rotation of the compaction planes. A finite strain generalisation can be developed in a relatively straightforward manner, but is omitted here. The work conjugate stress measures are the normal stress $\sigma^{(\alpha)}$ and two shear stresses $\tau_s^{(\alpha)}$ and $\tau_t^{(\alpha)}$, such that the work rate per unit volume $w^{(\alpha)}$ is

$$w^{(\alpha)} = \sigma^{(\alpha)}\dot{\varepsilon}^{(\alpha)} + \tau_s^{(\alpha)}\dot{\gamma}_s^{(\alpha)} + \tau_t^{(\alpha)}\dot{\gamma}_t^{(\alpha)} \tag{2.1}$$

where no sum is performed over the index α unless explicitly stated by a summation sign. Suppose the aggregate is comprised of N compaction planes. Then, in terms of the Cartesian reference frame x_i, the macroscopic plastic strain rate $\dot{E}_{ij}^p$ is related to the strain rate $\left(\dot{\varepsilon}^{(\alpha)}, \dot{\gamma}_s^{(\alpha)}, \dot{\gamma}_t^{(\alpha)}\right)$ on each active compaction plane α by

$$\dot{E}_{ij}^p = \sum_{\alpha=1}^N \left[P_{ij}^{(\alpha)}\dot{\varepsilon}^{(\alpha)} + Q_{ij}^{(\alpha)}\dot{\gamma}_s^{(\alpha)} + R_{ij}^{(\alpha)}\dot{\gamma}_t^{(\alpha)} \right] \tag{2.2}$$

where the orientation factors $P_{ij}^{(\alpha)}$, $Q_{ij}^{(\alpha)}$ and $R_{ij}^{(\alpha)}$ are defined by

$$P_{ij}^{(\alpha)} \equiv n_i^{(\alpha)}n_j^{(\alpha)}, \quad Q_{ij}^{(\alpha)} \equiv \frac{1}{2}\left(n_i^{(\alpha)}s_j^{(\alpha)} + s_i^{(\alpha)}n_j^{(\alpha)}\right), \quad R_{ij}^{(\alpha)} \equiv \frac{1}{2}\left(n_i^{(\alpha)}t_j^{(\alpha)} + t_i^{(\alpha)}n_j^{(\alpha)}\right). \tag{2.3}$$

The macroscopic stresses on each compaction plane $\left(\sigma^{(\alpha)}, \tau_s^{(\alpha)}, \tau_t^{(\alpha)}\right)$ are related to the macroscopic stress Σ by substituting (2.2) into the work statement

$$\Sigma_{ij}\dot{E}_{ij}^p = \sum_{\alpha=1}^N \left[\sigma^{(\alpha)}\dot{\varepsilon}^{(\alpha)} + \tau_s^{(\alpha)}\dot{\gamma}_s^{(\alpha)} + \tau_t^{(\alpha)}\dot{\gamma}_t^{(\alpha)} \right] \tag{2.4}$$

to get

$$\sigma^{(\alpha)} = P_{ij}^{(\alpha)}\Sigma_{ij}, \qquad \tau_s^{(\alpha)} = Q_{ij}^{(\alpha)}\Sigma_{ij} \quad \text{and} \quad \tau_t^{(\alpha)} = R_{ij}^{(\alpha)}\Sigma_{ij}. \tag{2.5}$$

We assume that a compaction plane α suffers plastic flow when the yield function

$$f^{(\alpha)} \equiv \sigma_e^{(\alpha)}\left(\sigma^{(\alpha)}, \tau_s^{(\alpha)}, \tau_t^{(\alpha)}\right) - \sigma_Y^{(\alpha)} = 0 \qquad (2.6)$$

is satisfied. Here, $\sigma_e^{(\alpha)}$ is an effective stress measure and $\sigma_Y^{(\alpha)}$ is a scalar measure of the current magnitude of the yield surface. The shape of the yield surface is dependent upon the local contact law between particles, including the shear strength and the cohesive strength. For example, a yield surface of elliptical shape is given by

$$\sigma_e^{(\alpha)} = \sqrt{\left(\sigma^{(\alpha)}\right)^2 + \left(a^{(\alpha)}\tau_s^{(\alpha)}\right)^2 + \left(b^{(\alpha)}\tau_t^{(\alpha)}\right)^2} \qquad (2.7)$$

where $a^{(\alpha)}$ and $b^{(\alpha)}$ are constants defining the ellipticity of the yield surface. Here, we shall continue to work in terms of the general form (2.6) rather than the particular form (2.7).

For simplicity, we assume that plastic flow occurs in a direction normal to the yield surface for each compaction plane, giving

$$\dot{\varepsilon}^{(\alpha)} = \lambda^{(\alpha)}\frac{\partial\sigma_e^{(\alpha)}}{\partial\sigma^{(\alpha)}}, \quad \dot{\gamma}_s^{(\alpha)} = \lambda^{(\alpha)}\frac{\partial\sigma_e^{(\alpha)}}{\partial\tau_s^{(\alpha)}} \quad \text{and} \quad \dot{\gamma}_t^{(\alpha)} = \lambda^{(\alpha)}\frac{\partial\sigma_e^{(\alpha)}}{\partial\tau_t^{(\alpha)}}. \qquad (2.8)$$

The magnitude of the plastic multiplier $\lambda^{(\alpha)}$ is determined from a work hardening statement, as follows. Introduce an effective strain rate $\dot{\varepsilon}_e^{(\alpha)}$ by the work statement

$$w^{(\alpha)} \equiv \sigma_e^{(\alpha)}\dot{\varepsilon}_e^{(\alpha)} = \sigma^{(\alpha)}\dot{\varepsilon}^{(\alpha)} + \tau_s^{(\alpha)}\dot{\gamma}_s^{(\alpha)} + \tau_t^{(\alpha)}\dot{\gamma}_t^{(\alpha)}. \qquad (2.9)$$

Then, upon substituting (2.8) into (2.9), $\lambda^{(\alpha)}$ is expressible in terms of $\dot{\varepsilon}_e^{(\alpha)}$ as

$$\lambda^{(\alpha)} = C^{(\alpha)}\dot{\varepsilon}_e^{(\alpha)} \qquad (2.10)$$

where

$$C^{(\alpha)} \equiv \frac{\sigma_e^{(\alpha)}}{\sigma^{(\alpha)}\dfrac{\partial\sigma_e^{(\alpha)}}{\partial\sigma^{(\alpha)}} + \tau_s^{(\alpha)}\dfrac{\partial\sigma_e^{(\alpha)}}{\partial\tau_s^{(\alpha)}} + \tau_t^{(\alpha)}\dfrac{\partial\sigma_e^{(\alpha)}}{\partial\tau_t^{(\alpha)}}}. \qquad (2.11)$$

(It is noted that $C^{(\alpha)}=1$ when $\sigma_e^{(\alpha)}$ is homogeneous and of degree one in $\left(\sigma^{(\alpha)}, \tau_s^{(\alpha)}, \tau_t^{(\alpha)}\right)$ such as given by (2.7).) The macroscopic plastic strain rate can now be expressed in terms of the effective strain rate on each compaction plane $\dot{\varepsilon}_e^{(\alpha)}$ by rewriting (2.2) with the aid of (2.8) and (2.10) as

$$\dot{E}_{ij}^p = \sum_{\alpha=1}^{N}\left[C^{(\alpha)}T_{ij}^{(\alpha)}\dot{\varepsilon}_e^{(\alpha)}\right] \qquad (2.12)$$

where

$$T_{ij}^{(\alpha)} \equiv \frac{\partial\sigma_e^{(\alpha)}}{\partial\sigma^{(\alpha)}}P_{ij}^{(\alpha)} + \frac{\partial\sigma_e^{(\alpha)}}{\partial\tau_s^{(\alpha)}}Q_{ij}^{(\alpha)} + \frac{\partial\sigma_e^{(\alpha)}}{\partial\tau_t^{(\alpha)}}R_{ij}^{(\alpha)}. \qquad (2.13)$$

The overall hardening law is specified by

$$\dot{\sigma}_e^{(\alpha)} = \sum_{\beta=1}^{N}\left[h_{\alpha\beta}\dot{\varepsilon}_e^{(\beta)}\right]. \qquad (2.14)$$

In general, the hardening matrix $h_{\alpha\beta}$ can be homogeneous and of degree zero in the effective strain rates $\dot{\varepsilon}_e^{(\alpha)}$; here, only hardening laws for which the $h_{\alpha\beta}$ are independent of the effective strain rates are employed.

In the foregoing we have assumed that $\sigma_e^{(\alpha)}$ is a function of $\varepsilon_e^{(\beta)}$. An alternative work hardening hypothesis is to assume normal hardening with no shear hardening, such that $\sigma_e^{(\alpha)}$ depends only upon $\varepsilon^{(\beta)}$. For the case of independent hardening, the hardening matrix is then given by

$$h_{\alpha\alpha} = \frac{\sigma^{(\alpha)}}{\sigma_e^{(\alpha)}} \frac{d\sigma_e^{(\alpha)}}{d\varepsilon^{(\alpha)}}$$

$$h_{\alpha\beta} = 0 , \quad \alpha \neq \beta . \tag{2.15}$$

The above structure remains unchanged with this minor modification to the hardening rule. This form of hardening has been used by Schofield and Wroth (1969) in their Cam Clay model and by Calladine (1971) in his microstructural view of clay.

3. Calibration of crystal plasticity law

Ashby and co-workers (Helle et al. (1985)) have developed accurate relations for the hydrostatic stage I compaction of a powder aggregate. They assume that spherical particles are composed of elastic, perfectly-plastic material of yield strength σ_y. The yield pressure p_y for the aggregate is dependent upon its relative density D (D= density of aggregate/ full density) according to

$$p_y = 3D^2 \frac{(D-D_o)}{(1-D_o)} \sigma_y \tag{3.1}$$

where D_o is the initial relative density corresponding to random packing of the particles. For example, for dense random packing $D_o=0.64$. We calibrate the hardening matrix $h_{\alpha\beta}$ in the crystal plasticity model against (3.1) in the hydrostatic limit.

4. Bishop-Hill calculation of yield surface for hydrostatic compaction

So far we have dealt with the 'single crystal response' of a finite set of compaction planes for a representative particle. Now consider the case of an aggregate comprising randomly oriented particles, bonded at their mutual contacts. The macroscopic 'polycrystalline' limit yield surface for the aggregate of compaction planes can be estimated using the upper bound method laid down by Bishop and Hill (1951). Elastic deformation of the particles is ignored and a work calculation is performed to determine the collapse response in stress space for the random aggregate of particles. We restrict ourselves to in-plane biaxial straining of the aggregate, and assume that the deformation response for each particle is adequately described by a pair of orthogonal compaction planes, oriented at an angle ω as defined in Fig. 2a. The aggregate is assumed to be isotropic, with the compaction planes distributed uniformly over all orientations.

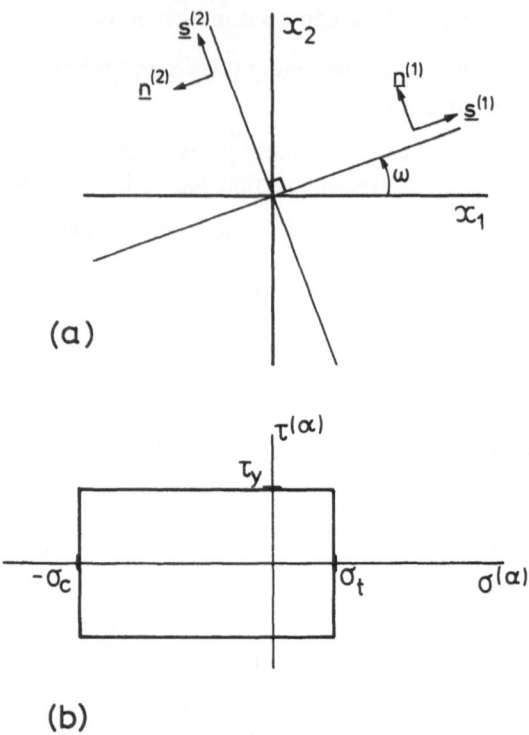

Figure 2. (a) A pair of orthogonal compaction planes, with orientation ω.
(b) Assumed yield surface for each compaction plane.

For simplicity, we assume that the yield surface for each compaction plane is rectangular in shape as shown in Fig. 2b, and is characterised by a compressive yield strength $\sigma_c^{(\alpha)}$, a tensile yield strength $\sigma_t^{(\alpha)}$ and a shear yield strength $\tau_y^{(\alpha)}$. The magnitude of the yield surface is taken to scale with the normal strain $\varepsilon^{(\alpha)}$, such that

$$\sigma_c^{(\alpha)}\left(\varepsilon^{(\alpha)}\right) = \sigma_c^{(\alpha)} .$$

(4.1)

Consider the case where an aggregate has been compacted hydrostatically from an initial density D_o to a current density D. Then, the magnitude of $\sigma_c^{(\alpha)}$ for all compaction planes follows from (3.1) as

$$\sigma_c^{(\alpha)} = p_y = 3D^2\frac{(D-D_o)}{(1-D_o)}\sigma_y .$$

(4.2)

Gurson (1977) has shown that the macroscopic stress Σ on the aggregate, corresponding to a plastic strain rate $\dot{E}^p$ is given by

$$\Sigma_{ij} = \partial W\left(\dot{E}^P\right)/\partial \dot{E}^P_{ij} \qquad (4.3)$$

where W is the plastic dissipation rate of the aggregate per unit volume. It remains to estimate $W(\dot{E}^P)$. For a pair of orthogonal compaction planes as shown in Fig. 2a, the plastic dissipation per unit volume w is

$$w = \sigma^{(1)}\dot{\varepsilon}^{(1)} + \sigma^{(2)}\dot{\varepsilon}^{(2)} + \tau^{(1)}\dot{\gamma}^{(1)} + \tau^{(2)}\dot{\gamma}^{(2)} \qquad (4.4)$$

where we have dropped the subscript s from the shear terms, to keep notation compact. A simple connection exists between the strain rates on the compaction planes and the macroscopic strain rate; after some manipulation, (2.2) leads to

$$
\begin{pmatrix} \dot{\varepsilon}^{(1)} \\ \dot{\varepsilon}^{(2)} \\ \dot{\gamma}^{(1)} - \dot{\gamma}^{(2)} \end{pmatrix} =
\begin{pmatrix} \sin^2\omega & \cos^2\omega & -\sin 2\omega \\ \cos^2\omega & \sin^2\omega & \sin 2\omega \\ -\sin 2\omega & \sin 2\omega & 2\cos 2\omega \end{pmatrix}
\begin{pmatrix} \dot{E}^P_{11} \\ \dot{E}^P_{22} \\ \dot{E}^P_{12} \end{pmatrix} . \qquad (4.5)
$$

Since we are dealing with an isotropic aggregate we can consider principal stresses and principal strains and, without loss of generality, we can set $\dot{E}^P_{12}=0$. For a given $\left(\dot{E}^P_{11}, \dot{E}^P_{22}\right)$ the stress state for each of the two compaction planes is at a vertex, and both $\dot{\varepsilon}^{(1)}$ and $\dot{\varepsilon}^{(2)}$ are determined uniquely from (4.5). The values of $\dot{\gamma}^{(1)}$ and $\dot{\gamma}^{(2)}$ follow from (4.5) and from the minimum plastic work hypothesis of Bishop and Hill (1951): the strain rates are selected to minimise w. This optimisation gives $\dot{\gamma}^{(1)} > 0$, $\dot{\gamma}^{(2)}=0$ for $(\dot{\gamma}^{(1)} - \dot{\gamma}^{(2)}) > 0$, and $\dot{\gamma}^{(1)}=0$, $\dot{\gamma}^{(2)}>0$ for $(\dot{\gamma}^{(1)} - \dot{\gamma}^{(2)}) < 0$.

The macroscopic stress Σ is calculated by volume averaging the response for a pair of compaction planes at all orientations, that is,

$$\Sigma_{ij} = \frac{2}{\pi}\int_0^{\pi/2} \frac{\partial w}{\partial \dot{E}^P_{ij}}\, d\omega . \qquad (4.6)$$

Upon substituting into (4.6) the expression (4.4) for w, and (4.5) for the strain rates in each compaction plane, we obtain a specification for the macroscopic limit yield surface. The yield surface is plotted in Fig. 3 for the cases $\sigma_t/\sigma_c = 0, 1$ and $\tau_y/\sigma_c = 0, 1/(2+\pi)$. The value $\tau_y/\sigma_c = 1/(2+\pi)$ corresponds to perfectly sticking contacts as discussed in section 3. We conclude from Fig. 3 that the cohesive strength ratio σ_t/σ_c has a more major effect on the size and shape of the yield surface than has the shear strength ratio τ_y/σ_c. For a wide range of strain rate directions ($\dot{E}^P_{11}>0$ and $\dot{E}^P_{22}>0$; $\dot{E}^P_{11}<0$ and $\dot{E}^P_{22}<0$) the macroscopic stress lies at a vertex close to the hydrostatic axis. Akisanya and Cocks (1994) observed a similar behaviour in their analysis of the compaction of a hexagonal array of cylindrical particles.

5. Bishop-Hill calculation of yield surface for closed die compaction: the development of anisotropy

The above Bishop-Hill calculation can be repeated for the case of closed die compaction where, without loss of generality we take $E^P_{11}=0$, $E^P_{22}<0$. Again, we consider the

108

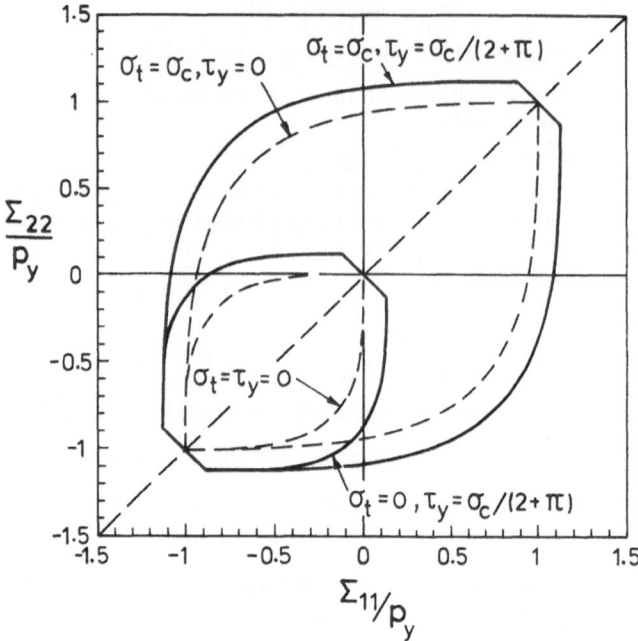

Figure 3. Yield surface for a uniform distribution of an orthogonal pair of compaction planes, subjected to hydrostatic compaction.

response for an isotropic aggregate, composed of a uniform distribution of pairs of orthogonal compaction planes as shown in Fig. 2a. The yield surface is taken to be rectangular in shape, see Fig. 2b, and the magnitude of the yield surface scales with the normal strain $\varepsilon^{(\alpha)}$ on that compaction plane, with $\sigma_c^{(\alpha)} = \sigma_e^{(\alpha)}\left(\varepsilon^{(\alpha)}\right)$.

Consider a compaction plane with orientation $\omega=0$ such that the unit normal $n^{(1)}$ is aligned with the x_2 axis. The complementary compaction plane is orthogonal with a normal $n^{(2)}$ aligned with the x_1 axis. Then, after a small amount of closed die compaction (say D=0.7 from an initial value of $D_0 = 0.64$), we have $\sigma_c^{(1)} = 2p_y$ and $\sigma_c^{(2)} = 0$, where $p_y(D)$ is given by (3.1). For a pair of compaction planes with orientation ω, the normal strain on the compaction planes follows from (4.5) as

$$\varepsilon^{(1)} = E_{22}^p \cos^2 \omega , \qquad \varepsilon^{(2)} = E_{22}^p \sin^2 \omega \qquad (5.1)$$

We assume that the degree of compaction is small (D increases by less than 10%) and so the yield strengths $\left(\sigma_c^{(\alpha)}, \sigma_t^{(\alpha)}, \tau_y^{(\alpha)}\right)$ and $\sigma_e^{(\alpha)}$ increase linearly with $\varepsilon^{(\alpha)}$ for all compaction planes, giving

$$\sigma_c^{(1)} = 2p_y \cos^2 \omega , \qquad \sigma_c^{(2)} = 2p_y \sin^2 \omega \qquad (5.2)$$

The shape of the yield surface for each compaction plane is taken to be constant with $\sigma_t^{(\alpha)}/\sigma_c^{(\alpha)}$ fixed at zero (for cohesionless aggregate) and at unity (for full cohesive strength). Perfectly sticking contacts are modelled by putting $\tau_y^{(\alpha)}/\sigma_c^{(\alpha)} = 1/(2+\pi)$, while the choice $\tau_y^{(\alpha)}/\sigma_c^{(\alpha)} = 0$ is appropriate for frictionless contacts.

The macroscopic yield surface is evaluated from (4.6), with w given by (4.4), and the strain rate for each compaction plane specified by (4.5). Again, the relative magnitudes of $\dot{\gamma}^{(1)}$ and $\dot{\gamma}^{(2)}$ are selected to minimise w, subject to the constraint on $(\dot{\gamma}^{(1)} - \dot{\gamma}^{(2)})$ given by (4.5).

The yield surface for closed die compaction is plotted in Fig. 4 for $\sigma_t/\sigma_c = 0, 1$ and $\tau_y/\sigma_c = 0, 1/(2+\pi)$. The main features are the same as for isostatic compaction, as shown in Fig. 3: the degree of cohesive strength has a major influence and the shear strength at the contacts has a minor influence upon the yield surface. As a result of preferential hardening of compaction planes aligned with the direction of compaction, significant anisotropy develops for closed die compaction. The compact is about three times stronger in the compaction direction x_2 than in the transverse x_1 direction.

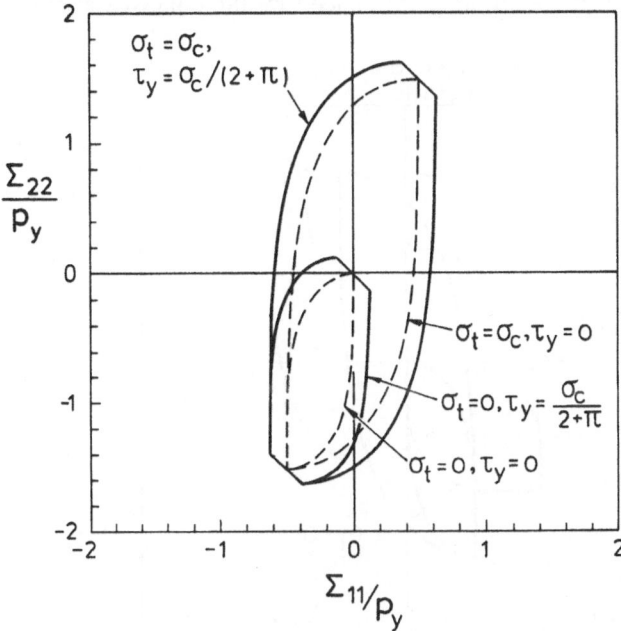

Figure 4. Yield surface for a uniform distribution of an orthogonal pair of compaction planes, subjected to closed die compaction, $E_{11}^p = 0$, $E_{22}^p < 0$.

The yield surface for closed die compaction is compared with that for isostatic compaction in Fig. 5, for frictionless contacts ($\tau_y/\sigma_c=0$) and for the two limiting cases of full cohesive strength ($\sigma_t/\sigma_c=1$) and vanishing cohesive strength ($\sigma_t/\sigma_c=0$). The comparison is made at the same value of relative density D slightly greater than D_0. The development of anisotropy under closed die compaction is obvious: the yield strength along the x_2 direction is greater after closed die compaction than after isostatic compaction. Conversely, the yield strength in the transverse direction is less for closed die compaction than for isostatic compaction.

6. Concluding discussion

Fleck et al. (1992) have previously used the Bishop-Hill method to estimate the macroscopic yield locus for stage I compaction of a powder aggregate. They assume that plastic flow occurs in accordance with Green's (1954) slip line field solution at all contacts on the surface of a representative spherical particle: all contacts are active for an arbitrary macroscopic strain rate. The contacts are assumed to be perfectly sticking with a cohesive strength equal to the indentation strength. More recently, Fleck (1994) has repeated the calculation for a range of shear strength and cohesive strength. He finds that the macroscopic yield surface is influenced to a minor extent by the level of shear strength, and much more strongly influenced by the cohesive strength. These

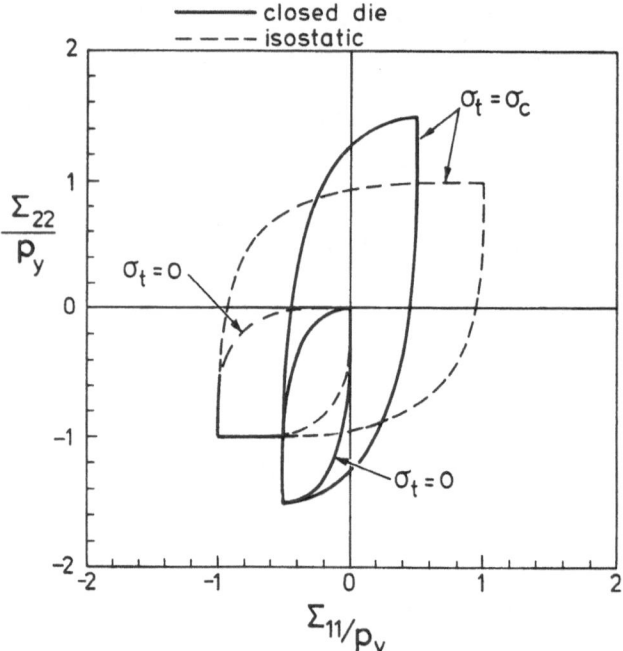

Figure 5. Comparison of yield surface for isostatic compaction and for closed die compaction. The yield surfaces are calculated for the same small increment in relative density D above the initial density D_0. Contacts are frictionless with $\tau_y = 0$.

conclusions are fully supported by the findings of the current study.

The predictions of Fleck (1994) are compared with those of the current study in Fig. 6 for the case of isostatic compaction. Yield surfaces are given for frictionless contacts, with either zero cohesion or perfect cohesion between particles. The results are presented in terms of the deviatoric stress measure $\Sigma \equiv \Sigma_{22} - \Sigma_{11}$ and the mean stress measure $\Sigma_m \equiv \frac{1}{2}(\Sigma_{11} + \Sigma_{22})$, for the case of an isotropic distribution of two orthogonal compaction planes as described in section 4 above. The calculation by Fleck (1994) was done for axisymmetric loading of an aggregate with $\Sigma_{33} = \Sigma_{11}$; the appropriate deviatoric stress measure remains $\Sigma \equiv \Sigma_{22} - \Sigma_{11}$, and the mean stress is defined by $\Sigma_m \equiv \frac{1}{3}(\Sigma_{11} + 2\Sigma_{22})$. We conclude from Fig. 6 that the yield surface given by the plane strain crystal plasticity calculation of section 4 and Fleck's (1994) axisymmetric calculation (assuming plastic dissipation at up to twelve contacts per particle) give closely similar results. This is not surprising since both calculations assume that all contacts are active, and predictions have been calibrated to give the result (3.1) for hydrostatic loading.

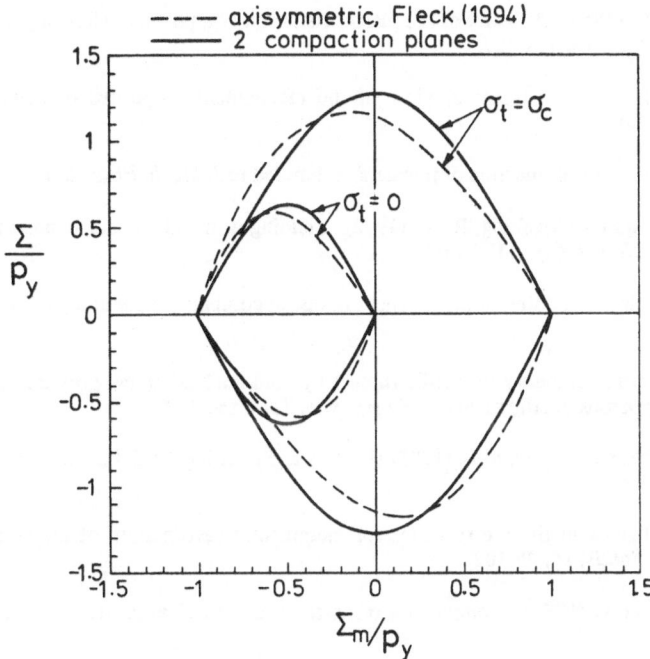

Figure 6. Predictions of the yield surface after a small amount of hydrostatic compaction, for the two compaction plane model of the current study, and the axisymmetric model of Fleck (1994). Frictionless contacts.

Akisanya and Cocks (1994) have recently examined the densification of a hexagonal array of cylinders under in-plane loading, and Ogbonna and Fleck (1994) have performed an approximate calculation of axisymmetric compaction using a cell model. In both cases it is found that as densification proceeds plastic flow occurs throughout each particle and is no longer confined to each contact. Independent collapse of each contact is replaced by a discrete set of collapse mechanisms for the representative particle: these collapse mechanisms involve plastic dissipation at several contacts. Latent hardening occurs between one collapse mechanism and the next. The crystal plasticity framework has adequate flexibility to include this cross-hardening between mechanisms via the off-diagonal terms of the hardening matrix $h_{\alpha\beta}$. Indeed, the cell calculations may be used to calibrate the off-diagonal terms of $h_{\alpha\beta}$. Further work is needed to further develop this calculation scheme.

References

Akisanya, A. R. and Cocks, A. C. F. (1994) Stage I compaction of cylindrical patricles under non-hydrostatic loading, submitted to *J. Mech. Phys. Solids*.

Bishop, J.W.F. and Hill, R. (1951) A theory of the plastic distortion of a polycrystalline aggregate under combined stresses, *Phil. Mag.* **42**, 414-427.

Calladine, C. R. (1971) A microstructural view of the mechanical properties of saturated clay, *Geotechnique*, **21**(4), 391-415.

Fleck, N. A. (1994) On the cold compaction of powders, submitted to *J. Mech. Phys. Solids*.

Fleck, N. A., Kuhn, L. T. and McMeeking, R. M. (1992) Yielding of metal powder bonded by isolated contacts, *J. Mech. Phys. Solids*, **40**(5), 1139-1162.

Green, A. P. (1954) The plastic yielding of metal junctions due to combined shear and pressure, *J. Mech. Phys. Solids*, **2**, 197-211.

Gurson, A. L. (1977) Continuum theory of ductile rupture by void nucleation and growth: part I - yield criteria and flow rules for porous ductile media, *J. Engng. Mat. Tech.* **99**, 2-15.

Helle, A.S., Easterling, K.E. and Ashby, M.F. (1985) Hot-isostatic pressing diagrams: new developments, *Acta metall.*, **33**(12), 2163-2174.

Hill, R. (1966) Generalized constitutive relations for incremental deformation of metal crystals by multislip, *J. Mech. Phys. Solids*, **14**, 95-102.

Ogbonna, N. and Fleck, N.A. (1994) Compaction of an array of spherical particles, *Acta Metall. et Mater.*, in press.

Schofield, A. N. and Wroth C. P. (1968) *Critical state soil mechanics*, McGraw Hill, London.

CRACKED LAMINATES WITH IMPERFECT INTERLAMINAR INTERFACE

Z. HASHIN

Dept. of Solid Mechanics, Materials and Structures
Faculty of Engineering
Tel-Aviv University
Tel-Aviv, Israel

Abstract. Variational analysis of cracked laminates with imperfect inter-laminar interface is developed on the basis of a generalized extemum principle of thermoelastic complementary energy. Closed form results for effective Young's modulus, thermal expansion coefficient, shear modulus and internal stresses are developed for cracked cross-ply laminates. The results provide an assessment of the significance of interlaminar imperfection.

1. Introduction

The present paper is concerned with the effect of intralaminar crack (IC) accumulation on the thermomechanical properties of fiber composite laminates and the resulting internal stress distributions. Such cracks develop in the matrix along fibers due to load or temperature change. They are thus parallel crack distributions within the layers which propagate very rapidly until the laminate edges. Therefore,the formation of a typical IC is not viewed as a crack propagation phenomenon but as a **fracture event** which occurs instantaneously. Thus, the concern is with a laminate which contains IC distributions which are quantitatively described by crack density, the number of IC per unit length. The problems are then to determine deterioration of thermoelastic peoperties in terms of crack density, laminate internal geometry and ply properties, internal stresses resulting from crack accumulation and their relation to failure mechanisms, and more ambitiously - to predict crack density due to load or temperature.

113

R. Pyrz (ed.), IUTAM Symposium on Microstructure-Property Interactions in Composite Materials, 113–128.
© 1995 *Kluwer Academic Publishers.*

The problems outlined have been the subject of a large number of research papers over the last 15 years. There are two major approaches : the first may be termed the micromechanics approach and the second, the continuum damage approach. In the first approach it is attempted to carry out analysis recognizing the cracks as defects on which the tractions must vanish. The advantage of this approach is physical realism and information about internal (micro) stresses, which is important for failure considerations. The disadvantage is analytical difficulty and for this reason the micromechanics approach has to date been confined to cross-plies.

In the second approach effect of IC on a layer is modeled by an abstract damage function whose form is not unique and which invariably contains unknown coefficients. The disadvantage is that such coefficients must be backed out from experiment on the laminate and it is not clear whether such coefficients qualify as ply material parameters or are fitting parameters which change from laminate to laminate. The advantage of the approach is that it can be applied to practical laminates, more complicated than cross-plies.

The present work is concerned with the micromechanics approach for cross-ply laminates. Review of the voluminous literature is not within the present scope. It is recalled that initial analytical efforts were based on the shear-lag approximation e.g. Reifsnider and Jamison (1982),Laws and Dvorak (1988). This method requires the determination of a so-called shear lag parameter on the basis of the fracture toughness of the ply materials. Analysis in terms of a displacement formulation represented, arbitrarily, by hyperbolic functions was given by Tsai et al. (1990). Work of similar nature with the choice of different form displacement functions has been done by Lee et al. (1990). A variational method based on the principle of minimum complementary energy has been developed by Hashin (1985) with application to stiffness reduction and stress analysis of cross-ply laminates with one layer cracked. This has been extended to the case of all layers cracked in Hashin (1987) and to evaluation of thermal expansion coefficients in Hashin (1988). The only assumption made in the variational anslysis is that-in plane stresses in the ply are constant over the thickness. Analysis based on similar assumptions has been given by McCartney (1992). Analysis for more general in-plane stresses has been given by Varna and Berglund (1994). Nairn et al. (1993) have successfully used the variational analysis for prediction of crack density resulting from in plane loading of cross-ply laminates.

The purpose of the work presented here is to extend the variational analysis to the evaluation of thermoelastic properties and internal stresses of cross-ply laminates when there is imperfect interlaminar bond between the layers.

2. Thermoelastic Extremum Principle for Imperfect Interface

Perfect interface between two solid constituents implies continuity of traction and displacement vectors at the interface. When the interface displacement vector is discontinous, while the traction vector remains continuous for reasons of equilibrium, the interface is called imperfect. Let the displacement jump at interface S_{12} be denoted

$$[\mathbf{u}] = \mathbf{u}^2 - \mathbf{u}^1. \tag{1}$$

Then the simplest imperfect interface condition is

$$\begin{aligned} T_n &= D_n\,[u_n] \\ T_s &= D_s\,[u_s] \\ T_t &= D_t\,[u_t], \end{aligned} \tag{2}$$

where n, s, t are normal and tangential components of the interface normal n, assumed here as pointing into phase 2, and D_n, D_s, D_t are spring constant type interface parameters. With respect to a fixed cartesian coordinate system, (2) assumes the forms

$$\mathbf{T} = \mathbf{D}.[\mathbf{u}] \quad [\mathbf{u}] = \mathbf{R}.\mathbf{T} \quad \mathbf{R} = \mathbf{D}^{-1}, \tag{3}$$

where the Cartesian components of D and its inverse R now vary along the interface. It has been shown in Hashin (1990) that the effect of a thin and very compliant interphase between constituents can actually be expressed in the form (2) and that the interface parameters can be expressed in terms of interphase thickness and stiffness.

In the variational analysis to be employed here the generalization of the extremum principle of minimum complementary energy for imperfect interface conditions will be needed, Hashin (1992). This will here be further generalized to the thermoelastic case and will be stated for the case when tractions are prescribed over the entire external surface S. Let σ be the actual stress field and $\tilde{\sigma}$ an admissible stress field for a body with surface load $\mathbf{T}\,(S)$ and imperfect interface S_{12}. Define

$$\begin{aligned} W &= \tfrac{1}{2}\sigma : \mathbf{S} : \sigma \\ \widetilde{W} &= \tfrac{1}{2}\tilde{\sigma} : \mathbf{S} : \tilde{\sigma}, \end{aligned} \tag{4}$$

where S is the compliance tensor. Here W is the stress energy density while $\widetilde{W}$ has no physical meaning. Next define the functionals

$$U = \int_V \left[W + \boldsymbol{\alpha}.\sigma\ \theta - c_p\,(\theta^2/2\theta_o) \right] dV + \tfrac{1}{2} \int_{S_{12}} \mathbf{T} : \mathbf{R} : \mathbf{T} dS$$

$$\tilde{U} = \int_V \left[\widetilde{W} + \boldsymbol{\alpha}.\tilde{\sigma}\ \theta - c_p\,(\theta^2/2\theta_o) \right] dV + \tfrac{1}{2} \int_{S_{12}} \tilde{\mathbf{T}} : \mathbf{R} : \tilde{\mathbf{T}} dS \tag{5}$$

Here α is the thermal expansion tensor, c_p the specific heat at constant pressure and θ is the (known) temperature relative to a reference temperature θ_o. Then the thermoelastic principle of minimum complementary energy is expressed by the inequality

$$\tilde{U} \geq U, \tag{6}$$

equality occurring if, and only if, $\tilde{\sigma} = \sigma$.

For composite materials applications it is of importance to consider the case of constant temperature and so-called homogeneous traction boundary conditions which are defined as

$$\mathbf{T}\left(S\right) = \sigma^{o}.\mathbf{n}\left(S\right) \tag{7}$$

Where σ^{o} is a constant stress tensor. Then σ^{o} is the average stress tensor and it can be shown that the first of (5) is, rigorously

$$U = \frac{1}{2}\left[\sigma^{o} : \mathbf{S}^{*} : \sigma^{o} + \alpha^{*}.\,\sigma^{o} - c_{p}^{*}\left(\theta^{2}/2\theta_{o}\right)\right] V \tag{8}$$

where $\mathbf{S}^{*}$,α^{*} and c_{p}^{*} are the effective elastic compliance tensor, thermal expansion tensor and specific heat, respectively.

In the following the variational principle will be exploited to analyze approximately thermo-elastic properties and internal stresses in cracked laminates.

3. Cross-Ply Laminates with One Ply Family Cracked

The case to be considered here is a $[0_m^0, 90_n^0]_s$ laminate in which either the 0^0 or the 90^0 plies are cracked, fig. 1a. The variational method will be employed to obtain strict lower bounds for the effective Young's modulus E_x^* and the effective shear modulus G_{xy}^* and approximations for the effective thermal expansion coefficient α_x^* and internal stresses, for the case of imperfect interlaminar interface as defined by a damaged interphase between plies.

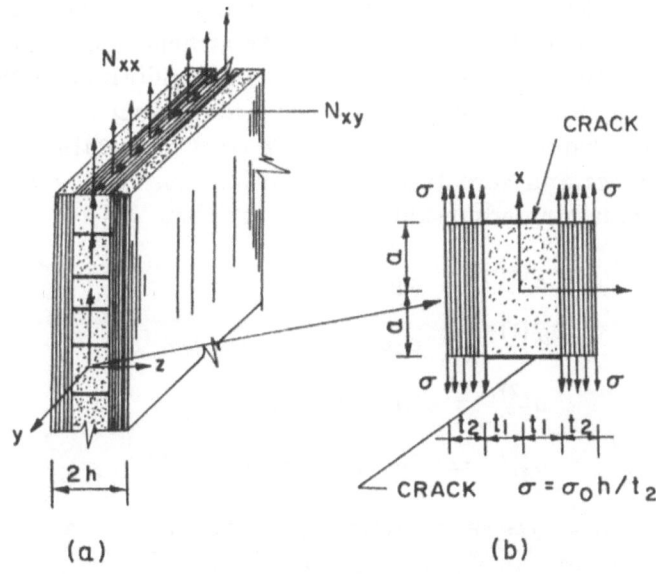

Figure 1. Cracked laminate

3.1. EFFECTIVE YOUNG'S MODULUS

Let it be assumed that the laminate is subjected to a constant tensile membrane force N_{xx} and let the σ_{xx} stresses in the layers of the **uncracked** laminate be denoted σ_1 and σ_2, respectively, where from now on the label 1 refers to the 90^0 ply and 2 refers to the 0^0 ply. As is well known, these stresses are constant throughout the layers . The actual stress state in the cracked laminate is described by generalized plane strain in reference to the y axis, and therefore all stresses are functions of x,z only. It is at present assumed that only the 90^0 ply is cracked. The admissible stress state in the cracked laminate will be constructed on the basis of the simplification that the σ_{xx} stresses are functions of x and not of z. Thus these stresses may be written in the form

$$\sigma_{xx}^{(1)} = \sigma_1[1 - \phi_1(x)]$$

$$\sigma_{xx}^{(2)} = \sigma_2[1 - \phi_2(x)]$$

(9)

where ϕ_1 and ϕ_2 are unknown functions. These functions are, however, related since the stress pairs σ_1, σ_2 and (9) are each in equilibrium with the same N_{xx}. Therefore

$$\sigma_1 t_1 \phi_1 + \sigma_2 t_2 \phi_2 = 0$$

Consider a typical region between two adjacent cracks at distance 2a, fig.1b. It is emphasized that the cracks do not have to be equidistant. For example, the interdistance 2a could be a random variable. For a typical intercrack region as shown in fig.1b the admissible stress field is constructed by integration of the two dimensional equations of equilibrium in the xz plane with the stresses (9). The integration produces residual unknown functions which are determined by satisfaction of traction continuity conditions at layer interfaces and zero traction conditions on the external laminate surface. The remaining admissible stresses are then

$$\sigma_{xz}^{(1)} = \sigma_1 \phi'(x) z$$

$$\sigma_{zz}^{(1)} = \sigma_1 \phi''(x)(ht_1 - z^2)/2$$

$$\sigma_{xz}^{(2)} = (\sigma_1/\lambda) \phi'(x)(h - z)$$

$$\sigma_{zz}^{(2)} = (\sigma_1/\lambda) \phi''(x)(h - z)^2/2$$

(10)

where $\phi = \phi_1$, $\lambda = t_2/t_1$ and prime superscript denotes x differentiation. These stresses have already been given in Hashin (1985). On the crack surfaces $\sigma_{xx}^{(1)}$ and $\sigma_{xz}^{(1)}$ must vanish and therefore

$$\phi(\pm a) = 1 \qquad \phi'(\pm a) = 0 \tag{11}$$

The aggregate of the stress fields (9-11) for all intercrack regions are the admissible stress field. The stress energy densities for the layers 1 and 2 are:

$$2W_1 = \sigma_{xx}^{(1)^2}/E_T - 2\sigma_{xx}^{(1)}\sigma_{zz}^{(1)}\nu_T/E_T + \sigma_{zz}^{(1)^2}/E_T + \sigma_{xz}^{(1)^2}/G_T$$

$$2W_2 = \sigma_{xx}^{(2)^2}/E_A - 2\sigma_{xx}^{(2)}\sigma_{zz}^{(2)}\nu_A/E_A + \sigma_{zz}^{(2)^2}/E_T + \sigma_{xz}^{(2)^2}/G_A$$

(12)

where the elastic ply properties in (12) are : E_A, ν_A - axial Young's modulus and associated Poisson's ratio ; E_T, ν_T - transverse Young's modulus and associated Poisson's ratio ; G_A, G_T - axial and transverse shear moduli.

For reasons of symmetry it is sufficient to evaluate the complementary energy functional for the regions $-a_m \le x \le a_m$; $0 \le y \le 1$; $0 \le z \le h$. Then for such a region and for an isothermal state, the second of (5) assumes the form

$$\tilde{U}_{C_m} = \int_{-a_m}^{a_m} \int_0^{t_1} W_1 dz dx + \int_{-a_m}^{a_m} \int_{t_1}^h W_2 dz dx +$$

$$\frac{1}{2} \int_{-a_m}^{a_m} (\sigma_{zz}^2(x, t_1)/D_n + \sigma_{xz}^2(x, t_1)/D_s) \, dx$$

(13)

where D_n and D_s are normal and shear interface parameters, and for the entire laminate

$$\tilde{U}_C = \sum \tilde{U}_{C_m} \tag{14}$$

The actual complementary energy is given for the present case by

$$U_C = \sigma^{0^2} V / 2E_x^* \qquad \sigma^0 = N_{xx}/2h \qquad (15)$$

where V is the volume of the entire cracked laminate of thickness 1 in y direction. Introduction of the admissible stress field with functions ϕ_m for each region into (14) results, according to the principle of minimum complementary energy, in an upper bound on (15). The lowest upper bound is obtained by minimizing the resulting functional with respect to the functions ϕ_m. This is a standard problem in the calculus of variations resulting in Euler equations and boundary conditions for the minimizing functions which have the form

$$\frac{d^4\phi_m}{d\xi^4} + p\frac{d^2\phi_m}{d\xi^2} + q\phi_m = 0 , \qquad \phi_m(\pm\rho_m) = 1, \qquad \frac{d\phi_m}{d\xi}_{(\xi=\pm\rho_m)} = 0 \quad (16)$$

where

$$\xi = x/t_1 \qquad \rho_m = a_m/t_1 \qquad p = (C_{02} - C_{11})/C_{22} \qquad q = C_{00}/C_{22}$$

$$\begin{aligned}
C_{00} &= 1/E_T + 1/\lambda E_A \qquad C_{02} = \frac{\nu_T}{E_T}(\lambda + \frac{2}{3}) - \frac{\nu_A}{3E_A}\lambda \\
C_{22} &= (\lambda+1)(3\lambda^2 + 12\lambda + 8)/60E_T + \lambda^2/4D_n t_1 \qquad (17) \\
C_{11} &= \frac{1}{3}(1/G_T + 1/\lambda G_A) + 1/D_s t_1
\end{aligned}$$

Evaluation of the complementary energy functional in terms of the functions ϕ_m as was done in Hashin(1985), introduction of the result into the complementary inequality (6) with (15) as the actual complementary energy gives the result

$$1/E_x^* \le 1/E_x^0 + (\frac{\sigma_1}{\sigma_0})^2 \frac{C_{22}}{\lambda+1} \frac{\langle \chi(\rho) \rangle}{\langle \rho \rangle} , \qquad \chi_m(\rho) = -\frac{d^3\phi_m}{d\xi^3}_{(\xi=\rho_m)} \qquad (18)$$

where the brackets denote average with respect to the random variable a_m, half of the intercrack spacing. The form of χ depends on the nature of the roots of the characteristic equation of (16). When these roots are of the form $\pm(\alpha + i\beta)$, where $i = \sqrt{-1}$, then the solution of (16) is of the form

$$\phi_m = A_m Cosh(\alpha\xi)\cos(\beta\xi) + B_m Sinh(\alpha\xi)\sin(\beta\xi) \qquad (19)$$

where the constants are determined by the boundary conditions in (16). Then the associated χ_m is

$$\chi_m = 2\alpha\beta(\alpha^2 + \beta^2)\frac{Cosh(2\alpha\rho_m) - \cos(2\beta\rho_m)}{\alpha\sin(2\beta\rho_m) + \beta Sinh(2\alpha\rho_m)} \qquad (20)$$

If the roots are real, thus of the form $\pm\alpha, \pm\beta$,then

$$
\begin{aligned}
\phi_m &= A_m Cosh(\alpha\xi) + B_m Cosh(\beta\xi) \\
\chi_m &= \frac{\beta^2 - \alpha^2}{Coth(\alpha\rho)/\alpha - Coth(\beta\rho)/\beta}
\end{aligned}
\tag{21}
$$

3.2. IMPERFECT INTERFACE MODEL

A physical interpretation of the interface parameters D_n and D_s has been given in Hashin (1990). If there is a thin elastic isotropic interphase of thickness t_i between the phases, then

$$
D_n = (K_i + \frac{4}{3}G_i)/t_i \qquad D_s = D_t = G_i/t_i \tag{22}
$$

where K_i and G_i are the bulk and shear moduli of the interphase. It is easily shown that if the interphase is orthotropic, with material axes n,s and t , (22) becomes

$$
D_n = C_{nn}/t_i \qquad D_s = G_{ns}/t_i \qquad D_t = G_{nt}/t_i \tag{23}
$$

where C_{nn} is the normal stiffness and G_{ns} and G_{nt} are shear moduli. The first of (22,23) is strictly valid only when the interphase elastic moduli are much smaller than those of the constituents, but there is no such restriction with respect to D_s and D_t. If the interphase moduli are of the order of constituent moduli then the thin interphase effect is negligible and is equivalent to a perfect interface with displacement continuity. Consider a thin interphase in-between the layers of a cross-ply, fig.2. A relevant example is an oxidation protection layer between the laminae of a ceramic composite. Such a layer may develop many transverse cracks due to thermal stresses produced by manufacturing cooldown, fig. 2. These cracks are roughly orthogonal in fiber directions of the layers. In a ceramic fiber composite laminate, for example a SiC matrix reinforced by graphite fibers, the stiffness of the interphase layer is of the order of the stiffness of the layer material. For large crack density in the layer the shear moduli decrease very significantly, but not C_{nn}. Therefore such a cracked layer can be considered as an interface which is perfect for normal contact , $[u_n] = 0$, but is imperfect for shear. In that event the term in C_{22}, equ.(17), containing D_n, is negligible.

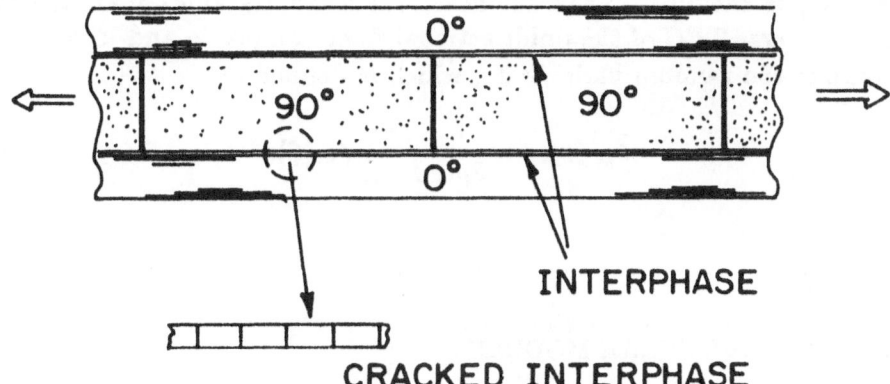

Figure 2. Laminate with damaged interphase

3.3. THERMAL EXPANSION

To evaluate the effective thermal expansion coefficient (TEC) of a cracked laminate it is very convenient to use the Levin relation as has been done for perfect interlaminar interface in Hashin (1988). For this purpose consider any elastic composite which is subjected to the homogeneous traction boundary conditions (7) and let the internal stresses due this loading be $\sigma^M(x)$. Denoting the local TEC $\alpha(x)$ and the effective TEC α^*, the Levin relation , Levin (1967), is expressed as

$$\int_V \alpha . \sigma^M dV = \alpha^* . \sigma^0 V \qquad (24)$$

Levin's original derivation of (24) is based on displacement continuity, but it may be shown that it remains valid for interface displacement continuities which obey the relations (3) and therefore (24) may be employed in the present case with the stresses (9-10) and the functions ϕ_m to give an approximate expression for the TEC. For the loading N_{xx} the only surviving component of σ^0_{ij} in (7) is $\sigma^0_{xx} = \sigma^0$ as defined by (15). Then from (24)

$$\alpha^*_{xx}\sigma^0 V = \int_0^L \left[\int_0^{t_1} \alpha_T(\sigma^{(1)}_{xx} + \sigma^{(1)}_{zz})dz + \int_{t_1}^h (\alpha_A\sigma^{(2)}_{xx} + \alpha_T\sigma^{(2)}_{zz})dz \right] dx \quad (25)$$

where L is the length of the laminate of unit thickness in y direction. Insertion of the stresses into (25) with use of the boundary conditions of (16) yields

$$\alpha^*_{xx} = \alpha^0_{xx} + \frac{\sigma_1}{\sigma_0}\frac{1}{1+\lambda}(\alpha_A - \alpha_T) < \phi > \qquad (26)$$

where α_{xx}^0 is the TEC of the uncracked laminate, α_A and α_T are the axial and transverse TEC of the unidirectional fiber composite and $< \phi >$ is the average of the random variable $\bar{\phi}_m$ which is defined as

$$\bar{\phi}_m\ (\rho_m) = \frac{1}{2\rho_m} \int_{-\rho_m}^{\rho_m} \phi_m(\xi)d\xi \tag{27}$$

Also

$$\alpha_{yy}^* \cong \alpha_{yy}^0 \tag{28}$$

3.4. EFFECTIVE SHEAR MODULUS

Let the cracked laminate shown in fig. 1 be subjected to constant shear membrane load N_{xy} which defines the average applied shear stress

$$\tau^0 = N_{xy}/2h \tag{29}$$

In this case the laminate is in a state of antiplane stress with respect to the y axis. Admissible stresses are defined as in Hashin (1985) by

$$\begin{aligned}
\sigma_{xy}^{(1)} &= \tau^0[1 - \psi(x)] & \sigma_{yz}^{(1)} &= \tau^0\psi'(x)z \\
\sigma_{xy}^{(2)} &= \tau^0[1 + \tfrac{1}{\lambda}\psi(x)] & \sigma_{yz}^{(2)} &= \tfrac{\tau^0}{\lambda}\psi'(x)(h - z)
\end{aligned} \tag{30}$$

Then a variational optimization as done above and in Hashin (1985) yields the results

$$\psi_m(\xi) = Cosh(\mu\xi)/Cosh(\mu\rho_m) \quad \mu^2 = \frac{3(1+1/\lambda)}{1+\lambda G_A/G_T+3G_A/t_1D_s} \tag{31}$$

$$G_{xy}^* \geq \frac{G_A}{1+ < Tanh(\mu\rho) > /\lambda\mu < \rho >}$$

In the case of equidistant cracks $a_m = a$, $\rho_m = a/t_1$, and all of the averages in all of the expressions above reduce to simple functions of ρ which are defined by removal of the brackets.

3.5. STRESS ANALYSIS AND CRACK OPENING DISPLACEMENTS

The optimal functions ϕ_m and ψ_m define optimal admissible stresses by the relations (9,10,30). The effective properties E_x^* and G_{xy}^* based on these are strict lower bounds which also agree quite well with experimental data. The status of the stresses associated with the optimal functions is less clear which is a typical situation for any variational field approximation. It is, however, believed that these stresses are of qualitative importance, at least, and the numerical results obtained, some of which are shown below, support this belief.

The results obtained in this work can be easily used to estimate the crack opening displacements (COD) when the cracks are equidistant. It is rigorously true that the stress energy U of a cracked elastic body, homogeneous or non-homogeneous, and the stress energy U_0 of the same uncracked body, under same load, are related by

$$U = U_0 + \frac{1}{2} \sum^{m} \int_{S_m} T^0.[u]dS \qquad (32)$$

where S_m is the surface of the mth crack, T^0 is the traction on same surface in the **uncracked** body and $[u]$ is the COD. In the case of simple tension discussed above U is given by (15) and $U_0 = (\sigma^{0^2}/2E_x^0)V$ for the uncracked laminate. Also, the only surviving T_i^0 is $T_x^0 = \sigma_1$. The COD is now estimated in the form of two equal and oppositely joined second order parabolas. Thus

$$[u_x] = 2\delta[1 - (\frac{z}{t_1})^2]$$

where 2δ is the maximum COD. It then follows easily that

$$\delta/t_1 = \frac{3}{2}\frac{\rho(1+\lambda)}{k_1}\sigma^0(1/E_x^* - 1/E_x^0) \qquad (33)$$

4. Results

Illustrative results are presented for a $[0^0, 90^0]_s$ laminate in which the layers of equal thickness are T300/SiC ceramic unidirectional composites with fiber volume fraction 0.45. The relevant properties of the layer material are:

$$
\begin{aligned}
E_A &= 431.5\,GPa & E_T &= 113.6\,GPa \\
G_A &= 90.8\,GPa & G_T &= 39.3\,GPa \\
\nu_A &= 0.182 & \nu_T &= 0.446 \\
\alpha_A &= 2.39\ 10^{-6}(C^0) & \alpha_T &= 5.49\ 10^{-6}(C^0)
\end{aligned}
$$

It is assumed that in between the layers there is a thin oxidation protection interphase of isotropic B_4C material with thickness 0.02 of the layer thickness and with properties

$$E_i = 380\,GPa \qquad G_i = 159.7\,GPa \qquad \nu_i = 0.19$$

Due to thermal treatment the interphase may develop many through cracks. As has been explained above, this creates an orthotropic interphase which may be considered perfect in normal z direction but imperfect in shear. Increase of crack density of the interphase may be considered as decrease of the effective shear stiffness of the interphase. All of the results given below are in the form of plot families where each plot is associated with a shear modulus value G_i/m , m=1,5, 10, 20, 50, 100, 200. Figs. 3–5 show plots of Young's modulus, TEC and shear modulus, all normalized with respect to their values for the uncracked laminate, as functions of crack density for the case of equidistant cracks, expressed by the parameter ρ. For large values of ρ the properties of the uncracked laminate are attained while for small values of ρ the properties reduce to those of a laminate in which the 90^0 layer has vanishing E_T and G_A but retains it's E_A value. Such stiffness loss is associated with the concept of laminate netting analysis. The values of the properties decrease with decreasing interphase shear modulus. Thus for each property the uppermost plot is for undamaged interphase which may be regarded as a perfect interface while the lowest plot is for m=200. It is seen that the effect of interface imperfection, i.e. interphase damage, is not very significant for Young's modulus and TEC but is very significant for the shear modulus. It is also seen that the plots for normalized Young's modulus and TEC are very similar and indeed these normalized quantities are almost the same numerically. It should be realized that interphase damage has no effect on an uncracked laminate under in-plane loading since there are no interlaminar stresses in this case.

Figs. 6–7 show internal stresses as functions of x ,for load N_{xx}, when the intercrack distance is $2a = 5t_1$. Fig.6 shows the in-plane stresses $\sigma_{xx}^{(1)}$ in the cracked layer and $\sigma_{xx}^{(2)}$ in the uncracked layer functions of x for the case when the intercrack distance is $2a = 5t_1$, the family of plots being defined by the sequence G_i/m. It is seen that the tensile stress in the cracked 90^0 layer decreases with increasing interphase damage amd therefore the stress in the uncracked 0^0 layer increases with damage. Fig. 7 shows similar plots for the interlaminar stresses $\sigma_{xz}(x, t_1)$ and $\sigma_{zz}(x, t_1)$. These stresses decrease with increasing interphase damage and the effect is significant.

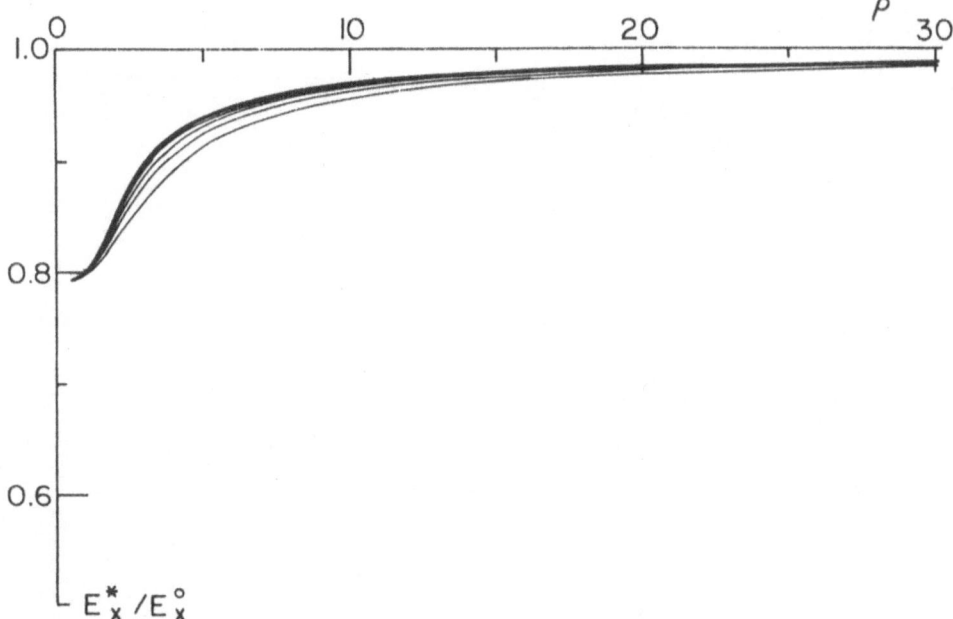

Figure 3. Effective Young's modulus versus crack spacing

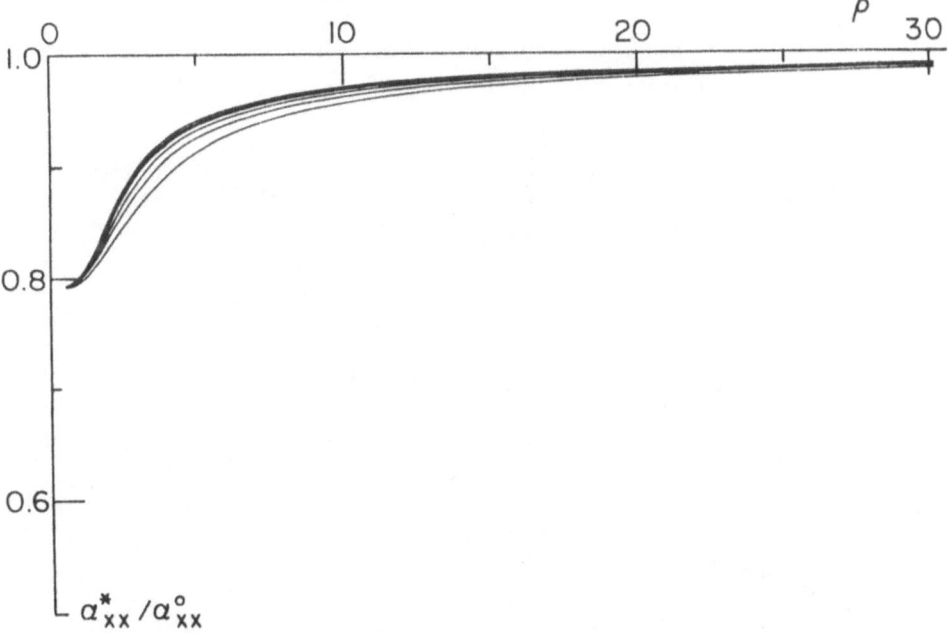

Figure 4. Effective thermal expansion coefficient versus crack spacing

126

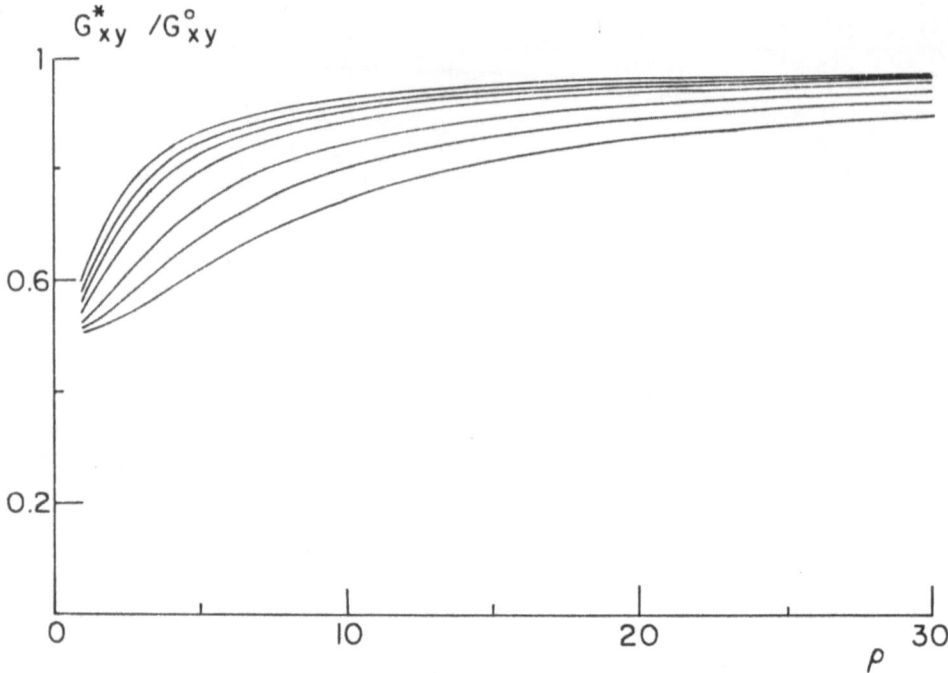

Figure 5. Effective shear modulus versus crack spacing

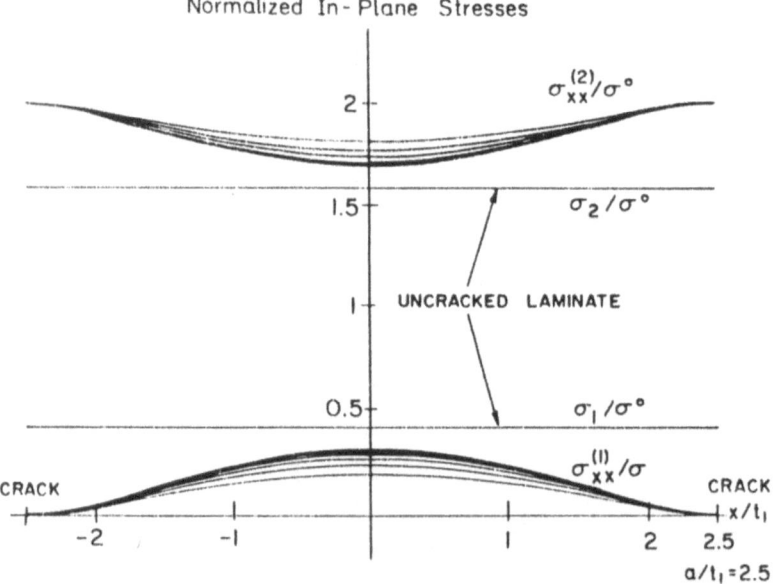

Figure 6. In-plane stresses in layers

Normalized Interface Stresses

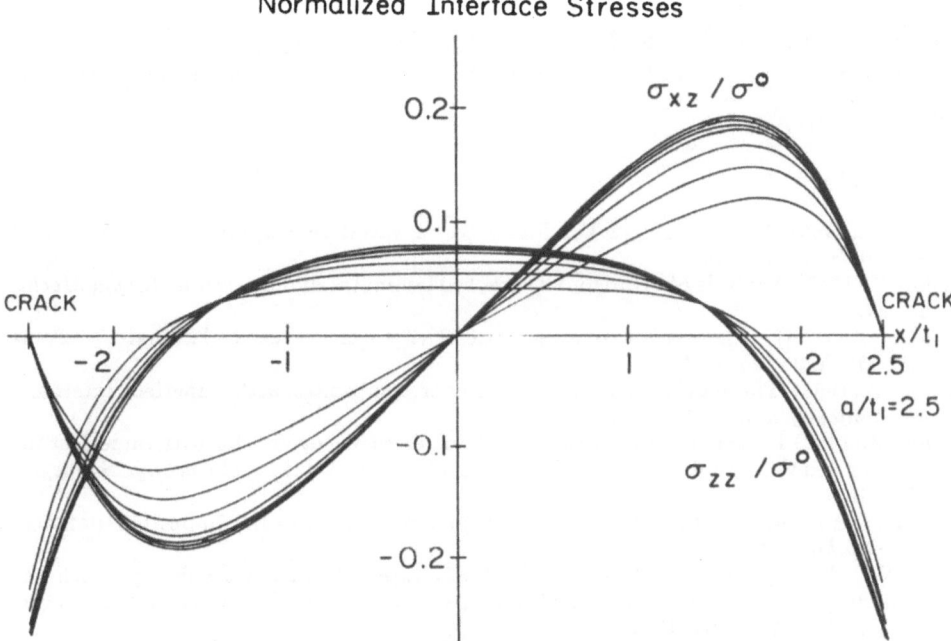

Figure 7. Interface stresses

5. Conclusion

Previously developed variational analysis of cracked laminates has been extended to the case of imperfect interlaminar interface by use of a generalized thermo-elastic variational principle for imperfect interface. In the present work detailed analysis has been confined to a cross-ply laminate in which only the 90^0 layer is cracked. But there is no difficulty to carry out similar analysis with imperfect interface for the case when all layers are cracked, on the basis of the admissible stress system which has been constructed in Hashin (1987) for orthogonally cracked laminates. It is also a straightforward matter to analyze mechanical and thermal stresses.

Present analysis has shown that for a cracked laminate with interlaminar interface which is imperfect in shear only, the quantitative effect of interface imperfection on effective Young's modulus and TEC is not drastic, but there is significant effect on the effective shear modulus and internal interlaminar stresses.

Acknowledgement

Support of the Air Force Office of Scientific Research, Dr. Walter Jones contract monitor, through the Materials Sciences Corporation is gratefully acknowledged.

References

Hashin, Z.(1985) Analysis of cracked laminates: a variational approach, *Mechanics of Materials*, 4, 121-136

Hashin, Z.(1987) Analysis of orthogonally cracked laminates under tension, *J.Appl.Mech.*, 54, 872-879

Hashin, Z.(1988) Thermal expansion coefficients of cracked laminates, *Comp.Sci.Tech.*,31, 247-260

Hashin, Z.(1990) Thermoelastic properties of fiber composites with imperfect interface, *Mechanics of Materials*,8, 333-348

Hashin, Z.(1992) Extremum principles for elastic heterogeneous media with imperfect interfaces and their application to bounding of effective moduli, *J.Mech.Phys.Solids*,40, 767-781

Laws, N. and Dvorak, G.J.(1988) Progressive transverse cracking in composite laminates, *J.Comp.Mat.*,22, 900-916

Lee, J.W., Allen, D.H. and Harris, C.E.(1989) Internal state variable approach for predicting stiffness reductions in fibrous laminated composites with matrix cracks, *J.Comp.Mat.*,23, 1273-1291

Levin, V.M. (1967) On the coefficients of thermal expansion in materials, *Mechanics of Solids* (translation of *Mekhanika Tverdogo Tela*, 2, 58-61.

McCartney. L.N.(1992) Theory of stress transfer in a $0^0 - 90^0 - 0^0$ cross-ply laminate containing a parallel array of transverse cracks, *J.Mech.Phys.Solids*,40, 27-68

Nairn, J.A., Hu, S. and Jong, S.B., (1993) A critical evaluation of theories for predicting microcracking in composite laminates, *J.Mat.Sci.* 28, 5099-5111

Reifsnider, K.L. and Jamison,R.(1982) Fracture of fatigue-loaded composite laminates, *Int.J.Fatigue*,6, 187-197

Tsai, C.L., Daniel, I.M. and Lee, J.W.(1990) Progressive matrix cracking of crossply composite laminates under biaxial loading, in G.J.Dvorak and D.C.Lagoudas (eds.), *Microcracking-Induced Damage in Composites*, AMD-Vol.111, ASME,New York,pp.9-18

Varna,J. and Berglund,L.A.(1994) Thermo-elastic properties of composite laminates with transverse cracks, *J.Comp.Tech.Res.*, 16, 77-87

NUMERICAL MODELLING OF CRACK GROWTH IN MATERIAL MODELS OF FIBROUS COMPOSITES

K.P. HERRMANN and F. FERBER
Laboratorium für Technische Mechanik
Paderborn University
Pohlweg 47-49, D-33098 Paderborn, Germany

ABSTRACT

Experimental investigations of fracture phenomena in thermomechanically loaded fibrous composites demonstrate the appearance of different failure mechanisms, like matrix and interface cracks as well as of fiber breakings. In addition, the existence of branched crack systems consisting of a combination of those elementary failure mechanisms has also been observed several times. Therefore, in this paper the numerical modelling of complicated crack systems has been performed, consisting of a combination of matrix and fiber cracks where the latter are originated by local asymmetrical interface cracks and according to experimental results arise in a single layer of a thermomechanically loaded fibrous composite structure. The mathematical modelling of such branched crack systems in thermomechanically loaded two-phase compounds leads to mixed boundary value problems of the thermoelasticity. The corresponding solutions were obtained by using a closed finite element program capable of an automatic mesh generation. Further, special emphasis has been given to the crack path prediction of thermal cracks initiated in a plastic matrix/glass fiber reinforced composite structure by using a newly established crack growth criterion based on the total energy release rate of a quasistatic mixed-mode crack extension. This numerical simulation of the crack growth process in appropriate material models should allow a better understanding of the fracture behaviour of fibrous composites on a micromechanical level.

1. INTRODUCTION

The failure behaviour of fibrous composites differs considerably from that of homogeneous solids due to the large number of possible failure mechanisms arising in thermomechanically loaded composite materials. Thus, for example, the fracture behaviour of a laminate will be influenced by local failure mechanisms in a single layer such as fiber breaks, matrix and interface cracks, fiber pull-outs as well as the plastification of the matrix material. These failure modes existing on a microscale depend heavily from the orientation of the fibers, the individual ply thickness as well as on the constitutive equations describing the mechanical properties of the fibers, the matrix as well as the fiber-matrix interface. In addition to these elementary failure mechanisms having the size of the microstructure, i.e. the dimension of a fiber diameter, other failure modes arise, for instance intralaminar transverse cracks as well as extended delaminations

R. Pyrz (ed.), IUTAM Symposium on Microstructure-Property Interactions in Composite Materials, 129–140.
© 1995 *Kluwer Academic Publishers.*

between single plies with dimensions of several orders of magnitude larger than a fiber diameter. Because of the variety of parameters needed to be considered for a formulation of strength characteristics for composites based on experimental results only, today's composite research prefers the application of appropriate analytical models showing the essential details of the failure physics. There exist two different methods for a description of the failure behaviour of composites by using fracture mechanics; known in the literature as the micromechanical and macromechanical stress analysis. A review concerning the essential peculiarities of these two distinct continuum mechanical methods for a characterization of the fracture behaviour of composites has been given by Rosen et al. [1] and Mahishi [2,3]. Within the last two decades a considerable number of publications dealing with different aspects of the strength and fracture behaviour of composites has been accumulated. For instance, Goree and Groß [4] gave an analytical solution concerning the determination of stress and strain fields in a unidirectionally fiber-reinforced composite containing an arbitrary number of broken fibers as well as a plastified matrix material. The corresponding analytical model based on the shear-lag assumption as well as on a shear stress fracture criterion allows, for instance, the prediction of the characteristic strength and fracture properties of a boron fiber/aluminum matrix composite in agreement with experimental results. Further, Tvardovsky [5] considered an appropriate material model of an anisotropic layered composite containing isolated collinear and double periodic cracks, respectively. Thereby by assuming a remote constant loading of the composite structure as well as by consideration of the inherent boundary and continuity conditions the associated boundary value problem could be reduced to a singular integral equation by applying the finite Fourier transform. Furthermore, there already exist several micromechanical models which allow for an analysis as well as for a prediction of the overall behaviour of composites. These methods are known in the literature as the dilute approximation, the self-consistent scheme, the Mori-Tanaka and the differential scheme. In Aboudi's book [6] a description of these composite models with their advantages and disadvantages has been given. Aboudi himself developed a micromechanical composite model based on the study of interacting periodic cells. Thereby due to the assumed periodic microstructure, a representative volume element only needs to be considered consisting of the fiber and the matrix subcells. By using a homogenization procedure a set of continuum equations can be produced allowing the transition to an equivalent homogeneous continuum. The important advantage of Aboudi's model consists in the establishment of a unified approach in the prediction of the overall behaviour of composites.

Moreover, an interesting problem concerning the failure behaviour of thermomechanically loaded composite structures consists in the prediction of the prospective paths of microcracks which already exist in the heterogeneous microstructure depending on the geometrical configuration as well as on the applied thermomechanical load distribution belonging to a given composite. Several possible failure criteria have been discussed in this respect in the literature, e.g. in a review article by Rosen [7]. Besides, there exist some stochastic models for a description of the fracture behaviour of unidirectionally reinforced composites. Elementary failure mechanisms like fiber breaking, debonding as well as matrix cracking due to special loading and overloading were simulated by appropriate computer experiments, Kopyov et al. [8]. The corresponding activation criteria concerning the initiation of such damage mechanisms were obtained by an analysis of the stress redistribution due to rising microcracks in the fibrous composites.

2. BRANCHED CRACK SYSTEMS IN MATERIAL MODELS OF FIBROUS COMPOSITES

2.1. DISK-LIKE BIMATERIAL SPECIMEN
In this paper, branched crack systems consisting of a combination of curved matrix and interface cracks as well as of fiber breaks and arising in different material models of fibrous composites are considered.

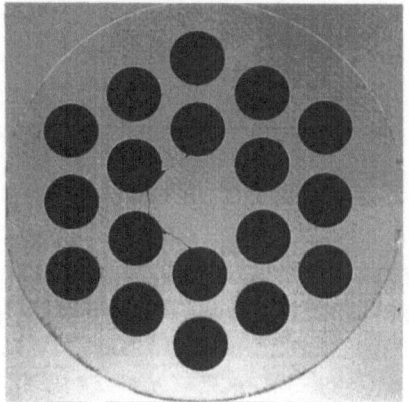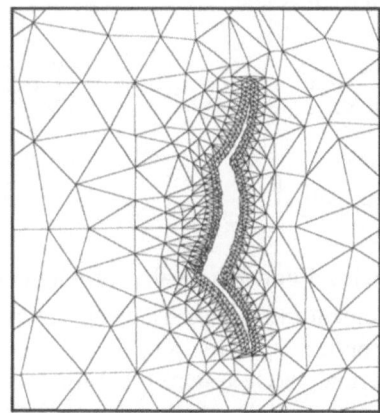

Figure 1. Branched thermal crack system in a disk-like material model of a fibrous composite and associated finite element discretization

Figure 1 shows the cross section of a cracked disk-like two-phase solid (matrix: Araldite F, fibers: steel) containing three matrix as well as two interface cracks at the fiber-matrix interfaces of two neighbouring fibers due to thermal loading after a special casting process. The thermal loading of the composite structure took place due to a cooling from the temperature $T_0 = 60 \deg C$ of the unstressed initial state to a loading temperature $T_1 = -7,5 \deg C$ according to a special cooling curve. The latter as well as the associated material properties of the two-phase compound and also the geometrical parameters of the disk-like bimaterial specimen can be taken from reference [9].

Further, an appropriate nomenclature of different possible model geometries of thermally loaded two-phase composite structures has been introduced in order to be able to formulate the following boundary value problems of the plane thermoelasticity [10]

$$\sigma_{rr}^m(r_m,\phi) = \sigma_{r\phi}^m(r_m,\phi) = 0; \qquad\qquad (|\phi| \le \pi) \tag{1}$$

$$\sigma_{\rho\rho}^j(\rho_f,\psi) = \sigma_{\rho\psi}^j(\rho_f,\psi) = 0; \qquad\qquad (|\psi| < \psi_i),\,(j=f,m) \tag{2}$$

$$\left[\sigma_{\rho\rho}(\rho,\psi)\right]_{\rho_i=\rho_f} = \left[\sigma_{\rho\psi}(\rho,\psi)\right]_{\rho_i=\rho_f} = 0; \qquad (\psi_i \le |\psi| \le \pi) \tag{3}$$

$$\left[u_\rho(\rho,\psi)\right]_{\rho_i=\rho_f} = \left[u_\psi(\rho,\psi)\right]_{\rho_i=\rho_f} = 0; \qquad (\psi_i \le |\psi| \le \pi) \tag{4}$$

Besides, global plane polar coordinates r,ϕ with respect to the centers of the composite structures (with the outer radius r_m) as well as local coordinates ρ,ψ, with the center in each fiber (with the radius ρ_f) have been introduced into the boundary and continuity conditions $(1)-(4)$. By applying the basic equations of the stationary thermoelasticity for a plane stress state

$$\sigma_{ij} = \frac{E}{(1+v)}\left\{\varepsilon_{ij} + \frac{v}{1-2v}\varepsilon_{kk}\delta_{ij} - \frac{1+v}{1-2v}\alpha\Delta T\delta_{ij}\right\} \tag{5}$$

$$\sigma_{ji,j} = 0 \tag{6}$$

$$\varepsilon_{ij} = \frac{1}{2}\left\{u_{i,j} + u_{j,i}\right\} \tag{7}$$

the associated boundary value problems $(1)-(4)$ can be solved either by means of the finite element method or by using the experimental methods of the photoelasticity and shadow optics. By using a refined finite element mesh in the neighbourhood of a branched thermal crack system (cf. Fig. 1) strain energy release rates at the tips of matrix and interface cracks were calculated. Furthermore, by implementing an appropriate crack growth criterion based on a maximum energy release rate principle a theoretical prediction of the experimentally observed branching phenomenon of curvilinear thermal cracks in disk-like material models of fibrous composites could be performed. Figure 2 shows a summary of experimentally obtained results concerning the crack velocity v and the opening-mode stress intensity factor K_I at the tip of a curvilinear thermal matrix crack propagating in a disk-like material model of a self-stressed composite structure. Moreover, the radius r_0 of the initial curve of the caustic surrounding the crack tip as function of the crack length a is given which was used for a determination of the diameter of the caustic. Finally, Fig. 2 clearly demonstrates that the values of the stress intensity factor K_I obtained experimentally by the method of caustics in transmission and reflection for a curvilinear thermal matrix crack show a very good coincidence with the corresponding numerical K_I-values determined by associated finite element calculations especially in the region of stable crack propagation. More details about these investigations are given in reference [10].

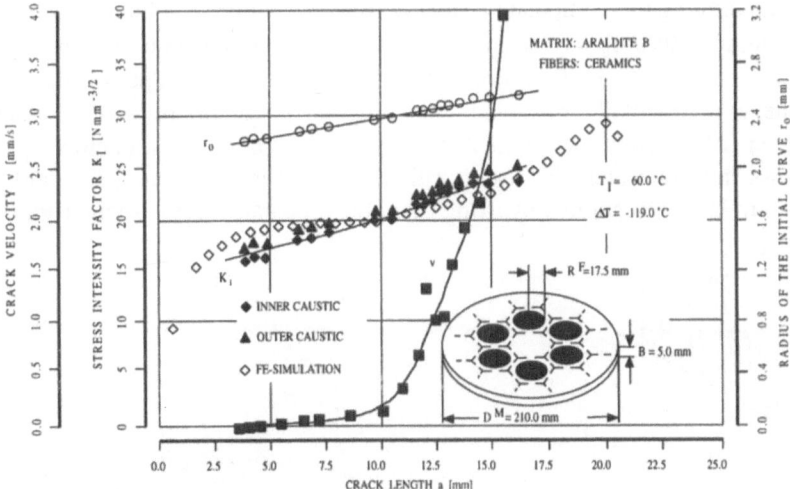

Figure 2. Stress intensity factor K_I, crack velocity v, and radius r_0 of the initial curve
of the caustic as function of crack length a

2.2. DETERMINATION OF FRACTURE MECHANICAL DATA AND CRACK PATH PREDICTION OF A BRANCHED CRACK SYSTEM IN A SINGLE UD-LAYER OF A LAMINATE

2.2.1. Crack growth criterion

Figure 3 shows the cross section of a cracked single UD-layer laminate model (matrix: Araldite F, fibers: glass) containing matrix and interface cracks as well as fiber breaks due to a thermal loading after a special casting process.

Figure 3. Material model of a cracked UD-layer of a laminate

The thermal loading of the composite structure took place due to a homogeneous cooling from the temperature $T_0 = 60$ deg C of the unstressed initial state to a loading temperature of $T_1 = -5.0$ deg C. Further, finite element calculations were performed in order to predict the experimentally observed branching phenomenon of cracks arising in single UD-layers of different thermomechanically loaded composite structures.

Figure 4 shows the finite element discretization of a UD-layer where a standard finite element program has been applied by using triangular constant strain 3-node elements.

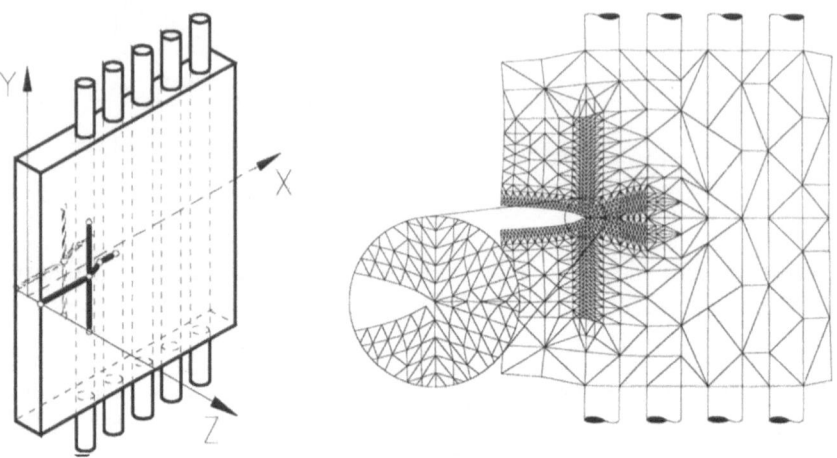

Figure 4. Finite element discretization of a cracked UD-layer of a two-phase compound

The directional criteria for the description of crack growth in brittle solids proposed in the past, like the criteria of principal stress [11], maximum of strain energy release rate [12,13], minimum of strain energy density [14,15] require the knowledge of the near-tip stress and displacement fields in the vicinity of the original crack tip characterized for a general plane loading situation by the stress intensity factors K_I and K_{II}, respectively.

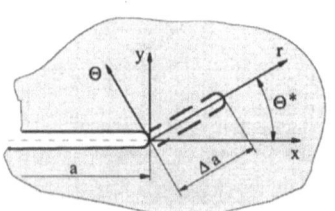

Figure 5. Global und local coordinates at a kinked crack tip

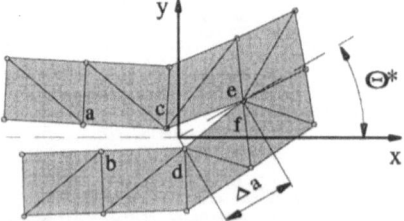

Figure 6. Finite element mesh at a crack tip with a new crack lengthening Δa

Moreover, the application of these criteria to a cracked brittle solid delivers equations for the determination of the angle ϑ^* describing the direction of further crack growth (cf. Fig. 5). By combining the essentials from the principal stress and the maximum energy release rate criterion, an appropriate crack growth criterion based upon the numerical calculation of energy release rates at crack tips has been proposed by Herrmann and Dong [16 – 18].

Moreover, it was shown earlier by Herrmann and Grebner [19] that the calculation of strain energy release rates at the tips of quasistatically extending curvilinear thermal cracks in self-stressed two-phase solids could be reached by using a method originated by Rybicki and Kanninen [20]. This method is founded on the evaluation of Irwin's crack closure integral

$$G(a,\vartheta) = G_I(a,\vartheta) + G_{II}(a,\vartheta) = \lim_{\Delta a \to 0} \frac{1}{2\Delta a} \int_0^{\Delta a} \left[\sigma_{\vartheta\vartheta}(r,\vartheta) \cdot \bar{u}_\vartheta(r,\vartheta) \right] dr$$

$$+ \lim_{\Delta a \to 0} \frac{1}{2\Delta a} \int_0^{\Delta a} \left[\sigma_{r\vartheta}(r,\vartheta) \cdot \bar{u}_r(r,\vartheta) \right] dr \qquad (8)$$

where the quantities $\sigma_{\vartheta\vartheta}, \sigma_{r\vartheta}$ represent the near-tip stresses in the local coordinate system at the crack tip prior to crack extension, whereas the quantities $\bar{u}_r, \bar{u}_\vartheta$ stand for the corresponding normal and tangential displacements between opposite points of the crack surface after crack extension and Δa represents the crack lengthening.

By using a finite element mesh according to Fig. 6 a numerical calculation of the average energy release rates G, G_I, G_{II} related to the global coordinate system x,y as indicated can be performed where the displacements along the new crack surfaces are approximated by a linear interpolation function in the related finite elements, i.e. usually 3-node triangular elements. The corresponding formulae read as follows [17]

$$G\left(a + \frac{\Delta a}{2}, \vartheta\right) = G_I\left(a + \frac{\Delta a}{2}, \vartheta\right) + G_{II}\left(a + \frac{\Delta a}{2}, \vartheta\right) \qquad (9)$$

$$G_I\left(a + \frac{\Delta a}{2}, \vartheta\right) \approx G_I(a \to a + \Delta a, \vartheta) = A \sin^2 \vartheta - 2B \sin \vartheta \cos \vartheta + C \cos^2 \vartheta \qquad (10)$$

$$G_{II}\left(a + \frac{\Delta a}{2}, \vartheta\right) \approx G_{II}(a \to a + \Delta a, \vartheta) = A \cos^2 \vartheta + 2B \sin \vartheta \cos \vartheta + C \sin^2 \vartheta \qquad (11)$$

with

$$A = \frac{1}{2t\Delta a} F_x^c \left(u_x^d - u_x^c\right); \quad B = \frac{1}{4t\Delta a} \left[F_x^c \left(u_y^d - u_y^c\right) + F_y^c \left(u_x^d - u_x^c\right) \right] \qquad (12a)$$

$$C = \frac{1}{2t\Delta a} F_y^c \left(u_y^d - u_y^c\right) \qquad (12b)$$

where t denotes the specimen thickness, F_x^c, F_y^c mean the components of the nodal forces in the global coordinate system x,y before the crack extension Δa, and $u_x^i, u_y^i (i = c,d)$ stand for the components of the corresponding nodal displacements of the new crack surfaces after the crack extension Δa. Besides, it should be mentioned that the coefficient B in the equations $(10)-(12)$ differs from the former coefficient B introduced in the references $[16-18]$ by a factor of 2.

By using the equations $(9)-(11)$ the strain energy release rates G, G_I, G_{II} can be determined if the new crack extension direction ϑ is known. However, the latter quantity is just wanted. Therefore, an approximate method for the determination of the new crack extension direction has been developed, based on the physical mechanisms of brittle fracture. Thus, it is assumed that the new direction of crack extension Δa is given by the direction in which $G_{II} = 0$ holds true and crack extension occurs for $G_I \geq G_{Ic}$ (critical crack extension force). Further, an iteration scheme concerning the determination of the new crack extension direction has been stated which is based on this criterion $G_{II} = 0$ and on the numerical calculation of the strain energy release rates $G_i (i = I,II)$.

2.2.2. Iteration scheme

Step 1: Select a start angle ϑ^* of the prospective crack growth.

Step 2: Let $\vartheta = \vartheta^*$ by arranging a local finite element mesh in the vicinity of the crack tip. By using this preliminary extension angle calculate a set of coefficients A,B,C according to the equations $(12a,b)$ as well as the associated strain energy release rates, equations $(9)-(11)$. If $G_{II} = 0$ is valid, then ϑ^* already gives the desired new crack extension direction and the iteration is finished. However, usually G_{II} becomes not be zero and then step 3 has to be carried out.

Step 3: By taking $G_{II} = 0$ it follows

$$A \cos^2 \vartheta + 2B \sin \vartheta \cos \vartheta + C \sin^2 \vartheta = 0 \tag{13}$$

from which a new angle ϑ^* for the desired crack extension direction can be calculated, namely

$$\vartheta^* = \arctan\left[-\frac{B}{C} \pm \sqrt{\left(\frac{B}{C}\right)^2 - \frac{A}{C}}\right] \tag{14}$$

Step 4: After obtaining this new crack extension direction ϑ^*, the iteration scheme can be started again. The procedure can be stopped if after a few reiterations of the steps 2 and 3 the new angle ϑ^* nearly equals ϑ, the prior angle used to calculate the coefficients A,B,C. Then the strain energy release rate G_{II} reaches approximately zero value. Thus, this calculated angle then corresponds to the new crack extension angle.

2.2.3. Crack path prediction in a UD-layer of a two-phase compound

The crack growth criterion stated in the previous session has been applied for the prediction of a combined crack system extending in an epoxy matrix/glass fiber reinforced composite model due to well-defined stress fields caused by uniform temperature changes and additional mechanical loadings. Figure 3 gives the cross section of the two-phase compound consisting of homogeneous isotropic and linearly elastic materials with different thermoelastic properties according to Table 1.

Table 1. Material properties of a two-phase compound

Material	Matrix: Araldite F	Fibers: Glass (SF11)
Young's modulus E [N/mm^2]	2600	74850
Poisson's ratio v [1]	0.39	0.17
Thermal expansion coefficient α $\left[10^{-7} K^{-1}\right]$	530	5.0

The numerical modelling of the crack path starts with the development of a straight matrix crack, initiated in the origin of the coordinate system shown in Fig. 4, and growing quasistatically towards a neighbouring fiber. By striking the nearest fiber-matrix interface a material defect has been modelled by a short asymmetrical interface crack. Further, the following prediction of the prospective crack path has been performed by applying the crack growth criterion already mentioned. The graphs 7-8 show the crack extension direction angles ϑ^* as functions of the projected crack lengths a_x as well as the prospective crack paths by applying a thermomechanical loading system consisting of a homogeneous cooling of $\Delta T = -65^0 \deg C$ and variable mechanical loads to the composite structure. It can be seen that for two of the additional mechanical loads only cracks exist up to the end of the prospective crack path. The other mechanical loads deliver either a further crack extension or a crack arrest in the neighbouring fiber. The latter result can also be taken from Fig. 9 showing the strain energy release rate G_I in dependence on the projected crack length a_x.

Besides, it can been from the same graph that the G_I-values at the tip of the first straight matrix crack reach certain maximum values near to the fiber-matrix interface. In addition, Fig. 10 shows that for all cases of a thermomechanical loading a mixed-mode loading situation exists along the fiber-matrix interface. Moreover, the program developed for an automatic mesh generation was not able to handle the large crack deviation angles arising in the fiber due to the applied crack growth criterion. Therefore, in these cases the further crack extension in the fiber has not been investigated. Finally, it should be mentioned that a second UD-layer composite model (SSKN5 glass matrix/SF11-glass fiber) has been studied in reference [21].

138

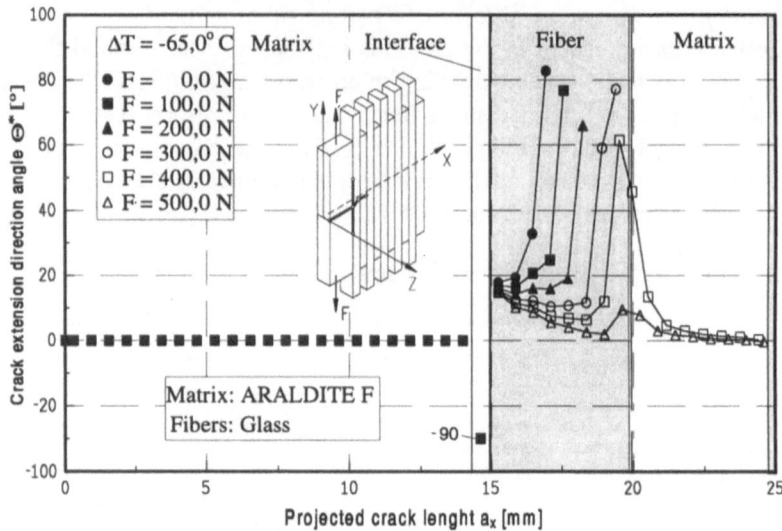

Figure 7. Crack extension direction angle ϑ^* as function of the projected crack length a_x

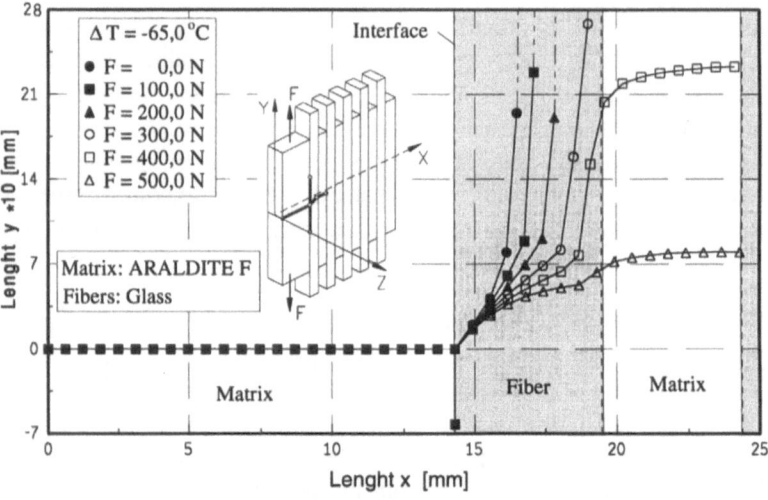

Figure 8. Predicted crack paths in a UD-layer laminate model

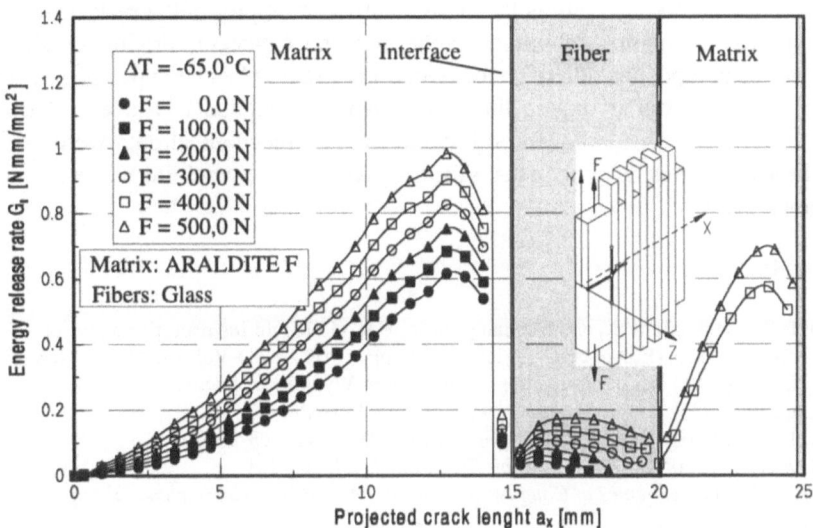

Figure 9. Strain energy release rate G_I for a kinked crack system

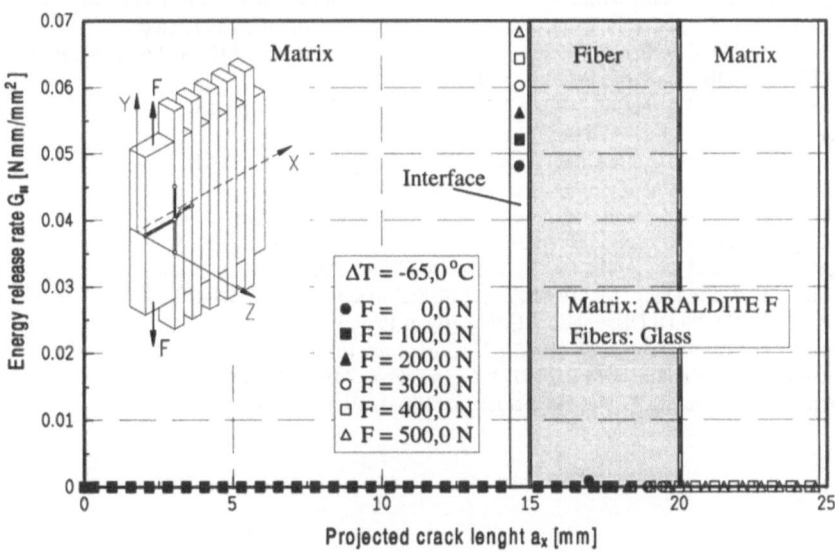

Figure 10. Strain energy release rate G_{II} for a kinked crack system

3. CONCLUSIONS

A numerical modelling of branched crack systems arising in thermomechanically loaded material models of fibrous composites is given. By taking the substructure technique of

140

the finite element method as well as by implementing an appropriate crack growth criterion a numerical simulation of branched thermal crack systems arising in disk-like models of fibrous composites due to a steady cooling process has been performed. Besides, kinked crack systems arising in thermomechanically loaded UD-layers of fibrous composites have been considered. In both cases fracture mechanical data were calculated where these quantities have been used for the prediction of crack initiation as well as of further crack growth.

REFERENCES

[1]   Rosen, B.W., Kulkarni, S.V., and Mc Laughlin Jr., P.V. (1975), In *Inelastic Behaviour of Composite Materials*, (Edited by C.T. Herakovich) pp. 17-72, AMD-Vol. 13, ASTM, New York.
[2]   Mahishi, J.M. (1984), Ph.D. dissertation, University of Wyoming, Laramie.
[3]   Mahishi, J.M. (1986), *Engng Fracture Mechanics* 25, 197-228.
[4]   Goree, J.G. and Gross, R.S. (1979), *Engng Fracture Mech.* 13, 563-578.
[5]   Tvardovsky, V.V. (1992), *Theoret. Appl. Fract. Mech.* 17, 205-231.
[6]   Aboudi, J. (1991) *Mechanics of Composite Materials. A Unified Micromechanical Approach*, Elsevier, Amsterdam.
[7]   Rosen, B.W. (1983), In *Mechanics of Composite Materials. Recent Advances*, (Edited by Z. Hashin and C.T. Herakovich) pp. 105-134, Pergamon Press, New York.
[8]   Kopyov, I.M., Ovtchinsky, A.S., and Bilsagayev (1982), In *Fracture of Composite Materials* (Edited by G.C. Sih and V.P. Tamuzh) pp. 45-52, Martinus Nijhoff, The Hague.
[9]   Herrmann, K.P. and Ferber, F. (1986), In *Brittle Matrix Composites 1* (Edited by A.M. Brandt and I.H. Marshall) pp. 49-68, Elsevier, London.
[10] Herrmann, K.P. and Ferber, F. (1992), *Computers & Structures* 44, 41-53.
[11] Erdogan, F. and Sih, G.C. (1963), *J. Basic Engng* 85, 519-527.
[12] Hussain, M.A., Pu, S.L., and Underwood, I. (1972), *ASTM STP* 560, 2-28.
[13] Strifors, H.C. (1974), *Int. J. Solids Structures* 10, 1389-1404.
[14] Sih, G.C. (1973), In *Mechanics of Fracture* 1 (Edited by G.C. Sih) pp. 21-45.
[15] Sih, G.C. (1974), *Int. J. Fract.* 10, 305-321.
[16] Herrmann, K.P. and Dong, M. (1990), *ZAMM* 70, T 292-294.
[17] Herrmann, K.P. and Dong, M. (1992), *Int. J. Solids Structures* 29, 1789-1812.
[18] Dong, M. (1993), Ph.D. Dissertation, University of Paderborn.
[19] Herrmann, K.P. and Grebner, H. (1982), *Theoret. Appl. Fract. Mech.* 2, 133-155.
[20] Rybicki, E.F. and Kanninen, M.F. (1977), *Engng. Fract. Mech.* 9, 931-938.
[21] Ferber, F., Herrmann, K.P., and Hoppstock, J. (1993), *DVM-Mitteilungen*, 495-504.

ANALYTICAL ESTIMATE OF INTERACTION AMONG ELLIPSOIDAL INCLUSIONS: UPPER AND LOWER BOUNDS FOR STRAIN ENERGY DUE TO INTERACTION

M. HORI
Department of Civil Engineering,
University of Tokyo,
Tokyo 113, Japan

1. Introduction

Recent advancement of material and computer sciences enables one to analyze behavior of members and structures made of composite materials with superior properties, using a large numerical computation. The developed computational mechanics is being applied to highly heterogeneous and inelastic geomaterials to predict behavior of foundations or underground structures. It is essential in such computation to implement the constitutive relations of the material which exhibits anisotropy, inelasticity, nonlinearity or path-dependence. Micromechanical analysis is effective in predicting the constitutive relations since such responses are often due to irreversible deformation, failure or evolution of microconstituents. To reduce the required computational efforts, the analysis ought to lead to a closed-form expression of the constitutive relations.

The evaluation of interaction effects among microconstituents is a key issue in the micromechanical analysis. Eshelby (1957) succeeded to analytically estimate interaction between one inclusion and the surrounding matrix, solving a problem of an infinite body containing an ellipsoidal inclusion (a *single-inclusion problem*). The solution is expressed in a closed-form, and has been applied to various averaging schemes which predict the effective properties; see, for example, Nemat-Nasser and Hori (1993) for references. To evaluate the interaction among plural inclusions, one may need a solution to a problem of a body containing many ellipsoidal inclusions (a *many inclusion problem*), though it is an open question whether this problem can be solved analytically or not. The prediction of the effective properties, however, may not need an exact solution of the problem. An approximate but sufficiently accurate solution is preferable if it is given in a simple closed-form. For such a solution, the error due to the approximation should be estimated.

To predict the effective properties of a heterogeneous material, Nemat-Nasser and Hori (1993) has found *universal bounds* that give a range of the average strain energy

R. Pyrz (ed.), IUTAM Symposium on Microstructure-Property Interactions in Composite Materials, 141–151.

stored in any arbitrary finite body when subjected to various boundary conditions. The universal bounds are rigorously computed by using a functional for eigenstress fields. It is shown that one can set the functional such that it has either the maximum or the minimum value, which equals the average strain energy of the body.

Applying the universal bounds to the many inclusion problem, we seek to evaluate upper and lower bounds for the strain energy caused by the interaction which acts among the ellipsoidal inclusions. To this end, first, a functional which bounds the strain energy of the many inclusion problem is derived from the universal bounds in Section 2. To compute the functional analytically, we solve the single inclusion problem in Section 3, using an equivalent inclusion method which is formulated in a more general setting. The solution of the single inclusion problem is expressed in terms of a characteristic function for the ellipsoidal inclusion and a function which is defined outside of the inclusion and decays in the farfield. In Section 4, using these functional and the solution, we obtain suitable approximate solutions of the many inclusion problem. It is shown that upper and lower bounds for the strain energy can be analytically computed from the approximate solutions.

2. Universal Inequalities

A many inclusion problem considered here is as follows: an infinite elastic body, denoted by B, contains plural ellipsoidal inclusions, Ω^α's, embedded in matrix M. This body is subjected to farfield strains or stresses ($\epsilon^\infty = C^M : \sigma^\infty$). Each Ω^α or M is uniform, and has a distinct elasticity tensor, C^α or C^M, respectively; see Fig. 1. $C(x) = \sum_\alpha \varphi^\alpha(x)C^\alpha + \varphi^M(x)C^M$ is the elasticity tensor field of B, where $\varphi^{(\cdot)}(x)$ is the characteristic function of domain (.). Superscript (.) designates that the quantity is associated with domain (.). Note that (.) is used to denote a domain as well as its volume or shape. The displacement, traction, strain and stress fields are denoted by $u(x)$, $t(x)$, $\epsilon(x)$ and $\sigma(x)$, respectively.

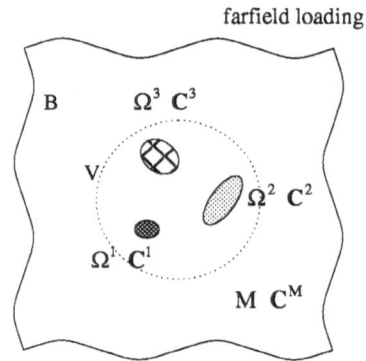

Fig. 1: Many Inclusion Problem

For simplicity, we take a finite region, V, which contains all Ω^α in it, and regard B as the limit of V expanding infinitely. The boundary conditions for V are homogeneous stress or uniform traction ($t(x) = \nu(x) \cdot \sigma^\circ$ on ∂V with constant σ°). According to the *universal inequalities* established by Nemat-Nasser and Hori (1993), the strain field of this problem, denoted by $\epsilon^{exact}(x)$, produces the minimum total strain energy among all compatible strain fields that have the same volume average as the average of $\epsilon^{exact}(\boldsymbol{x})$. That is,

$$0 \geq \int_V \frac{1}{2} \, \epsilon^{exact} : \boldsymbol{C} : \epsilon^{exact} \, dV - \int_V \frac{1}{2} \, \epsilon : \boldsymbol{C} : \epsilon \, dV, \tag{1}$$

where $\epsilon(\boldsymbol{x})$ is a compatible strain field which satisfies $< \epsilon >_V =< \epsilon^{exact} >_V$.

To obtain a suitable strain field, we apply the equivalent inclusion method. Introducing an infinite homogeneous body, denoted by B^*, we consider its subregion V^* (which is the same shape as V). The elasticity tensor of B^* or V^* is $\boldsymbol{C}^o$ and an eigenstress field $\boldsymbol{s}^*(\boldsymbol{x})$ is prescribed in B^* such that the constitutive relations become $\boldsymbol{\sigma}(\boldsymbol{x}) = \boldsymbol{C}^o : \epsilon(\boldsymbol{x}) + \boldsymbol{s}^*(\boldsymbol{x})$. For given $\boldsymbol{s}^*(\boldsymbol{x})$, we can formally solve this subsidiary problem. Using the Green function of B^*, we define an integral operator, $\boldsymbol{\Gamma}^o(\boldsymbol{x}; \boldsymbol{s}^*)$, which gives the disturbance strain field produced by $\boldsymbol{s}^*(\boldsymbol{x})$, and denote the strain field in B^* by $\epsilon(\boldsymbol{x}) = \epsilon^\infty + \boldsymbol{\Gamma}^o(\boldsymbol{x}; \boldsymbol{s}^*)$. The consistency condition is to choose an eigenstress field that makes the fields of the subsidiary problem coincide with those of the original problem. Denoting this eigenstress field by $\boldsymbol{\sigma}^*(\boldsymbol{x})$, we can write the condition as $\epsilon^\infty + \boldsymbol{\Gamma}^o(\boldsymbol{x}; \boldsymbol{\sigma}^*) - (\boldsymbol{C}(\boldsymbol{x}) - \boldsymbol{C}^o)^{-1} : \boldsymbol{\sigma}^*(\boldsymbol{x}) = \boldsymbol{O}$; see Nemat-Nasser and Hori (1993). The strain field of the original many inclusion problem is then given as $\epsilon^{exact} = \epsilon^\infty + \boldsymbol{\Gamma}^o(\boldsymbol{x}; \boldsymbol{\sigma}^*)$.

Using $\boldsymbol{C}^-$ which makes $\boldsymbol{C}(\boldsymbol{x}) - \boldsymbol{C}^-$ negative-semi-definite for all $\boldsymbol{x}$, we can obtain the following inequality from the consistency condition:

$$0 \geq \int_V \frac{1}{2} \, (\epsilon^\infty + \boldsymbol{\Gamma}^- - \hat{\boldsymbol{\epsilon}}) : (\boldsymbol{C} - \boldsymbol{C}^-) : (\epsilon^\infty + \boldsymbol{\Gamma}^- - \hat{\boldsymbol{\epsilon}}) \, dV \tag{2}$$

where $\hat{\boldsymbol{\epsilon}}(\boldsymbol{x}) = (\boldsymbol{C}(\boldsymbol{x}) - \boldsymbol{C}^o)^{-1} : \boldsymbol{s}^*(\boldsymbol{x})$. Superscript $-$ of $\boldsymbol{\Gamma}^-$ emphasizes that the reference elasticity tensor of B^* is $\boldsymbol{C}^-$. Equality in Eq. (2) holds only for $\boldsymbol{\sigma}^*(\boldsymbol{x})$. Note that $\epsilon^\infty + \boldsymbol{\Gamma}^-(\boldsymbol{x}; \boldsymbol{s}^*)$ in Eq. (2) is compatible for other eigenstress fields, and can be used in Eq. (1).

In the many inclusion problem, the change of the total strain energy due to the presence of Ω^α's is

$$W = \lim_{V \to \infty} \int_V \frac{1}{2} \, \epsilon^{exact} : \boldsymbol{C} : \epsilon^{exact} - \frac{1}{2} \, \epsilon^\infty : \boldsymbol{C}^M : \epsilon^\infty dV. \tag{3}$$

Taking the sum of Eq. (1) and Eq. (2) and its limit as $V \to \infty$, we can obtain $0 \geq W + J(\boldsymbol{s}^*; \boldsymbol{C}^-)$, where J is a functional for $\boldsymbol{s}^*(\boldsymbol{x})$ defined as

$$J(\boldsymbol{s}^*; \boldsymbol{C}^o) = \lim_{V \to \infty} \int_V \frac{1}{2} \, \boldsymbol{s}^* : \left((\boldsymbol{C} - \boldsymbol{C}^o)^{-1} : \boldsymbol{s}^* - \boldsymbol{\Gamma}^o(\boldsymbol{s}^*) - 2\epsilon^\infty \right)$$

$$- \epsilon^\infty : \boldsymbol{C}^o : \boldsymbol{\Gamma}^o(\boldsymbol{s}^*) - \frac{1}{2} \, \boldsymbol{\Gamma}^o(\boldsymbol{s}^*) : (\boldsymbol{C}^o : \boldsymbol{\Gamma}^o(\boldsymbol{s}^*) + \boldsymbol{s}^*)$$

$$+ \frac{1}{2} \, \epsilon^\infty : (\boldsymbol{C}^M - \boldsymbol{C}^o) : \epsilon^\infty \, dV. \tag{4}$$

144

Inequality $0 \geq W + J(\boldsymbol{s}^*; \boldsymbol{C}^-)$ is *universal* in the sense that it applies to V of any arbitrary microstructure[1]. Although J is the essentially same form as the one used in the Hashin-Shtrikman variational principle (1962), it does not assume the statistical homogeneity[2] for V.

Another universal inequality, $0 \leq W + J(\boldsymbol{s}^*; \boldsymbol{C}^+)$, can be derived in a similar manner. The major difference is that we use the change of the complementary strain energy,

$$\lim_{V \to \infty} \int_V \frac{1}{2} \boldsymbol{\sigma}^{exact} : (\boldsymbol{C})^{-1} : \boldsymbol{\sigma}^{exact} - \frac{1}{2} \boldsymbol{\sigma}^\infty : (\boldsymbol{C}^M)^{-1} : \boldsymbol{\sigma}^\infty dV,$$

where $\boldsymbol{\sigma}^{exact}(\boldsymbol{x})$ is the stress field when V is subjected to homogeneous strain boundary or linear displacement conditions ($\boldsymbol{u}(\boldsymbol{x}) = \boldsymbol{x} \cdot \boldsymbol{\epsilon}^o$ on ∂V with constant $\boldsymbol{\epsilon}^o$), and that $\boldsymbol{C}^+$ makes $\boldsymbol{C}(\boldsymbol{x}) - \boldsymbol{C}^+$ positive-semi-definite; see Nemat-Nasser and Hori (1993). In the limit as $V \to \infty$, the solution of the two boundary-value problems coincide if $\boldsymbol{\sigma}^\infty = \boldsymbol{C}^M : \boldsymbol{\epsilon}^\infty$, and the change of the complementary strain energy becomes W. Therefore, we obtain

$$- J(\boldsymbol{s}^{*+}; \boldsymbol{C}^+) \leq W \leq -J(\boldsymbol{s}^{*-}; \boldsymbol{C}^-), \tag{5}$$

where superscript $+$ or $-$ for $\boldsymbol{s}^*$ emphasizes that it is considered in B^* of $\boldsymbol{C}^-$ or $\boldsymbol{C}^+$, respectively. As J uses an eigenstress field which is *closer* to the one that satisfies the consistency condition, $J(\boldsymbol{s}^{*+}; \boldsymbol{C}^-)$ or $J(\boldsymbol{s}^{*-}; \boldsymbol{C}^+)$ becomes a sharper bound for W.

3. Eshelby's Solution and Complementary Eshelby's Solution

For the computation of J, we need an eigenstress field which is close to the solution of the consistency condition and is given in a closed form such that the resulting fields can be analytically computed. To this end, we solve a single inclusion problem, applying the equivalent inclusion method which is formulated in the manner as presented in the last section. A class of eigenstress fields that satisfy the above requirements are then obtained by considering a subsidiary problem which uses B^* of $\boldsymbol{C}^o \neq \boldsymbol{C}^M$; see Fig. 2. To simplify the expression in the following discussion, we omit the arguments of a field or a function if it does not make confusion.

The original problem is an infinite domain B which contains one ellipsoidal inclusion, Ω, and the subsidiary problem is an infinite homogeneous domain B^* which has the elasticity tensor $\boldsymbol{C}^o$ and an eigenstress field $\boldsymbol{s}^*$. Noting that a strain field in B^* is $\boldsymbol{\epsilon}^\infty + \boldsymbol{\Gamma}^o$, we write the consistency condition as

$$\boldsymbol{C}(\boldsymbol{x}) : (\boldsymbol{\epsilon}^\infty + \boldsymbol{\Gamma}^o(\boldsymbol{x}; \boldsymbol{\sigma}^*)) = \boldsymbol{C}^o : (\boldsymbol{\epsilon}^\infty + \boldsymbol{\Gamma}^o(\boldsymbol{x}; \boldsymbol{\sigma}^*)) + \boldsymbol{\sigma}^*(\boldsymbol{x}), \tag{6}$$

---

[1] One may obtain the same functional as J when a unit cell of a periodic structure is considered instead of V in B; see Nemat-Nasser and Hori (1993).

[2] See Francfort and Murat (1986), Milton and Kohn (1988), and Torquato (1991) for the statistical homogeneity and the related variational technique.

single inclusion problem

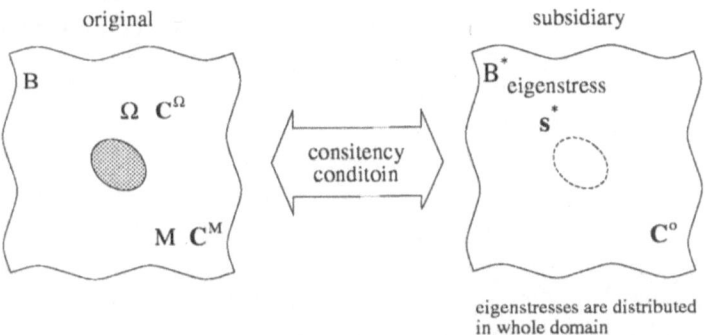

Fig. 2: Equivalent Inclusion Method Applied to Single Inclusion Problem

where $C(\boldsymbol{x}) = \varphi^{\Omega}(\boldsymbol{x})C^{\Omega} + \varphi^{M}(\boldsymbol{x})C^{M}$, and M denotes the domain outside of Ω.

In terms of an eigenstress, Eshelby's solution can be expressed as follows: a uniform distribution of eigenstresses, $\boldsymbol{\sigma}^{*}$, in the ellipsoidal Ω (i.e., $\varphi^{\Omega}(\boldsymbol{x})\boldsymbol{\sigma}^{*}$) produces

$$\boldsymbol{\Gamma}^{o}(\boldsymbol{x};\varphi^{\Omega}\boldsymbol{\sigma}^{*}) = \left(\varphi^{\Omega}(\boldsymbol{x})\boldsymbol{p}^{\Omega}(C^{o}) + \varphi^{M}(\boldsymbol{x})\boldsymbol{q}^{\Omega}(\boldsymbol{x};C^{o})\right) : \boldsymbol{\sigma}^{*}, \qquad (7)$$

where $\boldsymbol{p}^{\Omega}(C^{o})$ and $\boldsymbol{q}^{\Omega}(\boldsymbol{x};C^{o})$ are symmetric fourth-order tensors which give uniform strain field in Ω and a decaying strain field out of Ω. They can be explicitly determined from the ellipsoidal shape of Ω and the reference elasticity C^{o}; see Mura (1987) for the detailed explanation of Eshelby's solution[3]. This solution was found for an anisotropic C^{o} as well as an isotropic one; see, for example, Willis (1964).

Taking advantage of Eq. (7), we can solve Eq. (6) for the case of $C^{o} = C^{M}$, and obtain $\boldsymbol{\sigma}^{*} = \varphi^{\Omega}\boldsymbol{A} : \boldsymbol{\epsilon}^{\infty}$ with $\boldsymbol{A} = ((C^{\Omega} - C^{M})^{-1} - \boldsymbol{p}^{\Omega}(C^{M}))^{-1}$. In the original problem, the strain field due to the presence of Ω is then given by

$$\boldsymbol{\epsilon}^{d}(\boldsymbol{x}) = \boldsymbol{\Gamma}^{M}(\boldsymbol{x};\varphi^{\Omega}\boldsymbol{A} : \boldsymbol{\epsilon}^{\infty}),$$

where superscript M for $\boldsymbol{\Gamma}$ emphasizes $C^{o} = C^{M}$.

For the case of $C^{o} \neq C^{M}$, due to the uniqueness of the solution, the eigenstress field that satisfies Eq. (6), $\boldsymbol{\sigma}^{*}(\boldsymbol{x})$, must produces the same strain field as $\boldsymbol{\epsilon}^{\infty} + \boldsymbol{\epsilon}^{d}$, i.e., $\boldsymbol{\Gamma}^{o}(\boldsymbol{x};\boldsymbol{\sigma}^{*}) = \boldsymbol{\epsilon}^{d}(\boldsymbol{x})$. Therefore, $\boldsymbol{\sigma}^{*}(\boldsymbol{x})$ can be expressed in terms of $\boldsymbol{\epsilon}^{d}$ as $\boldsymbol{\sigma}^{*} = (C - C^{o}) : (\boldsymbol{\epsilon}^{\infty} + \boldsymbol{\epsilon}^{d})$. This eigenstress field[4] consists of the following three: a homogeneous

---

[3]In the two dimensional setting, the solution of an elliptical inclusion problem can be used instead of Eshelby's solution, and $\boldsymbol{p}^{\Omega}$ and $\boldsymbol{q}^{\Omega}$ are computed by using Airy's stress function or complex stress functions.

[4]Recall that $\boldsymbol{\epsilon}^{d}$ is constant in Ω and decays in M.

part, $(C^M - C^o) : \epsilon^\infty$, a constant in Ω, $\varphi^M(C^\Omega - C^M) : \epsilon^\infty + (C^\Omega - C^o) : \epsilon^d$, and a decaying part in M, $(C^M - C^o) : \epsilon^d$. Since the homogeneous part does not produce strains and the constant in Ω produces a field given by Eq. (7), a contribution of the decaying part to the disturbance strain field is determined. Therefore, we can explicitly compute the strain field due to a decaying eigenstress field in M (i.e., $\varphi^M(\boldsymbol{x})(C^M - C^o) : q^\Omega(\boldsymbol{x}; C^M) : \sigma^*$) as

$$\boldsymbol{\Gamma}^o(\boldsymbol{x}; \varphi^M(C^M - C^o) : q^\Omega(C^M) : \sigma^*) = \left(\varphi^\Omega(\boldsymbol{x})(p^\Omega(C^M) - B^\Omega : p^\Omega(C^o)) \right.$$
$$\left. + \varphi^M(\boldsymbol{x})(q^\Omega(\boldsymbol{x}; C^M) - B^\Omega : q^\Omega(\boldsymbol{x}; C^o)) \right) : \sigma^* \quad (8)$$

where $B^\Omega = I + (C^M - C^o) : p^\Omega(C^M)$. If Eq. (8) is used as well as Eq. (7), we can directly solve Eq. (6) for any arbitrary C^o. In this sense, we call[5] Eq. (8) *complementary Eshelby's solution*; see Fig. 3.

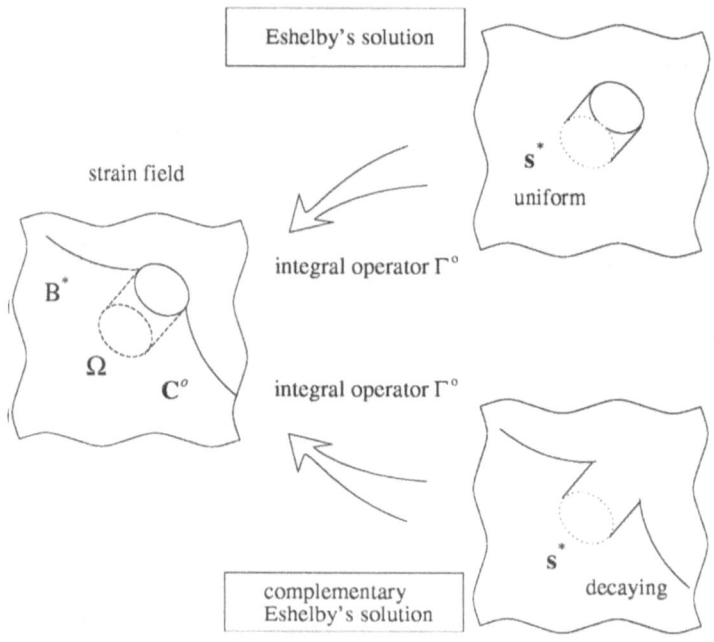

Fig. 3: Eshelby's Solution and Complementary Eshelby's Solution

---

[5]The complementary Eshelby solution uses p^Ω's and q^Ω's which are measured for C^o and C^M. A simpler equation is derived by setting $C^M = 2C^o$ and using $p^\Omega(2C^o) = p^\Omega(C^o)/2$ in Eq. (8). It is expressed as $\boldsymbol{\Gamma}^o(\varphi^M C^o : q^\Omega : \sigma^*) = 2(I + C^o : p^\Omega) : (\varphi^\Omega p^\Omega + \varphi^M q^\Omega) : \sigma^*$ only in terms of p^Ω and q^Ω measured for C^o.

4. Analytical Estimate of Interaction

Making use of Eshelby's solution and complementary Eshelby's solution, we seek to obtain an approximate solution of the many inclusion problem and to estimate the interaction among the inclusions using J given by Eq. (4). The key issue is the computation of J. If J is analytically computed for a class of eigenstress fields which are expressed in terms of these solutions we can determine the eigenstress field that stationarizes J. This eigenstress field gives the most suitable approximate solution and yields the sharpest bounds for W among the class when C^- or C^+ is used.

4.1. COMPUTATION OF J FOR TRIAL EIGENSTRESS FIELD

First, we consider a class of trial eigenstress fields. In view of the form of the exact eigenstress field for the single inclusion problem, the form of a trial eigenstress field should be $(C^M - C^o) : \epsilon^\infty + \sum_\alpha \varphi^\alpha s^{*\alpha 1} + (1 - \varphi^\alpha)(C^M - C^o) : q^\alpha (C^M) : s^{*\alpha 2}$ with a set of (constant) eigenstresses, $\{s^{*\alpha 1}, s^{*\alpha 2}\}$. To reduce the number of eigenstresses half, we set $s^{*\alpha 1} = B^\alpha : s^{*\alpha 2}$. Rewriting $s^{*\alpha 12} = s^{*\alpha}$, we have an eigenstress field of the following form:

$$s^* = (C^M - C^o) : \epsilon^\infty + \sum_\alpha \left(\varphi^\alpha B^\alpha + (1 - \varphi^\alpha)(C^M - C^o) : q^\alpha \right) : s^{*\alpha}. \qquad (9)$$

The strain and stress fields produced by this trial eigenstress field are

$$\epsilon = \epsilon^\infty + \sum_\alpha (\varphi^\alpha p^\alpha + (1 - \varphi^\alpha) q^\alpha) : s^{*\alpha}, \qquad (10)$$

$$\sigma = C^M : \epsilon^\infty + \sum_\alpha \left(\varphi^\alpha (C^M : p^\alpha + I) + (1 - \varphi^\alpha) C^M : q^\alpha \right) : s^{*\alpha}. \qquad (11)$$

As is seen, these fields[6] are analytically expressed in terms of p^α's and q^α's which are measured for C^M.

Next, we compute J for this class of eigenstress fields, taking advantage of the following two properties of Γ^o: 1) $\int_B \Gamma^o(s^*) : (C^o : \Gamma^o(s^*) + s^*) dV = 0$ for $s^*(x)$ which vanishes sufficiently fast at the farfield; and 2) $\int_V \Gamma^o(s^*) dV = (finite)$ for any sufficiently large ellipsoidal V (Nemat-Nasser and Hori (1993)). Substituting a trial eigenstress field into the right side of Eq. (4) and computing the integration of the second and third terms, we can obtain

$$\sum_\alpha \int_{\Omega^\alpha} \frac{1}{2} s^* : \left((C^\alpha - C^o)^{-1} : s^* - \epsilon^d - 2\epsilon^\infty \right) + \frac{1}{2} \epsilon^\infty : (C^M - C^o) : \epsilon^\infty dV$$

$$+ \frac{1}{2} \epsilon^\infty : (C^M + C^o) : \lim_{V \to \infty} \left(\int_V \epsilon^d dV \right),$$

where $\epsilon^d = \sum_\alpha (\varphi^\alpha p^\alpha + (1 - \varphi^\alpha) q^\alpha) : s^{*\alpha}$.

---

[6]Note that $\sigma = C^M : \epsilon$ in $M = V - \sum_\alpha \Omega^\alpha$ and $\sigma = C^M : \epsilon + s^*$ in Ω^α.

The integration in the last term is analytically computed by applying the Tanaka-Mori theorem[7] (1972), which are stated in terms of p^Ω and q^Ω as follows: for any ellipsoidal V including Ω, $\int_V \varphi^\circ p^\Omega + (1 - \varphi^\circ)q^\Omega dV = \Omega p^V$. Therefore, the last term becomes $p^V : \sum_\alpha s^{*\alpha}$, if the shape of V is ellipsoidal and remains the same as it expands. Functional J is now reduced to the following function of $\{s^{*\alpha}\}$:

$$J^s(\{s^{*\alpha}\}; C^\circ) = \sum_\alpha \int_{\Omega^\alpha} \frac{1}{2} s^* : \left((C^\alpha - C^\circ)^{-1} : s^* - \epsilon^d - 2\epsilon^\infty\right)$$

$$+ \frac{1}{2} \epsilon^\infty : (C^M - C^\circ) : \epsilon^\infty dV$$

$$+ \frac{1}{2} \epsilon^\infty : (C^M + C^\circ) : p^V : (\sum_\alpha \Omega^\alpha s^{*\alpha}). \tag{12}$$

4.2. APPROXIMATE SOLUTION AND BOUNDS FOR STRAIN ENERGY

The derivative of J^s with respect to $s^{*\alpha}$'s is computed from Eq. (12). The eigenstresses that stationarize J^s, denoted by $\{\sigma^{*\alpha}\}$, then satisfy the following set of tensorial equations:

$$\sum_\beta \int_{\Omega^\beta} (\varphi^\alpha B^\alpha + (1 - \varphi^\alpha)q^\alpha : (C^M - C^\circ)) : \left((C^\alpha - C^\circ)^{-1} : s^* - \epsilon^d - \epsilon^\infty\right) dV$$

$$- \frac{1}{2} \Omega^\alpha p^V : (C^M + C^\circ) : \epsilon^\infty = 0, \tag{13}$$

for $\alpha = 1, 2, \dots$. Terms in Eq. (13) can be analytically computed except for the volume integration[8] of q^β's over Ω^α. Substituting $\{\sigma^{*\alpha}\}$ into Eqs. (10,11), we can obtain the most suitable approximate strain and stress fields among trial eigenstress fields.

It follows from Eq. (5) that when $C^\circ = C^-$ or C^+, J^s computed for $\{\sigma^{*\alpha}\}$ yields the sharpest upper or lower bound for W, respectively. That is,

$$- J^s(\{\sigma^{*\alpha+}\}; C^+) \leq W \leq -J^s(\{\sigma^{*\alpha-}\}; C^-). \tag{14}$$

This inequality can be applied to a many inclusion problem for any set of ellipsoidal inclusions of arbitrary shape and elasticity which are arranged in an arbitrary manner. As mentioned, if $\{\sigma^{*\alpha+}\}$ or $\{\sigma^{*\alpha-}\}$ is closer to the exact eigenstress field, J^s provides a sharper bound for W. Therefore, $J^s(\{\sigma^{*\alpha+}\}; C^+) - J^s(\{\sigma^{*\alpha-}\}; C^-)$ can be used as a measure of the error of the approximate solutions that are given by substituting $\{\sigma^{*\alpha-}\}$ and $\{\sigma^{*\alpha+}\}$ into Eqs. (10,11).

Since p^V depends on V, the last term in Eq. (13) represents the effects of the surrounding ellipsoidal matrix on the inclusions. If this term is neglected, Eq. (13)

---

[7]See Nemat-Nasser and Hori (1993) for the detailed derivation of the Tanaka-Mori theorem.
[8]One may evaluate this integral using the mean value theorem.

becomes

$$\int_{\Omega^\alpha} \boldsymbol{B}^\alpha : \left((\boldsymbol{C} - \boldsymbol{C}^o)^{-1} : \boldsymbol{s}^* - \boldsymbol{\epsilon}^d - \boldsymbol{\epsilon}^\infty \right) dV$$

$$+ \sum_{\beta \neq \alpha} \int_{\Omega^\beta} \boldsymbol{q}^\alpha : (\boldsymbol{C}^M - \boldsymbol{C}^o) : \left((\boldsymbol{C} - \boldsymbol{C}^o)^{-1} : \boldsymbol{s}^* - \boldsymbol{\epsilon}^d - \boldsymbol{\epsilon}^\infty \right) dV = \boldsymbol{O}.$$

This is the volume average of the consistency conditions, $(\boldsymbol{C} - \boldsymbol{C}^o) : \boldsymbol{s}^* - \boldsymbol{\epsilon} = \boldsymbol{O}$, weighted by $\boldsymbol{B}^\alpha$ in Ω^α and by $\boldsymbol{q}^\alpha : (\boldsymbol{C}^M - \boldsymbol{C}^o)$ in other Ω^β's. Denoting by $\{\overline{\boldsymbol{\sigma}}^{*\alpha}\}$ the eigenstresses that satisfy the above set of equations[9], we can compute $J^s(\{\overline{\boldsymbol{\sigma}}^*\}; \boldsymbol{C}^o)$ as

$$-\frac{1}{2} \boldsymbol{\epsilon}^\infty : \left(\sum_\alpha \int_{\Omega^\alpha} (\boldsymbol{B}^\alpha - (\boldsymbol{C}^M + \boldsymbol{C}^o) : \boldsymbol{p}^V) : \overline{\boldsymbol{\sigma}}^{*\alpha} + \sum_{\beta \neq \alpha} \boldsymbol{q}^\beta : \overline{\boldsymbol{\sigma}}^{*\beta} dV \right).$$

Although $J^s(\{\overline{\boldsymbol{\sigma}}^{*\alpha}\}; \boldsymbol{C}^o)$ is not optimum, it still provides an upper or lower bound for W when $\boldsymbol{C}^o = \boldsymbol{C}^-$ or $\boldsymbol{C}^+$.

4.3. CASE OF INCLUSIONS OF COMMON SHAPE AND ELASTICITY

As an illustrative example, we consider a simple case when all Ω^α's are of common shape and elasticity tensor, denoted by Ω and $\boldsymbol{C}^\Omega$. Setting $\boldsymbol{s}^{*\alpha} = \boldsymbol{s}^*$ for all α's, we can rewrite J^s given by Eq. (12) as a function of $\boldsymbol{s}^*$, and reduce Eq. (13) to a tensorial equation for $\boldsymbol{s}^*$. If the term which includes $\boldsymbol{p}^V$ is neglected, the solution of this equation is

$$\overline{\boldsymbol{\sigma}}^* = \left(\sum_\alpha \int_{\Omega^\alpha} \boldsymbol{Y}^\alpha : (\boldsymbol{A}^{-1} - \boldsymbol{Q}^o) dV \right)^{-1} : \left(\sum_\beta \int_{\Omega^\beta} \boldsymbol{Y}^\beta dV \right) : \boldsymbol{\epsilon}^\infty,$$

where $\boldsymbol{Q}^\alpha = \sum_{\beta \neq \alpha} \boldsymbol{q}^\beta$ and

$$\boldsymbol{Y}^\alpha = \left(\boldsymbol{B}^\Omega + \sum_{\beta \neq \alpha} \boldsymbol{Q}^\beta : (\boldsymbol{C}^M - \boldsymbol{C}^o) \right) : (\boldsymbol{C}^\Omega - \boldsymbol{C}^o)^{-1} : (\boldsymbol{C}^\Omega - \boldsymbol{C}^M).$$

If $\boldsymbol{Q}^\alpha$'s are omitted, $\overline{\boldsymbol{\sigma}}^*$ coincides with the exact solution of the single inclusion problem. Therefore, $\boldsymbol{Q}^\alpha$ represents the strains which are produced in Ω^α by the existence of other inclusions, and the contribution of $\boldsymbol{Q}^\alpha$'s on J^s represents the interaction effects among the inclusions.

Substituting $\overline{\boldsymbol{\sigma}}^*$ into Eqs. (10,11), we can obtain the approximate strain and stress fields as

$$\boldsymbol{\epsilon} = \boldsymbol{\epsilon}^\infty + \sum_\alpha (\boldsymbol{p}^\Omega + \boldsymbol{Q}^\alpha) : \overline{\boldsymbol{\sigma}}^*,$$

$$\boldsymbol{\sigma} = \boldsymbol{C}^M : \boldsymbol{\epsilon}^\infty + \sum_\alpha (\boldsymbol{C}^M : (\boldsymbol{p}^\Omega + \boldsymbol{Q}^\alpha) + \boldsymbol{I}) : \overline{\boldsymbol{\sigma}}^*.$$

---

[9] Instead of the weighted average consistency condition, one may use the average consistency condition over each Ω^α to determine the eigenstresses.

Function J^s for $\overline{\sigma}^*$ becomes

$$-\frac{1}{2}\,\epsilon^\infty : \left(\sum_\alpha \int_{\Omega^\alpha} (B^\Omega - (C^M + C^o) : p^V) + Q^\alpha dV\right) : \overline{\sigma}^*.$$

As mentioned, $J^s(\overline{\sigma}^*; C^o)$ provides an upper or lower bound for W when $C^o = C^-$ or C^+, although it is not the sharpest among trial eigenstresses. From $J^s(\overline{\sigma}^{*+}; C^+) - J^s(\overline{\sigma}^{*-}; C^-)$, we can measure the error of the approximate strain and stress fields produced by $\overline{\sigma}^{*+}$ and $\overline{\sigma}^{*-}$.

5. Conclusion

For the many inclusion problem, a functional which gives the strain energy is derived from the universal bounds. This functional can be used to bound the change of the strain energy caused by the interaction effects among the ellipsoidal inclusions. Applying the equivalent inclusion method in a general setting, we find complementary Eshelby's solution, and compute the functional analytically. It is shown that a suitable approximate solution of the many inclusion problem is obtained by stationarizing the functional and the bounds for the strain energy are analytically computed.

We briefly mention the application of the many inclusion problem and its approximate solution. The problem is suitable to model a microstructure of a material which contains microconstituents of various kinds, and the effective elasticity of the material can be predicted from the approximate solutions. The advantage of this prediction is 1) the required computation is analytical except for the volume integration of functions; and 2) the error of the estimate can be measured from the difference of the strain energy bounds which are computed from the approximate solutions.

6. References

Eshelby, J. D. (1957), The determination of the elastic field of an ellipsoidal inclusion, and related problems, *Proc. Roy. Soc.*, Vol. A241, 376-396.

Nemat-Nasser, S. and Hori, M. (1993), *Micromechanics: Overall Properties of Heterogeneous Materials*, North-Holland, London.

Francfort, G. A. and Murat, F. (1986), Homogenization and optimal bounds in linear elasticity, *Archive Rat. Mech. and Analysis*, Vol. 94, 307-334.

Hashin, Z. and Shtrikman, S. (1962), On some variational principles in anisotropic and nohomogeneous elasticity, *J. Mech. Phys. Solids*, Vol. 10, 335-342.

Hori, M. and Nemat-Nasser, S. (1993), Double-Inclusion Model and overall moduli of multi-phase composites, *Mech. Mat.*, Vol. 14, 189-206.

Miloton, G. W. and Kohn, R. (1988), Variational bounds on the effective moduli of anisotropic composites, *J. Mech. Phys. Solids*, Vol. 43, 63-125.

Mura, T. (1987), *Micromechanics of Defects in Solids*, Martinus Nijhoff Publisher, Dordrecht.

Tanaka, K. and Mori, T. (1972), Note on volume integrals of the elastic field around an ellipsoidal inclusion, *J. Elasticity*, Vol. 2, 199-200.

Torquato, S. (1991), Random heterogeneous media: microstructure and improved bounds on effective properties, *Appl. Mech. Rev.*, Vol. 42, No.2, 37-76.

Willis, J. R., (1964), Anisotropic elastic inclusion problem, *Q. J. Mech. Appl. Math.*, Vol. 17, 157-174.

EFFECTIVE MODULI OF CONCENTRATED PARTICULATE SOLIDS

B. L. KARIHALOO AND J. WANG
School of Civil and Mining Engineering
The University of Sydney
NSW 2006
Australia

ABSTRACT. The overall elastic constants of a particulate composite material are theoretically estimated. The composite consists of a high concentration of randomly arranged spherical particles embedded in an isotropic elastic matrix. Because of the high concentration of particles (volume fraction close to the maximum possible), the load transfer occurs mainly at the regions of near contact between neighbouring particles. The self-consistent approach is therefore unlikely to give an accurate prediction. It is now necessary to estimate the load transfer between two neighbouring particles separated by a thin layer of matrix material. This has been done in the present paper without placing any restrictions on the rigidity of the particles or on the length of contact zone between them. The latter two limiting cases have been previously solved by Batchelor and O'Brien (1977), Phan-Thien and Karihaloo (1982), and Dvorkin, Mavko and Nur (1991). The results of the present study are applicable in particular to cemented granular materials.

1. Introduction

The determination of the bulk properties of particulate solids has been pursued in two ways. In the first, the emphasis has been placed on improving the accuracy of the results with respect to the volume concentration of the particles. Many results have been obtained which are accurate up to the order $0(\phi)$ or to $0(\phi^2)$, where ϕ is the volume concentration of the embedded particles. These were deduced by ignoring the interaction effect between the particles. If the volume fraction is small enough, the results of Smallwood (1944), Dewey (1947), Walpole (1971) are all accurate up to terms of order $0(\phi)$. Further analysis using the variational approach and the self-consistent

R. Pyrz (ed.), IUTAM Symposium on Microstructure-Property Interactions in Composite Materials, 153–164.
© *1995 Kluwer Academic Publishers.*

method can be found in the works of Hashin and Shtrikman (1963), and Willis (1977).

The second approach is to consider the interaction effects between the particles as for instance by Laws and McLaughlin (1979) for the analogous fibre-reinforced problem who used the self-consistent method developed by Hill (1963). This method approximates the interaction effects between particles by solving the elasticity problem of one particle enclosed in a spherical shell of the effective composite. Chen and Acrivos (1978), based upon the elasticity solution of two particles in an infinite region, gave the effective elastic moduli accurate up to terms of order $0(\phi^2)$.

When the particle concentration is high, the load transfer occurs mainly at the regions of near contact between the neighbouring particles. In this case, the self-consistent method is unlikely to give an accurate prediction. Dvokin, Mavko and Nur (1991) considered the load transfer problem of two spherical particles with a layer of cement between them. They gave the elastic moduli of the cemented system when the length of the cement layer was prescribed and remained constant. However, the method developed by Batchelor and O'Brien (1977) that makes explicit use of the load transfer characteristics is expected to be more accurate for the problem of high concentration particles close to the maximum volume fraction and when the length of contact zone is a variable. In this paper an attempt is made to determine the overall elastic constants of a particulate material. The material consists of a high concentration of randomly arranged particles embedded in an isotropic elastic matrix. In view of the high concentration, the particles are assumed to be nearly in contact with each other. We shall confine our attention to the situation in which both the ratio of Young's modulus of the particles to that of the matrix and the volume concentration of the particles are high. The method adopted in this paper closely follows the procedure used by Phan-Thien and Karihaloo (1982) in their study of the limiting case when the modulus ratio tends to infinity.

2. Mathematical Preliminaries

As is customary in the suspension mechanics of materials with random structure, we shall assume that the particulate solid is statistically homogeneous so that ensemble averages can be replaced by corresponding volume averages. Thus, the bulk stress tensor is given by

$$< \sigma_{ij} >= \frac{1}{V} \int_{V_m} \sigma_{ij} \, dV + \frac{1}{V} \sum_\alpha \int_{V_\alpha} \sigma_{ij} \, dV \qquad (1)$$

where V is a representative volume element (Hill, 1963), V_m, the matrix volume in V and V_α is the volume of particle α in V, such that

$$V_m + \sum_\alpha V_\alpha = V \tag{2}$$

In eqn (1) σ_{ij} are the local stresses, namely

$$\sigma_{ij}(\mathbf{x}) = \begin{cases} \lambda \varepsilon_{kk} \delta_{ij} + 2\mu \varepsilon_{ij}, & \mathbf{x} \text{ inside the matrix material,} \\ \lambda^* \varepsilon_{kk} \delta_{ij} + 2\mu^* \varepsilon_{ij}, & \mathbf{x} \text{ inside the particles,} \end{cases} \tag{3}$$

where λ, μ and λ^*, μ^* are the Lamé constants of the matrix and the particle, respectively.

Based upon the replacement of the ensemble averages by the volume averages of the stress and the strain, the integrals on the right hand side of eqn (1) can be converted to the following expression

$$<\sigma_{ij}> \ = \ (1 - \xi\phi)\lambda < \varepsilon_{kk} > \delta_{ij} + 2(1 - \xi\phi)\mu < \varepsilon_{ij} > +$$
$$+ \frac{1}{V} \sum_\alpha \int_{A_\alpha} x_i \sigma_{jk} n_k \, dA - \Re_{ij} \tag{4}$$

where ϕ is the particle volume fraction $(= \sum_\alpha V_\alpha / V)$ and A_α is the surface of a particle. In eqn (4), the remainder term is given by (with $V_p = \sum_\alpha V_\alpha$)

$$\Re_{ij} = \frac{1}{V} \int_{V_p} \{\lambda(\varepsilon_{kk} - \xi < \varepsilon_{kk} >)\delta_{ij} + 2\mu(\varepsilon_{ij} - \xi < \varepsilon_{ij} >)\} \, dV \tag{5}$$

If $\Re_{ij}$ can be made as small as possible by choosing ξ properly, then it may be neglected from (4) to give

$$<\sigma_{ij}> \ = \ (1 - \xi\phi)\lambda < \varepsilon_{kk} > \delta_{ij} + 2(1 - \xi\phi)\mu < \varepsilon_{ij} > +$$
$$+ \frac{1}{V} \sum_\alpha \int_{A_\alpha} x_i \sigma_{jk} n_k \, dA \tag{6}$$

It now remains to show how $\Re_{ij}$ can be made as small as possible by a proper choice of ξ. It is noted that the stresses at the interfaces between the matrix and the particles are continuous. We approximate this continuity condition by

$$\int_{V_m} \varepsilon_{ij} \, dV = \frac{\mu}{\mu^*} \frac{1 - \phi}{\phi} \int_{V_p} \varepsilon_{ij} \, dV \tag{7}$$

With this approximation, it can be shown that

$$\xi = \frac{1}{(1 - \phi)\frac{\mu}{\mu^*} + \phi} \tag{8}$$

leads to $\Re_{ij} = 0$. In the following we shall use eqn (8) in the average stress (6).

In the limit of a dilute concentration of particle when $\phi \to 0$, the stress field within a particle is unaffected by the presence of other particles. For simple particle shapes, the particle contribution to the stress can be explicitly obtained. For a sphere of radius a we have (Walpole, 1972),

$$\frac{\lambda_c}{\lambda} = 1 + \frac{1}{\lambda}\left[\left(K + \frac{4}{3}\mu\right)\gamma_1 - 10\mu(1-\nu)\gamma_2\right]\phi + 0(\phi^2) \qquad (9)$$

$$\frac{\mu_c}{\mu} = 1 + 15(1-\nu)\gamma_2\,\phi + 0(\phi^2), \qquad (10)$$

where the subscript c denotes a composite property; K is the bulk modulus $(= \lambda + \frac{2}{3}\mu)$ and ν Poisson's ratio. Other parameters in eqns (9) and (10) are defined as follows

$$\gamma_1 = \frac{3K^* - 3K}{3K^* + 4\mu}, \qquad \gamma_2 = \frac{\beta - 1}{2\beta(4 - 5\nu) + 7 - 5\nu}, \qquad \beta = \frac{\mu^*}{\mu} \qquad (11)$$

Accurate expressions for the composite Lamé constants (up to $0(\phi^2)$) have been obtained by Chen and Acrivos (1978) and have been confirmed by O'Brien (1979).

In this paper we are concerned with the opposite limit, namely when the particles are in high concentration and ϕ is close to its maximum value (almost touching particles). We adopt the procedure of a previous paper by Phan-Thien and Karihaloo (1982).

We consider here the simple case of identical elastic particles of radius a embedded in an elastic matrix (Fig. 1). At high concentrations of particles, the load transfer in the composite occurs at the regions of near contact of neighbouring particles. As the magnitude of the transfer load will be large in the vicinity of the regions of near contact, the integrals in (6) can be approximated by the sum of contributions from these regions:

$$\int_{A_\alpha} x_i \sigma_{jk} n_k \, dA = \sum_\beta x_i H_j \qquad (12)$$

where $\{x_i\}$ is the position vector of the contact region β and

$$H_j = \int_{A_{\alpha\beta}} \sigma_{jk} n_k \, dA \qquad (13)$$

in which $A_{\alpha\beta}$ is the area of the contact region β of particle α.

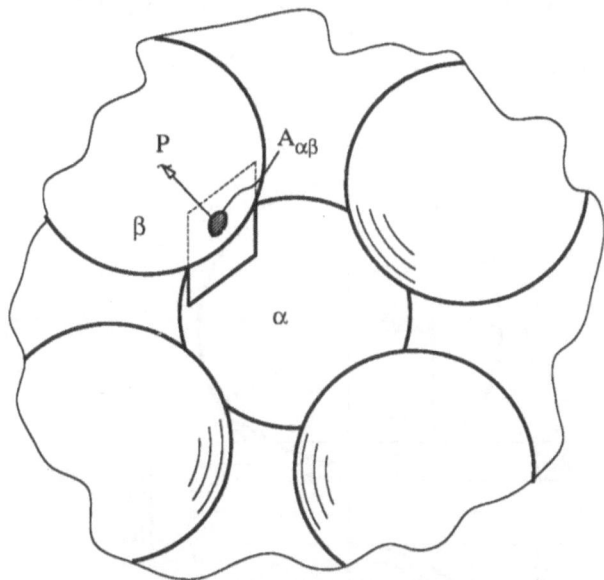

Fig.1 A Typical Microstructure of Particulate Solids

At high concentrations of particles we may replace V by the volume of one particle. Thus

$$< \sigma_{ij} > \;=\; (1 - \xi\phi)\lambda < \varepsilon_{kk} > \delta_{ij} + 2(1 - \xi\phi)\mu < \varepsilon_{ij} > +$$

$$+\frac{3}{4\pi a^3} \sum_{\beta} p_i H_j, \tag{14}$$

where $\{p_i\}$ is unit vector defining the contact region β of the generic particle. Thus, to evaluate the bulk stress tensor, $< \sigma_{ij} >$, we need to evaluate $\{H_j\}$, the load transfer vector in the contact region β. We consider this question in the next section.

3. Load Transfer between two Particles nearly in Contact

We consider here the contact problem of two identical spheres of high Young's modulus (compared to the matrix modulus), see Fig. 2. When the particles are nearly in contact, the thickness of the matrix layer, $2h$, can be approximated by a quadratic function of the radial coordinate r:

$$h = h_0 + r^2/2a + 0(r^4/a^3) \tag{15}$$

where $2h_0$ is the minimum thickness of the matrix layer between the particles.

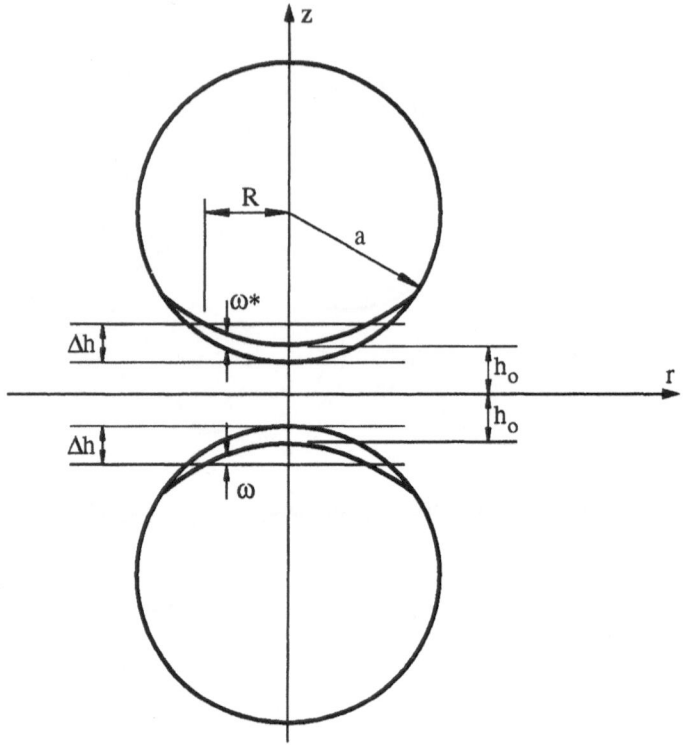

Fig.2 The Load Transfer Problem between two Generic Particles

If the relative displacement of the centres of the spherical particles is $2\Delta h$, we have (Fig. 2)

$$\omega^* + \omega = \Delta h - r^2/2a \qquad (16)$$

where ω^* and ω are given by the Hertzian contact theory (Landau and Lifshitz, 1959):

$$\omega = \frac{1-\nu^2}{\pi E} \int\int_{A_{\alpha\beta}} \frac{q(r')}{(r^2 + r'^2 - 2rr'\cos\theta)^{1/2}} r' \, d\theta dr', \qquad (17)$$

$$\omega^* = \frac{1-\nu^{*2}}{\pi E^*} \int\int_{A_{\alpha\beta}} \frac{q(r')}{(r^2 + r'^2 - 2rr'\cos\theta)^{1/2}} r' \, d\theta dr'. \qquad (18)$$

Here, $A_{\alpha\beta}$ is the contact area; $q(r)$ the normal loading and, as before, an asterisk denotes a particle property.

Let $I\left(\frac{r'}{r}\right)$ represent the following function

$$I\left(\frac{r'}{r}\right) \equiv \frac{1}{\pi} \int_0^{2\pi} \frac{r'}{(r^2 + r'^2 - 2rr'\cos\theta)^{1/2}} \, d\theta = \frac{4r'}{\pi(r+r')} K\left(\frac{2\sqrt{rr'}}{r+r'}\right),$$
(19)

where $K(m)$ is the complete elliptical integral of the first kind ($K(m)$ used here is identical to $K(m^2)$ used by Batchelor and O'Brien (1977)), then from (16)–(18), we get

$$\left(\frac{1-\nu^2}{E} + \frac{1-\nu^{*2}}{E^*}\right) \int_0^R q(r') I\left(\frac{r'}{r}\right) dr' = \Delta h - \frac{r^2}{2a},$$
(20)

where R is the contact radius.

This problem has been studied by Tu and Gazis (1964). However, in calculating the load transfer they regarded h to be constant and thus their treatment does not reflect the physical problem being considered here.

4. Thin Plate Approximation

By Fourier transform techniques, Tu and Gazis (1964) found that, if the matrix layer between the particles was thin enough, the displacement field in the layer corresponded to plane stress deformation. This can also be seen independently from the order-of-magnitude analysis of Phan-Thien and Karihaloo (1982). It can be shown that if the matrix layer between the spherical particles is thin, the integral (17) reduces to

$$\omega(r) = \frac{2(1-\nu^2)}{(2-\nu)E} h \, q(r),$$
(21)

so that eqn (20) becomes

$$\frac{1}{(2-\nu)} khq + \frac{1}{2}k^* \int_0^R q(r') I\left(\frac{r'}{r}\right) dr' = \Delta h - \frac{r^2}{2a},$$
(22)

where

$$k = \frac{2(1-\nu^2)}{E}, \quad k^* = \frac{2(1-\nu^{*2})}{E^*}.$$
(23)

We shall endeavour to find the solution of $q(r)$ by transforming (22) to an expression similar to that obtained by Batchelor and O'Brien (1977). To do this, let us define $f(r)$ by

$$\frac{1}{(2-\nu)} khq = \Delta h[1 - f(r)],$$
(24)

whereupon the integral equation (22) is transformed to

$$f(r) = \frac{(2-\nu)}{2}\frac{k^*}{k}\int_0^R \frac{1-f(r')}{h_0 + r'^2/2a} I\left(\frac{r'}{r}\right) dr + \frac{r^2}{2a\Delta h}. \qquad (25)$$

Nondimensionalizing all linear dimensions by a and denoting $\varepsilon = \frac{k^*}{k}$, eqn (25) can be written as

$$f(r) = (2-\nu)\varepsilon \int_0^{R/a} \frac{1-f(r')}{2h_0 + r'^2} I\left(\frac{r'}{r}\right) dr' + \frac{r^2}{2\Delta h}. \qquad (26)$$

The magnitude of the load transfer vector is

$$H = 2\pi \int_0^R q(r) r\, dr = \frac{2\pi(2-\nu)a^2}{k}\int_0^{R/a} \frac{2r[1-f(r)]}{2h_0 + r^2} dr. \qquad (27)$$

From (24) and (27), the nondimensional load transfer is

$$\aleph = \frac{Hk}{2\pi(2-\nu)a\Delta h} = \int_0^{R/a}\frac{2r[1-f(r)]}{2h_0 + r^2} dr. \qquad (28)$$

5. Overall Moduli of Composite

The bulk stress tensor of the composite, eqn (14), can now be written as

$$\begin{aligned} <\sigma_{ij}> \; = \; & (1-\xi\phi)\lambda <\varepsilon_{kk}>\delta_{ij} + 2(1-\xi\phi)\mu<\varepsilon_{ij}> + \\ & +\frac{3(2-\nu)}{4(1-\nu)}E\sum_\beta \frac{\Delta h}{a}\aleph\, p_i p_j, \end{aligned} \qquad (29)$$

where $\aleph$ is given by (28).

Following the procedure of Phan-Thien and Karihaloo (1982), it can be shown that the constitutive equations of the composite are

$$\begin{aligned} <\sigma_{ij}> \; = \; & (1-\xi\phi)\lambda<\varepsilon_{kk}>\delta_{ij} + 2(1-\xi\phi)\mu<\varepsilon_{ij}> + \\ & +\frac{(2-\nu)}{20(1-\nu^2)}NE\aleph(<\varepsilon_{kk}>\delta_{ij} + 2<\varepsilon_{ij}>), \end{aligned} \qquad (30)$$

or

$$\begin{aligned} <\sigma_{ij}> \; = \; & \left[(1-\xi\phi)\lambda + \frac{2-\nu}{20(1-\nu^2)}NE\aleph\right]<\varepsilon_{kk}>\delta_{ij} + \\ & +2\left[(1-\xi\phi)\mu + \frac{2-\nu}{20(1-\nu^2)}NE\aleph\right]<\varepsilon_{ij}>, \end{aligned} \qquad (31)$$

where N is a constant which depends on the packing mode of the particles. For an isotropic random packing, $N = 6.5$.

The ratios of the moduli of the composite to those of the matrix are

$$\frac{\lambda_c}{\lambda} = (1 - \xi\phi) + \frac{(2 - \nu)(1 - 2\nu)}{20(1 - \nu)\nu} N\aleph \tag{32}$$

$$\frac{\mu_c}{\mu} = (1 - \xi\phi) + \frac{(2 - \nu)}{10(1 - \nu)} N\aleph. \tag{33}$$

In order to obtain the overall moduli of the composite, it is necessary to solve the integral eqn (26). The kernel of this equation has a weak and integrable singularity. It can be solved by iteration.

For an isotropic matrix and particle both of which have a Poisson's ratio of 0.3, we have calculated the ratio of the shear moduli from eqn (33). The results are shown in Figs 3 and 4. Since the primary interest is in the case of high particle concentration and high particle modulus, large values of ϕ and small values of ε are chosen. a/h_0 is approximated by $\phi^{1/3}/(\phi_m^{1/3} - \phi^{1/3})$, where ϕ_m is the maximum possible particle fraction (=0.63) for random packing.

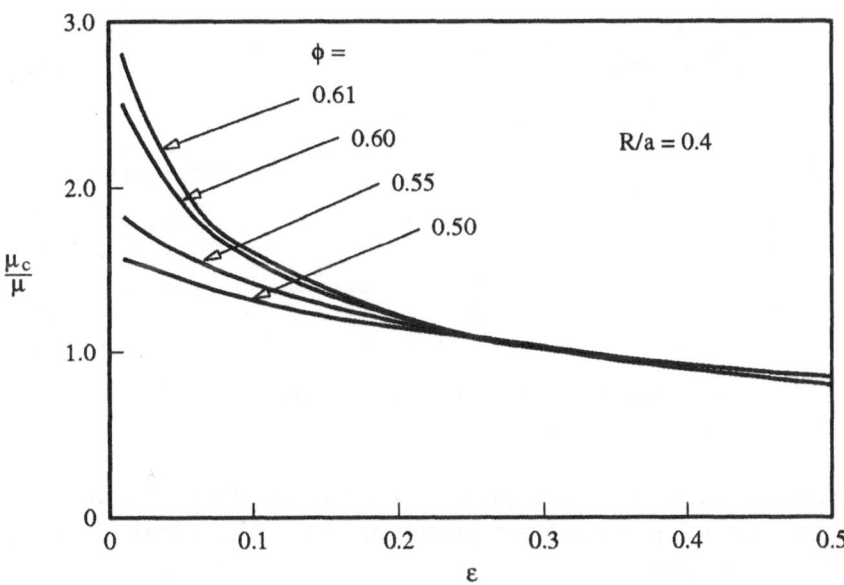

Fig.3 Ratio of the Shear Moduli when R/a = 0.4

Because of the non-linear nature of the contact problem, the elastic properties obtained above depend upon the contact radius R, or the relative displacement Δh, besides the properties of the constituent phases. Figs 3 and 4 show the results for $R/a = 0.4$ and $R/a = 0.5$, respectively. The shear modulus of the composite material increases with the chosen contact radius R. In practice this will mean that the stiffness of the composite will increase with deformation. It is expected that the present calculation will be more accurate when ε is low and ϕ is high. When the stiffnesses of the matrix and the particle approach each other, it is seen that the predicted μ_c/μ can drop below 1.0. The reason for this is that we only take into account the load transferred at the near contact areas between the particles and ignore the load transferred through the matrix surrounding the particle. When the stiffnesses of the matrix and particle are close, load transferred through the matrix will become significant.

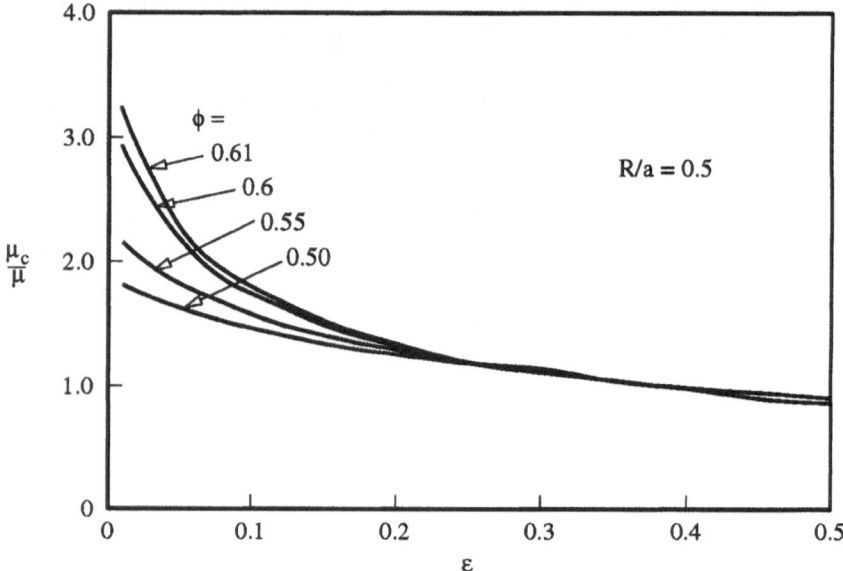

Fig.4 Ratio of the Shear Moduli when R/a = 0.5

The contact model can be approximately linearized by assuming $R/a = 1.0$. In this case, $\omega^* + \omega = \Delta h$ everywhere in eqn (16), so that the integral equation (22) becomes

$$\frac{1}{2-\nu} khq + \frac{1}{2} k^* \int_0^R q(r') I\left(\frac{r'}{r}\right) dr' = \Delta h, \qquad (34)$$

The results obtained from the solution of the above integral equation together with eqn (33) are shown in Fig. 5. Also shown are those calculated using the formula given by Willis and Acton (1976) which is accurate up to the term of $0(\phi^2)$

$$\frac{\mu_c}{\mu} = 1 + \phi \left(\frac{\mu^* - \mu}{\mu^* + \mu'}\right) \left(1 + \frac{\mu'}{\mu}\right) + \phi^2 \left(\frac{\mu^* - \mu}{\mu^* + \mu'}\right)^2 \left(1 + \frac{\mu'}{\mu}\right) +$$

$$+ \frac{1}{20}\phi^2 \big\{(\beta + \gamma)[6\alpha(3\hat{\kappa}) + 8\beta(2\hat{\mu})] + (2\hat{\mu})(28\gamma^2 - \beta^2)\big\}(2\hat{\mu})^2 \left(\frac{\mu^*}{\mu} - 1\right) \quad (35)$$

where α, β, γ, μ', $\hat{\kappa}$ and $\hat{\mu}$ are constants related to the moduli of the matrix and particle.

It is seen that the results calculated from eqns (34) and (35) are fairly close when ϕ is moderate, e.g. $\phi = 0.4$. However, the results diverge for large ϕ. Therefore, it seems that at low particle concentrations the self-consistent model (e.g. Willis and Acton, 1976) gives a more accurate result, but at moderate to high concentrations the present linearized contact model produces a reasonable prediction of the elastic properties of the particulate composite.

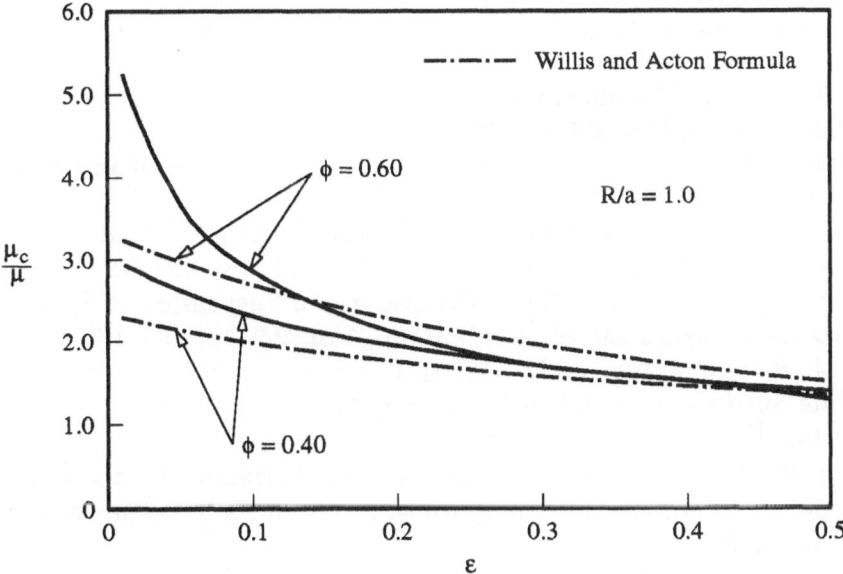

Fig.5 Ratio of the Shear Moduli when R/a = 1.0

6. References

Batchelor, G. K. and O'Brien, R. W. (1977) Thermal or Electrical Conduction through a Granular Material, *Proc. R. Soc. Lond.*, **A355**, pp. 313–333.

Chen, H. S. and Acrivos, A. (1978) The Effective Elastic Moduli of Composite Materials Containing Spherical Inclusions at Non-Dilute Concentrations, *Int. J. Solids Struct.*, **Vol. 14**, pp. 349–364.

Dewey, J. M. (1947) The Elastic Constants of Materials Loaded with Non-Rigid Fillers, *J. Appl. Phys.*, **Vol. 18**, pp. 578–581.

Dvorkin, J., Mavko, G. and Nur, A. (1991) The Effect of Cementation on the Elastic Properties of Granular Material, *Mech. Mater.*, **Vol. 12**, pp. 207–217.

Hashin, Z. and Shtrikman, S. (1963) A Variational Approach to the Theory of the Elastic Behaviour of Multiphase Materials, *J. Mech. Phys. Solids*, **Vol. 11**, pp. 127–140.

Hill, R. (1963) Elastic Properties of Reinforced Solids: Some Theoretical Principles, *J. Mech. Phys. Solids*, **Vol. 11**, pp. 357–372.

Landau, L. D. and Lifshitz, E. M. (1959) *Theory of Elasticity*. Pergamon.

Laws, N. and McLaughlin, R. (1979) The Effect of Fibre Length on the Overall Moduli of Composite Materials, *J. Mech. Phys. Solids*, **Vol. 27**, pp. 1–13.

O'Brien, R. W. (1979) A Method for the Calculation of the Effective Transport Properties of Suspensions of Interacting Particles, *J. Fluid Mech.* **Vol. 91**, pp. 17–39.

Phan-Thien, N. and Karihaloo, B. L. (1982) Effective Moduli of Particulate Solids, *ZAMM*, **Vol. 62**, pp. 183–190.

Smallwood, H. M. (1944) Limiting Law of the Reinforcement of Rubber, *J. Appl. Phys.*, **Vol. 15**, pp. 758–766.

Tu, Y. O. and Gazis, D. C. (1964) The Contact Problem of a Plate Pressed between Two Spheres, *J. Appl. Mech.*, **Vol. 31**, pp. 659–666.

Walpole, L. J. (1971) THe Elastic Behaviour of a Suspension of Spherical Particles, *Quart. J. Mech. Appl. Math.*, **Vol. 25**, pp. 153–160.

Willis, J. R. and Acton, J. R. (1976) The Overall Elastic Moduli of a Dilute Suspension of Spheres, *Quart. J. Mech. Appl. Math.*, **Vol. 29**, pp. 163–177.

Willis, J. R. (1977) Bounds and Self-Consistent Estimates for the Overall Properties of Anisotropic Composites, *J. Mech. Phys. Solids* **Vol. 25**, pp. 185–202.

MICROMECHANICAL ANALYSIS OF THE VISCOELASTIC BEHAVIOR OF COMPOSITES

Patrick Kim, Staffan Toll, and Jan-Anders E. Månson
Laboratoire de Technologie des Composites et Polymères
Ecole Polytechnique Fédérale de Lausanne
CH–1015 Lausanne

1. Introduction

In two-phase materials such as fiber-reinforced composites, the effective behavior of the material is determined not only by the properties of the constituents, but also by their geometry and their arrangement in the composite – the microstructure of the material [1]. When calculating effective properties and local stress or strain fields, it is therefore important to consider the effect of modelling assumptions concerning the microstructure. In analytical approaches, the information on the composite's microstructure is contained implicitly in certain model parameters. For example, in the Halpin-Tsai equations, the empirical fitting parameter is a function of the Poisson's ratio and of the reinforcing phase's geometry [2]. For analyses by finite elements (FE), the microstructure is defined explicitly by the FE mesh. A certain degree of idealization is necessary to keep the problem computationally tractable. This leads to the definition of a model material that is usually not representative of reality. On the other hand, a composite with a "random" fiber distribution is hard to define and a given distribution can be fairly arbitrary. A further difficulty lies in describing local concentrations of fibers or resin. However, certain models do describe experimental data better than others. This was shown by Adams and Tsai, who studied random fiber packings based on periodic arrays of possible fiber positions [1]. Coming closer to observed microstructures, Pyrz has explored ways to quantitatively describe the microstructure of unidirectional composites and the influence it can have on the material properties [3].

The linear viscoelastic behavior of composites can be modelled micromechanically based on the viscoelastic correspondence principle (VCP) [4, 5]. Using the correspondence principle, one can derive the viscoelastic response from the elastic solution by an inverse Laplace transform into the time domain. One can thus derive the composite properties from the constituent properties by micromechanics. However the inverse Laplace transform usually requires a numerical procedure. A useful approximation is the so-called pseudoelastic assumption, which simply states that elastic phase moduli in the elastic solution can be replaced by the corresponding relaxation moduli to obtain the viscoelastic response. It will be shown later on that this approximation can be quite accurate. Unidirectional composites are often modelled using pseudoelastic versions of the Kerner or Halpin-Tsai equations [6] or other micromechanics models [7]. This is mostly for transverse properties, whereas the longitudinal response is usually considered to be approximatively elastic for continuous fibers.

R. Pyrz (ed.), IUTAM Symposium on Microstructure-Property Interactions in Composite Materials, 165–176.
© *1995 Kluwer Academic Publishers.*

166

The direction of loading with respect to the fiber axis in unidirectional laminates has been found to be important when analyzing composites with a stress-dependent matrix behavior in that it affects the mean stress in the matrix [8]. Thus, factors such as the choice of array, which also affect the mean stress, as will be shown here, are also expected to affect the calculated viscoelastic response.

The viscoelastic behavior of composites has been recognized to be too complex to lend itself to an exact micromechanical analysis [9]. Many researchers accordingly rely on the method of cells [10] or FE methods to solve, at the micromechanics level, problems of composite creep or relaxation with various constituent material properties [11] and microstructures [12]. Since the effective properties are influenced by the local microstructure, it is important to understand the dependence of the results on the fiber arrangement chosen for the model.

This paper examines the evolution of the stresse state in a unidirectional composite under transverse tension and the influence of the FE model's unit cell on the calculated effective viscoelastic properties. The viscoelastic response is shown to be related to the initial, elastic stress field in the matrix, and is thus sensitive to parameters such as fiber content and packing. A hybrid (COMP) unit cell is introduced that combines features of the square (SQ) and hexagonal (HCP) arrays. It identifies the elements of elastic solutions which are significant for the viscoelastic behavior and evaluates various analytical approaches based on modified elastic models with respect to the FE results.

2   The influence of stress state on stress relaxation

Consider a linear elastic material

$$\varepsilon'_{ij} = \frac{\sigma'_{ij}}{2G} \quad \text{and} \quad \varepsilon_{kk} = \frac{\sigma_{kk}}{3K} \qquad \text{Eqs. 1}$$

where the primes denote deviatoric components of stress and strain. For a material consisting of two phases m and f of volume fractions $(1-\phi)$ and ϕ, respectively, the total strain is

$$\varepsilon_{ij} = (1-\phi)\overline{\varepsilon}^m_{ij} + \phi\overline{\varepsilon}^f_{ij} \qquad \text{Eq. 2}$$

where the overbars denote volume average. Substituting Eqs. (1) one obtains

$$\varepsilon_{ij} = \frac{1-\phi}{2G_m}\overline{\sigma}'^m_{ij} + \frac{1-\phi}{9K_m}\delta_{ij}\overline{\sigma}^m_{kk} + \frac{\phi}{2G_f}\overline{\sigma}'^f_{ij} + \frac{\phi}{9K_f}\delta_{ij}\overline{\sigma}^f_{kk} \qquad \text{Eq. 3}$$

Material compliances can be obtained by differentiating (3) with respect to the total stress σ_{kl}:

$$S_{ijkl} = \frac{1-\phi}{2G_m}A^m_{ijkl} + \frac{1-\phi}{9K_m}\delta_{ij}B^m_{kl} + \frac{\phi}{2G_f}A^f_{ijkl} + \frac{\phi}{9K_f}\delta_{ij}B^f_{kl} \qquad \text{Eq. 4}$$

where A_{ijkl} and B_{kl} are influence coefficients relating the average deviatoric and isotropic stress in a phase to the total applied stress,

$$\overline{\sigma}'^m_{ij} = A_{ijkl}\sigma_{kl} \ , \quad \overline{\sigma}^m_{kk} = B_{ij}\sigma_{ij} \qquad \text{Eqs. 5}$$

and similarly for the fiber phase. Equation (4) is exact in the elastic case, and applies to stress relaxation within the pseudoelastic approximation, by replacing G_m by $G_m(t)$.

Consider a homogeneous material whose bulk modulus is elastic (constant) and shear modulus is linear viscoelastic. If this material is subjected to a relaxation test in simple tension, the total stress relaxes over time but the ratios between different stress components stay constant, and thus the stress coefficients A_{ijkl} and B_{kk} are temporally constant. In any type of stress state where stress ratios are kept constant the relaxation behavior is entirely dictated by the stress coefficients. The situation in a composite is similar: the phase arrangement influences the relaxation behavior simply through changing the average stress state in the matrix. However, in a composite, the stress state itself may change as the material relaxes, making the stress coefficients time dependent.

A crude relaxation model can be constructed by ignoring this time dependence and assuming the average stress state in the relaxing phase to be constant, i.e. only its magnitude is time dependent. For a two-phase composite with only G_m relaxing this means setting the last three terms in (4) constant, to obtain

$$S_{ijkl}(t) - S^o_{ijkl} = \frac{1-\phi}{2}\left(\frac{1}{G_m(t)} - \frac{1}{G^o_m}\right)A^m_{ijkl} \qquad \text{Eq. 6}$$

giving the modulus

$$\frac{1}{E_1(t)} = \frac{1}{E^o_1} + \frac{1-\phi}{2}\left(\frac{1}{G_m(t)} - \frac{1}{G^o_m}\right)A^m_{1111} \qquad \text{Eq. 7}$$

Other components, such as the Poisson's ratio, may be similarly obtained.

In the particular case of rigid fibers, the stress coefficients can be simply, and exactly, related to overall composite moduli. A rigid fiber phase means that all strain comes from the matrix. Using (1), we obtain

$$\varepsilon'_{ij} = \frac{1-\phi}{2G_m}\bar{\sigma}'^m_{ij} \qquad \text{Eq. 8}$$

differentiation with respect to σ_{kl} yields the stress coefficients

$$A_{ijkl} = \frac{2G_m}{1-\phi}\left(\frac{\partial\varepsilon_{ij}}{\partial\varepsilon_{mn}} - \frac{1}{3}\delta_{ij}\delta_{mn}\right)S_{mnkl} \qquad \text{Eq. 9}$$

In the case of continuous rigid fibers aligned with x_3 (plane strain), $\nu_{31} = 0$ and we can write the particular components

$$A_{1111} = \frac{2G_m(2+\nu_{21})}{3(1-\phi)E_1} \quad \text{and} \quad A_{2211} = -\frac{2G_m(1+2\nu_{21})}{3(1-\phi)E_1} \qquad \text{Eqs. 10}$$

These can be substituted into, e.g., (7) to yield a simple model for relaxation in terms of the initial Poisson's ratio:

$$\frac{E_1^o}{E_1(t)} = 1 + \frac{2 + v_{21}^o}{3}\left(\frac{G_m^o}{G_m(t)} - 1\right)$$

Eq. 11

The same can be done for other components, such as the time-dependent Poisson's ratio.

This result (Eq. 11) suggests that the composite's elastic Poisson's ratio carries the essential information about the matrix's ability to relax, and may thus be used as an indicator of the phase arrangement effect. It will be shown below that the relaxation behavior is indeed a function of v_{21}^o.

3. Finite element models

3.1. FIBER ARRAYS

The most common modelling configurations for transverse properties are the square array, with a 0° (SQ) or 45° (SFC) orientation [13] and the hexagonal (HCP) array [14]. In this study, two regular arrays with a total of four loading configurations were examined (Fig. 1): square (SQ), face-centered square (SFC), hexagonal loaded along the densest axis (HCPy), and hexagonal loaded along the least dense axis (HCPx). Furthermore, a hybrid (COMP) array consisting of a combination of features from the HCP and SQ elements was analyzed (Fig. 1). This model has a broader distribution of interfiber spacings than the SQ and HCP models, since the assembly of the different constituent cells leads to much smaller as well as significantly larger interfiber distances. It is thus more "random" than a simple linear series of different unit cells. The main objective of the comparison between these arrays was to clarify the influence of the packing on the stresses in the matrix and thus on the calculated elastic and viscoelastic properties. The finite element program ABAQUS was used for the solution of the associated boundary value problems.

The FE meshes consist of six-node, triangular plane-strain elements with a quadratic interpolation function. A mesh refinement of four elements through the thickness of the matrix between two neighboring fibers was found to be sufficient for convergence on a stable solution. The following assumptions were made in all cases:
- A perfect bond between the fiber and the matrix, no interphase
- All edges subjected to kinematic conditions of periodic symmetry

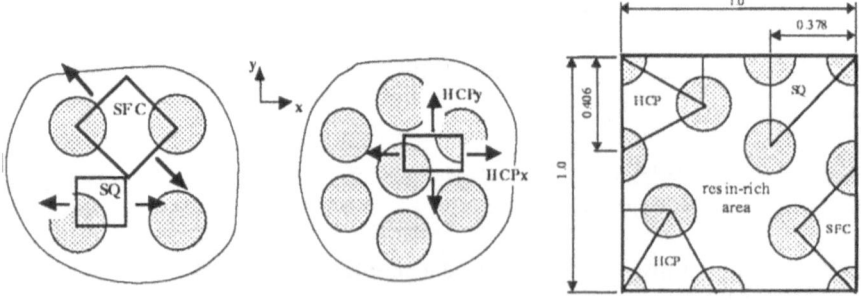

Figure 1. Arrays for the finite element analysis

- Linear elastic fibers and linear viscoelastic matrix
- Plain strain loading conditions, since the fibers are much stiffer than the matrix

3.2. FIBERS AND MATRIX: CONSTITUENT PROPERTIES

The variety of commonly used fibers and matrices in current composites applications calls for an approach to the problem that is as general as possible. For modelling the elastic case, this can be done by using ratios of constituent properties. For a typical glass fiber-reinforced polypropylene, for example, the ratio between the moduli is about 70, as is the case for a carbon fiber-reinforced epoxy system. Here the composite was modeled with a modulus ratio of 70 and also with rigid fibers in order to clarify the contribution of simple geometric effects and that of the interplay between deformations in the fibers and the matrix. Poisson's ratios of 0.22 and 0.35 for the fibers and the matrix, respectively, were used. Only isotropic constituents were considered.

A Prony series, which gives the modulus of a generalized Maxwell model (Eq. 12), was used for modelling the viscoelastic behavior of the composite.

$$G(t) = G_o \sum_{i=1}^{n} g_i \exp\left(-\frac{t}{\tau_i}\right)$$ Eq. 12

The relaxing shear and bulk components are given by two distinct series, and one element per decade is usually sufficient to describe very accurately a stress relaxation curve. For the present study a model matrix with a single relaxation time $\tau = 100s$ and a relaxed shear modulus $G_\infty = 0.5\ G_0$, where G_0 is the unrelaxed modulus, was used. The bulk modulus K was defined as non-relaxing, which has been shown to be the case for neat polymers [15]. As a result, the Poisson's ratio increases with time [16] and can be calculated from the elastic relationship (Eq. 13).

$$v(t) = \frac{3K - 2G(t)}{2(3K + G(t))}$$ Eq. 13

With few exceptions, such as aramid, the fibers can be assumed to be linear elastic over the time spans considered.

4. FE results: stress distributions and linear viscoelastic response

It has been shown [1] that the calculated elastic properties of the composite depend strongly on the array that is used for the calculation. This array dependence is, however, strongly reduced when results are compared on the basis of effective volume fraction ϕ / ϕ_{max}. The different effective stiffness values obtained for the composite at the same actual fiber content (Fig. 2), are attributed to the elastic stress distribution, which varies from one array to the next. Such a distribution, or stress spectrum, is shown in Fig. 3 in the form of a frequency distribution of the stress in loading direction for a fiber content of 30%. The fraction of matrix at a certain stress level is plotted as a function of the local stress normalized by the stress applied to the composite. The mean value of the distributions over the volume of the matrix, which is the stress coefficient A_{1111}, varies from 0.89 for the SQ array to 1.02 for the SFC array. The SFC array, with the highest mean stress value, also gives the lowest calculated modulus for the composite, while the SQ array, with the lowest mean, results in the highest modulus. Similar variations are obtained in plots of the stress distribution with varying fiber contents for a given array,

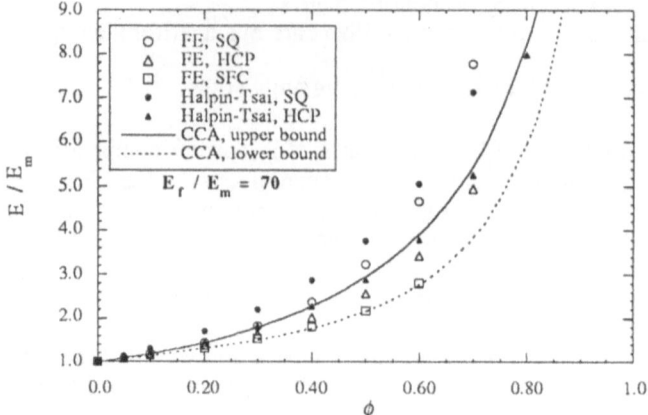

Figure 2. Effective elastic reponse of the composite for various arrays

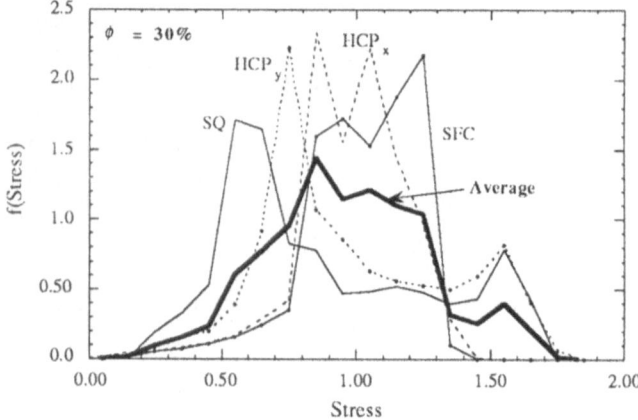

Figure 3. Distribution of $A^m_{1111} = \bar{\sigma}^{\prime m}_{11}/\sigma_{11}$ for various arrays, elastic analysis

with mean stress values going down to 0.49 for the SQ array with a fiber volume content of 70%.

The calculated effective viscoelastic response of the composite with a fiber volume content of 30% is plotted in Fig. 4 for different arrays . All curves are normalized by the initial, unrelaxed modulus. Since only the shear modulus G relaxes from 1 to 0.5, the tensile modulus E of the matrix decreases from 1 to 0.56 in plane strain through the relaxation. All curves thus lie above the relaxation curve of the matrix, but no shift or extension of the relaxation time is observed. A shift toward longer relaxation times was described in [17], but this was attributed to a modified structure of the polymeric matrix material as a result of varying local curing conditions near the filler/matrix interface. A change in the relaxation times could, however, also be related to the geometric constraint which the fibers present if the matrix material is nonlinear viscoelastic. Such effects are discussed below.

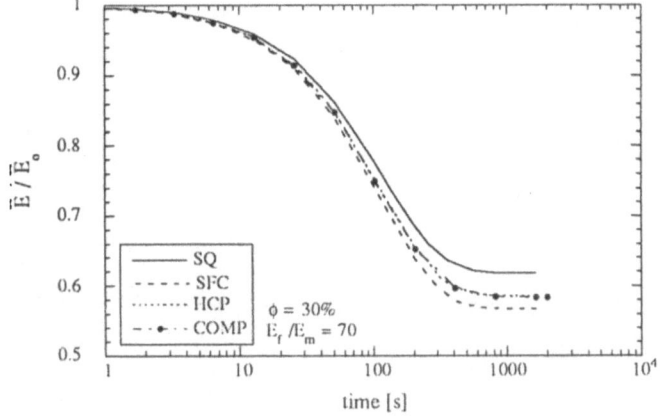

Figure 4 Normalized effective viscoelastic reponse of the composite for various arrays, ϕ=30%

As for the elastic case, the normalized relaxed modulus can be seen to depend on the array used for the calculation. Clearly this discrepancy occurs only for a non-zero relaxed modulus of the matrix, and it increases with an increasing value of the latter. The SQ array gives the least amplitude of relaxation, i.e. the highest normalized relaxed modulus. On the other hand, the calculated relaxation curve for the SFC array is only slightly higher than that for the matrix. The curves for the HCP and COMP models coincide and lie between those for the SQ and the SFC models. The COMP model however gives an initial, unrelaxed modulus which is lower than that obtained with both the SQ and the HCP array. The agreement of the results from the COMP model in the two loading directions despite the differences in fiber arrangement along the two axes fits in part the condition of transverse isotropy desired of a micromechanics model for an arbitrary unidirectional composite.

The appropriate choice of an arrangement of fibers in a monodisperse array is one difficulty encountered in modelling unidirectional composites. It is especially important to account correctly for the degree of alignment and the homogeneity of the fiber packing in the viscoelastic case, since the long-term response, in particular the relaxed modulus, is affected by the stress distribution, and thus by the fiber arrangement and spacing. Local variations in fiber content can affect the global, effective properties of the composite [3].

The relaxation curves for the SQ and the SFC arrays may be seen as an envelope within which are the results for an arbitrary composite at the given fiber volume content, since they represent loading along the densest and least dense line packing, respectively. The results of the COMP model supports this: although it has a local regularity of fiber packing, it contains resin-rich as well as fiber-rich areas, and yet gives a relaxation curve that corresponds very closely to that of the most densely packed array, the HCP. This is further supported by an examination of the influence of the fiber content. An increasing deviation from the behavior of the neat model matrix occurs with increasing fiber contents of the composite (Fig. 5). Such an effect has been shown to exist in the dynamic-mechanical behavior of two-phase composites, where the storage modulus at low frequencies decreases to a smaller extent with increasing filler contents [18]. The variation is related to the changing von Mises, or octahedral shear, stress spectrum in the matrix, which have been shown to determine the effective relaxation properties of

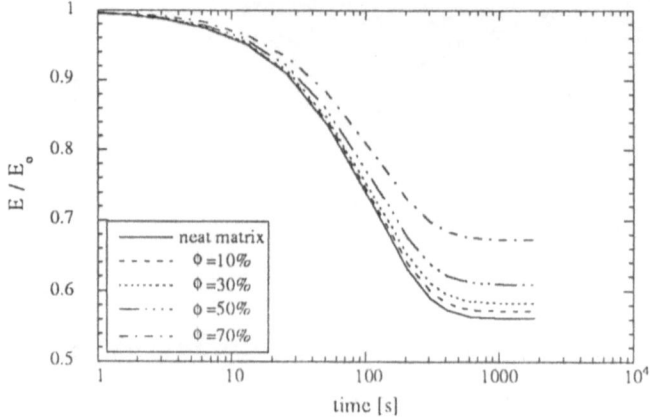

Figure 5. Normalized effective viscoelastic reponse of the composite for various fiber contents, HCP array

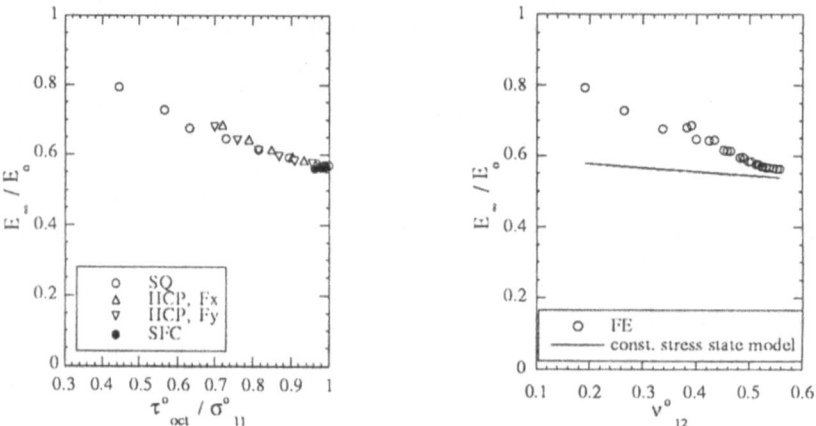

Figure 6. Normalized strength of relaxation as a function of mean initial elastic octahedral shear stress in the matrix (left) and transverse Poisson's ratio v_{12} (right)

unidirectional laminates [8]. The effect of a higher fiber content is to raise the normalized relaxed modulus of the composite, shown plotted as a function of fiber content for different models in Fig. 7. This graph confirms that the similarity between the COMP model and the HCP array extends to fiber contents up to at least 50% for the linear viscoelastic case. The results of the analytical models, discussed below, are also given in Fig. 7.

It was found that the mean octahedral shear stress in the matrix decreases with increasing fiber content. Different mean stress values were obtained for various arrays and fiber contents. The relaxation amplitude is found to be monotonically related to the mean octahedral shear stress (Fig. 6a). This applies both to variations of the fiber content and to the choice of different arrays for the model.

The constant stress state model (Fig. 6b) works only in very dilute conditions, where the elastic plane strain Poisson's ratio of the composite is close to that of the matrix.

Nevertheless, v_{12}^o seems to be a useful indicator of the relaxation strength to be expected.

Two effects must be considered in the relaxation of the composite: the change of modulus and Poisson's ratio of the matrix material and the associated redistribution of stresses in the matrix. The shear modulus of the neat model matrix used in the present study decreases by half through the relaxation, while the bulk modulus remains constant. As a result, the Poisson's ratio of the matrix increases from 0.35 to 0.42. For the linear viscoelastic analysis, the composite relaxation modulus at any time can be obtained by the pseudoelastic approximation, i.e. from an elastic FE calculation with the corresponding matrix property, $G(t)$, at the given time for the linear viscoelastic case. This hints to a possibility of using simple analytical micromechanics models for approximate calculations of the viscoelastic response of the unidirectional composites.

5. Evaluation of analytical models

Pseudoelastic micromechanical models have been used in which the matrix elastic modulus E is simply replaced by the relaxation modulus $E(t)$, among them the Halpin-Tsai or related equations [6, 17]. These models use fitting parameters that take into account the shape of the inclusions and are usually a function of the inclusion aspect ratio. Comparable results are obtained for both square (SQ) and hexagonal (HCP) arrays when the fiber content ϕ is replaced by an effective fiber content such as the fiber content normalized by the maximum content for the given array. This approach gives fairly good results for the viscoelastic behavior of particulate composites with relatively low filler contents.

For fibrous materials and higher filler contents, however, these models become inaccurate. This comes from the fact that they implicitly assume that Poisson's ratio is constant, while the better assumption usually is that the bulk modulus K is constant. This can be helped by finding an empirical, time-varying expression for the fitting parameters as a function of $v(t)$. A better approach however is a solution such as the composite cylinder assemblage (CCA) model [19], which includes explicitly both $G(t)$ and $v(t)$ of the matrix, i.e. in which the full time-dependence of the behavior of the matrix is accounted for.

Figure 7 shows the normalized relaxed modulus as a function of fiber content for different arrays modelled by FE and for CCA. The pseudoelastic Halpin-Tsai equation gives a normalized relaxed modulus practically equal to that of the matrix for all fiber contents and has thus been omitted from Fig. 7. This shows its inapplicability to composites with higher fiber contents, as it does not reproduce the influence of the fiber arrays on the relaxation behavior of the composite. The CCA bounds for the normalized relaxed modulus of the composite are slightly broader than the range given by the FE results, but the average of the CCA upper and lower bounds is nearly the same as the average of the FE solutions. It should be noted that the COMP array, which combines features of the four regular arrays, gives results that are very close to those obtained with the HCP arrays.

The differences between the models originate in part in the assumptions made on fiber microstructure in the composite. While the FE models allow an explicit definition of the array and the results of the Halpin-Tsai model can be made to correspond closely to the FE solutions at least in the elastic case for certain values of the fitting parameter, the CCA model is based on a quite different premise. In the latter, cylindrical units of varying sizes are assembled to completely fill an arbitrary space that is subjected to a

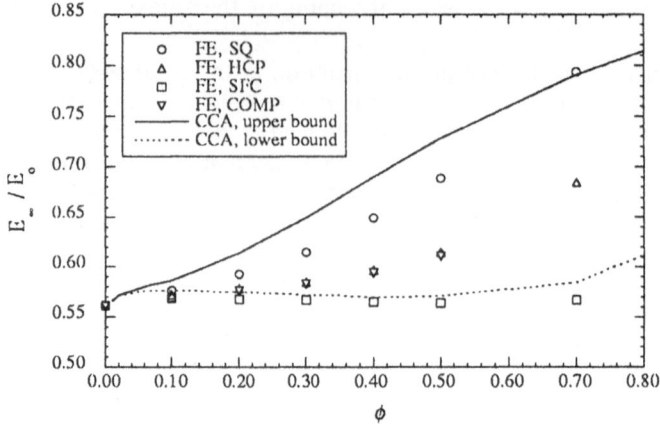

Figure 7. Normalized relaxed modulus E_∞/E_0 as a function of fiber content

mean displacement field. As was seen from the FE results, this affects very significantly the stress state in the matrix and thus the calculated relaxation response.

Since the curves in Fig. 7 are predominantly concave upward, any pseudo-elastic model can be used as an approximation, with the error staying low up to a fiber content of about 5-10%. This is the range over which experimental studies on the influence of fiber content on the relaxation behavior exist [17]. For higher fiber contents, there is still a need for experimental data to confirm the effects of the fiber-related constraints on the relaxation behavior of composites.

6. Extension to non-linear viscoelasticity

Nonlinearities in the viscoelastic response of polymer matrix composites can be caused by several factors: time-dependent effects, such as physical aging; changes of polymer structure due to uptake of humidity or solvents; changes of microstructure due to flaw or damage formation and accumulation in the matrix or at the fiber-matrix interface; and sensitivity to the stress or strain level. With respect to the latter, it is therefore clear that the choice of fiber array, which determines the local stress distribution, has an influence on the calculated nonlinear viscoelastic properties of the composite. It is well known that the geometric constraint presented by relatively stiff fibers results in a strain amplification in the matrix. The associated local stress concentration leads to a nonlinear behavior of the composite at low composite stress levels [20]. In such cases, the only current way of dealing with the non-linear properties of the matrix is by FE analysis [11]. It has been shown that even in the linear case the geometry chosen for the unit cell affects the results of the analysis. Given the considerable differences between the stress distributions in the matrix for the different arrays, nonlinearity of the matrix material can be expected to accentuate the range of responses obtained depending on the array used.

No shift in the relaxation times is observed for the present, linear case. The characteristic time of the relaxation remains the same as that of the matrix for all arrays and fiber contents. Furthermore, the composite response is linear viscoelastic, with a relaxation that is proportional to the applied stress. However, when the matrix behaves in a nonlinear viscoelastic way, the highly inhomogenous stress distribution in the matrix could lead to a significant shift and broadening of the relaxation time spectrum for the composite. The broadening would be less important in the HCP array, where the stresses in the matrix are more homogenous than for the SQ array. On the other hand, the SFC array, with a higher mean stress, can be expected to have shorter relaxation times than the others. Since the relaxation times of the matrix are usually shifted toward shorter times with increasing stress, the more highly stressed areas in the matrix will relax more rapidly and contribute to a shortening of the composite's relaxation times, while the less stressed areas will extend them.

7. Conclusions

This paper examined the interrelation between fiber arrangement, stress distribution in the matrix, and relaxation behavior. The normalized amplitude of relaxation is smaller for the composite than for the matrix. The characteristic relaxation times are not changed by the presence of the fibers in the linear viscoelastic case. The relaxation behavior of unidirectional composites under transverse loading in the linear viscoelastic domain can therefore be predicted by pseudo-elastic analytical models, provided these explicitly include the evolution of the shear and bulk moduli of the matrix. It was shown that the average stress state in the matrix is determinant for the effective relaxation behavior of the composite, and in particular the relaxed modulus. The reinforcing phase effectively reduces the relaxation of the matrix by imposing a geometrical constraint that depends on the packing geometry and fiber content. The different unit cells thus lead to considerable differences in the calculated effective viscoelastic response of the composite under transverse loading. The constraint effect is reflected in the stress spectrum of the matrix as well as in the Poisson's ratio obtained from an elastic solution.

Acknowledgements

The authors thank the Fonds National Suisse de la Recherche Scientifique for its financial support of the present work.

References

1. Adams, D. F. and Tsai, S. W., *The influence of random filament packing on the transverse stiffness of unidirectional composites,* Journal of Composite Materials, Vol. 3, p. 368-381, 1969.

2. Halpin, J. C. and Kardos, J. L., *The Halpin-Tsai equations: a review,* Polymer Engineering and Science, Vol. 16, No. 5, p. 344-352, 1976.

3. Pyrz, R., *Correlation of microstructure variability and local stress field in two-phase materials,* Materials Science and Engineering, Vol. A177, p. 253-259, 1994.

4. Hashin, Z., *Viscoelastic fiber reinforced materials,* AIAA Journal, Vol. 4, No. 8, p. 1411-1417, 1966.

5. Hashin, Z., *Complex moduli of viscoelastic composites II: Fiber reinforced materials,* Solids Structures, Vol. 6, p. 797-807, 1970.

6. Ek, C.-G., Kubat, J., and Rigdahl, M., *Stress relaxation, creep, and internal stresses in high density polyethylene filled with calcium carbonate,* Rheologica Acta, Vol. 26, p. 55-63, 1987.

7. Horoschenkoff, A., *Characterization of the creep compliances J22 and J66 of orthotropic composites with PEEK and epoxy matrices using the nonlinear viscoelastic response of the neat resins,* Journal of composite Materials, Vol. 24, p. 879-891, 1990.

176

8. Lou, Y. C. and Schapery, R. A., *Viscoelastic characterization of a nonlinear fiber-reinforced plastic,* Journal of Composite Materials, Vol. **5**, p. 208-234, 1971.

9. Hashin, Z., Humphreys, E. A., and Goering, J. , *Analysis of thermoviscoelastic behavior of unidirectional fiber composites,* Composites Science and Technology, Vol. **29**, p. 103-131, 1987.

10. Aboudi, J., *Micromechanical characterization of the non-linear viscoelastic behavior of resin matrix composites,* Composites Science and Technology, Vol. **38**, p. 371-386, 1990.

11. Schaffer, B. G. and Adams, D. F., *Nonlinear viscoelastic analysis of a unidirectional composite material,* Journal of Applied Mechanics, Vol. **48**, p. 859-865, 1981.

12. Dragone, T. L. and Nix, W. D., *Geometric factors affecting the internal stress distribution and high temperature creep rate of discontinuous fiber reinforced metals,* Acta Metallica et Materialia, Vol. **38**, No. 10, p. 1941-1953, 1990.

13. Adams, D. F. and Doner, D. R., *Transverse normal loading of a unidirectional composite,* Journal of Composite Materials, Vol. **1**, p. 152-164, 1967.

14. Devries, F. and Léné, F., *Homogénéisation à contraintes macroscopiques imposées: inplémentation numérique et application,* La Recherche Aérospatiale, Vol. **1**, p. 33-51, 1987.

15. Bertilsson, H., Delin, M., Klason, C., Kubat, M. J., Rychwalski, W. R., and Kubat, J., *Volume changes during flow of solid polymers,* Relaxations in Complex Systems, Alicante, 1993.

16. Rigbi, Z., *The value of Poisson's ratio of viscoelastic materials,* Applied Polymer Symposia, Vol. **5**, p. 1-8, 1967.

17. Nicolais, L., Guerra, G., Migliaresi, C., Nicodemo, L., and Benedetto, A. T. D., *Viscoelastic Behavior of Glass-Reinforced Epoxy Resin,* Polymer Composites, Vol. **2**, No. 3, p. 116-120, 1981.

18. Brinson, L. C. and Knauss, W. G., *Finite element analysis of multiphase viscoelastic solids,* Journal of Applied Mechanics, Vol. **59**, p. 730-737, 1992.

19. Rosen, B. W. and Hashin, Z., *Analysis of Material Properties,* Engineered Materials Handbook, Vol. 1: Composites, ASM International: Metals Park, OH, USA, p. 185-205, 1987.

20. Howard, C. M. and Hollaway, L., *The characterization of the nonlinear viscoelastic properties of a randomly orientated fibre/matrix composite,* Composites, Vol. **18**, No. 4, p. 317-323, 1987.

CONSTRUCTION OF A STOCHASTIC MACRO FAILURE MODEL OF UNIDIRECTIONAL FIBER REINFORCED COMPOSITES BASED ON DYNAMIC FAILURE PROCESS SIMULATIONS OF MICRO FAILURE MODELS

ISAO KIMPARA* and TSUYOSHI OZAKI**

*Department of Naval Architecture and Ocean Engineering,
The University of Tokyo,
7-3-1 Hongo, Bunkyo - ku, Tokyo 113, Japan

**Material and Electronic Device Laboratory,
Mitsubishi Electric Corporation,
1-1-57 Miyashimo, Sagamihara , Kanagawa 229, Japan

Abstract

A new failure process simulation model of unidirectional fiber reinforced composite materials is introduced considering the effect of matrix shear failure as well as fiber breaks based on a shear - lag theory in which a failure can occur randomly not only in fiber elements but also in matrix elements. A stochastic static tensile failure process of unidirectional composite materials is simulated by means of a Monte Carlo method based on a repeated increment scheme using a finite difference technique.

Then a new dynamic failure process simulation model is proposed in which an additional time variable is also incorporated by taking into account the mass of fiber and matrix elements. An exact time - dependent stress redistribution process in a composite failure model is evaluated by means of a finite difference scheme based on a time - increment method. A statistical analysis is made on the dynamic failure model and compared with the static one in terms of the different strength parameters, which shows that the dynamic simulation gives a better estimation on the dispersion of strength data and on the actual failure pattern.

Finally a macro model is proposed which is a new cumulative failure model composed of elements of micro models based on the previous dynamic failure process simulation models. This method is successfully used in estimating the statistical nature of strength properties of composite materials of actual size, and it is shown that the strength of composites depends clearly on the length but not on the width of a specimen.

R. Pyrz (ed.), IUTAM Symposium on Microstructure-Property Interactions in Composite Materials, 177–188.
© 1995 *Kluwer Academic Publishers.*

1. Introduction

The failure process of fiber reinforced composites is a very complicated accumulation process of damage due to random failure of fibers, matrix and interface, which leads to a catastrophic fracture. It should be necessary to introduce a reliability assessment system to evaluate effectively a decrease in strength due to cumulative damage and defects taking every aspect of variation into consideration in order to understand thoroughly the statistical nature of strength properties of composite materials.

A Monte Carlo simulation is one of the most effective methods to analyze such a complicated probabilistic phenomenon as a failure process of composites and several studies have so far been carried out including the author's earlier work [1,2,3,4]. In most of the past investigations, however, a simple failure model, which is called Rosen - Zweben's model [5,6], has been applied in which only random fiber breaks and stress concentration in the nearest fiber to the broken one are taken into account. This simple model leads to no more than a flat cleavage plane of a specimen.

In examining a tensile failure process of unidirectional fiber reinforced composites, not only fiber break but also marked interfacial debonding between fibers and matrix and pull - out of fibers are frequently observed especially in single and hybrid fiber composites with a medium - to - high volume fraction of fibers. Such a complex failure mechanism often leads to a complicated zigzag cleavage plane of a specimen. Although a few investigations taking into account interfacial debonding have been carried out, the interfacial failure criterion has not been clearly defined [7]. Therefore, it is important to take interfacial strength into consideration in modelling a basic failure model of composites.

For these reasons, the present paper aims at establishing a general assessment system to predict a decrease in reliability of composite materials due to cumulative damage, with a main system to simulate a stochastic failure process of composite materials considering the effect of a scatter in strength of fibers and matrix and defects mixed in during a fabrication process, accompanied with a subsystem of statistical analysis.

For this purpose, new static and dynamic failure process simulation models of unidirectional composites are introduced considering the effect of matrix shear failure as well as fiber breaks, based on a repeated increment scheme using a finite difference technique. A macro model is also proposed, which is a new cumulative failure model composed of elements of micro models based on the dynamic failure process simulation. The statistical nature of strength of unidirectional composites is characterized and discussed on the basis of these simulation models.

2. Basic Failure Model

Consider a lamina of unidirectional composite as shown in Fig. 1. Not only each reinforcing fiber arranged parallel to the direction of load has a varied fiber strength, but also every part of a fiber has a different strength. So each of k reinforcing fibers arranged at the same intervals is assumed to be composed of n links of "ineffective length", δ, where ineffective length means the minimum length in which a fiber does not break at more than two places. And it is assumed that each fiber should break at the center of a fiber element. Then nodal displacements, u_i^{j-1}, u_i^j, are set up above or below fiber element, $F(i, j)$, which is the i - th fiber from the leftside and the j - th element from the top as shown in Fig. 2. And as for the matrix, the matrix element, $M(i, j)$, is introduced between nodes u_i^j and u_{i+1}^j, which is a characteristic of this model.

Considering a stress transmission mechanism in the tensile failure of unidirectional composite, the mechanism is approximated by the shear - lag theory, which means that fibers transmit only axial force and matrix only shear force. Therefore tensile strength is allocated at $n \times k$ fiber elements and shear strength is allotted at $(n-1) \times (k-1)$ matrix elements. Namely, random numbers based on a statistical distribution are generated about fibers and matrices. $\sigma (i, j)$ is allocated at the fiber element, $F(i, j)$, and $\tau (i, j)$ is allotted at the matrix element, $M(i, j)$, respectively.

The strength distribution of fiber elements is given by the random number, $\sigma (i, j)$, based on the Weibull distribution. As for the matrix elements, the same distribution of shear strength should be assumed, but there are almost no measured data. So, supposing the effect of fiber strength distribution to be the most important factor for determining the strength of composites, a uniform shear strength, τ_{max}, is assumed in this study.

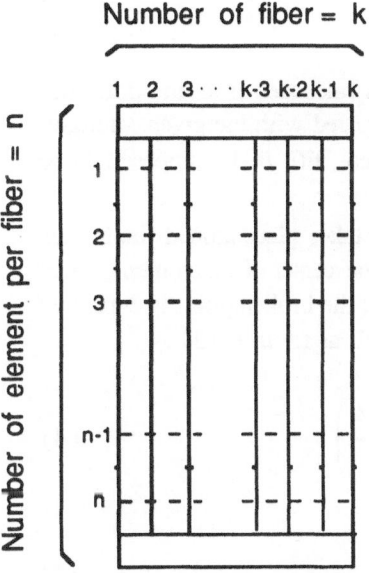

Figure 1. Simulation model

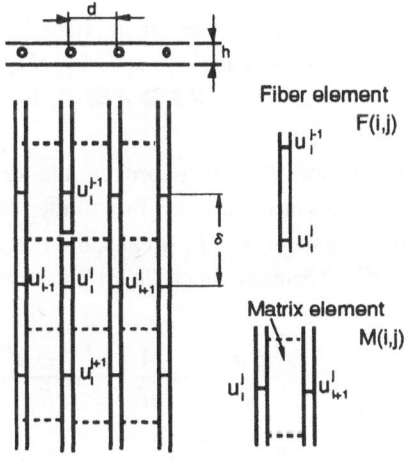

Figure 2. Fiber and matrix elements

From a shear - lag theory the equilibrium equation of force is given by:

$$
\left.
\begin{array}{ll}
EA\dfrac{d^2u_i}{dx^2}+\dfrac{Gh}{d}(u_{i-1}-2u_i+u_{i+1})=0 & (1<i<k)\\[2mm]
EA\dfrac{d^2u_i}{dx^2}+\dfrac{Gh}{d}(u_2-u_1)=0 & (i=1)\\[2mm]
EA\dfrac{d^2u_i}{dx^2}+\dfrac{Gh}{d}(u_{k-1}-u_k)=0 & (i=k)
\end{array}
\right\}
\tag{1}
$$

Where, u_i : axial displacement of i-th fiber, E : Young's modulus of fiber, A ; cross - sectional area of fiber, G : shear modulus of matrix, d : distance between fibers, h : thickness of lamina.

Approximating the second - order differential equations such as Eq. (1) by the following equation of finite difference :

$$
\frac{d^2u_i}{dx^2}=\frac{u_i^{j-1}-2u_i^j+u_i^{j+1}}{\delta^2}
\tag{2}
$$

the nodal displacement, u_i^j, is given by   [3] :

$$
u_i^j=\frac{u_i^{j-1}+u_i^{j+1}+\dfrac{Gh\delta^2}{EAd}\left(u_{i-1}{}^j+u_{i+1}{}^j\right)}{2\left(1+\dfrac{Gh\delta^2}{EAd}\right)}
\tag{3}
$$

The tensile stress caused in the fiber element, F (i, j), is calculated from the difference of nodal displacements, u_i^j, u_i^{j-1}, by $E(u_i^j-u_i^{j-1})/\delta$ and compared with the given strength, σ (i, j). If E $(u_i^j-u_i^{j-1})/\delta \geq \sigma$ (i, j), then the fiber element, F(i, j), is supposed to be broken.

When a fiber element is broken, the stress field disorder takes place around the broken element. As it is assumed that a fiber always breaks at the center of an element, when the fiber element, F(i, j), breaks, the nodal displacement of the broken point is expressed by $u_i^{j-1/2}$. Therefore when F(i, j) breaks as shown in Fig. 2, as for u_i^{j-1} [3] ,

$$
\frac{d^2u_i^{j-1}}{dx^2}=\frac{4}{3\delta}\left(\frac{u_i^{j-2}-u_i^{j-1}}{\delta}-\frac{u_i^{j-1}-u_i^{j-1/2}}{\delta/2}\right)
\tag{4}
$$

and since $u_i^{j-1/2}=u_i^{j-1}$,

$$
\frac{d^2u_i^{j-1}}{dx^2}=\frac{4}{3\delta^2}\left(u_i^{j-2}-u_i^{j-1}\right)
\tag{5}
$$

similarly, as for u_i^j,

$$\frac{d^2 u_i^j}{dx^2} = \frac{4}{3\delta^2}\left(u_i^{j+1} - u_i^j\right) \tag{6}$$

so that the equation of finite difference in the form of Eq.(2) is replaced by Eq. (5) for u_i^j and by Eq.(6) for u_i^{j-1} in this case.

On the other hand, shear stress caused in the matrix element, M(i, j), is calculated from the difference between nodal displacements, u_i^j, u_{i+1}^j, by $|\,G\tan^{-1}\{u_{i+1}^j\text{-}u_i^j\}\!/d\,|$, and compared with the shear strength of matrix, $\tau\,(i, j)$. If $|\,G\tan^{-1}\{u_{i+1}^j\text{-}u_i^j\}\!/d\,| \geq \tau$ (i, j), the matrix element, M(i, j), is supposed to be broken. In this case, the equilibrium equations become as follows :

$$\left. \begin{aligned} EA\frac{d^2 u_i^j}{dx^2} + \frac{Gh}{d}\left(u_{i-1}^j - u_i^j\right) = 0 \\ EA\frac{d^2 u_{i+1}^j}{dx^2} + \frac{Gh}{d}\left(u_{i+2}^j - u_{i+1}^j\right) = 0 \end{aligned} \right\} \tag{7}$$

A tensile failure process simulation is carried out on the basis of a repeated increment scheme [3] until it arrives at the ultimate fracture of a lamina.

3. Dynamic Failure Model

The basic failure model as described above is extended to a new dynamic failure model, in which an aditioinal time variable is also incorporated by taking into account the mass of fiber and matrix elements. From a shear-lag theory, the equilibrium equation of force is given in the finite difference form :

$$\frac{EA}{\delta^2}\left(u_i^{j-1} - 2u_i^j + u_i^{j-1}\right) + \frac{Gh}{d}\left(u_{i+1}^j - 2u_i^{j-1} + u_{i-1}^j\right) = m\frac{d^2 u_i^j}{dt^2} \tag{8}$$

where m is mass per unit length of a fiber including that of surrounding matrix region.

The dynamic problem is solved by means of Wilson's theta method. When the nodal displacement, u_{ij} , at the time, $t + \theta \Delta t$, is given by u_{ij} $(t + \theta \Delta t)$, Eq. (8) is combined with the following two sets of equations, from which the acceleration, $\ddot{u}_{ij}(t+ \theta \Delta t)$, is obtained.

$$u_i^j(t + \theta\Delta t) = \dot{u}_i^j(t) + \theta\Delta t u_i^j(t)$$

$$+ \frac{(\theta\Delta t)^2}{3}\ddot{u}_i^j(t) + \frac{(\theta\Delta t)^2}{6}\ddot{u}_i^j(t + \theta\Delta t) \tag{9}$$

$$\dot{u}_i^j(t+\theta\Delta t) = \dot{u}_i^j(t) + \frac{\theta\Delta t}{2}\left\{\ddot{u}_i^j(t) + \ddot{u}_i^j(t+\theta\Delta t)\right\} \tag{10}$$

Using $\ddot{u}_{ij}(t+\theta\Delta t)$, $u_{ij}(t+\Delta t)$ is obtained from the following equations of interpolation:

$$\ddot{u}_i^j(t+\Delta t) = \frac{1}{\theta}\left\{(\theta-1)\ddot{u}_i^j(t) + \ddot{u}_i^j(t+\theta\Delta t)\right\} \tag{11}$$

$$\dot{u}_i^j(t+\Delta t) = \dot{u}_i^j(t) + \frac{\Delta t}{2}\left\{\ddot{u}_i^j(t) + \ddot{u}_i^j(t+\Delta t)\right\} \tag{12}$$

$$u_i^j(t+\Delta t) = u_i^j(t) + \Delta t\,\dot{u}_i^j(t)$$
$$+\frac{(\Delta t)^2}{3}\ddot{u}_i^j(t) + \frac{(\Delta t)^2}{6}\ddot{u}_i^j(t+\Delta t) \tag{13}$$

It is known that $\theta \geq 1.37$ gives an unconditional stable condition in the Wilson's theta method. As a large value of θ, however, decrease the accuracy, Wilson recommends $\theta = 1.4$ as practical value [8], which is adopted in this calculation.

The failure criteria of fiber and matrix elements are the same as in the stastic simulation as described above. The equilibrium eqation is replaced in a similar way as in the static case when a fiber or matrix element fails. The strain increases regardless wether an element fails or not in the dynamic simulation, while the strain does not increase until a successive failure at a certain strain level comes to an end arriving at an equilibrium condition in the static simulation.

A statistical analysis is made of the strength and failure pattern distributions of typical unidirectioiial CFRP(carbon fiber reinforced plastics) and GFRP(glass fiber reinforced plastics) based on the dynamic failure model and compared with the static one in terms of different strength parameters. The dynamic model is tile same as the static one and the volume fraction of fiber is assumed to be 40 and 60 per cent. The strain rate is supposed to be 10^{-2} ms^{-1} and the interval of time increment is assumed to be $\Delta t = 2 \times 10^{-8}$ s. The number of simulations is 30 for each case for the purpose of statistical analysis.

TABLE 1 shows a comparison between the mean values and the coefficients of variation for maximum stress (σ_{max}), maximum strain (ε_{max}) and number of broken elements in fiber and matrix based on the static and dynamic simulations. Judging from these results, the dynamic simulation gives a larger σ_{max} than the static one by 37.5 MPa in GFRP of $V_f = 60$ %, while there is no significant diifference in GFRP of $V_f = 40$ %. On the other hand, the dynamic result gives a lower σ_{max} in CFRP (by 54.7 MPa in $V_f = 60$ %) and gives a larger coefficient of variation in either $V_f = 40$ % or 60 %. As for ε_{max}, the static simulation gives a little larger value in botlh CFRP and GFRP. It is noteworthy that there is a more significant difference, in the number of broken elements : the number of broken fiber elements is larger in the dynamic 'simulation, especially in GFRP, and the numbers of broken matrix elements are considerably different in CFRP.

In the static model, a "static" stress redistribution is made until a next equilibrium condition is attained, which causes a few "multiple fracture" as shown in TABLE 1 : almost no "multiple fracture" is observed especially in GFRP. The static model also tends to make the number of failures smaller in matrix in case of CFRP with a larger number of failures in matrix. Hence the active generation of failures in a microscopic region can be explained only by the dynamic failure model. Therefore, it is suggested that both static and dynamic simulations give a little difference as far as the advantage values of tensile strength and maximum strain are concerned, but that the dynamic simulations gives a better estimation in terms of the dispersion of strength data and the actual failure pattern.

TABLE 1. Comparison between static and dynamic simulations

		V_f %	σ max MPa	ε max %	Number of broken elements	
					Fiber	Matrix
CFRP	Static	60	1297.6 (7.8)	1.045 (10.5)	13.70 (16.6)	46.24 (23.2)
		40	845.5 (6.7)	0.959 (7.0)	15.02 (18.8)	22.04 (21.0)
	Dynamic	60	1224.9 (9.6)	1.017 (10.1)	17.13 (43.9)	55.67 (18.2)
		40	826.7 (9.4)	0.958 (7.7)	18.73 (40.4)	30.90 (23.3)
GFRP	Static	60	724.4 (10.0)	3.413 (12.6)	11.66 (7.6)	106.50 (9.8)
		40	502.3 (9.9)	3.046 (13.6)	11.82 (8.3)	70.24 (14.0)
	Dynamic	60	761.9 (11.6)	3.247 (15.0)	17.83 (47.7)	106.47 (10.9)
		40	502.6 (9.7)	2.920 (19.3)	16.03 (40.3)	75.23 (16.8)

Note : Figures in parentheses indicate coefficient of variation (%).

4. Macro Model and Size Effects in Strength

4.1 CONSTRUCTION OF MACRO MODEL

The basic failure model as described above is a "micro" model as indicated by the white broken line in Fig. 3 which is a typically observed final failure pattern of unidirectional CFRP [9] . Although such a micro model explains well the actual failure pattern of composites as described above, there remains a practical problem on what kind of statistical correlations exist between the simulated strength of "micro" models and the observed strength of "macro" region of composites of actual size. Consider, for examples, a rectangular region of 10 mm in width and 50 mm in gauge length in a tensile test specimen. As this region is equivalent to an assembly of about 5,000 micro models, composied of 165 ×5,000 elements, some kind of macroscopic considerations is required for simulating the failure process of such a "macro" region.

For this reason, a macro model is proposed which is a new cumulative failure model composed of elements of micro models based on the dynamic failure process simulations. Consider a macro model composed of $m \times n$ micro model elements (m rows in longitudinal direction and n columns in transverse direction) as shown in Fig. 4. A longitudinal strength, σ_{er} (i, j), is allotted at an element (i, j) according to the Weibull random numbers, which are given by the Weibull parameters deribed from σ_{max} data based on the dynamic simulations.

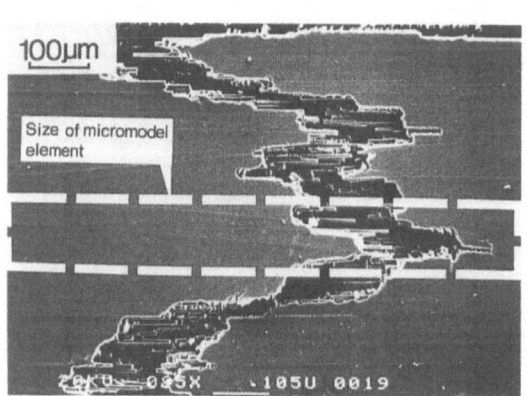

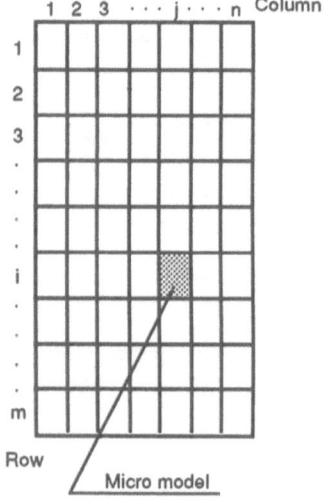

Figure 3. Micrography by SEM observation
- fracture surface -

Figure 4. Macro model

A tensile failure simulation is made on a macro model. As the number of elements is as many as several thousands, it is not practical to solve the equilibrium equations of a system successively based on a strain increment scheme by means of a finite element method. Hence only the maximum load (ultimate strength) of a macro model is simulated based on a load increment scheme, in which a uniform stress is first applied to each element and stress concentration factors are then given to surrounding elements around a broken one : a stress redistribution is repeated by judging a failure of elements on the basis of successive comparisons between stress and strength of an element.

The stress concentration factors are calculated in advance for various failure patterns by means of a finite element method and stored in a data file. Then, if a failure occurs, the failure pattern is searched and the corresponding stress concentration factors are given to the neighboring three elements around a broken one on both sides : $k(i, j \pm 1)$, $k(i, j \pm 2)$ and $k(i, j \pm 3)$, ane the element stress is increased to $k \cdot \sigma$ in a next step. A failure of a macro model can be assumed in the following two ways :i) at least one element fails in every column of n columns (series - parallel type), ii) all elements fail in either row out of m rows (parallel - series type). The former criterion is adopted in this study in order to allow a longitudinal crack to extend between micro model elements, which is often observed in a failure of unidirectional composites.

First, a macro model simulation is made on a small - scaled model composed of 10×10 elements and compared with a FEM simulation. A comparison is made between the resulting failure patterns of elements in macro model and FEM simulations on the same models with the identical strength distribution. It is shown that failure patterns are apparently similar in both cases, though a macro model simulation is not intended to express an exact failure pattern. The simulated ultimate strength is 872.7 MPa by FEM and 891.7 MPa by macro model, which shows the validity of a macro model simulation.

Then, a macro model simulation is carried out on unidirectional CFRP and GFRP (V_f=60%) of 50 mm in length and 10 mm in width corresponding to a region over gauge length in a tensile test specimen. The simulated results are shown in TABLE 2, in which the micro model simulations in TABLE 1 are also listed for the purpose of comparison. The mean values by macro model are 61-66% as large as those by micro models, which shows that the size effect in strength is remarkably large. On the other hand, the coefficients of variation by macro models are considerably smaller than those by micro models, which reflects a general statistical tendency that the variation becomes smaller as the number of elements increases.

The theoretical strength of macro models is expressed by the following Weibull analysis which is also shown in TABLE 2. The probability that an element does not fail under a stress σ , $R(\sigma)$, is given by :

$$R(\sigma) = \exp\left(-\frac{\sigma^\alpha}{\beta^\alpha}\right) \qquad (14)$$

where α is the shap parameter and β is the scale parameter of a Weibull distribution.

So that, the probability of failure of a macro model composed of m columns and n rows under the failure criterion (i), $F_1(\sigma)$, is expressed by :

$$F_1(\sigma) = \left[1 - \left\{ \exp\left(\frac{\sigma^\alpha}{\beta^\alpha} \right) \right\}^m \right]^n \tag{15}$$

and the probability of failure under the failure criterion (ii) is given by :

$$F_2(\sigma) = 1 - \left[1 - \left\{ 1 - \exp\left(-\frac{\sigma^\alpha}{\beta^\alpha} \right) \right\}^n \right]^m \tag{16}$$

The stress levels corresponding to the probability of failure : $F_1(\sigma)$ or $F_2(\sigma)$=0.1, 0.01, that is, the probability of survival : $R_1(\sigma)$ or $R_2(\sigma)$=0.90, 0.99, are shown in TABLE 2.

In comparing the simulated result with the theoretical one under the criterion (i), corresponding to the failure criterion of a macro model, the former is smaller than the latter by about 25%, since the latter is based on a simple probabilistic theory in which the effect of stress concentration is neglected. The failure criterion (ii) appears to be irrational, as it gives larger values of strength than the mean values of a macro model.

TABLE 2. Comparison between micro and macro model simulations

			CFRP	GFRP
Simulation	Micro Model		1224.9 (9.6)	761.9 (11.6)
	Macro Model		805.2 (2.9)	468.8 (4.8)
Weibull	Criterion (i)	R=0.90	928.1	573.6
		R=0.99	1000.5	598.1
	Criterion (ii)	R=0.90	1383.0	853.5
		R=0.99	1398.2	867.5

Notes : Units in MPa. Figures in parentheses indicate coefficient of variation (%).

4.2 DISCUSSIONS ON SIZE EFFECT IN STRENGTH

It is well recognized that there is an obvious size effect in strength of composite materials. A simulation is made on the size effect in strength of unidirectional CFRP

($V_f = 60\%$) by varying the number of elements in a macro model. The number of rows is varied as one - forth, one - half and two times in comparison to the standard one to examine the size effect in length. The mean values analyzed by assumed normal distributions based on 30 simulations in each case are shown in Fig. 5, together with the theoretical values based on the simple Weibull theory under the criterion (i). Similarly, the number of columns is varied as one-forth, one - half and two times in comparison to the standard one to examine the size effect in width. The simulated and analyzed results are shown in Fig. 6.

In comparing Fig. 5 with Fig. 6, the simulations show that there is an obvious tendency that the strength depends on the length : the strength is decreased but the rate of decrease becomes dull as the length is increased. The Weibull theory gives a similar tendency. On the other hand, the strength does not appear to depend on the wigth in the simulated results. The size effect in width is small also in the theoretical results : even if the number of columns is 10^4, the strength would be increased only to 1,063.1 MPa. As there is the effect of stress concentration in the simulations, the increase in reliability due to the increase in numbers of columns is cancelled by the decrease in reliability due to the increase in number of elements.

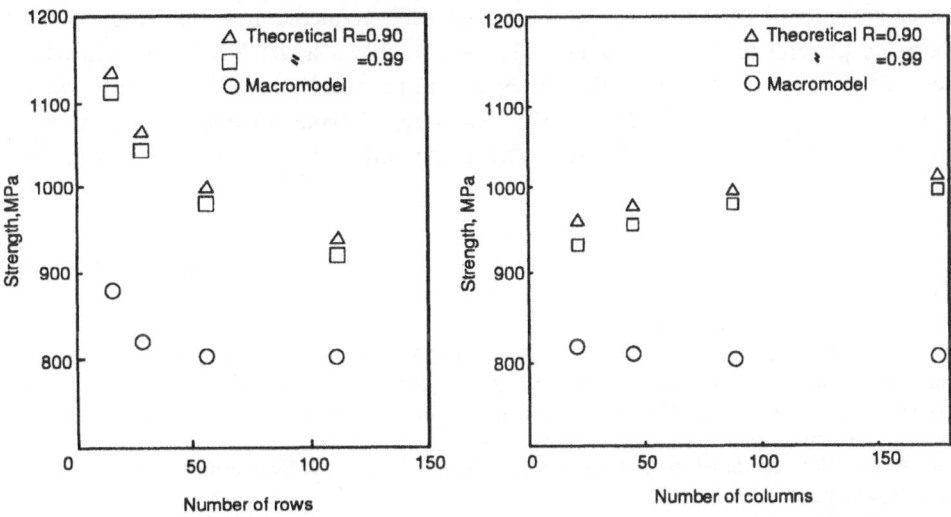

Figure 5.   Variation of stress with   Figure 6.   Variation of stress with
number of rows   number of columns

5. Conclusions

In this study, a reliability assessment system is proposed on the basis of static and dynamic failure process simulations of unidirectional lamina models in order to examine the statistical nature of strength properties of unidirectional composite materials. The main results are summarized as follows :

1) A successful simulation is carried out tracing faithfully an actual failure process by considering interfacial debonding between fibers and matrix as well as random fiber breaks. The reliability and damage tolerance levels of unidirectional composites are evaluated quantitatively based on the simulations.

2) An exact time - dependent stress redistribution process due to progressive failures is evaluated based on the dynamic simulations, which give a better estimation in terms of the dispersion of strength data and the active generation of failures than in the static simulations.

3) A macro model is proposed which is a new cumulative failure model composed of elements of micro models based on the dynamic simulations. This method is successful in estimating the statistical nature of strength of composite materials of actual size.

Acknowledgements

The authors are indebted to Mr. S. Takada, Mr. N. Tsuji and Mr. K. Inoue who were graduate students of the University of Tokyo for their cooperation in this study. The authors are grateful to Mr. T. Takabayashi, Mr. H. Yamada and Mr. Y. Morita who were students of the university for their assistance in the calculations. The authors wish to express their sincere thanks to Dr. N. Sato, Toyota Central Research and Development Laboratories, Inc., for having permitted us to use his valuable photographs.

References

1) Kimpara, I., Watanabe, I., Ohkatsu, T. and Ueda, N. (1974) A Simulation of Failure Process of Fiber Reinforced Materials, *Proc. 7th Sympos. on Compos. Mater.*, JUSE, 169-174 (in Japanese)

2) Fukuda, H. and Kawata, K. (1977) On the Strength Distribution of Unidirectional Fiber Composites, *Fibre Sci. Tech.*, 10, 53-63

3) Oh, K. P.(1979) The Strength of Unidirectional Fiber - Reinforced Composites, *Jour. Compos. Mater.*, 13, 311-328

4) Fukuda, H. and Chou, T.W. (1982) A Statistical Approach to the Strength of Hybrid Composites, *Proc. ICCM - IV*, Tokyo, 2, 1145-1151

5) Rosen, B.W. (1964) Tensile Failure of Fibrous Composites, *AIAA Jour.*, 2, 1985-1991

6) Zweben, C. (1968) Tensile Failure of Fiber Composites, *AIAA Jour.*, 6, 2325-2331

7) Okuno, S. and Miura, I. (1978) Analysis of Fracture Process and Strength in Fiber Reinforced Alloy by Monte Carlo Simulation, *Jour. Japan Soc. Metals*, 42, 736-742 (in Japanese)

8) Clough, R. W. and Bathe, K. J. (1972) *Finite Element Analysis of Dynamic Response*, UAH Press

9) Sato, N., Kurauchi, T. and Kamigaito, O. (1986) Fracture Mechanism of Unidirectional Carbon - Fibre Reinforced Epoxy Resin Composite, *Jour. Mater. Sci.*, 21, 1005-1010

ON IMPREGNATION QUALITY AND RESULTING MECHANICAL PROPERTIES OF COMPRESSION MOULDED COMMINGLED YARN BASED THERMOPLASTIC COMPOSITES

V. KLINKMÜLLER*, R. KÄSTEL*, L. YE⁺, AND K. FRIEDRICH*

*Institute for Composite Materials Ltd., University of Kaiserslautern
67633 Kaiserslautern, Germany
⁺Dept. of Mechanical Engineering, The University of Sydney,
Sydney, New South Wales 2006, Australia

Abstract

A glass fibre (GF) / polypropylene (PP) commingled yarn was selected to investigate the relationship between impregnation mechanisms and processing conditions during consolidation. Furthermore a carbon fibre (CF) / polyetheretherketone (PEEK) commingled yarn was studied. Laminates out of these material forms were fabricated by hot pressing. Microscopy of cross-sections and density measurements helped to examine the quality of impregnation and consolidation.

Based on microscopic observations, an impregnation model for the qualitative description of the consolidation behaviour was generated. It can be used to describe variations in void content over laminate thickness as a function of bundle geometry and combinations of processing parameters. The relationship between processing temperature, holding time, and applied pressure, required to reach full consolidation, were evaluated.

Results of transverse flexure tests were used to correlate the mechanical properties with the impregnation quality. For each kind of material the optimum processing window for manufacturing of laminates could be suggested.

1. Introduction

Thermoplastic resins as matrix materials for advanced composite materials have many advantages over thermosetting composites. For example, their fracture toughness is very high compared to thermosetting resins, they do not require extra time for chemical reaction after processing, and there is no need for sub-zero temperature storage [1]. On the other hand, thermoplastics at their processing temperature have viscosities of 500-5000 Pa·s compared to thermosets which possess values less than 100 Pa·s. The high viscosity imposes many problems in the manufacturing process of thermoplastic composites [2]. Along with poor dispersion of fibres in the thermoplastic matrices, the quality of impregnation has been one of the major concerns.

189

R. Pyrz (ed.), IUTAM Symposium on Microstructure-Property Interactions in Composite Materials, 189–201.
© *1995 Kluwer Academic Publishers.*

To make impregnation and consolidation easier, some intermediate material forms have been developed, such as powder impregnated fibre bundles and commingled yarns [3, 4]. These can be further consolidated into fully or partly impregnated, stiffer tapes, or directly processed into final part geometries during an on-line impregnation and consolidation process, e.g. by filament winding or pultrusion.

The present study is intended to provide a deeper insight into the impregnation and consolidation behaviour of commingled yarns during a compression molding process. For this propose, a GF/PP and a CF/PEEK system were selected. An impregnation model was generated to describe the consolidation process. Both approaches will help to predict under which conditions of pressure, time and temperature the material forms result in perfect composite macrostructures and good mechanical performance of the parts made out of them.

2. Materials and Evaluations

2.1 MATERIALS

The GF/PP - commingled yarn was supplied by Toyobo, Co. Japan. The yarn consisted of a 50 : 50 weight-% mixture of glass and polypropylene fibres. The melting peak, T_m, of the PP-polymer was determinated by DSC analysis as 162.9° C. In addition, CF/PEEK commingled yarn supplied by BASF, Germany was studied (exact description "PEEK/AS4 3k RC40"). It was composed of a 60 : 40 weight-% mixture of carbon and PEEK fibres. The melting peak, T_m, of the PEEK-polymer amounted to 345° C (Table 1).

TABLE 1. Properties of two commingled yarns studied

Properties	GF/PP	CF/PEEK
Weight of one bundle W_b	$8.0 \cdot 10^{-3}$ g/cm	$3.5 \cdot 10^{-3}$ g/cm
Real fibre volume fraction V_f*	0.26	0.53
Assumed fibre volume fractions in fibre rich areas V_f	0.05	0.53
Fibre radius r_f	8.5 μm	3.5 μm
Density of reinforcing fibres ρ_f	2.56 g/cm^3	1.78 g/cm^3
Density of matrix fibres ρ_m	0.905 g/cm^3	1.332 g/cm^3
Theoretical density ρ_t	1.337 g/cm^3	1.569 g/cm^3
Area of fully consolidated bundle A_b	0.6 mm^2	0.24 mm^2
Area of matrix in a consolidated bundle A_M	0.44 mm^2	0.11 mm^2
Average width of a fibre bundle y	2000 μm	1250 μm
Height of fibre bundle h	300 μm	192 μm
Distance x	55.5 μm	22.5 μm
Initial void content X_{vo}	15.6 %	10.5 %
Kozeny-Carman constant k_o	700	80

2.2   MANUFACTURING OF LAMINATES

Each laminate contained 16 layers of unidirectional "prepreg" sheets, which were made by winding individual bundles onto an aluminum plate, with subsequent two lines welding of bundles at both ends of the plate. Consolidation of the laminates was performed by using a small steel mould with a square cavity and a laboratory hot press. Once the mould reached the desired temperature (185 °C-220 °C for GF/PP; 380 °C-420 °C for CF/PEEK), pressure was applied. Different impregnation pressures (0.5, 1.5 and 3.0 MPa) and holding times (3, 5, 10, and 20 min) were selected to identify the impregnation mechanisms as a function of processing conditions. The composite panels were cooled rapidly to room temperature in order to avoid formation of voids in the resin-rich areas during cooling [5]. The average cooling rate was about 30 °C/min.

2.3   VOID CONTENT

Density measurements were carried out in order to correlate consolidation states with apparent void contents in relation to the processing conditions. The laminate density, ρ_l, under different processing conditions was determined according to ASTM-792. The theoretical density, ρ_t, of a fully consolidated composite part could be estimated by the following equation:

$$\rho_t = \frac{\rho_f \cdot \rho_m}{W_f \cdot \rho_m + W_m \cdot \rho_f} \tag{1}$$

where ρ_f and ρ_m are the densities and W_f and W_m the weight fractions of fibres and matrix, respectively. The apparent void content, X_v, was then determined by:

$$X_v = \frac{\rho_t - \rho_l}{\rho_t} \tag{2}$$

However, when a consolidated part with higher fibre volume fraction was obtained (due to a large matrix volume squeezed out of the processing mould, i.e. when $\rho_l > \rho_t$), X_v was set to be zero.

2.4   MECHANICAL CHARACTERIZATION

Characterization of mechanical properties as a function of impregnation conditions was carried out by using a small transverse flexure (three point bending) testing facility. The length of the span amounted to 40 mm; the width and thickness of the specimens were about 10 mm and 3.3 mm, respectively. The cross-head speed was set to 1mm/min. Transverse elastic constants and flexural strength were determined according to ASTM standard D-790.

3.   Consolidation of Laminates

Based on microscopic observations and previously developed processing theories, an impregnation model for the qualitative description of the consolidation behaviour in laminates made out of commingled yarn was generated. By examining the consolidation process in each fibre bundle, it can be found that although there existed differences in the states of impregnation and compaction between each fibre bundle, the basic procedure was almost the same for each of them. Hence, by assuming that all fibre bundles undergo impregnation simultaneously in a laminate and that all of them are identical in geometry, the consolidation of the entire laminate can be described by the inward impregnation in a representative single bundle.

It was observed that the initially commingled polymer fibres and the reinforcing fibres became unmingled when non-uniform tension was applied, because of the mismatch in stiffness in the fibre direction (Figure 1). This may result in both a non-uniform distribution of fibres in the final composite part and in insufficient impregnation of the reinforcing fibres and therefore poor load transfer between them [6]. It can locally lead to a higher fibre volume fraction in the fibre rich areas (V_f), in comparison to the given fibre volume fraction ($V_f *$).

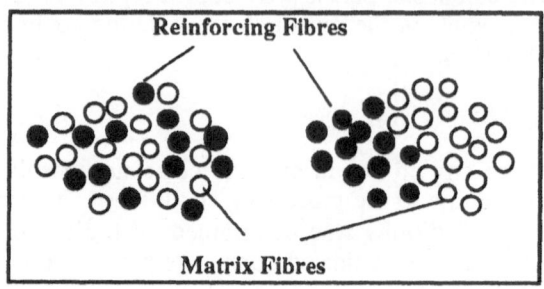

FIGURE 1. *Separation of the different fibres due to streching of the fibre bundle in fibre direction*

For the impregnation model it was assumed that the cross-sectional shape is a rectangle. Figure 2 shows this acceptance. Now, the mechanism can be described by a kind of film stacking process.

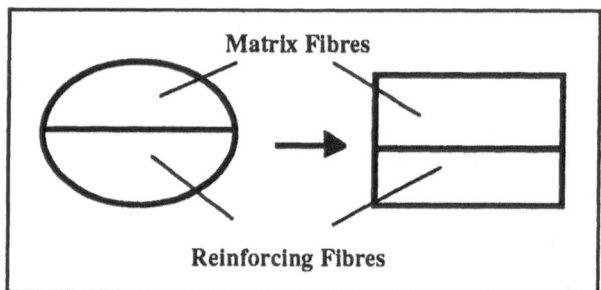

FIGURE 2. *Theoretical cross-sectional shape of a single commingled yarn bundle*

The processing time required for the consolidation can be evaluated by Darcy's equation. If it is assumed that the matrix impregnates the fibre network normal to the fibre axis, the rate of impregnation is given by

$$\frac{dz}{dt} = \frac{K_p}{\mu} \cdot \frac{dp}{dz}$$

(3)

where dp/dz is the pressure gradient, μ is the melt viscosity, and K_p the permeability of the fibre tow. Once the local fibre volume fraction V_f is known, the permeability can be estimated by the modified Carman-Kozeny equation [7, 8, 9]:

$$K_p = \frac{r_f^2}{4 k_o} \cdot \frac{(1 - V_f)^3}{V_f^2}$$

(4)

where r_f is the average radius of fibres and k_o is the Carman-Kozeny constant. In this case, the value of k_o will be determined during the evaluation of void content by fitting a set of selected experimental data. Assuming r_f, μ, k_o and V_f are constant, the time to reach a penetration distance z is

$$t = \frac{2 \cdot \mu \cdot k_o \cdot z^2}{r_f^2 \cdot P_a} \cdot \frac{V_f^2}{(1 - V_f)^3}$$

(5)

where p_a is the applied pressure. The penetration distance after a special time can then be expressed as:

$$z(t) = \sqrt{\frac{r_f^2 \cdot P_a}{2 \cdot \mu \cdot k_o} \cdot \frac{(1 - V_f)^3}{V_f^2}} \cdot \sqrt{t}$$

(6)

The cross-sectional area, A_b, of a fully consolidated (i.e. void free) commingled yarn bundle can be obtained from the theoretical density, ρ_t, and the weight of unit length bundle, W_b, according to

$$A_b = A_M + A_F = \frac{W_b}{\rho_t}$$

(7)

This cross-sectional area was assumed to be the initial area of the reinforcing fibres before pressure was applied. The presumed area of the matrix, A_m, could be calculated by

$$A_M = V_M \cdot A_b$$

(8)

and was imagined as a layer above the reinforcing fibre layer. The reinforcing fibres were spread equal about the area A_b. Figure 3 shows this arrangement.

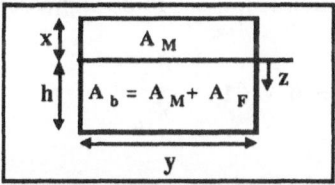

FIGURE 3. *The film stacking process in the impregnation model*

The distance "x" can be ascertained by the equation

$$x = V_M \cdot h$$

(9)

which leads to a void content of the laminate before applying pressure of

$$X_{vo} = \frac{A_M}{A_M + A_b} = \frac{x \cdot y}{x \cdot y + h \cdot y} = \frac{x}{x + h}$$

(10)

This value refers to the share in the voids of the whole laminate. The cross-sectional area, A_b, is the area without voids. Hence, the area of the voids is as large as the area of the matrix.

Based on the idea of such a representative bundle, the impregnation and consolidation behaviour in a real composite laminate can be described. For example, if the total number of fibre bundles, N, in the laminate manufacturing can be identified, the current laminate thickness, H, can be evaluated from the following equation:

$$H = \frac{N \cdot A_b (1 + X_v)}{B} = H_0 \cdot (1 + X_V)$$

(11)

where B is the width of the processing mould and $H_0 = N \cdot A_b/B$ is the thickness of the fully consolidated laminate.

In reality the matrix fibres are not completely unmingled and seperated from the reinforcing fibres, so that the penetration distance "h" is shorter. Figure 4 shows the initial situation when only a part of the fibres is unmingled.

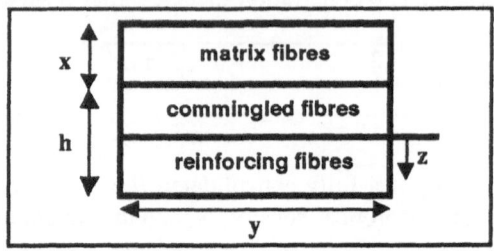

FIGURE 4. *Initial situation of the impregnation model*

Between the matrix layer and the layer with the reinforcing fibres is a layer with commingled matrix and reinforcing fibres. Hence the initial void content is lower, as well. Assuming three fourth of the matrix fibres are still commingled with the reinforcing fibres, this leads to a distance "x", of only

$$x = 1/4 \ V_M \cdot h$$

(12)

Equations (5) - (12) provide relationships between void content, degree of impregnation and processing variables in the consolidation process, namely, viscosity as a function of temperature $[\mu = \mu(T)]$, applied pressure, holding time and bundle geometry. To directly introduce the void content in equation (5), z has to be expressed in terms of the remaining void content:

$$z = h \cdot \left(1 - \frac{X_v}{X_{vo}} \right)$$

(13)

4. Results and Discussion

4.1 IMPREGNATION OF LAMINATES

The cross-sectional area, A_b, of a fully consolidated fibre bundle and the presumed area of the matrix, A_m, were calculated with equation (7) and (8) and are listed in Table 1. The values of k_0 were determined during the evaluation of void content by fitting a set of selected experimental data. Using the equations (3) and (4), the permeability of the fibre tow under different pressures can be estimated. Besides the applied pressure, the permeability also depends upon compaction of the fibre tow, i.e., it varies with the current fibre volume fraction during the impregnation process [3]. This effect is, however, neglected in the present approach.

The viscosity data of the polypropylene matrix were assumed to be the same as known from a comparable carbon fibre/polypropylene composite system [10]. The matrix viscosity, μ, is almost constant at low shear rates and can be fit to an equation as follows [8]:

$$\mu = 2.6 \cdot 10^{-3} \cdot \exp \left(\frac{5600}{T \, [K]} \right) \quad [Pa \cdot s]$$

(14)

where T is the processing temperature, expressed in Kelvin degrees.
With equation (15), the matrix viscosity, μ, of PEEK can be calculated:

$$\mu = 1.13 \cdot 10^{-10} \cdot \exp \left(\frac{19123}{T \, [K]} \right) \quad [Pa \cdot s]$$

(15)

The time to reach a laminate with a certain void content X_v can be calculated combining the equations (5) and (13):

$$t_{X_v} = \frac{2 \cdot \mu \cdot k_o \cdot h^2 \cdot \left(1 - \frac{X_v}{X_{vo}} \right)^2}{r_f^2 \cdot p_a} \cdot \frac{V_f^2}{(1 - V_f)^3}$$

(16)

Equation (17) calculates the resulting void content after a given time:

$$X_v (t) = X_{vo} \left(1 - \frac{r_f}{h \, V_f} \sqrt{\frac{p_a (1 - V_f)^3 \, t}{2 \mu \, k_o}} \right)$$

(17)

The Carman-Kozeny constant k_o was estimated by calculating the time to reach a fully or partly consolidated laminate, t_{X_v}, at different processing conditions and comparing it with the actual results. For the glass fibre/polypropylene in this study k_o amounted to $k_o = 700$. For the carbon fibre/PEEK system it was determined as $k_o = 80$.

In the case of the GF/PP yarn the fibre content in equations (16) and (17) is not the real fibre content of the yarn (V_f^*) but of an estimated loosely packing of glass fibres in the fibre rich areas and amounts to $V_f = 0.5$.

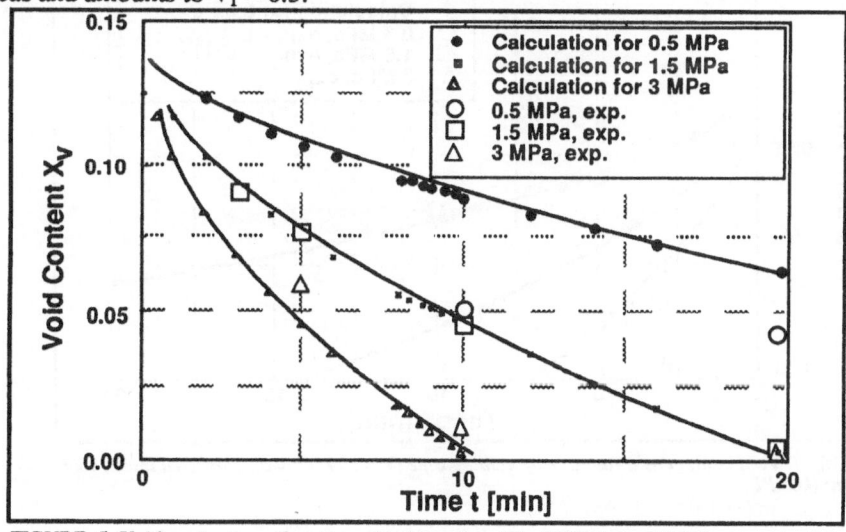

FIGURE 5. *Void content as a function of holding time (T = 185° C) for GF/PP - commingled yarn*

196

Next, the predicted void contents at different processing conditions were calculated. At first, a conservative initial void fraction was estimated from equations (10) and (12) to be $X_{vo} = 15\%$. For a temperature of 185°C this is a good assumption (Figure 5), but for higher temperatures such as 200°C it is more realistic to assume a value of $X_{vo} = 7\%$ (Figures 6 and 7). The model can well describe the trend of laminate consolidation, although there was significant scatter in the void content measured.

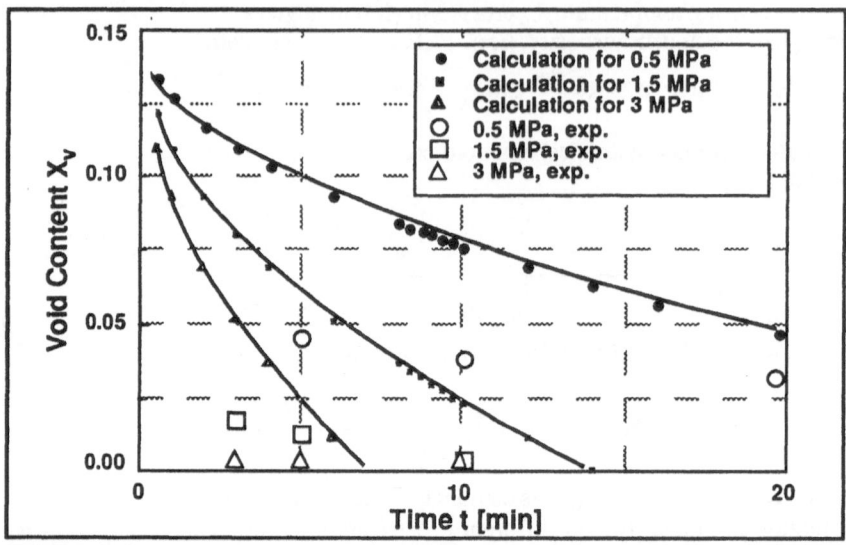

FIGURE 6. *Void content as a function of holding time (T = 200°C) for GF/PP - commingled yarn*

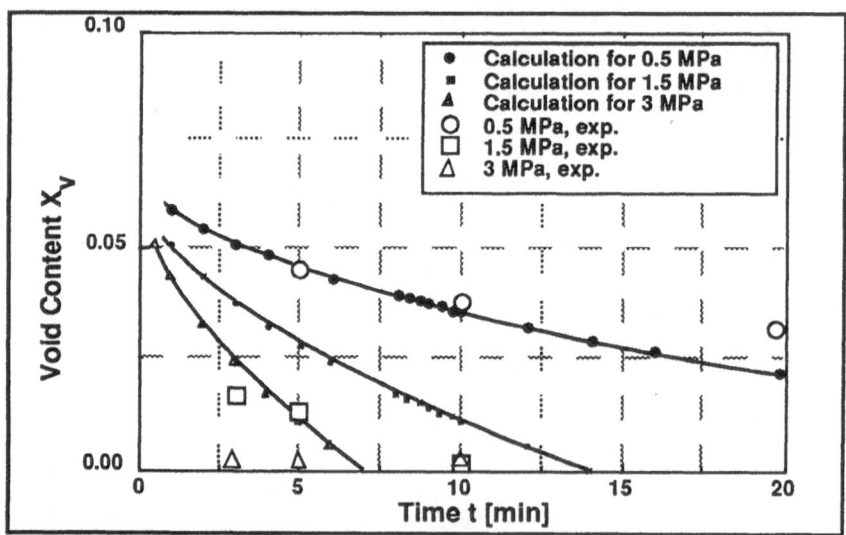

FIGURE 7. *Void content as a function of holding time (T = 200°C) with corrected initial void content (GF/PP)*

Figure 8 compares the predictions and the experimental measurements of the CF/PEEK system for a processing temperature of 380°C with X_{vo} = 10.5 %. It confirms that the basic trend of laminate consolidation is well characterized by the model used in these experiments.

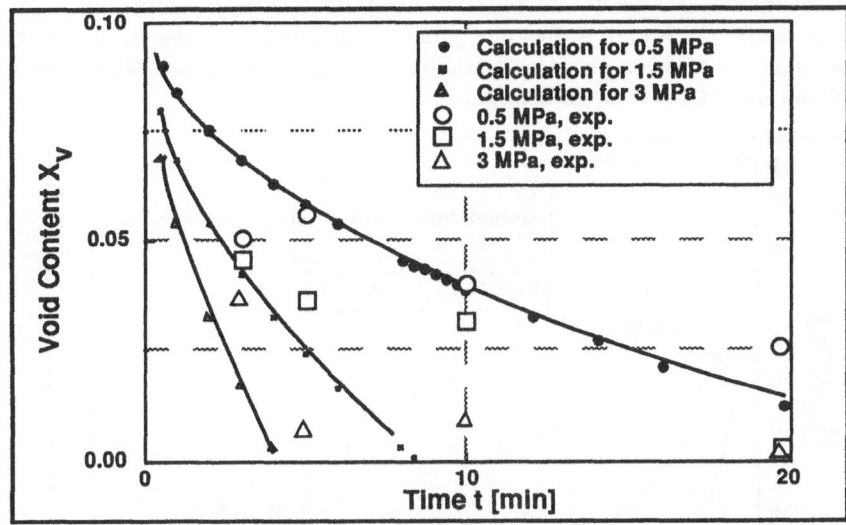

FIGURE 8. *Void content as a function of holding time at three different levels of applied pressure (T = 380°C) for CF/PEEK - commingled yarn*

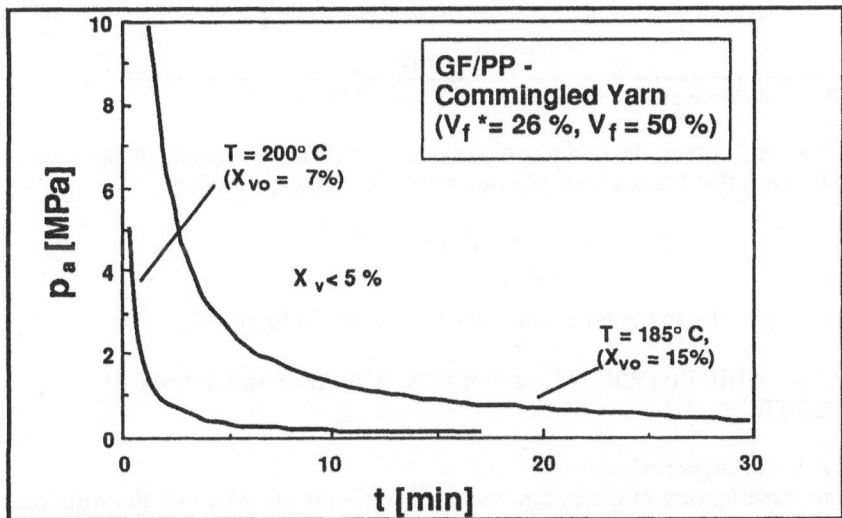

FIGURE 9. *Optimum processing window for commingled GF/PP fibres*

If the void content is set to be an indicative for the consolidation quality, the optimum processing window for manufacturing of laminates from commingled yarn can be evaluated from equation (16), based on a critical level of void content of e.g. X_v = 5 % [5]. For the GF/PP system the impregnation time amounts to

$$t = \left(1 - \frac{X_v}{X_{vo}}\right)^2 3.49 \ 10^6 \ \frac{\mu}{P_a}$$

(18)

considering, in addition, that the initial void content differs with temperature (X_{vo} = 15 % at 185°C and 7% at 200°C). Figure 9 illustrates, that on the left side of the curves the actual void content under the relevant processing conditions is still larger than 5%, whereas it is lower for pressure-time-conditions on the right.

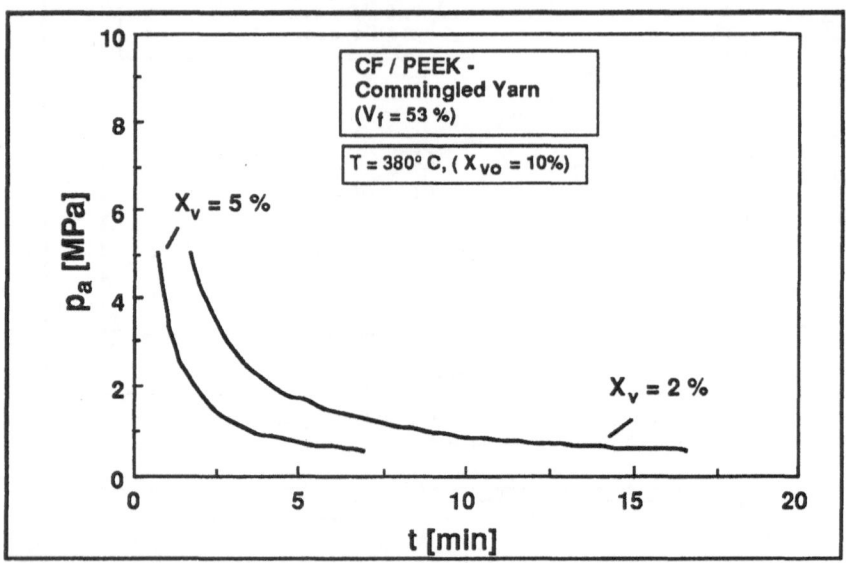

FIGURE 10. *Optimum processing window for commingled CF/PEEK fibres*

For the CF/PEEK system the relationship between temperature, applied temperature, and holding time to reach this desired level of void content can be expressed as:

$$t = \left(1 - \frac{X_v}{X_{vo}}\right)^2 1.29 \ 10^6 \ \frac{\mu}{P_a}$$

(19)

This leads for X_{vo} = 10% to the processing window shown in Figure 10.

4.2   RELATIONSHIP BETWEEN CONSOLIDATION AND MECHANICAL PROPERTIES

4.2.1 Glass Fibres/Polypropylene
From the transverse flexure stress-strain curves it becomes obvious that the responses of consolidated composite parts highly depend upon the processing temperature. There clearly exists a yielding point and a yielding period in the stress-strain curve for the laminates consolidated at T = 185° C. However, at high processing temperature (T = 220° C) the yielding period is significantly decreased. With increase in applied pressure and holding time, the yielding point is gradually reduced to a deviation point in the linearity of the stress-strain curve. In addition, the ultimate stress value is clearly higher in the case of both high temperature and pressure consolidation.

Figure 11 and 12 illustrate the effects of void content on the ultimate transverse flexure stress and the transverse elastic modulus, respectively. In both cases the mechanical properties get reduced with increasing amount of voids in the laminates. Due to difficulties in flexural testing of the rather small samples, only the trends are given, i.e. the absolute values were normalized to the average value measured for 5% void content.

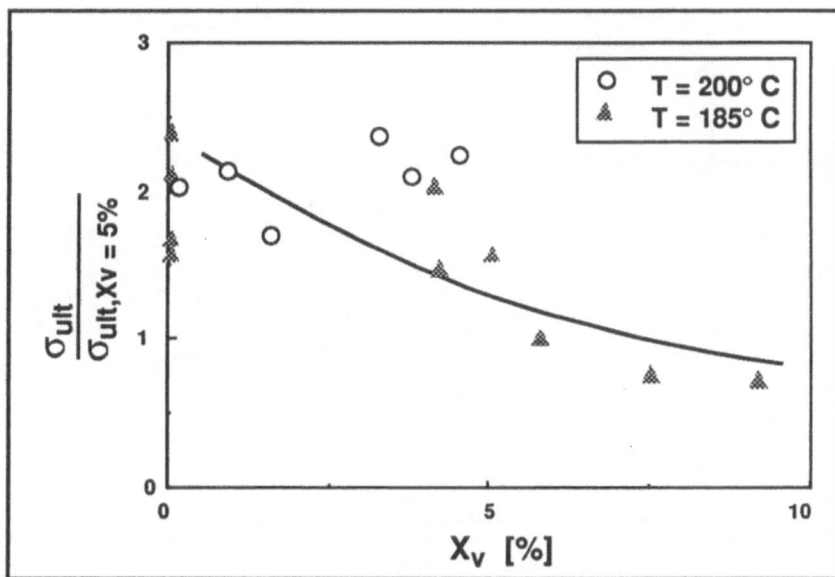

FIGURE 11. *Transverse strength properties as a function of void contents (GF / PP)*

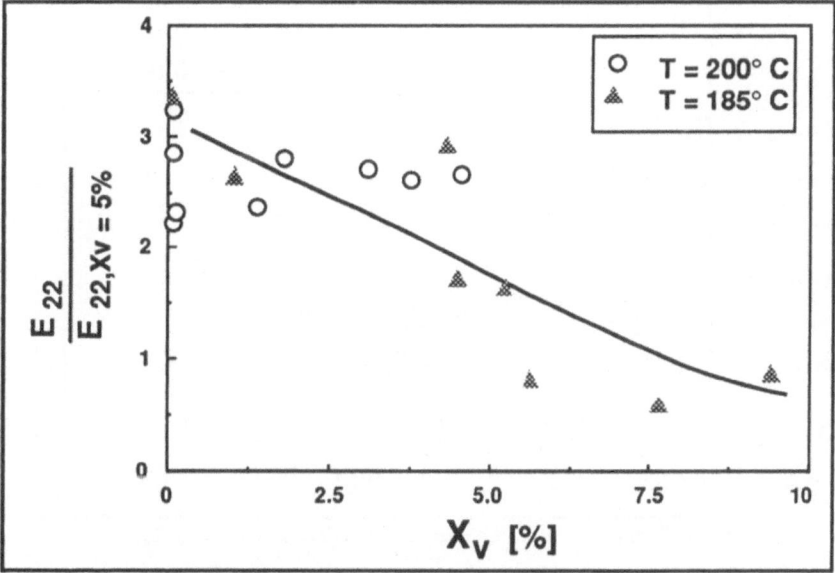

FIGURE 12. *Transverse elastic constants as a function of void contents (GF / PP)*

4.2.2 Carbon Fibres/Polyetheretherketone

Although the transverse flexure stress-strain curves of CF/PEEK looked slightly different from those of the GF/PP samples, the same trends with regard to the effects of processing parameters on mechanical properties were revealed (Figure 13 and 14).

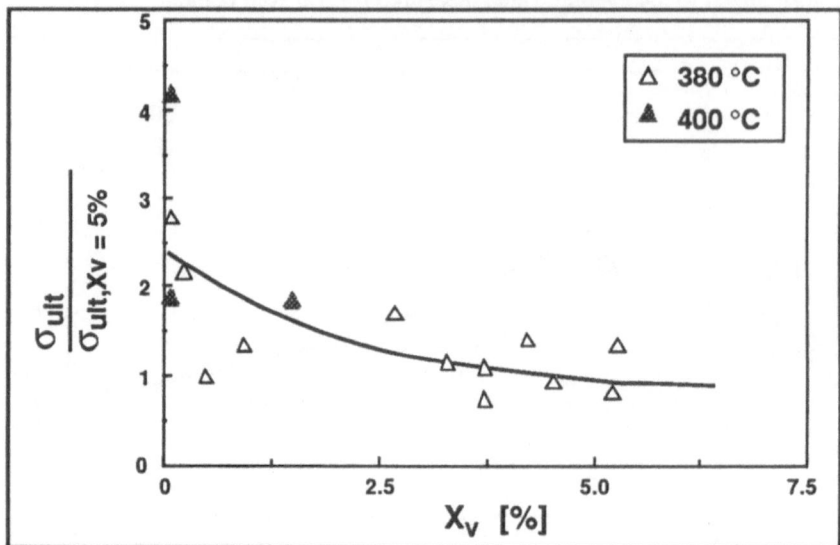

FIGURE 13. *Transverse strength properties as a function of void contents [CF / PEEK]*

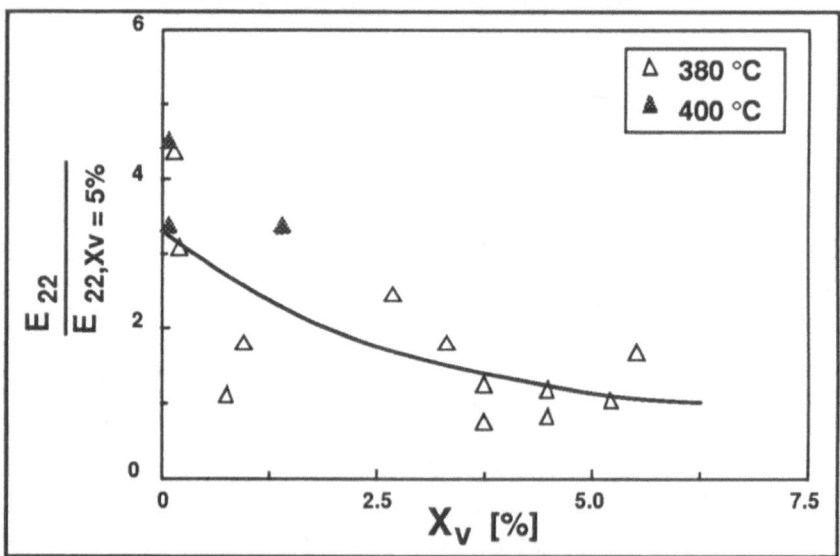

FIGURE 14. *Transverse elastic constants as a function of void contents [CF / PEEK]*

5. Conclusions

The impregnation and consolidation mechanisms in composites made out of commingled GF/PP and CF/PEEK fibre bundles were investigated. The consolidation process of this kind of material differs from other material forms because of the distribution of fibres and matrix in the unconsolidated states. A model has been developed to qualitatively describe the impregnation process during consolidation. Combined with the permeability model, this model predicts the current void content and laminate thickness etc. as a function of bundle geometry and processing variables (temperature, applied pressure and holding time). Good correlations with the experimental data indicate the success of this approach. Based on a desired, minimum level of void content (e.g. $X_V = 5\%$) in the laminates, optimum processing windows for manufacturing of composite parts from these materials are suggested. In practice, the present results will give important ideas to the user about how to optimize the manufacturing process, for example in case of laminate manufacturing, in order to obtain optimum structure-property relationships.

Acknowledgements

One of the authors, Dr. L. Ye would express his appreciation to the Alexander von Humboldt-Foundation (AvH) for the research fellowship at the University of Kaiserslautern. Prof. K. Friedrich acknowledges the financial help from the Fonds der Chemischen Industrie, Frankfurt, for his personal research activities in 1994. Further thanks are addressed to the Deutsche Forschungsgemeinschaft (DFG) for the support of this works in the frame of a German-Korean Research Project (DFG-FR-675-11-1).

References

[1] Bigg, D. M., Hiscock, D. F., Preston, J. R. and Bradbury, E. J. (1988) High Performance Thermoplastic Matrix Composites, *J. Thermoplastic Composite Materials*, 1, 146-160

[2] Chang, I. Y. and Lees, J. K. (1988) Recent Development in Thermoplastic Composites: A Review of Matrix Systems and Processing Methods, *Journal of Thermoplastic Composite Materials* 1, 277-295

[3] Ye, L., Klinkmueller, V., Friedrich, K. (1992) Impregnation and Consolidation in Composites Made of GF/PP Powder Impregnated Bundels, *Journal of Thermoplastic Composites* 5, 32-48

[4] Ye, L., Friedrich, K.,Cutolo, D., Savadori, A. (1994) Manufacturing of CF/PEEK Composites from Powder/Sheath-Fiber Preforms, *Composites Manufacturing* 5, *No.1*, 41 - 50

[5] Ye. L. and Friedrich, K. (1993) Mode I Interlaminar Fracture of Commingled Yarn Based Glass/Polypropylene Composites, *Composites Science and Technology* 46, 187-198

[6] Van West, B. P. Pipes, R. B. and Advani, S. G. (1991) The Consolidation of Commingled Thermoplastic Fabrics, *Polymer Composites* 12 No. 6, 417-427

[7] Gutowski, T. G., Cai. Z., Bauer, S., Boucher, D., Kingery, J. and Wineman, S. (1987) Consolidation Experiments for Laminate Composites, *Journal of Composite Materials* 21, 650-669

[8] Kim, W. T., Jun E. J. , Um, M. K. and Lee, W. I. (1989) Effect of Pressure on the Impregnation of Thermoplastic Resin into a Unidirectional Fibre Bundle, *Advances in Polymer Technology* 9, 275-279

[9] Greenkorn, R. A. (1983) *Flow Phenomena in Porous Media*, Dekker, New York

[10] Cutolo, D. (1991) private communication, Enichem, Italy

MORPHOLOGY/LOADING DIRECTION COUPLING ON THE TRANSVERSE BEHAVIOUR OF COMPOSITES

D. Kujawski, Z. Xia and F. Ellyin

Department of Mechanical Engineering
University of Alberta
Edmonton, Alberta, Canada T6G 2G8

Abstract

Experimental and numerical results are presented on the transverse stress-strain response of composite systems depending on the applied load direction. Coupon specimens of 6061-T0 aluminum alloy with square array of circular holes (hollow and with steel filament reinforcement) were used to simulate an ideal regular composite system. A different load direction was obtained by a rotation of the pattern of holes with respect to the longitudinal axis of the coupon sample. Numerical results were obtained by FEM analysis on unit cells. A change in the load direction results in the different unit cell to be used. The numerical results show a fair agreement with experimental data. The results of the perforated and reinforced periodic systems indicate the same trend in load direction dependency. This dependency is also affected by the presence of simulated voids. The effect of void pattern has a significant effect on the stress at failure. The variation in the failure stress due to change in load direction was also evaluated by preliminary macroscopic analysis which takes into account the observed failure modes.

1.   Introduction

The deformation characteristics of continuous fiber-reinforced composites depend, in general, on the constituent phases as well as concentration and arrangement of the fibers. However, when the composite is loaded in the reinforcement direction the fiber arrangement does not affect significantly the overall material stress-strain response. In contrast, when the load is applied transversely the fiber arrangement may have a primary effect on the transverse behaviour. Traditionally, this behaviour is predicted based on a unit cell modelling by assuming perfect periodicity of the fiber arrangement [e.g. see 1-4]. A two-dimensional view of three commonly used periodic arrangements of circular fibers are: square edge-packing, square diagonal-packing and triangle-packing. These are shown schematically in Fig. 1, where the dashed lines represent a unit cell. For the edge- (or diagonal-) packing and triangle-packing systems, the fibers are located in a square and a regular hexagonal array, respectively. In the square array arrangement, the distance between neighbouring fibers is not constant in the sense that the distance along the edge is different from that along the diagonal of the square. Therefore, this results in a significant direction-dependent behaviour of the square array morphology for transverse loading. Numerical results for composites [4] and experimental data for perforated materials [5] indicate that the square edge-packing system is the strongest whereas the square diagonal-packing is the softest one. The third system, i.e. the triangle-packing, exhibits the response which is somewhere between the above mentioned two arrangements.

R. Pyrz (ed.), IUTAM Symposium on Microstructure-Property Interactions in Composite Materials, 203–213.
© 1995 *Kluwer Academic Publishers.*

204

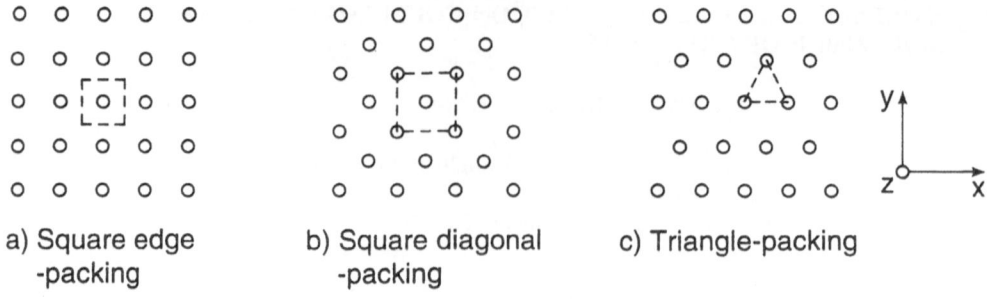

a) Square edge
-packing

b) Square diagonal
-packing

c) Triangle-packing

Fig. 1 A two-dimensional view of the different fiber arrangements
(a) square edge-packing, (b) square diagonal-packing, (c) triangle-packing

In general, for a specific periodic morphology the overall (macroscopic) transverse behaviour is direction-dependent. This dependency may also be manifested in different failure modes. Thus, there is a coupling between morphology and loading direction in the transverse behaviour of the unidirectional composite materials. However, this coupling effect is difficult to investigate experimentally for real composites due to variability of the fiber arrangements [6,7].

The objective of this work is to study experimentally the transverse behaviour of regular composite systems depending on the applied load direction. Coupon specimens of 6061-T0 aluminum alloy with square array of circular holes (hollow and reinforced) were used to simulate an ideal composite system. In addition, the effect of two different void arrangements on the overall response was considered, in which the voids were simulated as holes without reinforcement. Experimentally recorded stress-strain responses for perforated and reinforced systems are compared with the results obtained by FEM analysis of unit cells.

2.   Experimental and Numerical Approach

2.1   MATERIAL AND SPECIMENS

6061 aluminum alloy coupon specimens having a thickness of 3.175 mm with a width of 12.7 mm and a testing length of 50.8 mm were used in this investigation. A square array of holes, as depicted in Fig. 2, with a diameter $d \approx 0.985$ mm and a pitch $a = 1.25$ mm were drilled in the testing section of the sample. An angle α, shown in Fig. 2, specifies the inclination of the array pattern with respect to the longitudinal axis of the specimen. The specimens with $\alpha = 0°$, 26.6° and 45° were fabricated, as shown in Fig. 3. A change in the inclination angle α, results in a different unit cell. The corresponding unit cells are depicted in Fig. 2 by dashed lines. Note that for $\alpha = 0°$ and 45° one gets the square edge-packing and diagonal-packing systems, respectively. A high strength steel filament with a diameter $d_f = 0.99$ mm was used as the reinforcement material. The volume fraction of the filaments (fibers) was 0.47, which is a typical concentration for metal-matrix composites [6]. The tests were carried out on both types of specimens, i.e. unreinforced (perforated) and reinforced. Before testing all specimens (perforated and reinforced) were heat treated to obtain a fully annealed, T0, condition for the 6061 aluminum alloy matrix.

The mechanical properties of 6061-T0 matrix and the steel wire were obtained from tensile tests and are summarized in Table 1. Both materials, i.e. reinforcement and matrix, were assumed to be isotropic.

Table 1 Mechanical Properties

Material	Young's Modulus E(GPa)	Poisson's Ratio v	Yield Strength $\sigma_{0.2}$ (MPa)	Strain Hardening Exponent, n
6061-TO Alloy (fully annealed)	69	0.33[a]	43.5	0.333
High Strength Steel Wire	203	0.3[b]	1725	0.111

(a) From ref. [6], (b) estimated.

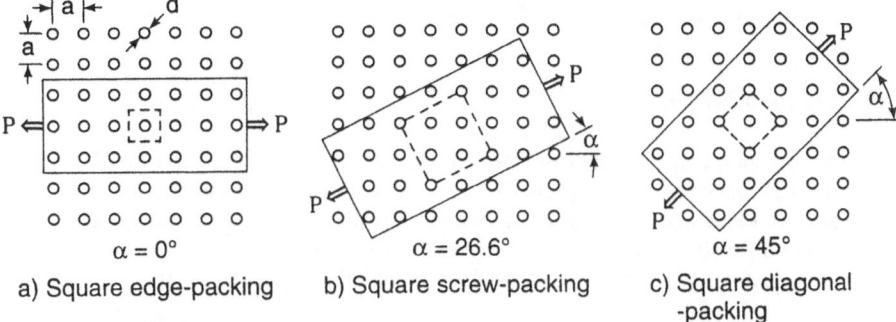

a) Square edge-packing   b) Square screw-packing   c) Square diagonal -packing

Fig. 2 Various orientations of coupon specimens relative to lattice direction for square array pattern

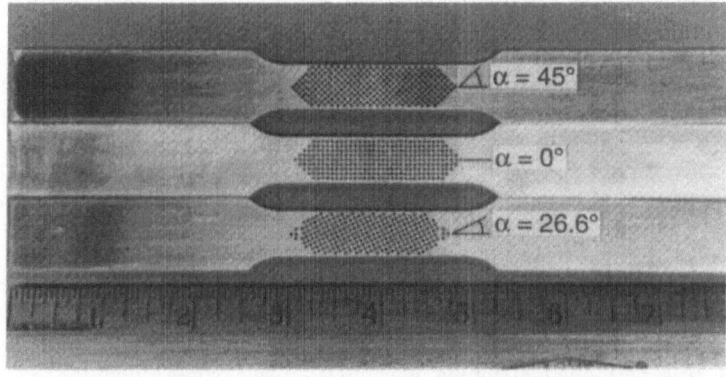

Fig. 3 Coupon specimens with various orientations of the lattice directions

The effect of void arrangement on the overall response of composite systems was also investigated in the case of transverse loading. For each angle $\alpha = 0°$, 26.6° and 45°, two types of void arrangements were included, viz. square-pattern (S-P) and rhomboidal pattern (R-P), as shown in Fig. 4. The voids were simulated as holes without reinforcement. The ratio of voids to fibers was 25:75.

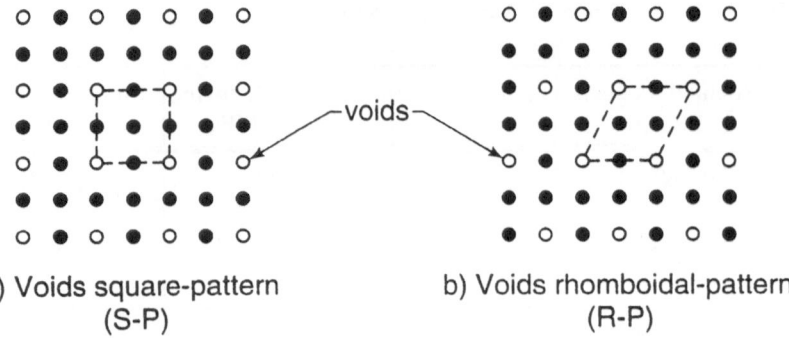

a) Voids square-pattern
(S-P)

b) Voids rhomboidal-pattern
(R-P)

Fig. 4 Arrangements of voids
(a) square-pattern (S-P), (b) rhomboidal-pattern (R-P)

2.2   TESTING PROCEDURE

All tests were carried out on an MTS servo-controlled system using ramp function for displacement. The displacement was measured by a clip-on 25.4 mm gauge length extensometer and converted to engineering strain. In the analysis two types of engineering stresses are used, a gross stress, S, and a net stress, σ, based on gross and net cross-section, respectively. An IBM PC computer was employed to provide a command signal. The data were recorded using both the IBM PC and an X-Y plotter.

2.3   NUMERICAL MODELLING

The numerical prediction of the stress-strain response of the perforated and reinforced systems was performed using a finite element code, ANSYS 5.0. A three-dimensional (3-D) unit cell with a circular fiber (or hole) was modelled for the three arrangements: square edge-packing (Fig. 2a), square screw-packing (Fig. 2b), and square diagonal-packing (Fig. 2c). Note that an appropriate unit cell selection (shown in Fig. 2) depends on the mutual orientation of the lattice with respect to the load direction which is specified by the angle α. For the square edge- (or diagonal-) packing system, due to symmetry, only one-fourth of the unit cell has to be considered in modelling. A finite element mesh in the transverse plane is shown in Fig. 5. The mesh was created using ANSYS mesh-generation program. In the direction of fiber/hole, only one layer of elements with unit thickness was modelled. For the composite with fibers, it is assumed that the fiber/matrix interfaces are debonded but contact each other. 3-D point to surface contact elements were used to model such interface conditions.

The FEM analysis was performed assuming generalized plane strain condition, i.e. boundary conditions prescribed for the unit cell satisfied the requirement that the right parallelepiped shape of the unit cell remains a right parallelepiped during deformation. The above boundary condition does not strictly apply to the square screw-packing arrangement because the edges of the unit cell need not necessarily remain straight during deformation. However, the straight edge assumption is an acceptable approximation for small macroscopic strains (up to 1%) considered in the numerical analysis. The validity of the above approximation has been checked experimentally using the perforated specimen. Before starting the test corresponding edge lines for the square screw-packing arrangement were marked on the sample. It was observed that these lines remained approximately straight for large macroscopic strains (2-3 percent).

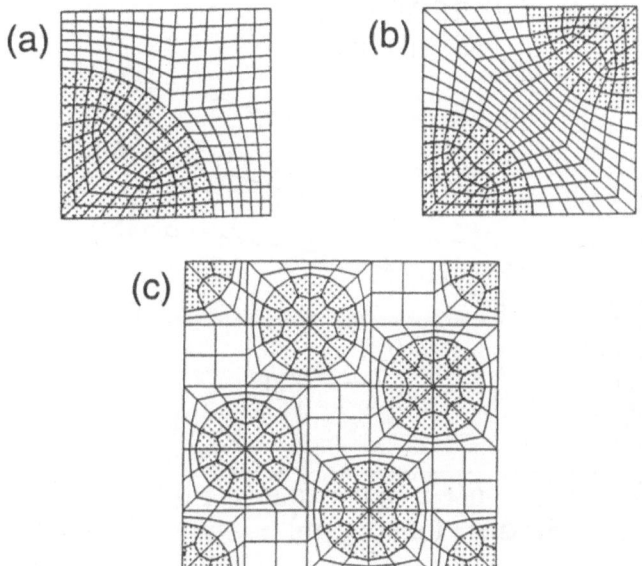

Fig. 5   A two-dimensional view of the finite element discretization
          (a) square edge-packing, (b) square diagonal-packing, (c) square screw-packing

The overall (macroscopic) stress-strain curves for transversely loaded composite systems were calculated based on the properties of the constituent materials given in Table 1. In the plastic regime the incremental (J_2 flow) theory of plasticity with kinematic hardening was used.

3.   Results

3.1   OVERALL STRESS-STRAIN RESPONSE

Figures 6a and 6b show the experimental results in terms of gross stress, S, versus overall strain, ε, for perforated (no fibers) and reinforced (no voids) materials systems, respectively. It is clear that the overall (macroscopic) stress-strain behaviour depends on the angle α specifying the orientation of the pattern with respect to the applied load direction. Both systems, i.e. perforated and reinforced, show the same trend in direction-dependent behaviour. The results indicate that the response for α = 0 is the strongest, whereas for α = 45 is the softest. The third system with α = 26.6° exhibits a response which is somewhere between the former two. The results obtained by FEM analysis are also depicted in Fig. 6 by dashed lines. A fairly good agreement between experiments and numerical analysis is observed, especially for the perforated systems. Further experimental results concerning the macroscopic transverse stress-strain behaviour, including an arrangement with voids, are shown in Fig. 7. The influence of void arrangement for each investigated angle α can be seen in this figure. The results indicate that material with voids arranged in the square-pattern (S-P) exhibit lower strain hardening than that with the rhomboidal-pattern (R-P). Consequently, for a given α lower stress at failure was recorded for the material with S-R void pattern in comparison to that of R-P.

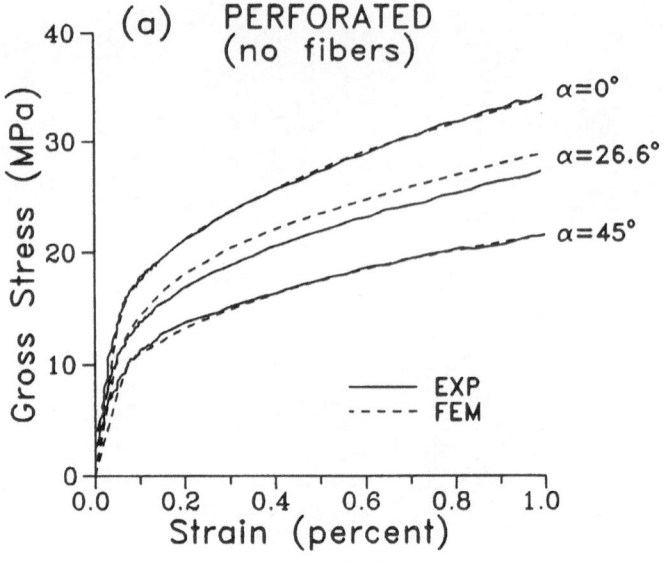

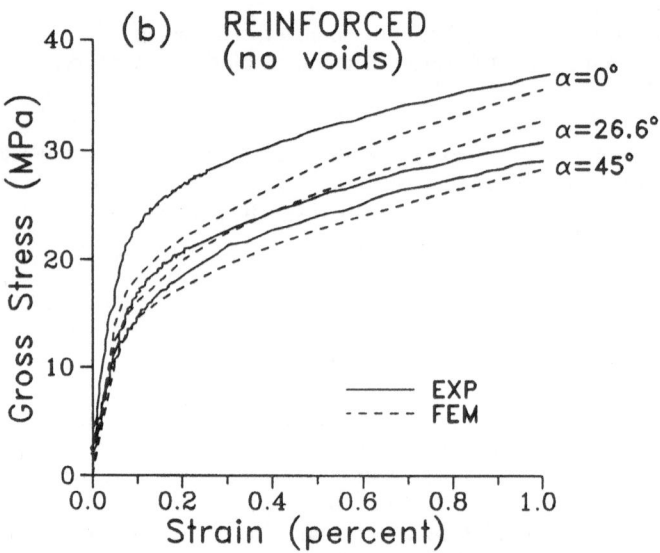

Fig. 6 Overall transverse stress-strain curves for
(a) perforated (no fibers), (b) reinforced (no voids) systems

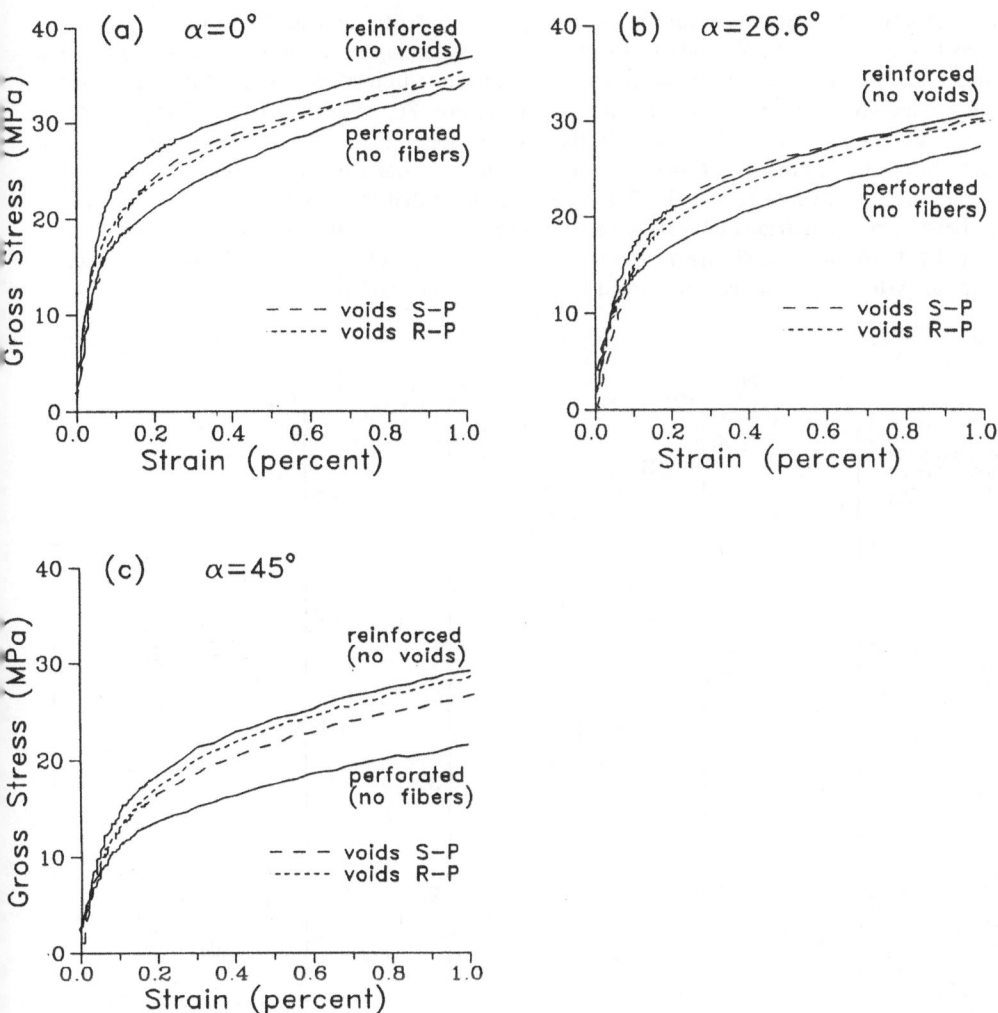

Fig. 7 Overall transverse stress-strain curves of composite systems for various loading directions
(a) α = 0°, (b) α = 26.6°, (c) α = 45°

3.2   MACROSCOPIC FAILURE MODES

Macroscopic failure modes and the corresponding stress and strain at failure were also investigated. Figure 8a presents the failure modes observed for the perforated (no fibers) material systems. It is seen that the failure modes are influenced by the orientation angle α i.e. a separation mode for α = 0°, a slip mode for α = 45°, and a combined separation and slip made for α = 26.6°. Similar failure modes (with respect to angle α) were observed for other material systems, i.e. reinforced (no void) and reinforced with voids arranged in the square- or rhomboidal-pattern. Figure 8b shows the recorded values of the macroscopic strain and gross stress at failure for perforated material (designated as ε_{fp} and S_{fp}, respectively) versus the orientation angle α. The results indicate that the gross stress at failure, S_{fp}, decreases with increasing angle α. By contrast, the reverse trend is seen with respect to the macroscopic strain at failure, ε_{fp}, i.e. ductility increases with the increasing α. The gross failure stresses, S_f, for the other three material systems, i.e. reinforced (no voids) and that with square and rhomboidal voids pattern, were normalized with respect to the failure stress for the perforated material, S_{fp}, and the results are shown in Fig. 8c. For $\alpha = 0$ and 26.6°, all three normalized failure stresses, S_f/S_{fp}, are less than 1, i.e. these material systems are weaker than perforated ones. In contrast, for $\alpha = 45°$, except for the material with square-pattern of voids, the normalized stresses are higher than 1. It is seen in Fig. 8c that the voids arranged in the square-pattern have a more detrimental effect on the stress at failure in comparison to that of the rhomboidal-pattern. In general, the overall stress at failure depends on both applied stress direction and material systems (perforated, reinforced and void contents and pattern).

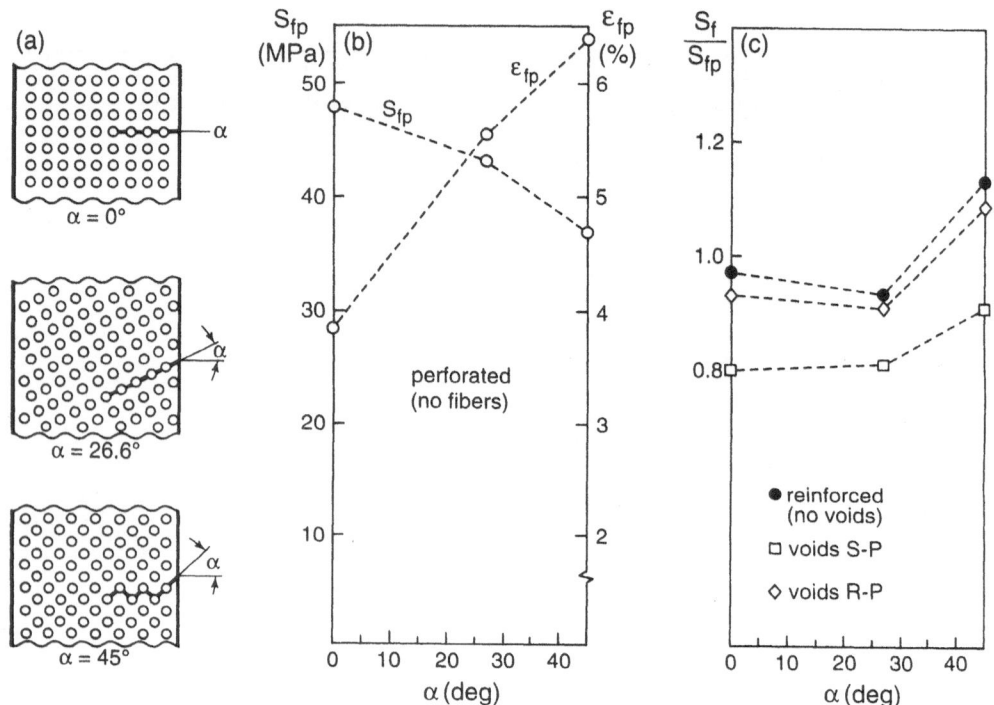

Fig. 8   Macroscopic (a) failure modes, (b) strain and gross stress at failure for perforated
system, and (c) normalized gross stress at failure versus loading direction angle

4.   Macroscopic Analysis

Using the failure modes, shown in Fig. 8a, the net stress at failure, $\sigma_{f\alpha}$, is calculated from

$$\sigma_{f\alpha} = P_{f\alpha}/A_\alpha^{net} \qquad (1)$$

where $P_{f\alpha}$ is the load at failure for a given angle α and A_α^{net} is the net cross-section of the corresponding failure plane. The variation of the normalized net stress at failure, defined as $\sigma_{f\alpha}$ divided by σ_{f0} (for $\alpha = 0$), versus the orientation angle α is shown in Fig. 9. The significant drop in the normalized net stress at failure, $\sigma_{f\alpha}/\sigma_{f0}$, is observed with the increasing angle α. To indicate a possible explanation for this variation, a preliminary macroscopic analysis is conducted. It is noted that at large plastic strains, the stress-strain relation for fully annealed matrix material shows relatively low strain hardening. At failure, the specimen cross-section is fully yielded which indicates that an elastic-perfectly plastic material model can be assumed. Therefore, this implies that the stress concentrations due to the hole/fiber can be neglected when calculating macroscopic stresses at failure.

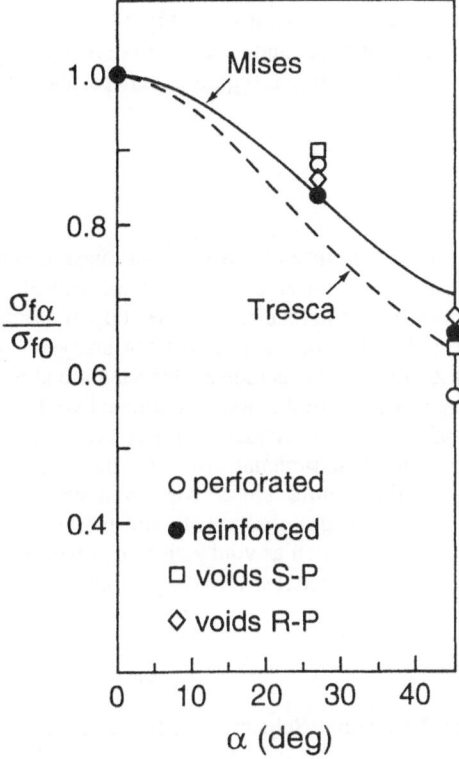

Fig. 9 Comparison of experimental and analytical net stresses at failure for different composite systems

Let us define the equivalent failure stress, $\bar{\sigma}_f$,

$$\bar{\sigma}_f = (\sigma_f^2 + \beta\tau_f^2)^{1/2} \qquad (2)$$

where σ_f and τ_f are the normal and shear stress components on the fracture plane. Equation (2) with $\beta = 3$ or 4 reduces to von Mises or Tresca equivalent stress, respectively. Using relations for $\sigma_f = \sigma_{f\alpha} \cos\alpha$ and $\tau_f = \sigma_{f\alpha} \sin\alpha$, eq. (2) yields

$$\bar{\sigma}_f = \sigma_{f\alpha}(\cos^2\alpha + \beta\sin^2\alpha)^{1/2} \qquad (3)$$

Noting that for $\alpha = 0°$, $\bar{\sigma}_f = \sigma_{f0}$, the above relation can be written as

$$\frac{\sigma_{f\alpha}}{\sigma_{f0}} = \frac{1}{(\cos^2\alpha + \beta\sin^2\alpha)^{1/2}} \tag{4}$$

Equation (4) describes the variation of the normalized net stress at failure depending on the angle α. The predictions of eq. (4) with $\beta = 3$ and 4 are depicted in Fig. 9. It is seen that the observed experimental variations of the failure stress with the angle α can be predicted using this preliminary macroscopic analysis. A more accurate prediction will require a detailed microscopic analysis and will be reported at a later date.

5. Conclusion

For regular composite systems, the transverse stress-strain response depends on the applied load direction relative to the arrangement pattern. A different load direction was obtained by a rotation of the periodic morphology with respect to longitudinal axis of the coupon specimen. A change in load direction results in the different unit cell to be used in an FEM analysis. The results for the perforated (no fibers) and reinforced (no voids) systems indicate the same trend for the load direction dependency. This dependency is altered by the introduction of simulated voids to the material. The effect of void pattern is not very significant at low stresses, but becomes more apparent at failure. The square-pattern of voids exhibits a more detrimental effect on the stress at failure than the rhomboidal-pattern. The variation of the failure stress was evaluated using a preliminary macroscopic analysis which takes into account the observed failure modes. The coupling effect between morphology of the fiber arrangement as well as void pattern relative to transversely applied load requires further investigation using a more detailed micromechanical analysis.

Acknowledgement

This work was supported, in part, by the Natural Sciences and Engineering Research Council of Canada.

References

[1]   Hashin, Z. (1983) Analysis of composite materials - A survey, *ASME Trans. J. Appl. Mech.* **50**, 481-505.

[2]   Aboudi, J. (1987) Closed form constitutive equations for metal matrix composites, *Int. J. Engng. Sci.* **25**, 1229-1240.

[3]   Teply, J.L. and Dvorak, G. (1988) Bounds on overall instantaneous properties of elastic-plastic composites, *J. Mech. Phys. Solids* **36**, 29-58.

[4]   Brockenbrough, J.R., Suresh, S. and Wienecke, H.A. (1991) Deformation of metal-matrix composites with continuous fibers: geometrical effects of fiber distribution and shape, *Acta Metall. Mater.* **39**, 735-752.

[5] Litewka, A. and Sawczuk, A. (1982) On a continuum approach to plastic anisotropy of perforated materials, in: J.P. Boehler (ed.), *Mechanical Behaviour of Anisotropic Solids*, Martinus Nijhoff, The Hague, 803-817.

[6] Backer, W., Pindera, M-J. and Herakovich, C.T. (1987) Mechanical response of unidirectional boron/aluminum under combined loading, Report CCMS-87-06 (VPI-E-87-17), VPI and SU, Blacksburg, VA.

[7] Pyrz, R. (1992) Stereological quantification of the microstructure morphology for composite materials, in: P. Peterson (ed.), *Optimal Design with Advanced Materials*, Elsevier, Amsterdam, 81-95.

BOUNDS FOR OVERALL NONLINEAR ELASTIC OR VISCOPLASTIC PROPERTIES OF HETEROGENEOUS SOLIDS

S. NEMAT-NASSER AND B. BALENDRAN
Center of Excellence for Advanced Materials
Department of Applied Mechanics and Engineering Sciences
University of California, San Diego
La Jolla, California 92093-0416, USA

MUNEO HORI
Department of Civil Engineering
University of Tokyo 7-3-1 Hongo
Bunkyo-ku, Tokyo 113, JAPAN

ABSTRACT: For a sample of a general heterogeneous nonlinearly elastic material, it is shown that, among all consistent boundary data which yield the same overall average strain (stress), the strain (stress) field produced by uniform boundary tractions (linear boundary displacements), renders the elastic strain (complementary strain) energy an absolute minimum. Similar results are obtained when the material of the composite is viscoplastic. Based on these results, universal bounds are presented for the overall potentials of a general, possibly finite-sized, sample of heterogeneous materials with arbitrary microstructures, subjected to any consistent boundary data with a common prescribed average strain (strain-rate) or stress. Statistical homogeneity and isotropy are neither required nor excluded.

1. Introduction

Composites consisting of nonlinearly elastic or nonlinearly viscoplastic constituents are considered. The strain (strain-rate) is denoted by $\varepsilon(x)$, where x measures position within the composite of volume V, bounded by ∂V. The corresponding stress field is denoted by $\sigma(x)$. It is further assumed that the constituent material admits stress and strain (strain-rate) potentials, $\phi(x,\varepsilon)$ and $\psi(x,\sigma)$, such that, at each point x in V,

$$\sigma = \partial\phi/\partial\varepsilon, \quad \varepsilon = \partial\psi/\partial\sigma, \qquad (1.1\text{a,b})$$

respectively. These potentials are related by a Legendre transformation,

$$\phi + \psi = \sigma : \varepsilon, \qquad (1.1\text{c})$$

215

R. Pyrz (ed.), IUTAM Symposium on Microstructure-Property Interactions in Composite Materials, 215–221.
© *1995 Kluwer Academic Publishers.*

216

where : denotes a double contradiction.

Define the overall stress and strain (strain-rate) potentials by the unweighted volume average of ϕ and ψ, respectively, i.e., set

$$\Phi \equiv <\phi> \equiv \frac{1}{V} \int_V \phi \, dV,$$

$$\Psi \equiv <\psi> \equiv \frac{1}{V} \int_V \psi \, dV, \tag{1.2a,b}$$

and seek to develop computable bounds for these potentials, for any consistent boundary data which may be prescribed on ∂V.

2. Averaging Theorems

Among all possible boundary data that may be prescribed on ∂V, the uniform-traction and the linear-displacement boundary data are of special consideration. Let E and Σ be given constant strain (strain-rate) and stress tensors. The linear-displacement boundary data are then defined by

$$u = x \cdot E \quad \text{on } \partial V, \tag{2.1a}$$

and the uniform-traction boundary data are given by

$$t = n \cdot \sigma = n \cdot \Sigma \quad \text{on } \partial V, \tag{2.1b}$$

where u is the displacement field and n is the unit exterior normal to ∂V. These boundary data are, in general, mutually exclusive, although, under restrictive circumstances they may coexist. When the boundary data are given by (2.1a), then it follows that (Hill, 1963)

$$\bar{\varepsilon}^E \equiv <\varepsilon> \equiv \frac{1}{V} \int_V \varepsilon(x; E) \, dV = E, \tag{2.2a}$$

where $\varepsilon = \varepsilon(x; E)$ is the strain (strain-rate) field. Denote the corresponding stress field by $\sigma = \sigma(x; E)$ and set

$$\bar{\sigma}^E \equiv <\sigma> = \frac{1}{V} \int_V \sigma(x; E) \, dV. \tag{2.3a}$$

The overall stress potential is then denoted by $\Phi^E(E) \equiv <\phi(x, \varepsilon(x; E))>$. It is easy to show that (Nemat-Nasser and Hori, 1990, 1993)

$$\bar{\sigma}^E = \partial \Phi^E(E)/\partial E. \tag{2.4a}$$

Similarly, when the boundary data are given by (2.1b), the average stress becomes

$$\bar{\sigma}^\Sigma \equiv <\sigma> \equiv \frac{1}{V} \int_V \sigma(x; \Sigma) \, dV = \Sigma, \tag{2.2b}$$

where $\sigma = \sigma(x; \Sigma)$ is the stress field in V. The corresponding strain (strain-rate) is denoted by $\varepsilon = \varepsilon(x; \Sigma)$, and its average is given by

$$\bar{\varepsilon}^\Sigma \equiv <\varepsilon> \equiv \frac{1}{V} \int_V \varepsilon(x; \Sigma) \, dV. \tag{2.3b}$$

The overall strain (strain-rate) potential then becomes $\Psi^\Sigma(\Sigma) = \langle\psi(x\,;\,\sigma(x\,;\,\Sigma))\rangle$. It can then be shown that

$$\bar{\epsilon}^\Sigma = \partial\Psi^\Sigma(\Sigma)/\partial\Sigma; \qquad (2.4b)$$

see Nemat-Nasser and Hori (1990, 1993).

In addition to the boundary data (2.1a,b), consider *any* general (but consistent) boundary data, and denote the corresponding strain (strain-rate), stress, and the associated potentials, respectively, by $\epsilon^G = \epsilon^G(x)$, $\sigma^G = \sigma^G(x)$, $\phi^G(x)$, and $\psi^G(x)$. Then, the overall average quantities are, $\bar{\epsilon}^G$, $\bar{\sigma}^G$, Φ^G, and Ψ^G, respectively. These are obtained by simple volume averaging. The aim is to obtain computable bounds for Φ^G and Ψ^G. This is done under the assumption that $\phi(x\,;\,\epsilon)$ and $\psi(x\,;\,\sigma)$ are convex functions of ϵ and σ, respectively. Hence, for any two strain (strain-rate) fields $\epsilon^{(1)}$ and $\epsilon^{(2)}$, and any two stress fields, $\sigma^{(1)}$ and $\sigma^{(2)}$, it follows that

$$\phi(x\,;\,\epsilon^{(1)}) - \phi(x\,;\,\epsilon^{(2)}) \geq (\epsilon^{(1)} - \epsilon^{(2)}) : \sigma^{(2)}, \qquad (2.5a)$$

and

$$\psi(x\,;\,\sigma^{(1)}) - \psi(x\,;\,\sigma^{(2)}) \geq (\sigma^{(1)} - \sigma^{(2)}) : \epsilon^{(2)}, \qquad (2.5b)$$

respectively.

Further results are obtained for restricted boundary data, as discussed below. First note the following identity which holds for any divergence-free stress field σ, and any (related or unrelated) strain (strain-rate) field ϵ which is obtained from a suitably smooth displacement (velocity) field u:

$$\langle\sigma:\epsilon\rangle - \bar{\sigma}:\bar{\epsilon} = \frac{1}{V}\int_{\partial V} (u - x \cdot \bar{\epsilon}) \cdot \{n \cdot (\sigma - \bar{\sigma})\}\, dS; \qquad (2.6a)$$

Hill (1963, 1967) and Mandel (1980). Then, for either linear-displacement (linear-velocity) or uniform-traction boundary data, the right-hand side of (2.6a) vanishes, leading to

$$\langle\sigma:\epsilon\rangle = \bar{\sigma}:\bar{\epsilon} \qquad (2.6b)$$

There are other boundary data for which (2.6b) holds. Denote the stress and strain (strain-rate) fields of these special boundary data by $\sigma^S = \sigma^S(x)$ and $\epsilon^S = \epsilon^S(x)$, respectively. The corresponding overall potentials are then denoted by Φ^S and Ψ^S; here the superscript S stands for "special."

3. Weakly Kinematically or Statically Admissible Fields

A self-compatible strain (strain-rate) field in V, with a prescribed finite average value, is called *weakly kinematically admissible*. A self-equilibrating stress field in V, with a prescribed finite average value, is called *weakly statically admissible*. These fields need not satisfy any specific boundary data. The following general results are then obtained for nonlinearly elastic (viscoplastic) composites of any convex constituents; see Nemat-Nasser and Hori (1990, 1993) for details.

Theorem I: Among all weakly kinematically admissible strain (strain-rate) fields, that which corresponds to uniform boundary tractions renders the overall stress potential Φ, an absolute minimum, i.e.,

$$\Phi^G - \Phi^\Sigma \geq 0 \quad \text{when } \overline{\epsilon}^G = \overline{\epsilon}^\Sigma. \tag{3.1a}$$

The proof follows directly from the convexity of the stress potential; take the volume integral of (2.5a) and use (2.6) with $\sigma^{(2)} = \sigma^\Sigma$. The following corallary is an immediate consequence of the above theorem.

Corollary I: Among all weakly kinematically admissible strain (strain-rate) fields with boundary data which satisfy (2.6b), those which correspond to uniform-traction and linear-displacement boundary data render the overall stress potential Φ an absolute minimum and an absolute maximum, respectively, i.e.,

$$\Phi^E \geq \Phi^S \geq \Phi^\Sigma \quad \text{when } \overline{\epsilon}^E (\equiv E) = \overline{\epsilon}^S = \overline{\epsilon}^\Sigma. \tag{3.2a}$$

The proof, again, follows from (2.5a) and the fact that the right-hand side of (2.6a) is identically zero for this class of boundary data.

Similar results are obtained for the strain (strain-rate) potential. Hence the following general theorem and its corollary can be stated.

Theorem II: Among all weakly statically admissible stress fields, that which corresponds to linear displacement (velocity) field, renders the overall strain (strain-rate) potential Ψ, an absolute minimum, i.e.,

$$\Psi^G \geq \Psi^E \quad \text{when } \overline{\sigma}^G = \overline{\sigma}^E. \tag{3.1b}$$

Corollary II: Among all weakly statically admissible stress fields with boundary data which satisfy (2.6b), those which correspond to linear-displacement (linear-velocity) and uniform-traction boundary data, render the overall strain (strain-rate) potential Ψ an absolute minimum and an absolute maximum, respectively, i.e.,

$$\Psi^\Sigma \geq \Psi^S \geq \Psi^E \quad \text{when } \overline{\sigma}^\Sigma (\equiv \Sigma) = \overline{\sigma}^S = \overline{\sigma}^E. \tag{3.2b}$$

As pointed out before, Φ^E and Ψ^Σ are the overall stress and strain (strain-rate) potentials, in the sense defined by (2.4a,b). In particular, for the special class of boundary data which satisfy (2.6b), it follows that

$$\overline{\sigma}^S = \partial\Phi^S/\partial\overline{\epsilon}^S, \quad \overline{\epsilon}^S = \partial\Psi^S/\partial\overline{\sigma}^S, \quad \Psi^S + \Phi^S = \overline{\sigma}^S : \overline{\epsilon}^S. \tag{3.3a~c}$$

4. Periodic Boundary Data

For periodically distributed inhomogeneities, V stands for a typical unit cell which may include any number of inhomogeneities with any desire distribution within the cell; e.g., the unit cell may even be isotropic in its overall average response. In the periodic case, the condition (2.6b) is always satisfied and hence results (3.1) and (3.3) are always valid; Nemat-Nasser and Hori (1993).

5. Calculation of Overall Potentials

The calculation of the overall stress potential for *an arbitrary heterogeneous solid* is outlined here. The calculation of the strain (strain-rate) potential then follows similar steps.

The calculation of the lower bound for Φ^Σ is based on the following general observation which is valid for any finite V, of any heterogeneity with pointwise convex material constituents; see (2.5). Suppose V_0 with boundary ∂V_0 is totally contained in V, and Ω_0 with boundary $\partial \Omega_0$ is totally contained in V_0. Assume the uniform stress $\hat{\Sigma}$ is prescribed in $V - V_0$, while Ω_0 is removed and instead, uniform tractions $-n \cdot \hat{\sigma}^0$ are applied on the boundary $\partial \Omega_0$, where n is the outward unit normal on $\partial \Omega_0$. Calculate $\hat{\sigma}^0$ (constant) such that the average "cavity strain" $\bar{\varepsilon}^C$ equals the average strain $\bar{\varepsilon}^{\Omega_0}$ when uniform tractions $n \cdot \hat{\sigma}^0$ are prescribed on $\partial \Omega_0$ of the isolated Ω_0, where

$$\bar{\varepsilon}^C \equiv \frac{1}{\Omega_0} \int_{\partial \Omega_0} \frac{1}{2} (n \otimes u + u \otimes n) \, dS. \tag{5.1}$$

Note that the displacement field u is not continuous across ∂V_0. Denote the stress and strain fields in V by $\hat{\sigma}(x)$ and $\hat{\varepsilon}(x)$, respectively, and observe that

$$\hat{\sigma} \equiv \begin{cases} \hat{\Sigma} & \text{in } V - V_0 \\ \hat{\sigma}(x\,;\,\hat{\Sigma}) & \text{in } V_0 - \Omega_0 \\ \hat{\sigma}(x\,;\,\hat{\sigma}^0), & \text{in } \Omega_0, \end{cases} \tag{5.2a}$$

where $\hat{\Sigma}$ and $\hat{\sigma}^0$ are constant tensors. Also, whatever the composition of $V - V_0$, $V_0 - \Omega_0$, and Ω_0, and whatever the $\hat{\sigma}^0$, the average stresses in V and Ω_0 are given by

$$<\hat{\sigma}>_V = \hat{\Sigma}, \quad <\hat{\sigma}>_{\Omega_0} = \hat{\sigma}^0, \tag{5.2b,c}$$

respectively, where the subscripts V and Ω_0 in the left-hand side of these equations denote the corresponding domain of averaging. Similarly, since ∂V_0 is subjected to uniform tractions $n \cdot \hat{\Sigma}$, it also follows that

$$<\hat{\sigma}>_{V_0} = \hat{\Sigma}. \tag{5.2d}$$

The corresponding strain (strain-rate) field depends on the composition of the composite. It is given by

$$\hat{\varepsilon} \equiv \begin{cases} \hat{\varepsilon}(x\,;\,\hat{\Sigma}) & \text{in } V - V_0 \\ \hat{\varepsilon}(x\,;\,\hat{\Sigma}) & \text{in } V_0 - \Omega_0 \\ \hat{\varepsilon}(x\,;\,\hat{\sigma}^0), & \text{in } \Omega_0, \end{cases} \tag{5.3a}$$

with the requirement that

$$<\hat{\varepsilon}>_{\Omega_0} = \bar{\varepsilon}^C \tag{5.3b}$$

which then yields the required $\hat{\sigma}^0$; here, $\bar{\varepsilon}^C$ is defined by (5.1).

Now, let $\varepsilon^{(1)}$ in (2.5a) be the strain (strain-rate) field in V, which is produced by uniform tractions $n \cdot \Sigma$ applied on ∂V. Denote this strain field by $\varepsilon^\Sigma \equiv \varepsilon(x\,;\,\Sigma)$; this is a *compatible* field. Let $\varepsilon^{(2)}$ and $\sigma^{(2)}$ in (2.5a) be the fields defined by (5.3) and (5.2), respectively. While the strain (strain-rate) field (5.3) is not compatible, the stress field (5.2) *is* self-equilibrating everywhere within V. Hence, taking the volume average of (2.5a) over V, and noting (5.3b), obtain

$$\Phi^{\Sigma} - \hat{\Phi}^{\hat{\Sigma}} \geq 0 \quad \text{when} \quad \bar{\varepsilon}^{\Sigma} = \bar{\hat{\varepsilon}}, \tag{5.4a}$$

where

$$\Phi^{\hat{\Sigma}} \equiv \frac{1}{V} \int_V \phi(x, \hat{\varepsilon}(x\,;\,\hat{\Sigma}))\, dV, \tag{5.4b}$$

and where the overall stresses, Σ and $\hat{\Sigma}$, are not, in general, the same. The overall stress $\hat{\Sigma}$ must be adjusted such that the corresponding average strain (strain-rate) $\bar{\hat{\varepsilon}}$ equals the average strain (strain-rate) $\bar{\varepsilon}^{\Sigma}$ which is produced by uniform tractions $n \cdot \Sigma$ applied on ∂V of the original composite.

From the above results it now follows that the given composite can be subdivided into a "matrix" M and a set of non-intersecting subregions, V_α, $\alpha = 1, 2, \cdots, n$, all totally contained within V, where $M = V - \bigcup_{\alpha=1}^{n} V_\alpha$. Each V_α may contain its own set of inhomogeneities. In M, a uniform stress $\hat{\Sigma}$ is prescribed such that the final overall strain (strain-rate) $\bar{\varepsilon}^{\Sigma}$ is attained. Each V_α can now be treated in the manner discussed before, and the corresponding stress potential $\hat{\Phi}^{\hat{\Sigma}}$ can then be computed, leading to a lower bound for Φ^{Σ}, and hence for Φ^G. The poorest bound is obtained when a uniform stress $\hat{\Sigma}$ is assumed for the entire composite. This then leads to the lower bound associated with the Reuss model; Reuss (1929). Similarly, if a uniform strain (strain-rate) $\hat{E}$ is used for the entire composite, then the poorest bound for $\Psi^{\hat{E}}$ is obtained. This corresponds to the Voigt model; Voigt (1889). All other bounds which may result from the procedure outlined in this work will be better than the Reuss and Voigt bounds.

The method outlined in this work has been applied by Balendran and Nemat-Nasser (1994) to linear composites. The advantage of these bounds is that they are *finite* for composites with cavities, rigid inclusions, or both cavities *and* rigid inclusions. The classical bounds obtained using methods which consider "comparison uniform solids" are generally zero and infinity when both cavities and rigid inclusions coexist.

6. References

Balendran, B. and Nemat-Nasser, S. (1994) Bounds on the Overall Moduli of Composites, submitted to *J. Mech. Phys. Solids*.

Hill, R. (1963) Elastic properties of reinforced solids: Some theoretical principles, *J. Mech. Phys. Solids* Vol. 11, 357-372.

Hill, R. (1967) The essential structure of constitutive laws for metal composites and polycrystals, *J. Mech. Phys. Solids*, Vol. 15, 79-95.

Mandel, J. (1980) Generalization dans R9 de la regle du potential plastique pour un element polycrystallin, *Compt. Rend. Acad. Sci. Paris*, Vol. 290, 481.

Nemat-Nasser, S. and Hori, M. (1990) Elastic solids with microdefects, in *Micromechanics and Inhomogeneity; The Toshio Mura Anniversary Volume*, Springer-Verlag, New York, 297-320, 305-350.

Nemat-Nasser, S. and Hori, M. (1993) *Micromechanics: Overall Properties of Heterogeneous Materials*,

Elsevier, North-Holland.

Reuss, A. (1929) Berechnung der Fliessgrenze von Mischkristallen auf Grund der Platizitätsbedingung für Einkristalle, *Z. angew. Math. Mech.*, Vol. 9, 49.

Voigt, W. (1889) Uber die Beziehungzwischen den beiden Elastizitätskonstanten isotroper Körper, *Wied. Ann.* Vol. 38, 573.

Anisotropic and Inhomogeneous Plastic Flow of Fibrous Composites

JIE NING* and E.C. AIFANTIS**

Center for Mechanics of Materials and Instabilities, Michigan Tech.,
Houghton, MI49931, USA
*On leave from Institute of Applied Mechanics, Southwestern Jiaotong
University, PRC, ** Also: Aristotle University of Thessaloniki,
Thessaloniki 54006, GREECE

Abstract. Anisotropic and inhomogeneous plastic flow are two basic deformation characteristics of fibrous composites. To include these effects into the constitutive relations, the scale invariance approach previously developed by the second author and his co-workers for finite plastic deformations of polycrystals is extended to composites. A rule of mixtures approach is then used to interpret the effect of fiber volume fraction. Finally, a strain gradient dependent flow stress is derived from self-consistent arguments to describe the heterogeneity of plastic flow in fibrous composites.

1. Introduction

Large deformation plasticity analyses are necessary for a comprehensive understanding of the deformation, instability and fracture behavior of composites. In particular, the issues of fiber anisotropy and heterogeneity development should be considered within a large deformation plasticity framework. Anisotropic effects were emphasized, for example, by Fares and Dvorak (1991) who indicated that the pronounced anisotropy of fibrous composites tends to magnify the effect of small rotations by causing relatively large changes in the resolved shear stress in the fiber direction. In this connection, non-Schmid type yielding effects may also be important in modelling the overall behavior of composites as was illustrated by Qing and Bassani (1993) for the case of polycrystals. The issue of deformation heterogeneity is also extremely important for composites. Thus, even though the overall deformation may be small, the local deformation at regions

223

R. Pyrz (ed.), IUTAM Symposium on Microstructure-Property Interactions in Composite Materials, 223–234.
© 1995 Kluwer Academic Publishers.

of microcrack nucleation and debounding are quite large. Moreover, patterning of deformation and cracking are routinely observed in composites and a gradient plasticity framework for considering such effects is required.

In this paper, the scale invariance approach previously developed by Aifantis (1984, 1987) and co-workers (Zbib and Aifantis 1988; Dafalias and Aifantis 1990; Shi et al. 1990; Ning and Aifantis 1994) for plastic polycrystals is extended to the case of fibrous composites. The microscopic yield condition is now allowed to account for non-Schmid (matrix) and inherent (fiber) anisotropic effects and a modified maximization procedure is developed to connect micro to macro scales and determine the corresponding plastic spins. To evaluate the effect of the volume fraction of the fibers, a rule of mixtures procedure is employed. Finally, a strain gradient dependent theory based on self-consistent arguments is formulated to describe heterogeneity development and deformation patterning in fibrous composites.

2.Plastic Flow in Fibrous Composites

In the scale invariance approach one starts from the microscopic configuration of a single slip system defined by two unit vectors $(\mathbf{n}, \mathbf{v})$ where $\mathbf{n}$ denotes the normal to the slip plane and $\mathbf{v}$ is in the slip direction. Then, the following basic equations hold for the plastic strain rate $\mathbf{D}^P$ and the dislocation or back (internal) stress $\mathbf{T}^D$ (see, for example, Aifantis 1987)

$$\mathbf{D}^P = \dot{\gamma}^P \mathbf{M}, \tag{1}$$

$$\mathbf{T}^D = t_m \mathbf{M} + t_n \mathbf{N}, \ \mathbf{T}^D = \sigma - \mathbf{T}^L, \tag{2}$$

where σ is the Cauchy stress and $\mathbf{T}^L$ is the lattice or effective stress. The parameters t_m and t_n are functions of the plastic strain history, commonly expressed in terms of the equivalent plastic strain γ^P ($\dot{\gamma}^P = \sqrt{2\mathrm{tr}\,(\mathbf{D}^P\mathbf{D}^P)}$). The orientation tensors $\mathbf{M}$ and $\mathbf{N}$ are defined by

$$\mathbf{M} = \frac{1}{2}(\mathbf{n} \otimes \mathbf{v} + \mathbf{v} \otimes \mathbf{n}), \ \mathbf{N} = \mathbf{n} \otimes \mathbf{n}. \tag{3}$$

For a single family of inextensible fibers in an incompressible material, yielding is assumed to be independent not only on superimposed hydrostatic stress but also on superimposed tension in the fiber direction. The effective stress $\mathbf{T}^L$ is thus decomposed into an effective reaction stress $\mathbf{r}$ and an effective extra-stress $\mathbf{R}$ (Spencer 1972)

$$\mathbf{T}^L = \mathbf{r} + \mathbf{R} \; ; \; \mathbf{r} = -p\mathbf{I} + t\mathbf{A} \tag{4}$$

$$\mathbf{R} = \mathbf{T}^L - \frac{1}{2}\text{tr}\,(\mathbf{T}^L - \mathbf{A}\mathbf{T}^L)\,\mathbf{I} + \frac{1}{2}\text{tr}\,(\mathbf{T}^L - 3\mathbf{A}\mathbf{T}^L)\,\mathbf{A} , \tag{5}$$

where $\mathbf{A} = \mathbf{a} \otimes \mathbf{a}$ with $\mathbf{a}$ denoting the unit vector in the fiber direction. In this case, the microscopic yield condition reads

$$\tau = \text{tr}\,(\mathbf{R}\mathbf{M}) = f\,(\mathbf{R}, \mathbf{A}, \mathbf{M}, \mathbf{N}, \gamma^p) , \tag{6}$$

with the following simple form for f being adopted

$$\tau = \eta\,\text{tr}\,(\mathbf{H}\mathbf{M}) + \kappa\,(\gamma^p) \; ; \; \mathbf{H} = (\mathbf{R}\mathbf{A} + \mathbf{A}\mathbf{R})/2 . \tag{7}$$

To obtain a macroscopic counterpart of the slip system $(\mathbf{n}, \mathbf{V})$, we postulate that macroscopic plastic flow occurs in the direction where the plastic work rate $w^p = \text{tr}\,(\mathbf{R}\mathbf{D}^p)$ is maximized (Aifantis 1984) subject to the constraints, $\text{tr}\mathbf{M} = 0$, $\text{tr}\mathbf{M}^2 = 1/2$, $\text{tr}\mathbf{N}\mathbf{M} = 0$, $\text{tr}\mathbf{N} = 1$, and the yield condition (7). This maximization procedure gives (Ning and Aifantis 1994-95)

$$\mathbf{M} = \frac{1}{2\lambda_2}\{\mathbf{R} - 2\lambda_3\mathbf{P}\} \; ; \; \mathbf{P} = \frac{1}{2}(\mathbf{R} - \eta\mathbf{H}) , \tag{8}$$

where $\lambda_2 = \sqrt{(2I_2 J_R - I_1^2)/(2I_2 - \kappa^2)}$ and $\lambda_3 = (I_1 - \kappa\lambda_2)/2I_2$. The invariant quantities I_1, I_2 and J_R are defined as $J_R = \text{tr}\,(\mathbf{R}\mathbf{R})/2$, $I_1 = \text{tr}\,(\mathbf{R}\mathbf{P})$, and $I_2 = \text{tr}\,(\mathbf{P}\mathbf{P})$.

Substitution of (8) into (7) gives

$$\frac{(1 - \lambda_3)}{2\lambda_2\kappa}\text{tr}\,(\mathbf{R}^2) + \frac{(2\lambda_3 - 1)\,\eta}{2\lambda_2\kappa}\text{tr}\,(\mathbf{A}\mathbf{R}^2) - \frac{\lambda_3\eta^2}{2\lambda_2\kappa}\text{tr}\,(\mathbf{H}^2) - 1 = 0 . \tag{9}$$

Mulhern, Rogers and Spencer (1967) proposed the following phenomenological yield condition

$$\alpha \mathrm{tr}\,(\mathbf{R}^2) + \beta \mathrm{tr}\,(\mathbf{A}\mathbf{R}^2) - 1 = 0,\tag{10}$$

which does not contain the effect of the third term in equation (9) derived on microscopic grounds by considering non-Schmid effects due to the presence of the fibers.

Experimental data for the variation of tensile stress with fiber orientation are shown in Figure 1, in relation with the predictions of yield conditions (9) and (10). In the calculation, the values of the parameters κ and η were taken as $\kappa = 6\,(\mathrm{ksi})$, $\eta = 0.6$. The predictions of the present approach are in excellent agreement with the corresponding experimental data and an improvement over those of the phenomenological yield condition (10).

From (1) and (8), the corresponding plastic flow rule reads

$$\mathbf{D}^{\mathrm{p}} = \frac{\dot{\gamma}^{\mathrm{p}}}{2\lambda_2} \{\mathbf{R} - 2\lambda_3 \mathbf{P}\},\tag{11}$$

$$\dot{\gamma}^{\mathrm{p}} = 2\mathrm{tr}\,(\overset{\circ}{\mathbf{P}}\mathbf{M}) / \{h + \eta'\mathrm{tr}\,(\mathbf{H}\mathbf{M})\},\tag{12}$$

where $h = d\kappa/d\gamma^{\mathrm{p}}$ and $\eta' = d\eta/d\gamma^{\mathrm{p}}$. The corotational rate $\overset{\circ}{\mathbf{P}}$ is defined in terms of the corresponding vorticity $\mathbf{W}$ and the plastic spin $\mathbf{W}^{\mathrm{p}}$ by

$$\overset{\circ}{\mathbf{P}} = \dot{\mathbf{P}} - \omega\mathbf{P} + \mathbf{P}\omega;\ \omega = \mathbf{W} - \mathbf{W}^{\mathrm{p}},\tag{13}$$

as it will be discussed in detail in the next section.

It is interesting to consider a simple J_2 - type flow model. If $\eta = 0$, the flow rule (11) becomes

$$\mathbf{D}^{\mathrm{p}} = \frac{\dot{\gamma}^{\mathrm{p}}}{2\sqrt{J_R}}\mathbf{R},\tag{14}$$

and the yield condition reads

$$J_R - \kappa^2 = 0.\tag{15}$$

It then turns out from (5) that

$$\frac{1}{2}\mathrm{tr}\,(\mathbf{T}^L{}'\mathbf{T}^L{}') - \frac{3}{4}\mathrm{tr}^2\,(\mathbf{T}^L{}'\mathbf{A}) - \kappa^2 = 0.\tag{16}$$

A widely used yield condition (Hill 1948) for transversely isotropic materials can be cast in the form (Dafalias 1987)

$$(F+2B)\,tr\,(T^{L'}T^{L'}) + 2\,(C-F-2B)\,tr\,(T^{L'^2}A) + (5F+B-C)\,tr^2\,(T^{L'}A) = \kappa^2 \,, (17)$$

It follows that our microscopically derived yield condition (16) is identical to Hill's phenomenological yield condition (17) when

$$F = 0, \quad B = 1/4, \quad C = 1/2. \tag{18}$$

If the coordinate x_1 is along the direction of the fibers, (16) can be written as

$$\frac{1}{4}\,(T_{22}^L - T_{33}^L)^2 + (T_{23}^L)^2 + (T_{12}^L)^2 + (T_{31}^L)^2 - \kappa^2 = 0 \,, \tag{19}$$

which, in the case of plane stress deformation, gives

$$\frac{1}{4}\,(T_{22}^L)^2 + (T_{12}^L)^2 - \kappa^2 = 0; \tag{20}$$

this being the surface of a cylinder with an elliptic cross section with T_{11}^L along its axis in the stress space. Comparison of the above results with phenomenological approaches (Dafalias 1987, Fares and Dvorak 1991) indicates that there are no unidentified parameters in the present micromechanics approach.

3.Plastic Spin of Fibrous Composites

In a complete finite deformation plasticity framework, constitutive relations must be provided not only for the stretching part of the deformation rate but also for the rate of rotation or plastic spin. In the representative single slip system, the exact microscopic expression for the plastic spin is given by the relation

$$W^p = \dot{\gamma}^p \Omega; \quad \Omega = \frac{1}{2}\,(v \otimes n - n \otimes v) \,, \tag{21}$$

with the orientation tensor Ω satisfying the conditions $tr\,(\Omega) = 0$ and $tr\,(\Omega^2) = -1/2$. It is noted that, the plastic spin does not enter into the expression for the plastic work rate which was maximized in the previous section to reduce the macroscopic counterpart of the orientation tensor M and therefore the stretching tensor D^p. It enters, however, the expression for the second order plastic work rate which has already been introduced by Hill (1968) and co-workers (Hill and Rice1972, Havner 1978) to discuss stability and uniqueness of plastic deformation. Motivated by the maximization procedure of the previous section for the plastic work rate, a similar maximization procedure for the second plastic work rate is developed to specify the macroscopic counterpart of the plastic spin by defining the corresponding Lagrangean

$$L = tr(\overset{\circ}{S}D^P) - l_1 tr\Omega - l_2\left(tr\Omega^2 + \frac{1}{2}\right), \tag{22}$$

with $l_{1,2}$ denoting Lagrange multipliers. It turns out that

$$W^P = -\frac{1}{t_\omega}(SD^P - D^PS) = -\frac{1}{t_\omega}\{(\underline{T}^LD^P - D^P\underline{T}^L) + (T^DD^P - D^PT^D)\}, \tag{23}$$

with

$$t_\omega = \sqrt{-2tr(SM - MS)^2} = \sqrt{-2tr\{(\underline{T}^LM - MT^L) + (T^DM - MT^D)\}^2}. \tag{24}$$

In view of the microscopic equation for the back stress and the scale invariance argument, (23) can be specified further to provide definite expressions for the plastic spin. For example, in the case of isotropic/kinematic hardening materials with no reinforcement ($A \equiv 0$), (1), (8) and (23) give

$$W^P = -\frac{1}{t_n}(T^DD^P - D^PT^D), \tag{25}$$

which is, indeed, the same expression as the one derived earlier by a simple scale invariance argument (Aifantis 1984, 1987).

In the present case of fiber-reinforced composites, (8) with ($A \neq 0$) should be used in connection with (1) and (23) to obtain the appropriate expression for W^P. This expression reads

$$W^P = \varphi_1(T^DS - ST^D) + \varphi_2(SA - AS) + \varphi_3(S^2A - AS^2)$$

$$+ \varphi_3(AT^DS - ST^DA) + \varphi_3(T^DAS - SAT^D) + \varphi_4(SA^2 - A^2S), \tag{26}$$

where φ_i are specific functions of the invariants of trT^L and $tr(AT^L)$ of the following form

$$\varphi_1 = -\frac{\dot{\gamma}^P}{t_\Omega}\frac{1-\lambda_3}{2\lambda_2}, \quad \varphi_2 = -\frac{\dot{\gamma}^P}{t_\Omega}\{\frac{1-\lambda_3}{2\lambda_2}R_2 - \eta\frac{\lambda_3}{2\lambda_2}R_1\}, \quad \varphi_3 = -\frac{\dot{\gamma}^P}{t_\Omega}\frac{\lambda_3}{4\lambda_2}\eta,$$

$$\varphi_4 = -\frac{\dot{\gamma}^p}{t_\Omega} \frac{\lambda_3}{2\lambda_2} \eta R_2; \quad t_\Omega = \sqrt{-2\text{tr}\{\frac{1-\lambda_3}{2\lambda_2}(SR-RS) + \eta \frac{\lambda_3}{2\lambda_2}(SH-HS)\}^2},$$

$$R_1 = \frac{1}{2}\{\text{tr}(T^L) - \text{tr}(AT^L)\} \ ; \ R_2 = \frac{1}{2}\{\text{tr}(T^L) - 3\text{tr}(AT^L)\} \ .$$

It is noted that this expression includes terms due to deformation induced anisotropy, inherent anisotropy, and their interaction. In the case of transversely isotropic deformation with no back stress effects, we have

$$W^p = \varphi_2(SA-AS) + \varphi_3(S^2A-AS^2) + \varphi_4(SA^2-A^2S) \ , \tag{27}$$

which is the same expression obtained through representation theorems (Dafalias 1985), but with the phenomenological coefficients $(\varphi_2, \varphi_3, \varphi_4)$ being now completely specified.

In the case of J_2 type plastic flow ($\eta = 0$), equation (26) reduces to

$$W^p = \varphi_2(SA-AS) \ , \tag{28}$$

which is similar to that given by Aravas and Aifantis (1991); [see also Zbib and Aifantis (1988)].

To interpret directly the effect of fiber volume fraction, it is not unreasonable to incorporate a rule of mixtures type of approach into the above continuum formulation. Generally, a composite may be regarded as a material made up by two phases: the matrix phase and the fiber phase. During deformation, plastic flow mainly occurs in the matrix phase. By extending the rule of mixtures argument to both stretching and rotating parts of deformation, we have

$$D = V_f D_f + V_m D_m; \quad W = V_f W_f + V_m W_m, \tag{29}$$

where D and W denote the total stretching and vorticity of the composite, while (D_f, W_f) and (D_m, W_m) denote respectively the corresponding stretching and vorticity of the fiber and matrix phases. The parameters V_f and V_m denote the volume fractions of the fibers and the matrix ($V_f + V_m = 1$). To proceed further we adopt simplifying assumptions for W_m^p and W_f^p in order to illustrate the effect of the fiber volume fraction. Based on the scale invariance approach (Aifantis 1987), the plastic spin associated with the matrix phase is given by

$$\mathbf{W}_m^p = -\frac{1}{t_n}\left(\mathbf{T}^D\mathbf{D}_m^p - \mathbf{D}_m^p\mathbf{T}^D\right). \tag{30}$$

In the fiber phase, the plastic deformation is negligible and the plastic spin $\mathbf{W}_f^p$ may be assumed (Zbib and Aifantis 1988, Aravas and Aifantis 1991) to be given by the expression

$$\mathbf{W}_f^p = \mathbf{A}\mathbf{D}_f - \mathbf{D}_f\mathbf{A}. \tag{31}$$

Next, we define the "**effective material spin**" tensor ω by the relation

$$\omega = \mathbf{W} - \mathbf{W}_{eff}^p, \tag{32}$$

with the "**effective plastic spin**" tensor given by

$$\mathbf{W}_{eff}^p = V_f\mathbf{W}_f^p + V_m\mathbf{W}_m^p = V_f(\mathbf{A}\mathbf{D}_f - \mathbf{D}_f\mathbf{A}) - \frac{V_m}{t_n}\left(\mathbf{T}^D\mathbf{D}_m^p - \mathbf{D}_m^p\mathbf{T}^D\right). \tag{33}$$

It is noted that equation (33) is a special case of equation (26). Figure 2 shows clearly that a remarkable difference exists between the anisotropy induced by plastic flow and the anisotropy due to fiber orientation. Moreover, the evolution of the plastic spin is quite sensitive to the volume fraction of the fibers. In the early stages of small to moderate strains, the direction of fibers has a strong effect on the evolution of plastic spin. However, the evolution of plastic spin is dominated by the anisotropy induced by the plastic flow at stages of large deformation (Ning and Aifantis 1993).

4. Inhomogeneous Deformation of Fibrous Composites

Because of the existence of fibers, the local deformation in composites is highly inhomogeneous. In order to describe heterogeneity and deformation patterning in composites, the strain gradient approach as developed by Aifantis (1984, 1987) and co-workers (Zbib and Aifantis 1989; Muhlhaus and Aifantis 1991; Vardoulakis and Aifantis 1991) is employed by utilizing the self-consistent framework (Eshelby 1956, Kroner 1958).

The self-consistent stress-strain relations for fibrous composites can be written as

$$\varepsilon = \mathbf{Q}^{-1}\sigma + \varepsilon^p + \mathbf{S}'(\varepsilon^p - \bar{\varepsilon}^p), \tag{34}$$

where $\bar{\sigma}$ and $\bar{\varepsilon}^p$ are the overall stress and plastic strain with $\varepsilon^e = \{\varepsilon_x^e, \varepsilon_y^e, \gamma^e\}^T = \mathbf{Q}^{-1}\sigma = \mathbf{Q}^{-1}\{\sigma_x, \sigma_y, \tau\}^T$ denoting the local stress-strain elastic relation in two dimensions and **Q** being the stiffness matrix. The tensor **S'** is given by the relation **S'** = **T⁻¹ST**, where **S** is the Eshelby tensor and the coordinate transformation (orthogonal tensor) **T** reads

$$
\mathbf{T} = \begin{bmatrix} \cos^2\theta & \sin^2\theta & \sin 2\theta \\ \sin^2\theta & \cos\theta & -\sin 2\theta \\ -\sin\theta\cos\theta & \sin\theta\cos\theta & \cos^2\theta - \sin^2\theta \end{bmatrix} \tag{35}
$$

with the angle θ denoting the orientation of the fibers with respect to axis $x_1 = x$.

It then follows that

$$
\sigma = \bar{\sigma} + \mathbf{Q}\,[\mathbf{S}'\Delta\varepsilon^p - \Delta\varepsilon^p]\;, \tag{36}
$$

where $\Delta\varepsilon^p = \varepsilon^p - \bar{\varepsilon}^p$.

In the case of simple shear, (36) reads

$$
\tau = \bar{\tau}(\gamma^p) - \beta\Delta\gamma^p\;, \tag{37}
$$

where $\beta = Q_{33}(1 - 2S'_{33})$.

Next, we note that the actual plastic shear strain γ^p is related to the "shear strain" γ_{12}^p in the direction normal to the fibers by the equation

$$
\gamma^p = \frac{1}{\cos^2\theta - \sin^2\theta}\gamma_{12}^p\;, \tag{38}
$$

where the coordinate $x_1 = x$ is in the direction of fibers. In the plane normal to the fiber, we may calculate the average shear strain in a characteristic volume element through the expression

$$
\gamma_{12}^p = \gamma_{12}^p + \frac{1}{\pi R^2}\int_0^R\int_0^{2\pi}\left[\partial_i\gamma_{12}^p n_i + \tfrac{1}{2}\partial_{ij}\left(\gamma_{12}^p n_i n_j\right) + \ldots\right]r\,dr d\theta \approx \gamma_{12}^p + \frac{R^2}{8}\nabla^2\gamma_{12}^p\;, \tag{39}
$$

where $R = R_f$ is the spacing between fibers, n_i denotes the outward unit normal of the circle $r = R_f$, and $\nabla^2 = \partial^2/\partial x_2^2$ since there is no strain variation (inextensibility assumption) along the fibers. It then turns out that

$$\gamma^p = \gamma^p + \frac{R_f^2}{8} \tilde{\nabla}^2 \gamma^p , \qquad (40)$$

where

$$\tilde{\nabla}^2 = \sin^2\theta \frac{\partial^2}{\partial x^2} - 2\sin\theta\cos\theta \frac{\partial^2}{\partial x \partial y} + \cos^2\theta \frac{\partial^2}{\partial y^2} . \qquad (41)$$

Substitution of (40) into (37) gives

$$\tau = \tau(\gamma^p) + (\beta + h) \frac{R_f^2}{8} \tilde{\nabla}^2 \gamma^p , \qquad (42)$$

with

$$\beta = \frac{1}{4(1-\nu_0)} \{ A + G_{12}(\sin^4\theta + \cos^4\theta) \} , \qquad (43)$$

$$A = \{ \frac{(1-\nu_{12})E_2 + (1-\nu_{21})E_1}{1-\nu_{12}\nu_{21}} - 2G_{12} \} \sin^2\theta\cos^2\theta , \qquad (44)$$

$$h = \frac{d\tau}{d\gamma^p} ; \quad \tilde{\nabla}^2 = n \cdot \nabla\nabla \cdot n , \qquad (45)$$

where n denotes the unite vector normal to the fibers, ν_0 is the Poisson's ratio of the matrix, and the parameters $(E_1, E_2, G_{12}, \nu_{12}, \nu_{21})$ are the overall elastic constants.

It is seen that the gradient effect is directly related to the orientation of the fibers. It is also easy to prove the directional gradient operator $\tilde{\nabla}^2$ is an objective operator (see also Oka and Aifantis 1993). Moreover the equation (42) suggests that the orientation of shear bands in fibrous composites is directly dependent on the directional part of the flow stress. The effect of the fiber direction on the strain gradient coefficient β is shown in Figure 3 for two different values of the overall elastic parameter $\zeta = \{ (1-\nu_{12})E_2 + (1-\nu_{21})E_1 \} / (1-\nu_{12}\nu_{21})G_{12}$.

Acknowledgments

The support of the US National Science Foundation (MSS-9310476) and Airforce Office of Scientific Research (AFOSR-91-0221) and the CEC (ER-BCHBGCT 920041) is gratefully acknowledged.

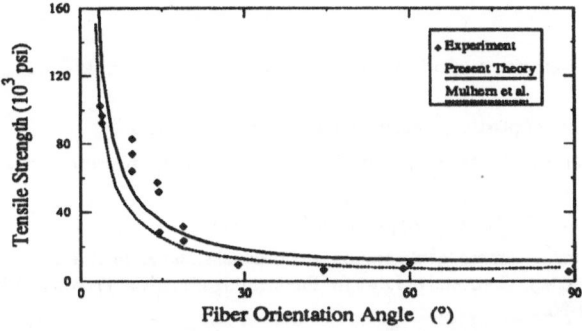

Fig. 1. Variation of tensile strength with fiber orientation in aluminium/silica composite.

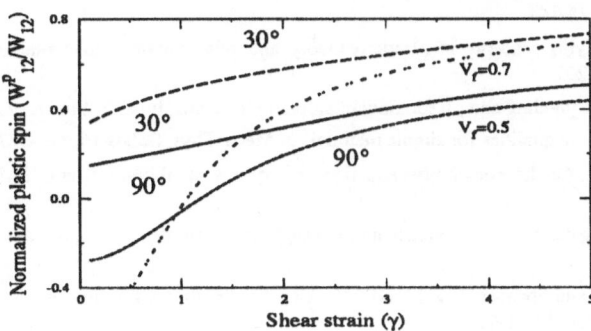

Fig. 2 Effect of the fiber volume fraction on the plastic spin

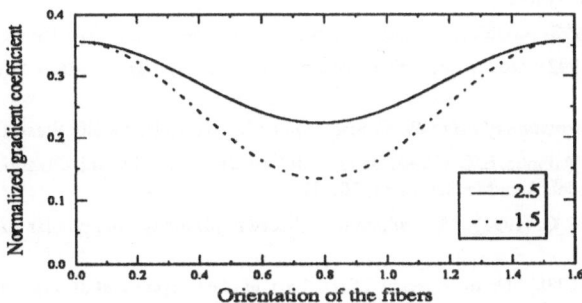

Fig. 3 Variation of normalized gradient coefficient β/G_{12} with fiber direction at different values of ζ.

234

References

Aifantis, E. C. (1984) On the microstructural original of certain inelastic models, ASME J. Eng. Mat. Tech. 106, 326-330.

Aifantis, E. C. (1987) The physics of plastic deformation, Int. J. Plasticity 3, 211-247.

Aravas, N. and Aifantis, E. C. (1991) On the geometry of slip and spin in finite plastic deformation, Int. J. Plasticity 7, 141-160.

Dafalias, Y. F. (1985) The plastic spin, ASME J. Appl. Mech. 52, 865-871.

Dafalias, Y. F. (1987) Anisotropy, reference configuration and residual stresses, in C. S. Desai, E. Krempl, P. D. Kiousis and T. Kundu (eds.), Constitutive Laws for Engineering Materials, Theory and Applications, Elsevier, Amsterdam, pp. 69-80.

Dafalias, Y. F. and Aifantis, E. C. (1990) On the microscopic origin of the plastic spin, Acta Mechanica 82, 31-48.

Eshelby, J. D. (1957) The determination of the elastic field of an ellipsoidal inclusion, and related problems, Proc. Roy. Soc. A241, 376-396.

Fares, N. and Dvorak, G. J. (1991) Large elastic-plastic deformations of fibrous metal matrix composites, J. Mech. Phys. Solids 39, 725-744.

Havner, K. S. (1978) On unifying concepts in plasticity theory and related matters in numerical analysis, Nucl. Eng. Design 46, 187-201.

Hill, R. (1948) A theory of the yielding and plastic flow of anisotropic metals, Proc. R. Soc. A193, 281-297.

Hill, R.(1968) On constitutive inequalities for simple material,. J. Mech. Phys. Solids 16, 315-322.

Hill, R. and Rice, J. R.(1972) On the constitutive equations of crystals at arbitrary strain, J. Mech. Phys. Solids 20, 401-413.

Kroner, E. (1958) Berechnung der elastischen konstanten des vielkristalls aus den konstanten des einkristalls, Z. Physik 151, 504-518.

Mulhern, J. F., Rogers, T. G. and Spencer, A. J. M. (1967) A continuum theory for fibre-reinforced plastic materials, Proc. R. Soc. A301, 473-492.

Muhlhaus, H. B. and Aifantis, E. C. (1991) A variational principle for gradient plasticity, Int. J. Solids Structures 89, 845-858.

Ning, J. and Aifantis, E. C. (1994) On anisotropic finite deformation plasticity: Part I. A two-back stress model, Acta Mechanica 106, 55-72.

Ning, J. and Aifantis, E. C. (1994-95) On the description of anisotropic plastic flow by the scale invariance approach, Int. J. Plasticity (In press).

Oka, F. and Aifantis, E. C. (1993) Instability of gradient dependent elastic-viscoplasticity for clay (Preprint).

Qing, Q. and Bassani, J. L. (1992) Non-Schmid yield behavior in single crystals, J. Mech. Phys. Solids 40, 813-833.

Spencer, A. J. M. (1972) *Deformation of Fibre-Reinforced Materials*, Oxford University Press, Oxford.

Shi, M. F., Gerdeen, J. C., and Aifantis, E. C. (1990) On finite deformation plasticity with directional softening: Part I. One -component model, Acta Mechanica 83, 103-117.

Vardoulakis, I. and Aifantis, E. C. (1991) A gradient flow theory of plasticity for granular materials, Acta Mechanica 87, 197-217.

Zbib, H. and Aifantis, E. C. (1988) On the concept of relative and plastic spins and its implications to large deformation theories: Part I & II, Acta Mechanica 101, 69-80.

Zbib, H. and Aifantis, E. C. (1989) A gradient-dependent flow theory of plasticity: Application to metal and soil instabilities, Appl. Mech. Rev. 42, S295-S304.

A FFT-BASED NUMERICAL METHOD FOR COMPUTING THE MECHANICAL PROPERTIES OF COMPOSITES FROM IMAGES OF THEIR MICROSTRUCTURES

H. MOULINEC, P. SUQUET

LMA/CNRS

31 Chemin Joseph Aiguier

13402. Marseille. Cedex 20. FRANCE

Abstract: The effective properties of composite materials are strongly influenced by the geometry of their microstructures, which can be extremely complex. Most of the numerical simulations known to the authors make use of two- or three-dimensional finite elements analyses which are often time consuming because of the complexity imposed by the requirement of extremely precise description of the reinforcements distribution. A numerical method is presented here that directly uses images of the microstructure - supposed to be periodically repeated - to compute the composite overall properties, as well as the local distribution of stresses and strains, without requiring further geometrical interpretation by the user. The linear elastic problem is examined first. Its analysis is based on the Lippmann-Schwinger's equation, which is solved iteratively by means of the Green operator of an homogeneous reference medium. Then the method is extended to non-linear problems where the local stress strain relation is given by an incremental relation.

Introduction

This study is devoted to a new numerical technique to compute the local and overall response of a nonlinear composite from *images* of its *real microstructure*. The need for developing these numerical simulations is twofold.

First, numerous studies have been devoted to nonlinear cell calculations using the Finite Element Method (FEM) and a list of comprehensive references (by no means exhaustive) include Adams and Donner (1967), Christman *et al* (1989), Tvergaard (1990), Brockenborough *et al* (1991), Böhm *et al* (1993), Michel and Suquet (1993), Nakamura and Suresh (1993). But the difficulties due to meshing and the large number of d.o.f.'s required by the analysis limit the complexity of the microstructures which can be investigated by means of the FEM.

The present method avoids the first difficulty (meshing), and makes use of fast Fourier transforms (FFT) to solve the unit cell problem, even in a nonlinear context. FFT algorithms require data sampled in a grid of regular spacing, allowing to use directly digital images of the microstructure. The second difficulty (size of the problem) is partially overcome by the use of an iterative method which does not require the formation of a stiffness matrix.

R. Pyrz (ed.), IUTAM Symposium on Microstructure-Property Interactions in Composite Materials, 235–246.
© *1995 Kluwer Academic Publishers.*

Second, the interest for numerical simulations of the nonlinear response of composites has recently been strengthened by the emergence of theoretical methods to predict analytically the nonlinear overall behavior of composites (Willis (1991), Ponte Castañeda (1992), Suquet (1993)). Part of the present study intends to give precise numerical results for uniaxial or multiaxial loading paths which could serve as guidelines for theoretical predictions.

1. Description of the method.

1.1. BOUNDARY CONDITIONS

The overall behavior of a composite is governed by the individual behavior of its constituents and by its microstructure. Its effective response to a prescribed path of macroscopic strains or stresses may be determined numerically via the resolution of the so-called "local problem" on a representative volume element (r.v.e.) V. In this study, the "representative" information on the microstructure is provided by an image (micrograph) of the microstructure of the composite of arbitrary complexity. The image contains $M \times N$ pixels and independent mechanical properties are assigned individually to each pixel.

The local problem consists in equilibrium equations, constitutive equations, boundary and interface conditions. All different phases are assumed to be perfectly bonded (hence displacements and tractions are continuous across interfaces). However, the displacements and tractions along the boundary of the r.v.e. are left undetermined and the local problem is ill-posed. We choose to close the problem with periodic boundary conditions which can be expressed as follows. The local strain field $\varepsilon(\mathbf{u}(\mathbf{x}))$ is split into its average $\mathbf{E}$ and a fluctuation term $\varepsilon(\mathbf{u}^*(\mathbf{x}))$:

$$\varepsilon(\mathbf{u}(\mathbf{x})) = \varepsilon(\mathbf{u}^*(\mathbf{x})) + \mathbf{E} \quad \text{or equivalently} \quad \mathbf{u}(\mathbf{x}) = \mathbf{u}^*(\mathbf{x}) + \mathbf{E}.\mathbf{x}$$

By assuming periodic boundary conditions it is assumed that the fluctuating term $\mathbf{u}^*$ is periodic (notation: $\mathbf{u}^*$ #), and the traction $\boldsymbol{\sigma}.\mathbf{n}$ is anti-periodic in order to meet the equilibrium equations on the boundary (notation: $\boldsymbol{\sigma}.\mathbf{n} - \#$).

1.2. PRELIMINARY PROBLEM.

First, the preliminary problem of an homogeneous linear elastic body, with stiffness $\mathbf{c}^0$, subject to a polarization field $\boldsymbol{\tau}(\mathbf{x})$, is considered

$$\left. \begin{array}{l} \boldsymbol{\sigma}(\mathbf{x}) = \mathbf{c}^0 : \varepsilon(\mathbf{u}^*(\mathbf{x})) + \boldsymbol{\tau}(\mathbf{x}) \quad \forall \mathbf{x} \in V \\ \mathbf{div}\boldsymbol{\sigma}(\mathbf{x}) = 0 \quad \forall \mathbf{x} \in V, \quad \mathbf{u}^* \#, \ \boldsymbol{\sigma}.\mathbf{n} - \# \end{array} \right\} \tag{1.1}$$

The solution of (1.1) can be expressed in real and Fourier spaces, respectively, by means of the periodic Green operator $\boldsymbol{\Gamma}^0$ associated with $\mathbf{c}^0$:

$$\varepsilon(\mathbf{x}) = -\boldsymbol{\Gamma}^0 * \boldsymbol{\tau}(\mathbf{x}) \quad \forall \mathbf{x} \in V, \quad \text{or} \quad \hat{\varepsilon}(\boldsymbol{\xi}) = -\hat{\boldsymbol{\Gamma}}^0(\boldsymbol{\xi}) : \hat{\boldsymbol{\tau}}(\boldsymbol{\xi}) \quad \forall \boldsymbol{\xi} \neq 0, \ \hat{\varepsilon}(0) = 0$$

The operator $\boldsymbol{\Gamma}^0$ is explicitly known in Fourier space and, when the reference material is isotropic (with Lamé coefficients λ^0 et μ^0), takes the form:

$$\hat{\Gamma}^0_{ijkh}(\boldsymbol{\xi}) = \frac{1}{4\mu^0|\boldsymbol{\xi}|^2}(\delta_{ki}\xi_h\xi_j + \delta_{hi}\xi_k\xi_j + \delta_{kj}\xi_h\xi_i + \delta_{hj}\xi_k\xi_i) - \frac{\lambda^0 + \mu^0}{\mu^0(\lambda^0 + 2\mu^0)}\frac{\xi_i\xi_j\xi_k\xi_h}{|\boldsymbol{\xi}|^4}.$$

1.3. THE LIPPMANN-SCHWINGER EQUATION.

The preliminary problem can be used to solve the problem of an inhomogeneous elastic composite material with stiffness $\mathbf{c}(\mathbf{x})$ at point $\mathbf{x}$ under prescribed strain $\mathbf{E}$. For simplicity $\mathbf{E}$ is assumed to be prescribed, although other average conditions could be considered as well (prescribed stresses).

$$\left.\begin{aligned} \boldsymbol{\sigma}(\mathbf{x}) &= \mathbf{c}(\mathbf{x}) : (\boldsymbol{\varepsilon}(\mathbf{u}^*(\mathbf{x})) + \mathbf{E}) \quad \forall \mathbf{x} \in V \\ \mathbf{div}\boldsymbol{\sigma}(\mathbf{x}) &= 0 \quad \forall \mathbf{x} \in V, \quad \mathbf{u}^* \;\#, \; \boldsymbol{\sigma}.\mathbf{n} - \# \end{aligned}\right\} \tag{1.2}$$

A reference material $\mathbf{c}^0$ is introduced and a polarization tensor $\boldsymbol{\tau}(x)$, which is unknown a priori, is defined as

$$\boldsymbol{\tau}(\mathbf{x}) = \boldsymbol{\delta}\mathbf{c}(\mathbf{x}) : \boldsymbol{\varepsilon}(\mathbf{u}(\mathbf{x})), \quad \boldsymbol{\delta}\mathbf{c}(\mathbf{x}) = \mathbf{c}(\mathbf{x}) - \mathbf{c}^0. \tag{1.3}$$

Thus, the problem reduces to the *periodic Lippmann-Schwinger equation* (Kröner (1972)), which reads, in real space and Fourier space respectively:

$$\left.\begin{aligned} \boldsymbol{\varepsilon}(\mathbf{u}(\mathbf{x})) &= -\boldsymbol{\Gamma}^0(\mathbf{x}) * \boldsymbol{\tau}(\mathbf{x}) + \mathbf{E}, \\ \widehat{\boldsymbol{\varepsilon}}(\boldsymbol{\xi}) &= -\widehat{\boldsymbol{\Gamma}}^0(\boldsymbol{\xi}) : \widehat{\boldsymbol{\tau}}(\boldsymbol{\xi}) \quad \forall \boldsymbol{\xi} \neq 0, \quad \widehat{\boldsymbol{\varepsilon}}(0) = \mathbf{E} \end{aligned}\right\} \tag{1.4}$$

where $\boldsymbol{\tau}$ is given by (1.3).

1.4. THE ALGORITHM.

The principle of the algorithm is to use alternately (1.3) and (1.4), in real space and Fourier space, respectively, in an iterative process, to solve (1.2):

$$\left.\begin{aligned} &\textit{Initialization}: \quad \varepsilon^0(\mathbf{x}) = \mathbf{E}, \quad \forall \mathbf{x} \in V, \\ &\textit{Iterate } i+1: \quad \varepsilon^i \text{ is known} \\ &\qquad a) \quad \sigma^i(\mathbf{x}) = \mathbf{c}(\mathbf{x}) : \varepsilon^i(\mathbf{x}). \text{ Convergence test} \\ &\qquad b) \quad \tau^i(\mathbf{x}) = \sigma^i(\mathbf{x}) - \mathbf{c}^0 : \varepsilon^i(\mathbf{x}), \\ &\qquad c) \quad \widehat{\tau}^i = \mathcal{F}(\tau^i), \\ &\qquad d) \quad \widehat{\varepsilon}^{i+1}(\boldsymbol{\xi}) = -\widehat{\boldsymbol{\Gamma}}^0(\boldsymbol{\xi}) : \widehat{\tau}^i(\boldsymbol{\xi}) \; \forall \boldsymbol{\xi} \neq 0 \text{ and } \widehat{\varepsilon}^{i+1}(0) = \mathbf{E}, \\ &\qquad e) \quad \varepsilon^{i+1} = \mathcal{F}^{-1}(\widehat{\varepsilon}^i) \end{aligned}\right\} \tag{1.5}$$

where $\mathcal{F}$ and $\mathcal{F}^{-1}$ denote the Fourier transform and the inverse Fourier transform.

The rate of convergence of the algorithm is governed by the choice of Lamé coefficients of the reference material. A good convergence rate was observed when λ^0 and μ^0 were prescribed to be the half sum of the minimum and maximum value of these coefficients in the composite (Moulinec and Suquet (1994)).

1.5. NONLINEAR BEHAVIOR.

The algorithm can be extended to the case where the individual constituents obey an incremental law (infinitesimal strains), *e.g.* phases with an elastic-plastic behavior with isotropic hardening. The loading is applied step by step. At each loading step n, the overall strain $\mathbf{E}_n$ is prescribed, and the local problem is solved for $(\boldsymbol{\sigma}_n, \boldsymbol{\varepsilon}_n, p_n)$ via the procedure summarized below:

$$
\left.
\begin{aligned}
&\textit{Iterate } i+1: \qquad \boldsymbol{\varepsilon}_n^i \text{ is known} \\
&a) \;\; \textit{Compute } \boldsymbol{\sigma}_n^i \textit{ and } p_n^i \textit{ from } (\boldsymbol{\varepsilon}_n^i, \; \boldsymbol{\sigma}_{n-1}, \; \boldsymbol{\varepsilon}_{n-1}, \; p_{n-1}) \\
&\quad \text{Convergence test} \\
&b) \;\; \boldsymbol{\tau}_n^i(\mathbf{x}) = \boldsymbol{\sigma}_n^i(\mathbf{x}) - \mathbf{c}^0 : \boldsymbol{\varepsilon}_n^i(\mathbf{x}), \\
&c) \;\; \widehat{\boldsymbol{\tau}}_n^i = \mathcal{F}(\boldsymbol{\tau}_n^i), \\
&d) \;\; \widehat{\boldsymbol{\varepsilon}}_n^{i+1}(\boldsymbol{\xi}) = -\widehat{\boldsymbol{\Gamma}}^0(\boldsymbol{\xi}) : \widehat{\boldsymbol{\tau}}_n^i(\boldsymbol{\xi}) \;\; \forall \boldsymbol{\xi} \neq 0 \text{ and } \widehat{\boldsymbol{\varepsilon}}_n^{i+1}(0) = \mathbf{E}, \\
&e) \;\; \boldsymbol{\varepsilon}_n^{i+1} = \mathcal{F}^{-1}(\widehat{\boldsymbol{\varepsilon}}_n^i)
\end{aligned}
\right\} \qquad (1.6)
$$

This procedure is similar to the one adopted in the linear elastic case. The main difference is due to the calculation of the stress field $\boldsymbol{\sigma}_n^i$ (step (a)).

2. Applications to unidirectional fiber reinforced composites

2.1. CONFIGURATIONS

The above numerical scheme was applied to predict the overall and local response of unidirectional fiber reinforced composites. Owing to the translation invariance along the axial direction (parallel to the unit vector e_3), the geometry and the material properties of these composites are completely specified by the same data on a cross section in the plane (e_1, e_2) transverse to the fibers' direction.

Several configurations were investigated. In all of them, the fibers cross sections were assumed to be impenetrable circular disks, with identical radii. In all the analyses presented below, the fiber volume fraction was 47.5% .

Random configurations. A prescribed number of identical impenetrable circular fibers were placed randomly in the unit cell. Fibers intersecting the boundary of the unit cell were treated modulo the periodic lattice, *i.e.* by moving the part of the fiber which would lie outside the unit cell to the opposite boundary (Figure 1a).

Standard configurations. Configurations which are classical in FEM modelling, namely circular fibers arranged at the nodes of a square or hexagonal lattice, were also considered (Figure 1b).

All calculations were performed under the *generalized plane strains* condition (Michel and Suquet (1993)).

2.2. CONSTITUTIVE BEHAVIOUR OF THE INDIVIDUAL PHASES

The individual phases are assumed to be isotropic and elastic plastic with isotropic hardening. More specifically, their elastic properties are given by a Young modulus E and a Poisson ratio ν (labelled by f or m for fibers and matrix respectively) and their plastic properties are governed by the flow rule and the Von Mises criterion with linear isotropic hardening,

$$
\sigma_{eq} = \sigma_0 + H\, p,
$$

where σ_0 is the initial flow stress ($\sigma_0 = +\infty$ when the phase is purely elastic), H is the hardening modulus ($H = 0$ when the phase is ideally plastic), p is the equivalent plastic strain. This general form is specialized below to several cases: isotropic linear elastic fibers in an elastic-ideally plastic matrix (section 3), isotropic linear elastic fibers in an elastic plastic matrix with isotropic linear hardening (section 3), elastic-ideally plastic matrix and fibers with different flow stresses (section 4). The specific values of the material constants are given in each section.

3. Uniaxial transverse tension

The use of generalized plane strains allows to follow arbitrary paths in the space of macroscopic stresses. In this section monotone uniaxial tension in a transverse direction is considered. Fibers are assumed to be purely elastic and the matrix is a Von Mises material

$$E^f = 400\ 000\ MPa, \quad \nu^f = 0.23,$$

$$E^m = 68\ 900\ MPa, \quad \nu^m = 0.35, \quad \sigma_0^m = 68.9\ MPa.$$

The hardening modulus of the matrix is either $H^m = 0$ (ideally plastic case) or $H^m = 1\ 710$ MPa (hardening case).

23 images at a spatial resolution of 1024×1024 points containing 64 fibers with a constant volume fraction (0.475) were considered. The choice of this spatial resolution (for a given number of fibers) stems from a study of the influence of spatial resolution on the accuracy of the results (reported elsewhere). The square and hexagonal configurations were also considered.

A uniaxial tension in the 0^0 direction was applied to each "random" configuration. The square and hexagonal configurations were submitted to a uniaxial tension at 0^0 and 45^0. The results are shown in Figure 2 (the average of the strain/stress response of the random configurations is the thick solid line) and summarized in Table 1.

Table 1. Uniaxial tension in the transverse plane.

Transverse Young modulus E_T^{hom}, flow stress σ_0^{hom} (ideally plastic matrix), hardening modulus H^{hom} (hardening matrix). Sample mean (s. mean) and sample standard deviations (ssd) over 23 random configurations (the sample standard deviations are expressed in percentage of the sample means of the corresponding constants). Hexagonal and square lattice.

Constant	Random config.		Hexag. lattice		Square lattice	
	s. mean	ssd	0^0	45^0	0^0	45^0
E_T^{hom} (MPa)	143 166	0.93%	139 655	139 580	153 190	128 600
σ_0^{hom} (MPa)	88.85	2.42%	87.95	79.55	98.01	79.56
H^{hom} (MPa)	10 002	6.54%	7 100	7 420	13 400	4 760

Comments 1. The square lattice has a marked transverse anisotropy, strengthened by the nonlinear behavior, which gives raise to different responses when the direction of tension makes an angle of 0^0 or 45^0 with one of the axis of the square lattice.

The low flow stress in the diagonal direction (45^0) is due to the presence of a shear plane passing through the matrix. Indeed, when a plane of shear can be passed through the weakest phase of a composite, the shear strength of the composite is exactly the strength of the weakest phase (Drucker (1959)). In uniaxial tension in a direction inclined at 45^0 on this plane, the transverse flow stress of the same composite cannot exceed $2\sigma_0^m/\sqrt{3}$. This is the flow stress observed on Figure 2 and Table 1 ($2\sigma_0^m/\sqrt{3} \simeq 79.56 MPa$). In conclusion, except at low volume fractions, the square lattice should not be used to investigate the transverse properties of transversely isotropic nonlinear composites.

2. The hexagonal lattice approaches transverse isotropy. When the matrix is a hardening material, the predictions obtained with the hexagonal lattice underestimate the stiffness of the composite, or at least are located below the average of the predictions for the random configurations in the range of overall deformations considered. Another computation, not reported here, has been performed up to 30% of transverse strain, with no modification in the conclusions. A similar observation was made by Brockenborough *et al* (1991) on another system. When the matrix is ideally plastic, the low flow stress in the diagonal direction (45^0) is again due to the presence of a shear plane passing through the matrix. In conclusion, the hexagonal lattice should be used with care to predict the transverse properties of nonlinear composite systems, even for hardening matrices.

3. The deviation from the average of the transverse Young's moduli computed on the different configurations is small. By contrast, the deviations in the other properties (flow stress, hardening modulus) are higher and might probably be attributed to the combined effects of nonlinearity and incompressibility.

4. The inspection of the local plastic strains reveals significant differences between the ideally plastic case and the hardening case. When the matrix is ideally plastic, the strain localizes in thin bands in the matrix. In most configurations, only a small percentage of the matrix contributes to the plastic dissipation. The overall flow stress of the composite is observed to be in direct relation with the "tortuosity" of these bands. This observation is consistent with Drucker's remark and one of the most meaningful geometrical parameter seems to be the length of the shortest path passing through the matrix at an angle of approximately 45^0 (in tension, or 0^0 in shear). When the matrix is a hardening material, the plastic strain spreads all over the matrix. The whole matrix contributes (although non homogeneously) to the plastic dissipation and, consequently, to the overall strengthening of the composite. In spite of these differences, it has been observed that the "stiffest" (respectively the "weakest") configurations in the ideally plastic case remain the stiffest (respectively the weakest) configurations in the hardening case.

4. Overall flow surface of unidirectional composites

4.1. OVERALL FLOW SURFACE

When the macroscopic stress state Σ is multiaxial instead of being uniaxial, the notion of *overall flow stress* can be generalized into the notion of *overall flow surface* of the composite. First, the *overall strength domain* of the composite is defined as the domain of overall stress states Σ which can be associated to a local stress field σ which is both in equilibrium with Σ and satisfies the strength condition (Suquet (1983) (1987)):

$$P^{hom} = \{\Sigma \in \mathbb{R}_s^{3\times 3}, \text{ such that there exists } \sigma(\mathbf{x}) \text{ with } \langle\sigma\rangle = \Sigma,$$
$$\mathbf{div}(\sigma(\mathbf{x})) = 0, \ \sigma_{eq}(\mathbf{x}) \leq \sigma_0(\mathbf{x}), \text{ for every } \mathbf{x} \text{ in } V\}.$$

The *overall flow surface of the composite* (or its *extremal surface* according to Hill (1967)) is the boundary of P^{hom}. It depends on the flow stress of each phase, on their volume fractions and on their arrangement but is independent of the behavior of its individual constituents prior to the flow stress.

The general properties of P^{hom} will not be discussed here (the interested reader is referred to Suquet (1983) (1987)). We limit our attention to the numerical calculation of P^{hom} in two-phase materials with two different flow stresses, one phase being under the form of cylindrical fibers with a circular cross section dispersed into the other phase. The boundary of P^{hom} is determined according to a procedure described in Marigo *et al* (1987) and Michel and Suquet (1993). In this procedure, each individual phase is assumed to be given an elastic ideally plastic behaviour and, for a prescribed radial direction in the space of stress, the response of the composite along this direction of loading is computed. The overall stress reaches an asymptotic value which is on the extremal surface.

The overall stresses under consideration consist of the superposition of a uniaxial tension and a transverse shear (this example was first considered by Ponte Castañeda and De Botton (1992))

$$\Sigma = \Sigma_1 \left(\mathbf{e}_1 \otimes \mathbf{e}_1 - \mathbf{e}_2 \otimes \mathbf{e}_2\right) + \Sigma_3 \ \mathbf{e}_3 \otimes \mathbf{e}_3.$$

The extremal yield surface lies in the plane (Σ_1, Σ_3).

4.2. NUMERICAL SIMULATIONS

Three contrast ratios between the strengths of the two phases σ_0^f/σ_0^m have been investigated: $\sigma_0^f/\sigma_0^m = 2, 5, 10$. For one of them, $\sigma_0^f/\sigma_0^m = 2$, the computations were performed with 11 of the 23 configurations used in section 3. For the other two ratios, the computations were performed one a single configuration representative of the average of the predictions over the whole set of configurations, when $\sigma_0^f/\sigma_0^m = 2$. More specifically this configuration approaches transverse isotropy and its overall strain/stress response is close to the mean response of all the configurations - as well under multiaxial loading, as under uniaxial tension. The results are shown in Figure 3 and summarized in Table 2.

In all three cases, the shape of the extremal surface ressembles a bimodal surface. Bimodal surfaces have been used by Hashin (1980), Dvorak (1988), De Buhan and Taliercio (1991), Ponte Castañeda and De Botton (1992) and several other authors to describe the first yield or the flow surface of unidirectional composites. The present calculations confirm the validity of this assumption or observation.

The inspection of the failure modes at the local level reveals that the unit cell can fail under two possible modes and confirms even further the bimodal shape of the overall flow surface. The first mode, observable for low values of the axial stress Σ_3, corresponds to shear bands in the matrix with an inclination of approximately $\pm45^0$ on the horizontal direction. When Σ_3 reaches a threshold corresponding to the vertex of the flow surface, the plastic zone spreads throughout the unit cell. As Σ_3 is increased, the plastic strain tends to be more and more homogeneous and becomes fully homogeneous when Σ_3 reaches $< \sigma_0 >$ (Figure 4).

These numerical calculations can be used to propose a closed form expression to the bimodal surface. Indeed, simple piecewise constant stress fields meeting both the requirements of equilibrium and of strength can be constructed and lead to the following **inner** approximation of P^{hom}

$$|\Sigma_3| \leq v_f \left((\sigma_0^f)^2 - 3\Sigma_1^2\right)^{1/2} + v_m \left((\sigma_0^m)^2 - 3\Sigma_1^2\right)^{1/2}. \tag{4.1}$$

The strength in shear predicted by (4.1) coincides with the strength in shear of the matrix, $\sigma_0^m/\sqrt{3}$, while the calculations show a small increase in strength due to the fibers. This increase can be taken into account by modifying (4.1) into

$$|\Sigma_3| \leq v_f \left((\sigma_0^f)^2 - 3\Sigma_1^2\right)^{1/2} + v_m \left((\sigma_0^m)^2 - (\frac{\sigma_0^m}{k_*}\Sigma_1)^2\right)^{1/2} \tag{4.2}$$

k_*, in-plane shear strength of the composite, is the only adjustable parameter contained in (4.2). The resulting simple expression (4.2), with k_* adjusted in pure in-plane shear, fits well with the results of numerical calculations performed on radial paths with arbitrary orientation.

Table 2. Overall flow surface under combined axial tension and in plane shear.

$\sigma_0^f/\sigma_0^m = 2$		$\sigma_0^f/\sigma_0^m = 5$		$\sigma_0^f/\sigma_0^m = 10$	
Σ_1/σ_0^m	Σ_3/σ_0^m	Σ_1/σ_0^m	Σ_3/σ_0^m	Σ_1/σ_0^m	Σ_3/σ_0^m
0.626	0.000	0.638	0.000	0.640	0.000
0.626	0.626	0.639	1.756	0.642	3.640
0.616	0.879	0.637	2.379	0.642	4.053
0.580	1.005	0.591	2.562	0.642	4.569
0.448	1.231	0.478	2.709	0.638	4.850
0.356	1.328	0.391	2.779	0.609	4.963
0.201	1.430	0.250	2.853	0.533	5.072
0.000	1.475	0.000	2.900	0.450	5.142
*	*	*	*	0.184	5.255
*	*	*	*	0.000	5.275

Acknowledgments.

This study is part of the "Eurhomogenization" project supported by the SCI-ENCE Program of the Commission of the European Communities under contract

ERB4002PL910092. Use of the supercomputer CRAY YMP at IMT was possible through a grant of the PACA region. Helpful discussions with J.C. Michel and P. Ponte Castañeda are gratefully acknowledged.

References.

Adams, D.F. and Doner, D.R. (1967) Tranverse normal loading of a unidirectional composite, *J. Comp. Mat.*, **1**, 152

Böhm, H.J., Rammerstoffer, F.G. and Weissenbeck,E. (1993) Some simple models for micromechanical investigations of fiber arrangements in MMCs, *Comput. Mat. Sc.*, **1**, 177-194.

Brockenborough, J.R., Suresh, S. and Wienecke, H.A. (1991) Deformation of Metal-Matrix Composites with continuous fibers: geometrical effects of fiber distribution and shape, *Acta Metall. Mater.*, **39**, 735-752.

Christman, T., Needleman, A. and Suresh, S. (1989) An experimental and numerical study of deformation in metal-ceramic composites, *Acta Metall. Mater.*, **37**, 3029-3050.

De Buhan, P. and Taliercio A. (1991) A homogenization approach to the yield strength of composite materials, *Eur. J. Mech. A/ Solids*, **10**, 129-150.

Drucker, D.C. (1959) On minimum weight design and strength of nonhomogeneous plastic bodies, in Olszak (ed.), *Nonhomogeneity in Elasticity and Plasticity*, Pergamon Press, 139-146.

Dvorak, G., Bahei-El-Din, Y.A., Macheret, Y. and Liu, C.H. (1988) An experimental study of elastic-plastic behavior of a fibrous Boron-Aluminum composite, *J. Mech. Phys. Solids*, **36**, 655-687.

Hashin, Z. (1980) Failure Criteria for Unidirectional Fiber Composites, *J. Appl. Mech.*, **47**, 329-334.

Hill, R. (1967) The essential structure of constitutive laws for metal composites and polycristals, *J. Mech. Phys. Solids*, **15**, 79-95.

Kröner, E. (1972) *Statistical Continuum Mechanics*, Springer Verlag, Wien, 117.

Marigo, J.J., Mialon, P., Michel J.C. and Suquet, P. (1987) Plasticité et homogénéisation : un exemple de prévision des charges limites d'une structure périodiquement hétérogène, *J. Méca. Th. Appl.*, **6**, 47-75.

Michel, J.C. and Suquet, P. (1993) On the strength of composite materials: variational bounds and numerical aspects, in M.P. Bendsoe, C. Mota-Soares (eds.), *Topology Design of Structures*, Kluwer Pub., Dordrecht pp. 355-374.

Moulinec, H. and Suquet, P. (1994) A fast numerical method for computing the linear and nonlinear properties of composites *C. R. Acad. Sc. Paris*, II, **318**, 1417-1423.

Nakamura, T. and Suresh, S. (1993) Effects of thermal residual stresses and fiber packing on deformation of metal-matrix composites, *Acta metall. mater.*, **41**, 1665-1681.

Ponte Castañeda, P. (1992) New variational principles in Plasticity and their application to composite materials, *J. Mech. Phys. Solids*, **40**, 1757-1788.

Ponte Castañeda, P. and De Botton, G. (1992) On the homogenized yield strength of two-phase composites, *Proc. Royal Soc. London A*, **438**, 1992, 419-431.

Suquet, P. (1983) Analyse limite et homogénéisation, *C. R. Acad. Sc. Paris*, II, **296**, 1355-1358.

Suquet, P. (1987) Elements of Homogenization for Inelastic Solid Mechanics, in E. Sanchez Palencia and A. Zaoui (eds.) *Homogenization Techniques for Composite Media*, Lecture Notes in Physics $N^0 272$, Springer-Verlag. Berlin, pp.193-278.

Suquet, P. (1993), Overall potentials and flow stresses of ideally plastic or power law materials, *J. Mech. Phys. Solids*, **41**, 981-1002.

Tvergaard, V. (1990) Analysis of tensile properties for a whisker–reinforced metal-matrix composite, *Acta Metall. Mater.*, **38**, 185-194.

Willis, J.R. (1991) On methods for bounding the overall properties of nonlinear composites, *J. Mech. Phys. Solids*, **39**, 73-86 and **40**, 441.

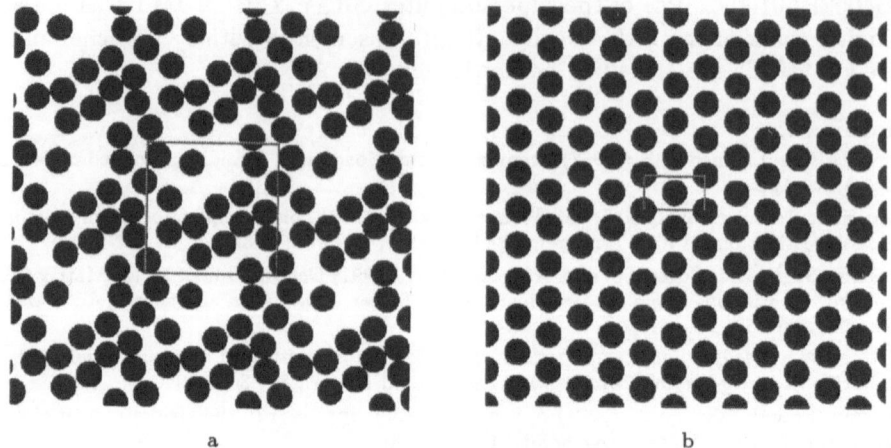

Figure 1. Configurations.

Periodic unit cell (framed area) containing 16 identical circular fibers placed randomly (Fig. 1.a). Hexagonal lattice (Fig. 1.b). The unit cell (framed area) contains $1 + 4 \times \frac{1}{4} = 2$ fibers.

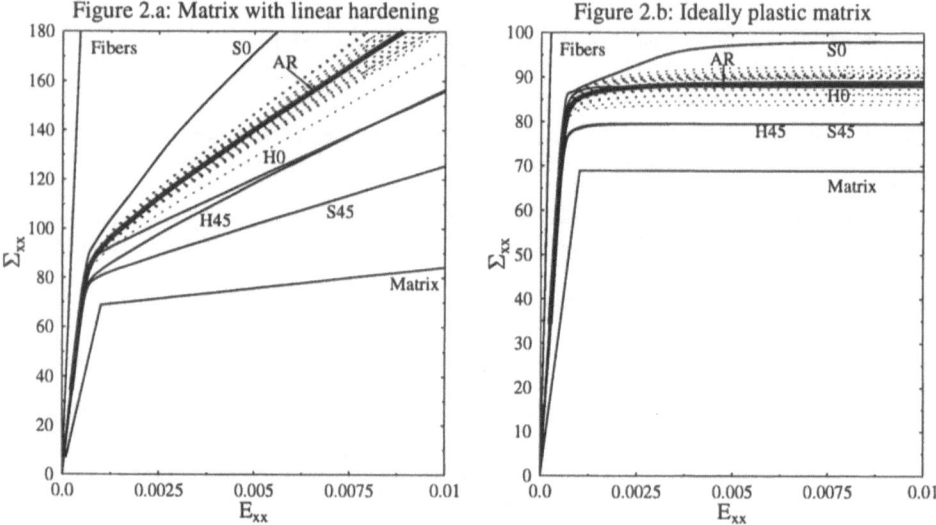

Figure 2. Uniaxial transverse tension.

Overall stress-strain response computed with the present method. Fibers volume fraction = 47.5%. Dotted lines: 23 configurations of 64 identical circular fibers placed randomly in the unit cell. Thick solid line: average of the random configurations (AR). S0 (resp. S45): fibers placed at the nodes of a square lattice, tension at $0°$ (resp. $45°$). H0 (resp. H45): fibers placed at the nodes of a hexagonal lattice, tension at $0°$ (resp. $45°$).

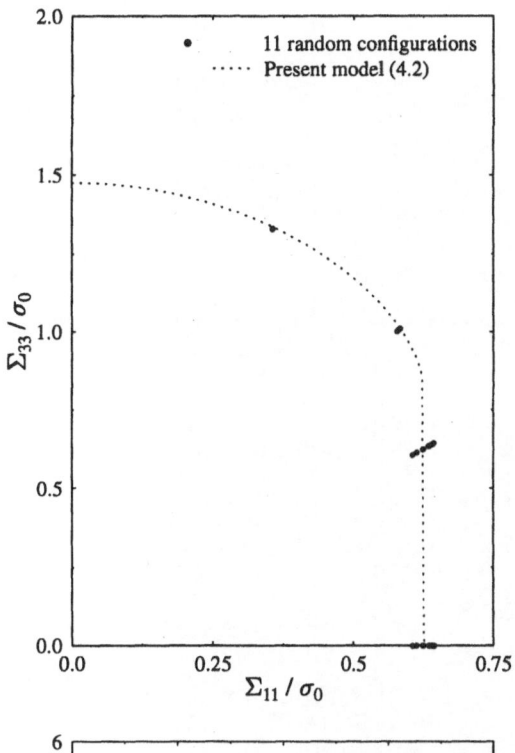

Figure 3. Overall flow surface of unidirectional composites.

Figure 3a: $\sigma_0^f/\sigma_0^m = 2$.

Numerical results for 11 random configurations (full circles).

Present model (4.2) (dotted line). The strength in shear k^* in (4.2) is computed on a representative configuration.

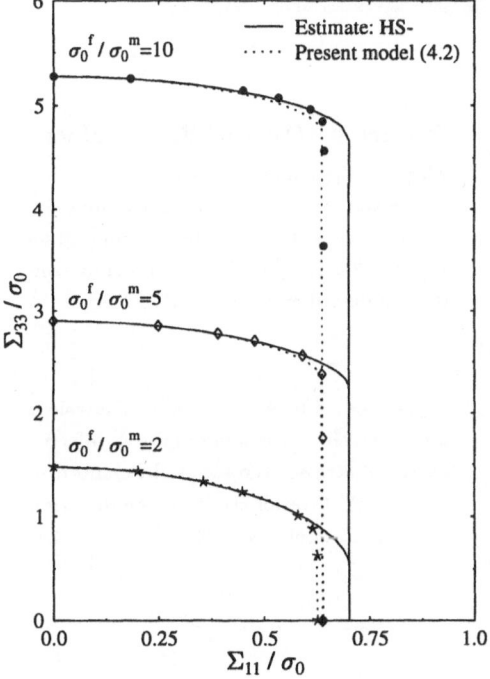

Figure 3b: Yield surface for a representative configuration.
- *: $\sigma_0^f/\sigma_0^m = 2$
- ◇: $\sigma_0^f/\sigma_0^m = 5$
- ●: $\sigma_0^f/\sigma_0^m = 10$

Present model (4.2) (dotted lines).

Prediction of nonlinear bounding theories (Ponte Castañeda and De Botton (1992), Suquet (1993)) using the lower Hashin-Shtrikman linear bound (solid line).

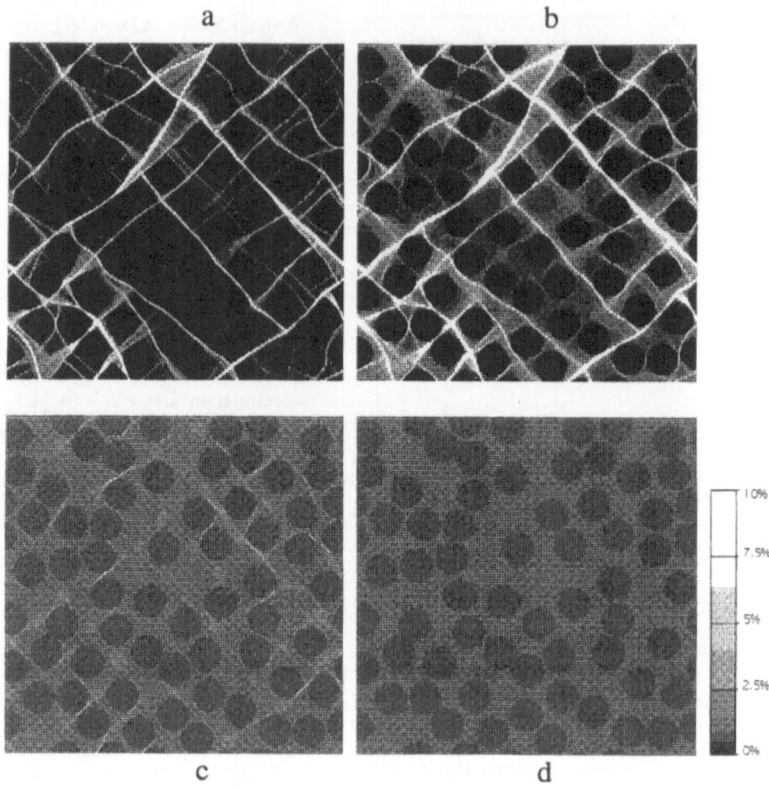

a   b

c   d

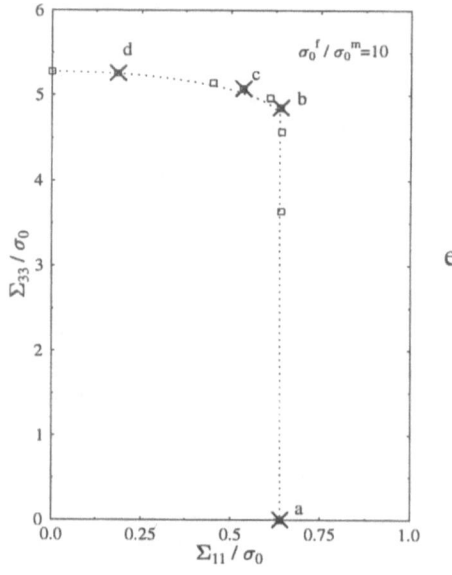

e

Figure 4. Overall flow surface.

Maps of the equivalent plastic strain in a composite reinforced by identical unidirectional circular fibers. Fibers volume fraction = 47.5%. $\sigma_0^f/\sigma_0^m = 10$. Loading conditions described in sect. 4.1.

Figure 4.a, 4.b, 4.c, and 4.d: equivalent plastic strains corresponding to 4 different extremal stresses (crosses a, b, c and d in figure 4.e). Overall strain in the direction of the applied stress= 1%.

HYSTERETIC EFFECTS AND PROGRESSIVE
DELAMINATION AT COMPOSITE INTERFACES

Z. MRÓZ AND S. STUPKIEWICZ
Institute of Fundamental Technological Research,
Polish Academy of Sciences
Świętokrzyska 21, 00-049 Warsaw, Poland

Abstract. Interface delamination is usually accompanied by friction slip at contacting interfaces under compressive normal traction. The present work is devoted to the analysis of friction slip phenomena for monotonic and cyclic loading involving progressive and reversal slips with account for memory of load reversal. The progressive delamination interacts with the friction slip by affecting hysteretic response. Both limit stress and potential energy conditions are used in specifying the delamination evolution. When friction slip conditions during cyclic loading are affected by surface wear with respective variation of contact parameters, the delamination growth may occur in the course of cyclic loading. Three specific examples are treated in detail: i) cyclic tension of an elastic strip adhering to the interface; ii) fiber pull-out problem for cyclically varying load; iii) monotonic and cyclic flexure of a beam sliding on the frictional interface.

1. Introduction

The propagation of interlayer cracks and the resulting delamination in composite structures induce significant degradation of their stiffness and strength. This topic is of fundamental interest in strength assessment and design of composites. A detailed survey of research in this area can be found, for instance, in articles by Garg [4], Storakers [18] or in the extensive essay by Hutchinson and Suo [6] on cracking in layered materials. The present work is concerned with the case when compressive normal tractions occur at the interface and the effect of frictional slip accompanies progressive delamination. When a composite material is subjected to cyclic loading, a

247

R. Pyrz (ed.), IUTAM Symposium on Microstructure-Property Interactions in Composite Materials, 247–264.
© *1995 Kluwer Academic Publishers.*

set of progressive and reverse slip zones develop at the interface and frictional hysteretic effects occur inducing contact dissipation and evolution of the state of delaminated inteface due to wear. The varying state of a delaminated portion may cause coupling between the propagation mode of delamination of surface and the hysteretic dissipation in the domain of frictional slip. This class of problems has not yet attracted sufficient interest of the researchers. The following major topics should be investigated in order to describe quantitatively frictional slip and delamination phenomena:

    i. formulation of accurate slip and wear rules at the delaminated interface portions with account for local dilatancy, compaction and interaction of asperities,
    ii. formulation of evolution rules of delamination with account for local stress state and friction dissipation during monotonic and cyclic loading,
    iii. study of localized temperature effects due to cyclic slip and interface dissipation.

Recently, the slip and memory rules in elastic, contact friction problems were discussed by Jarzębowski and Mróz [7] following the earlier phenomenological friction models by Michałowski and Mróz [11]. It was shown that progressive and reverse slip effects can be described by introducing active loading and memory surfaces in the space of contact tractions, similarly to cyclic plasticity for which multisurface hardening rules proved to be very efficient. The constitutive models for contact slip effects were discussed in a set of papers [13, 14, 15] by Mróz, Giambanco, Jarzębowski and Stupkiewicz. The account of asperity interaction was made by considering single or dual asperity models with resulting dilatancy phenomena and associated wear effects.

It is natural to expect that accurate modelling of friction and slip effects will constitute a substantial contribution to the analysis of coupled phenomena of delamination and slip under compressive normal tractions. This state of stress occurs in many cases when the lateral pressure or constraint is combined with tangential loading, so the separation and delamination buckling effects do not occur.

In the next section we shall present the general formulation of the problem. In Sections 3, 4, and 5, three simple cases will be disscussed, namely elastic strip, fiber pull-out or push-out problem and beam bonded to a rigid frictional foundation.

2. Problem formulation

Consider an interface between two elastic bodies Ω_1 and Ω_2, Fig. 1. Denote by t_1, t_2, n the local orthogonal reference system with the n-axis pointing

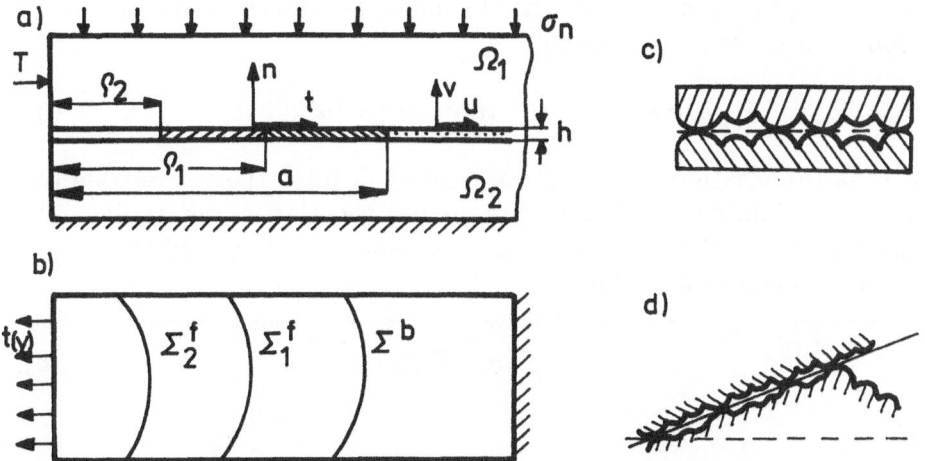

Figure 1. a) Interface between elastic materials, b) Progressive slip and delamination fronts in a plate resting on a frictional foundation, c) Spherical asperity model, d) Dual asperity model.

into the Ω_1-domain. The interface can be replaced by a contact layer of thickness h. The contact stress components acting on the layer are $\sigma_n = \mathbf{n} \cdot \boldsymbol{\sigma} \mathbf{n}$ and $\boldsymbol{\tau}_n = \boldsymbol{\sigma} - (\mathbf{n} \cdot \boldsymbol{\sigma} \mathbf{n}) \mathbf{n}$. For the plane problem with slip occuring along t_1, only the components σ_n and τ_{n1} are considered (τ_n identified with τ_{n1}). At the bonded interface portion, the displacement field is continuous. In view of continuity of interface traction components, on the bonded interface there is

$$[\tau_{n\alpha}] = [\sigma_n] = 0, \quad [u_\alpha] = [v] = 0, \qquad \alpha = 1, 2, \tag{1}$$

where [ ] denotes the discontinuity. At the debonded interface the traction components are continuous but the displacement components u_α, v may suffer discontinuity, thus

$$[u_\alpha] = u_\alpha^{(1)} - u_\alpha^{(2)}, \quad [v] = v^{(1)} - v^{(2)}, \tag{2}$$

where $u_\alpha^{(1)}$, $v^{(1)}$ and $u_\alpha^{(2)}$, $v^{(2)}$ are referred to bodies Ω_1 and Ω_2. The engineering strain components within the contact layer

$$\gamma_{n\alpha} = \frac{[u_\alpha]}{h}, \quad \varepsilon_n = \frac{[v]}{h}, \qquad \alpha = 1, 2 \tag{3}$$

are assumed constant within the layer. The following rate equality holds

$$\dot{W} = \sigma_n \dot{\varepsilon}_n h + \tau_{n\alpha} \dot{\gamma}_{n\alpha} h = \sigma_n [\dot{v}] + \tau_{n\alpha} [\dot{u}_\alpha], \tag{4}$$

so that σ_n, $\tau_{n\alpha}$ and $[\dot{v}]$, $[\dot{u}_\alpha]$ can be regarded as the conjugate stresses and strain rates at the interface. The strain components (3) can be used as measures of slip and dilatancy at the debonded interface. The additive decomposition is assumed, namely

$$\varepsilon = \varepsilon^e + \varepsilon^s \quad \text{or} \quad [u] = [u^e] + [u^s], \tag{5}$$

where ε^e is the elastic interface strain and ε^s denotes the irreversible (or siding) strain. Similarly, $[u^e]$ and $[u^s]$ denote the elastic and sliding displacements. The elastic interface strain components are related to contact stresses by the constitutive law

$$\sigma = E^e \varepsilon^e = \frac{1}{h} E^e [u^e], \tag{6}$$

where

$$\sigma = [\tau_{n\alpha}, \sigma_n]^T, \quad \varepsilon = [\gamma_{n\alpha}, \varepsilon_n]^T, \quad E^e = \begin{bmatrix} E^e_{t\alpha} & 0 \\ 0 & E^e_n \end{bmatrix}, \tag{7}$$

and $E^e_{t\alpha}$, E^e_n are the tangential and normal stiffness moduli. For the non-linear case, $E^e_{t\alpha}$ and E^e_n are the secant interface moduli.

The sliding mode along the interface occurs when the limit friction condition is reached. Assuming the classical Coulomb condition in the form

$$F_L(\sigma) = |\tau_n| - \sigma_n \tan \phi^b \le 0, \tag{8}$$

where $|\tau_n| = (\tau_{n\alpha} \tau_{n\alpha})^{1/2}$ and ϕ^b is the friction angle, the sliding rule is expressed as follows (compressive σ_n is assumed as positive)

$$\dot{\gamma}^s_{n\alpha} = \dot{\lambda} \frac{\tau_{n\alpha}}{|\tau_n|}, \quad \dot{\varepsilon}^s_n = 0, \quad \dot{\lambda} > 0, \quad \dot{\lambda} F_L = 0. \tag{9}$$

The rules (8) and (9) follow from the assumption of isotropic friction and tangential sliding with no dilatancy or compaction effects. The generalization to the case of orthotropic friction was discussed by Mróz and Stupkiewicz [15] who considered a class of non-associated rules predicting orientation of the sliding velocity vector with respect to the tangential traction vector.

A more general class of slip and sliding rules was recently discussed by Mróz and Giambanco [13] by assuming the interface to be composed of spherical asperities contacting each other and slipping under tangential tractions, Fig. 1c.

Considering two equal interacting spheres under normal and tangential forces, the classical solution of Mindlin and Deresiewicz [12] is used in order to generate contact slip rules and memory rules similar to plastic flow rules for hardening materials. For any distribution of spheres, the normal and tangential slip displacement can be determined and the incremental response is specified, namely

$$\dot{\sigma} = E^* \dot{\varepsilon}, \qquad (10)$$

where

$$E^* = \begin{bmatrix} E_{tt} & E_{tn} \\ 0 & E_{nn} \end{bmatrix}, \qquad (11)$$

and

$$E_{tt} = \frac{2G\beta}{2-\nu} \left(1 - \frac{\tan\theta}{\tan\phi^b}\right)^{\frac{p}{3}}, \quad E_{nn} = \frac{2G\beta}{1-\nu},$$
$$E_{tn} = \pm\frac{2G\beta}{1-\nu} \left[1 - \left(1 - \frac{\tan\theta}{\tan\phi^b}\right)^{\frac{p}{3}}\right]. \qquad (12)$$

Here $\beta = a/R = (4k\sigma_n)^{1/3}$, $k = 3(1-\nu^2)/(4E)$, a and R denote the contact radius and the sphere radius. The plus sign in (12) is applied for loading or reloading and minus sign is used for unloading. The parameter p specifies the non-linear stiffness variation due to non-simultaneous contact engagement (the value $p = 1$ corresponds to contact of two equal spheres). Further, at any instant there is an active slip zone of radius c progressing from the external contact boundary, and

$$\frac{c}{a} = \left(1 - \frac{\tan\theta}{\tan\phi^b}\right)^{\frac{1}{3}}. \qquad (13)$$

The typical contact response is depicted in Figs. 2a and 2b.

In order to predict realistically the dilatancy effect, the dual asperity model of Fig. 1d is assumed. Namely, the contact of spherical asperities is assumed to occur along the large asperity of linear or curvilinear profile. The dilatancy then occurs for progressive sliding and compaction for reverse sliding. Figure 2c presents the typical dilatancy curve for progressive sliding. For cyclic loading the consecutive wear of asperities will affect the dilatancy response and the state of contact will vary.

Having developed the constitutive models of frictional slip, the problem of progressive debonding accompanied by contact slip effect can be considered. Assume first the local or non-local stress condition of delamination progression, namely

$$F(\sigma_n, \tau_{n\alpha}) - c \leq 0, \qquad (14)$$

or

$$\frac{1}{\rho} \int_0^\rho F(\sigma_n, \tau_{n\alpha}) dx - \bar{c} \leq 0, \qquad (15)$$

where ρ denotes the non-locality range measured in the normal direction to the delamination front, and c, $\bar{c}$ are the local and non-local cohesion values.

252

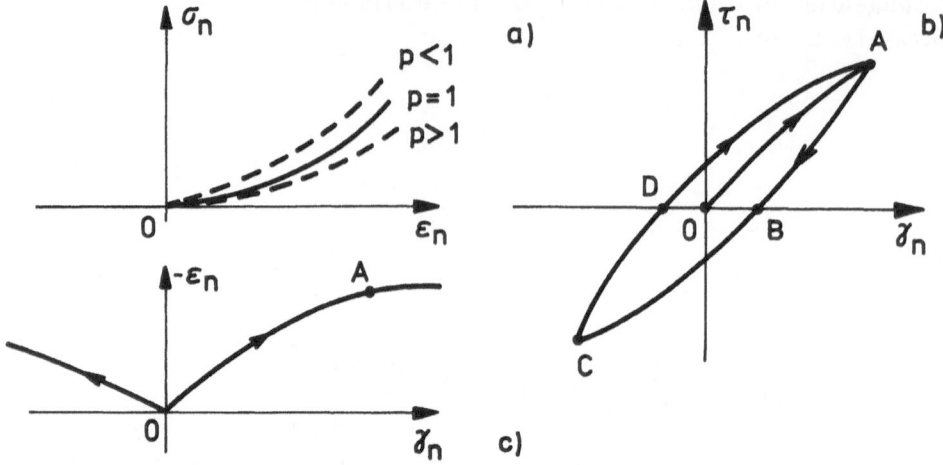

Figure 2. Spherical and dual asperity models: typical contact response: a) normal compression, b) shear response, c) dilatancy curves.

Non-local stress measures in delamination condition were successfully used by Brewer and Lagace [1], and recently the non-local condition (15) was applied by Seweryn and Mróz [17] to predict fracture of notched elements.

An alternative approach would follow the energy consideration for a dissipative system. Denote by $-d\Pi$ the elastic potential energy increment due to delamination propagation, by dD^f the variation of frictional dissipation and by dD^d the variation of specific delamination dissipation. Then, we can postulate

$$- d\Pi = dL - dU = dD^s + dD^d, \tag{16}$$

where dL is the increment of external work, dU is the increment of elastic energy.

The friction dissipation behind the delamination front depends on the loading history. In fact, for cyclic loading there may exist on the contact surface several distinct passive or active slip zones corresponding to progressive or reverse slip effects. The separating boundaries between zones are denoted by Σ_1^f, Σ_2^f, *etc.*, with the delamination boundary denoted by Σ^b, Fig. 1b. The evolution of slip zones occurs during varying loading and the delamination boundary evolves when the critical stress or energy conditions (14), (15) or (16) are satisfied. In the next sections we shall discuss in detail several simple examples of combined delamination and friction slip effects.

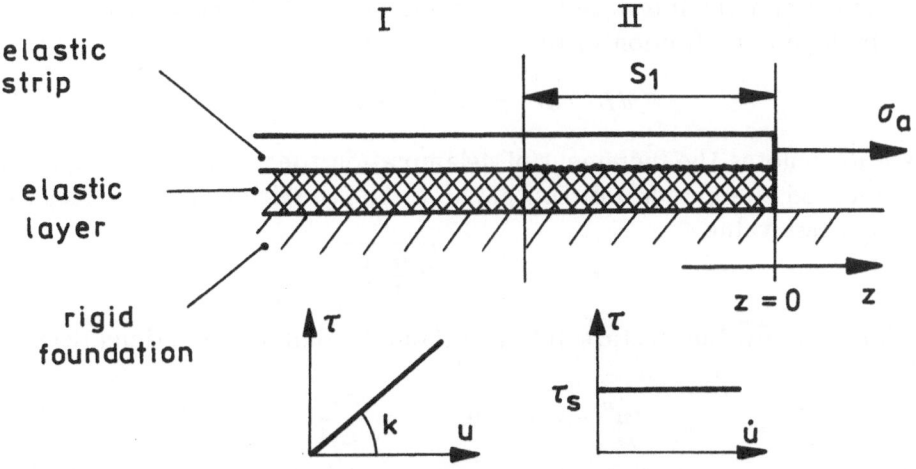

Figure 3. Elastic strip model (I—perfectly bonded zone, II—debonded (slip) zone).

3. Elastic Strip on a Frictional Foundation

Consider first a simple unidimensional case of an elastic strip bonded to a frictional foundation and acted on by a tensile stress σ_a, Fig. 3. The shear stress in the layer equals $\tau = k(u - u^s)$ where u is the strip displacement, u^s denotes the slip along the layer, and k is the layer shear modulus. Denote by A the strip cross section area and by a_c the contact area per unit length. The equilibrium equation provides

$$\sigma' - \frac{a_c}{A}\tau = 0 \tag{17}$$

and since $\sigma = E\varepsilon = Eu'$, there is

$$u'' - \frac{a_c}{EA}\tau = 0, \tag{18}$$

where prime denotes the differentiation with respect to z following the strip axis. The delamination propagation could be governed by the local contact stress condition

$$F(\sigma_n, \tau_n) - c^b = \tau_n - \sigma_n\mu - c^b = 0, \tag{19}$$

where c^b denotes the bonding cohesion and $\mu = \tan\phi^b$ is the friction coefficient. The friction slip occurs when the Coulomb condition is satisfied

$$f(\sigma_n, \tau_n) = \tau_n - \sigma_n\mu = 0. \tag{20}$$

The transition from (19) to (20) occurs by setting $c^b = 0$. Thus, there is a discontinuity of shear stress at the delamination front $z = -s$.

Assume first the interface layer as rigid and apply the energy condition (16). If there is no friction in the system, then

$$dL - dU = dD^d = a_c \Gamma ds, \tag{21}$$

where ds denotes the increment of delamination zone and Γ is the specific delamination energy per unit contact area. The condition (21) provides the critical stress value

$$\sigma_a = \sigma_{cr} = \sqrt{2 \frac{a_c}{A} E \Gamma}. \tag{22}$$

Consider now the elastic solution without debonding. Equation (18) now becomes

$$u'' - r^2 u = 0, \quad r^2 = \frac{k a_c}{E A} \tag{23}$$

with the boundary conditions: $\sigma = \sigma_a$ at $z = 0$ and $\sigma = 0$ at $z = -L$. The solution is

$$u = \frac{\sigma_a}{r E} \frac{e^{r(L+z)} + e^{-r(L+z)}}{e^{rL} - e^{-rL}}. \tag{24}$$

For an infinite strip, $L \to \infty$, we have

$$u = \frac{\sigma_a}{r E} e^{rz} = u_a e^{rz}, \quad \tau = \frac{k \sigma_a}{r E} e^{rz} = \tau_a e^{rz}, \quad \sigma = \sigma_a e^{rz}, \quad z < 0. \tag{25}$$

Consider now the strip with the slip zone $-s < x < 0$ and neglect the contact layer elasticity. The shear stress is constant within the shear zone, $\tau = \tau_s$, so the equilibrium equation provides

$$\sigma = \sigma_a + \frac{a_c}{A} \tau_s z, \quad u = \frac{\sigma_a}{E} (z + s) + \frac{a_c \tau_s}{2 E A} (z^2 - s^2), \quad -s < z < 0. \tag{26}$$

The propagation condition (16) provides the critical stress condition. We have

$$
\begin{aligned}
dL &= A \sigma_a du_a = \left(\frac{A \sigma_a^2}{E} - \frac{a_c \tau_s \sigma_a}{E} s \right) ds, \\
dU &= \frac{d}{ds} \left(\int_{-s}^{0} \frac{1}{2} E \varepsilon^2 A dz \right) = \left(\frac{A \sigma_a^2}{2E} - \frac{a_c \tau_s \sigma_a}{E} s + \frac{a_c^2 \tau_s^2}{2 E A} s^2 \right) ds, \\
dD^s &= \int_{-s}^{0} a_c \tau_s \left(\frac{\partial u}{\partial s} \right) ds dz = \left(\frac{a_c \tau_s \sigma_a}{E} s - \frac{a_c^2 \tau_s^2}{E A} s^2 \right) ds, \\
dD^d &= a_c \Gamma ds
\end{aligned}
\tag{27}
$$

and from (16) it follows that

$$\sigma_{a\,cr} = \frac{a_c}{A} \tau_s s + \sqrt{2 \frac{a_c}{A} E \Gamma}. \tag{28}$$

The first term of (28) presents the friction force, so the critical stress acting at the delamination front is again identical to (22). For the applied stress growing above σ_{cr}, we have

$$s = \frac{A(\sigma_a - \sigma_{cr})}{a_c \tau_s}, \quad u_a = \frac{A(\sigma_a^2 - \sigma_{cr}^2)}{2Ea_c \tau_s}, \quad \sigma_a > \sigma_{cr}. \tag{29}$$

It is seen that in this case the contact friction does not affect the value of the critical stress acting on the interface.

Consider now the elastic contact layer with the delaminated zone $-s < z < 0$. Assume first, according to (19), the local value of shear stress $\tau = \tau_m > \tau_s$ to govern the delamination propagation. We have in the elastic (bonded) zone $-\infty < z < -s$

$$\tau = \tau_m e^{r(z+s)}, \quad \sigma = \sigma_m e^{r(z+s)}, \quad u = u_m e^{r(z+s)} \tag{30}$$

and in the slip zone $-s < z < 0$

$$\tau = \tau_s, \quad \sigma = \frac{a_c}{A}\tau_s z + \sigma_a, \quad u = u_m + \frac{A(\sigma_a^2 - \sigma_m^2)}{2Ea_c\tau_s} + \frac{\sigma_a}{E}z + \frac{a_c\tau_s}{2EA}z^2. \tag{31}$$

The following relations now occur

$$\sigma_m = \frac{1}{r}\frac{a_c}{A}\tau_m = \text{const}, \quad u_m = \frac{1}{r}\frac{\sigma_m}{E}, \quad s = \frac{A(\sigma_a - \sigma_m)}{a_c\tau_s}, \quad \sigma_a > \sigma_m. \tag{32}$$

Let us now apply the energy condition (16). We have

$$
\begin{aligned}
dL &= A\sigma_a\left(\frac{\sigma_m}{E}r\frac{\tau_s}{k}\right)ds, \\
dU &= dU^l + dU^s \\
& \quad dU^l = \frac{a_c\tau_s^2}{2k} - \frac{rA\tau_s\sigma_m}{2k}, \quad dU^s = \frac{A\sigma_m^2}{2E} - \frac{rA\tau_s\sigma_m}{2k}, \\
dD^s &= \int_{-s}^{0} a_c\tau_s\left(\frac{\partial u}{\partial s}ds\right)dz = \left(\frac{a_c\tau_s\sigma_m}{E} - \frac{ra_c\tau_s^2}{k}\right)s\,ds, \\
dD^d &= a_c\Gamma ds
\end{aligned}
\tag{33}
$$

and from (16) it follows that

$$\sigma_m^2 = \sigma_{cr}^2 + \sigma_l^2, \tag{34}$$

where

$$\sigma_m = \sigma_{cr}^e = \sigma_a - \frac{a_c}{A}\tau_s s, \quad \sigma_{cr} = \sqrt{2\frac{a_c}{A}E\Gamma}, \quad \sigma_l = \frac{a_c\tau_s}{rA} \tag{35}$$

256

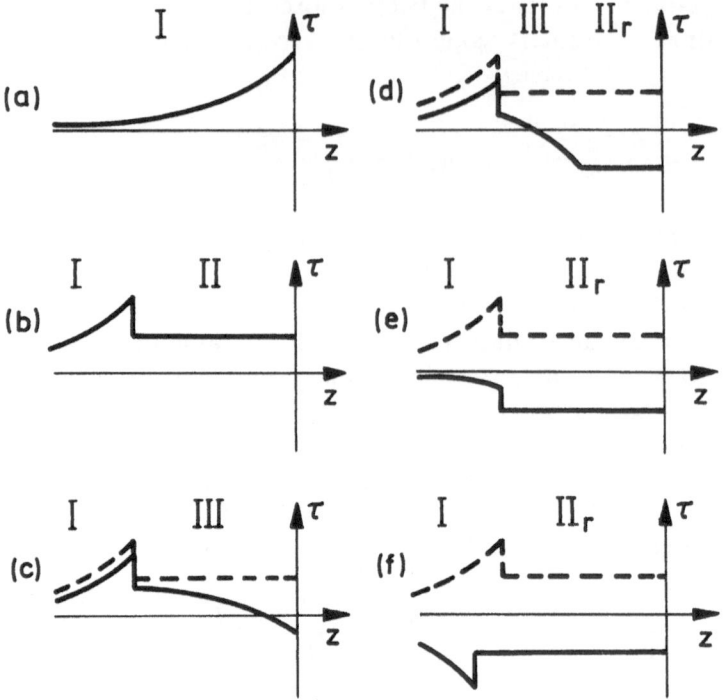

Figure 4. Evolution of the shear stress during loading and unloading (I—perfect bonding zone, II—debonded (slip) zone, II$_r$—reverse slip zone, III—stick zone).

and σ_l corresponds to the growth of the critical stress due to elastic energy of the contact layer. The critical load applied at the end of strip now equals

$$\sigma_{a\,cr} = \frac{a_c}{A}\tau_s s + \sqrt{\sigma_{cr}^2 + \sigma_l^2} = \frac{a_c}{A}\tau_s s + \sigma_{cr}^e \tag{36}$$

and the critical shear stress equals $\tau_m = k\sigma_{cr}^e/(rE)$. Thus, in this simple example the energy condition can be reduced to the critical stress condition $\tau_m = \tau_{cr}$.

Figure 4 presents the evolution of the shear stress during loading and unloading and Fig. 5 presents the hysteresis loop during cyclic loading. The portions 0–1, 2–3 and 5–3' of the loop correspond to elastic loading, the portion 1–2 corresponds to growth of delamination, the portions 3–4–5 and 3'–4'–2 correspond to frictional slip of the delaminated strip segment.

4. Fiber Pulling and Pushing with Interfacial Micro-Dilatancy

The model of the strip on the rigid foundation presented in Section 3 can be regarded as a simple (constant shear stress) model of fiber debonding and sliding in brittle matrix composites. However, the case of fiber sliding

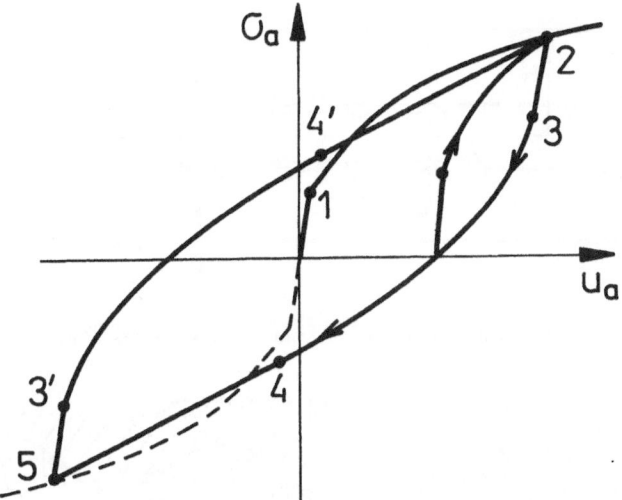

Figure 5. Load-displacement hysteresis loop.

is more complex due to fiber Poisson contraction and resulting change of the compressive normal stress on the interface. The detailed study of such unidimensional model can be found in Hutchinson and Jensen [5] and its application to measurements of interfacial properties of brittle matrix composites was presented by Marshall [9] and Marshall *et al.* [10]. The key role of interface sliding properties in such composites was emhasized by Evans and Marshall [2].

In this section the modification of this model will be briefly disscused, namely the effect of asperity interaction and resulting micro-dilatancy of the interface will be accounted for. We shall restrict ourselves to the case of a single isotropic fiber pulled out or pushed out from the isotropic matrix ($R_f/R = 0$ in Fig. 6), and only the stage of progressive debonding and relative sliding will be treated. The elastic compliance of the interface will be neglected, thus only the slip displacement occurs, $[u^e] = 0$, $[u] = [u^s]$. In contrast to the previous example, the axial stress in the fiber changes discontinously at the delamination front.

The residual stress state in the composite is induced during the curing process of the composite by the mismatch of fiber and matrix coefficients of thermal expansion

$$\sigma^+ = -a_2 E_m \epsilon_r^T, \quad p^+ = a_4 E_m \epsilon_r^T, \tag{37}$$

where σ^+ is the residual axial stress in the fiber, p^+ is the residual compressive stress on the interface and the radial misfit strain is denoted by ϵ_r^T. The non-dimensional parameters a_2, a_4 (introduced by Hutchinson and Jensen [5]) depend on the radial to axial misfit strain ratio $\lambda = \epsilon_r^T/\epsilon_z^T$ and

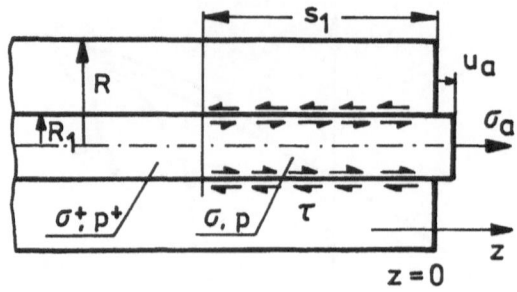

Figure 6. Axisimmetric model of fiber-matrix interaction.

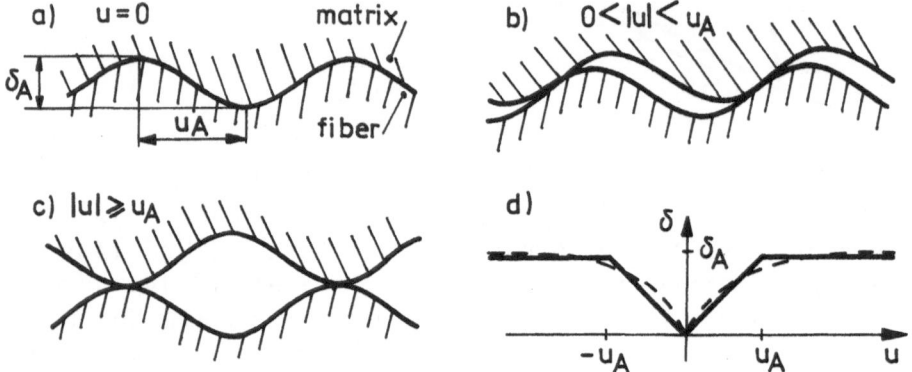

Figure 7. a)–c) Interaction of asperities after relative slip, d) Simplified dilatancy model.

the elastic properties of fiber and matrix: the Young's moduli E_f, E_m and Poisson's ratios ν_f, ν_m.

Implying common assumptions (*cf.* [5]), stress and strain state in the composite can be found from the Lamé solution of the axisimmetric rod-cylinder model (Fig. 6) and here the influence of the radial misfit δ/R_f (Fig. 7) is explicit

$$\Delta p = -b_1 \Delta\sigma + d_1 E_f (\delta/R_f), \quad \Delta\varepsilon = (b_2/E_f)\Delta\sigma + d_2(\delta/R_f), \qquad (38)$$

where the *changes* with respect to the initial state are defined as

$$\Delta\sigma = \sigma - \sigma^+, \quad \Delta p = p - p^+, \quad \Delta\varepsilon = \varepsilon - \varepsilon^+. \qquad (39)$$

The non-dimensional parameters of eqns (38) are given by

$$b_1 = \frac{\nu_f E_m}{(1+\nu_m)E_f + (1-\nu_f)E_m}, \quad b_2 = 1 - 2\nu_f b_1, \quad d_1 = \frac{b_1}{\nu_f}, \quad d_2 = 2b_1. \quad (40)$$

If the relative displacements of the fiber and matrix are of the same or-der as the characteristic asperity wave-length, the assumption of constant increase of the radial misfit due to irregularities of the contacting surfaces is no longer valid. Using the "push-back" experiments Jero et al. [8] showed that the frictional resistance of the interface increases when the fiber is moved from its original position reaching steady value when the relative displacement is of the order of the asperity wave-length. Moreover, when the fiber seats back in its original position the friction force decreases. The explanation of this phenonenon can be provided by assuming that the as-perities on both contacting surfaces match only when the relative displace-ment is equal to zero. Otherwise, due to non-periodic layout of asperities the radial misfit of fiber and matrix increases, Fig. 7. Following these ob-servations a very simple, piecewise-linear form of the dilatancy curve will be assumed allowing for analytical solution of governing equations, namely

$$\delta = \begin{cases} \eta \dfrac{\delta_A}{u_A} u & \text{for} |u| \le u_A \\ \delta_A & \text{for} |u| > u_A \end{cases} \tag{41}$$

Here $\eta = 1$ for pulling out and $\eta = -1$ for pushing out the fiber, u_A, δ_A are the model parameters and can be identified as characteristic asperity amplitude and half of the asperity wave-length, respectively.

In order to unify the notation for all loading events the frictional stress can be written as

$$\tau = \mu p \operatorname{sign} \dot{u} = \eta \vartheta \, \mu p, \quad \eta \vartheta = \pm 1, \tag{42}$$

where $\vartheta = 1$ in case of loading, reloading etc. and $\vartheta = -1$ for unloading etc. Combining the equilibrium equation (17), with eqns (38), (41) and (42) the set of ordinary differential equations is obtained:

$$\begin{Bmatrix} \Delta \bar{\sigma}' \\ \bar{u}' \end{Bmatrix} = \begin{bmatrix} -\eta \vartheta & \vartheta D_1 \\ 1 & \eta D_2 \end{bmatrix} \begin{Bmatrix} \Delta \bar{\sigma} \\ \bar{u} \end{Bmatrix} + \begin{Bmatrix} \eta \vartheta \\ 0 \end{Bmatrix} \tag{43}$$

where the non-dimensional form results from the following definitions

$$\bar{\sigma} = \frac{b_1 \sigma}{p^+}, \quad \bar{u} = \frac{b_1 E_f}{b_2 p^+} \alpha u, \quad \bar{z} = \alpha z, \quad \alpha = \frac{2\mu b_1}{R_f} \tag{44}$$

and parameters D_1, D_2 are given by

$$D_1 = \frac{b_2 d_1}{b_1} \frac{1}{\alpha R_f} \frac{\delta_A}{u_A}, \quad D_2 = d_2 \frac{1}{\alpha R_f} \frac{\delta_A}{u_A}. \tag{45}$$

Note that for $\delta_A = 0$ eqns (43) decouple and the classical model is obtained.

The boundary condition at the end of the debonded zone is

$$\bar{z} = -\bar{s}_1: \qquad \Delta\bar{\sigma} = \eta\bar{\gamma}, \quad \bar{u} = 0 \qquad (46)$$

where the axial stress jump γ can be found from the energy condition, eqn (16), and was given by Hutchinson and Jensen [5]:

$$\gamma = 2\sqrt{\frac{E_f \Gamma}{b_2 R_f}}. \qquad (47)$$

Similarly as in the case of the elastic strip the fiber axial stress at the debonded crack tip is constant and independent of the frictional properties of the interface. Equation (47) differs from (22) only by parameter b_2 (for fiber $a_c/A = 2/R_f$). Parameter b_2 is in fact close to unity and it accounts for the effect of radial fiber-matrix interaction.

The solution of eqns (43) can be found analytically giving the fiber stress and displacement in the debonded zone as a function of $\bar{z}$ coordinate and two integration constants C_i. Implying the appropriate boundary conditions for different loading events the integration constants can be found as functions of the lengths of the loading, unloading and reloading slip zones

$$C_i^{(1)} = C_i^{(1)}(\bar{s}_1), \quad C_i^{(2)} = C_i^{(2)}(\bar{s}_I, \bar{s}_2), \quad C_i^{(3)} = C_i^{(3)}(\bar{s}_I, \bar{s}_{II}, \bar{s}_3). \qquad (48)$$

where the superscript (1) denotes loading, *etc.* and roman subscripts of $\bar{s}_i$ denote frozen slip zones while arabic subscripts—the active ones. The explicit forms of $\Delta\bar{\sigma}$ and $\bar{u}$ will not be given here since they are quite lengthy. In fact the use of computer program handling symbolic operations is necessary when calculating $C_i^{(2)}$ and $C_i^{(3)}$.

In contrast to the classical model when the differential equations (43) are decoupled it is not possible to find direct relation $\bar{u}_a(\Delta\bar{\sigma}_a)$. Instead, the parametric forms $\bar{u}_a(\bar{s}_i)$, $\Delta\bar{\sigma}_a(\bar{s}_i)$ have to be resolved numerically in order to generate the relation $\Delta\bar{\sigma}_a(\bar{u}_a)$.

In the case when $\bar{\gamma} > \bar{\sigma}^+$ spontaneous debonding occurs with no load applied to the fiber. Description of this phenomenon can be found in Marshall [9]. In the same manner it can be incorporated into this model giving proper prediction of the measured load-displacement relation.

Some properties of the model are shown in Fig. 8. Load-displacement curves are plotted for varying $\bar{\gamma}$ (Fig. 8a) and $\bar{u}_A$ (Fig. 8b). It is seen that micro-dilatancy of the interface results in stiffer response.

The experimental data of Marshall *et al.* [10] was used to identify the model. Parameters of the model were found by means of curve-fitting of the monotonic loading path and the unloading-reloading cycle was then predicted. Resulting load-displacement curve is presented on Fig. 9 together

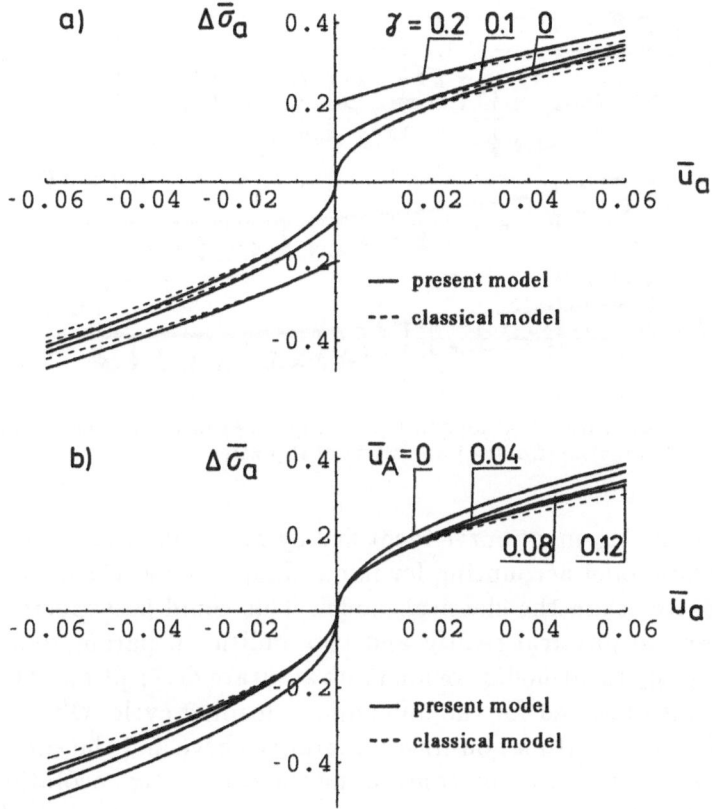

Figure 8. Effect of a) $\bar{\gamma}$ and b) $\bar{u}_A$ on load-displacement characteristics.

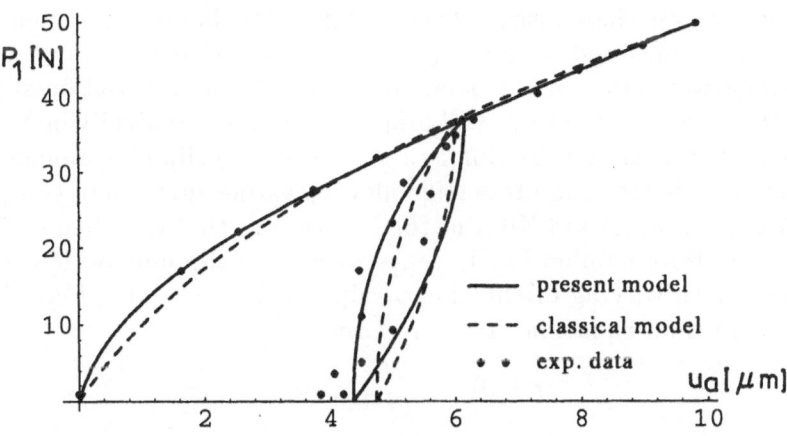

Figure 9. Identification of the experimental data of Marshall *et al.* [10]. For $\bar{\sigma}^+ = -0.123$, $\bar{\gamma} = 0$ and $\bar{u}_A > 10\,\mu$m best fit parameters are: $\delta_A/u_A = 0.018$, $\mu = 0.585$, $p^+ = 117\,$MPa.

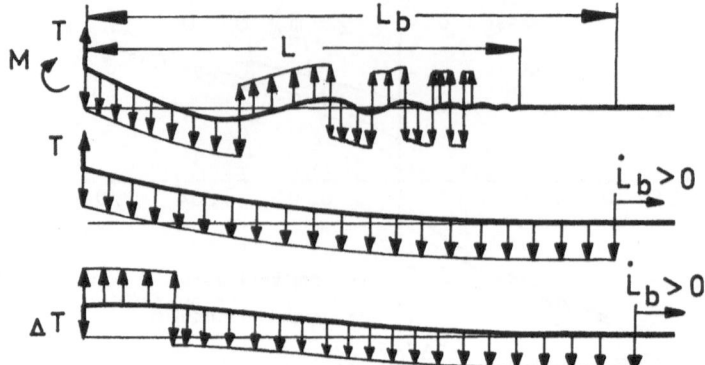

Figure 10. Beam on a frictional foundation: a) slip zones in the debonded domain; b) propagation of delamination front; c) unloading of slip zone.

with the load-displacement curve identified by Marshall *et al.* for the classical model. The model accounting for micro-dilatancy fits the experimental data much better than the classical model. This could be expected as this model is closer to physical reality and two additional parameters u_A and δ_A, when properly identified, provide more accurate description. Also a better prediction is obtained for the unloading-reloading cycle. Obviously, the model could be used with a non-linear dilatancy curve and the surface wear effect could be incorporated by considering the consecutive evolution of the initial dilatancy curve.

5. Frictional Slip and Delamination of a Beam

In this section, we shall discuss briefly slip and delamination effects for the case of a beam bonded to a rigid foundation and loaded tangentially to the foundation plane. In the debonded zone, the lateral friction slip occurs and the delamination front will propagate for monotonically increasing load. The case of Coulomb friction and the associated slip phenomena were studied by Stupkiewicz and Mróz [19] following earlier research of this problem by Fischer *et al.* [3] and Nikitin [16]. It turns out that for a beam loaded at its tip an infinite number of slip zones within the slip domain $0 \leq x \leq L$ is generated with varying orientation of slip rate in each zone, Fig. 10.

The equilibrium equation takes the form

$$EIw^{IV} = -q\,\text{sign}(\dot{w}) \tag{49}$$

with the boundary conditions at $x = 0$

$$EIw''(0,t) = M(t), \quad EIw'''(0,t) = T(t) \tag{50}$$

and the boundary conditions at $x = L$

$$w(L,t) = w'(L,t) = w''(L,t) = w'''(L,t) = 0, \qquad (51)$$

and with the initial conditions $w(x,0) = 0$. Here $w(x,t)$ denotes the lateral beam deflection within the foundation plane and q is the specific friction force per unit length.

When the slip domain interacts with the bonded domain, the consecutive slip zones are progressively annihilated and for monotonic loading eventually only one slip zone may remain, Fig. 10b. The boundary conditions at the interface between slip and bonded domains are

$$w(L_b,t) = w'(L_b,t) = 0 \qquad (52)$$

and the condition of propagation (16) must be satisfied when the delamination front propagates. During cyclic loading, only several slip zones will occur in the debonded domain. On the other hand, when the residual force state generated by friction slip is removed, an infinite number of slip zones will occur in the debonded domain for subsequent monotonic loading. A detailed study of these complex interaction will be presented separately.

6. Concluding Remarks

In the present paper we emphasized the importance of an accurate modelling of local interface friction and slip effects together with dilatancy phenomena occuring during monotonic or cyclic loading. The development of multiple slip zones in the contact area should be accounted for especially when the progression of the delamination interface is affected by prior loading and slip history. The wear of slipping interfaces will induce further coupling effect causing growth of delamination.

References

1. Brewer, J.C. and P.A. Lagace (1988), Quadratic stress criterion for initiation of delamination, *J. Composite Mat.*, **22**, 1141–1155.
2. Evans, A.G. and D.B. Marshall (1989), The mechanical behavior of ceramic matrix composites, *Acta Metall.* **37**(10), 2567–2583.
3. Fischer, F.D. and F.G. Rammerstorfer (1991), The thermally loaded heavy beam on a rough surface, in: F. Ziegler (ed.), *Proc. 8th Symposium on Trends in Applications of Mathematics to Mechanics*, Longman Higher Education & Reference, Burnt Mill, 10–21.
4. Garg, A.C. (1988), Delamination — a damage mode in composite structures, *Engn. Fract. Mech.*, **29**, 557–584.
5. Hutchinson, J.W. and H.M. Jensen (1990), Models of fiber debonding and pullout in brittle composites with friction, *Mech. Mater.* **9**, 139–163.
6. Hutchinson, J.W. and Z. Suo (1991), Mixed mode cracking of layered materials, in: J.W. Hutchinson and T.Y. Wu (eds.), *Adv. Appl. Mech.* **29**, Academic Press, 63–191.

264

7. Jarzębowski A. and Z. Mróz (1994), On slip and memory rules in elastic, friction contact problems, *Acta Mech.* **102**, 199–216.
8. Jero, P.D., R.J. Kerans and T.A. Parthasarathy (1991), Effect of interfacial roughness on the frictional stress measured using pushout tests, *J. Am. Ceram. Soc.* **74**(11), 2793–2801.
9. Marshall, D.B. (1992), Analysis of fiber debonding and sliding experiments in brittle matrix composites, *Acta Metall. Mater.* **40**(3), 427–441.
10. Marshall, D.B., M.C. Shaw and W.L. Morris (1992), Measurement of interfacial debonding and sliding resistance in fiber reinforced intermetallics, *Acta Metall. Mater.* **40**(3), 443–454.
11. Michałowski, R. and Z. Mróz (1978), Associated and non-associated sliding rules in contact friction problems, *Arch. Mech.*, **30**, 259–276.
12. Mindlin, R.D. and H. Deresiewicz (1953), Elastic spheres in contact under varying oblique forces, *J. Appl. Mech.*, **75**, 327–344.
13. Mróz, Z. and G. Giambanco (1994), An interface model for analysis of deformation behaviour of discontinuities, *Int. J. Num. Anal. Meth. Geom.* (submitted for publication).
14. Mróz, Z. and A. Jarzębowski (1994), Phenomenological model of contact slip, *Acta Mech.*, **102**, 59-72.
15. Mróz, Z. and S. Stupkiewicz (1994), An anisotropic friction and wear model, *Int. J. Sol. Struct.*, **31**, 1113–1131.
16. Nikitin, L.W. (1992), Bending of a beam on a rough surface, *Sov. Phys. Dokl.*, **37**(2), 98–100, American Inst. Physics.
17. Seweryn, A. and Z. Mróz (1994), A non-local stress failure condition for structural elements under multiaxial loading, *Eng. Fract. Mech.* (submitted for publication).
18. Storakers, B. (1989), Non-linear aspects of delamination in structural members, in: P. Germain *et al.* (eds.), *Proc. IUTAM Congress Theor. Appl. Mech.*, Grenoble 1988, Elsevier Sci. Publ., 315–336.
19. Stupkiewicz, S. and Z. Mróz (1994), Elastic beam on a rigid frictional foundation under monotonic and cyclic loading, *Int. J. Sol. Struct.*, (in print).

OBSERVATION OF INTERNAL DAMAGE AND INELASTIC DEFORMATION OF GRAPHITE/EPOXY LAMINATE TUBES UNDER CYCLIC TENSION-COMPRESSION

S. MURAKAMI, Y. KANAGAWA, T. ISHIDA,
A. KAWASAKI, and Q. S. BAI

Department of Mechanical Engineering
Nagoya University, Chikusa-ku, Nagoya, 464-01, Japan

1. Introduction

Composite laminates employed in practical engineering components are subject to various non-steady loading under combined states of stress. A number of papers[1]-[6] have been published to elucidate the mechanisms of damage and fracture of composites under combined quasi-static or repeated loading, together with the resulting reduction of stiffness and residual strength.

Laminates of these composites, however, show not only elastic but also salient inelastic deformation when they are subjected to the loading in off-fiber directions. The cyclic global loading and the resulting inter-lamina stress may induce significant internal damage of matrix cracking, delamination, fiber breakage, etc. The present paper is concerned with the observation of the evolution of internal damage and the related change of mechanical behavior of cross-ply graphite/epoxy laminate under cyclic tension-compression.

Fatigue tests of graphite/epoxy (CFRP) $[\pm 45°]_4$ laminate tubular specimens are first performed for the stress ratios $R = (\sigma_{max}/\sigma_{min}) = 0$ through $-\infty$ for a constant stress amplitude of $\sigma_a = 80$MPa. The mechanisms of the resulting inelastic behavior and the change of the mechanical properties were discussed in relation to the evolution of internal damage. Besides these global and macroscopic observations, we further perform quantitative measurement of matrix cracks and delaminations, and elucidate the interaction between the internal damage and the macroscopic mechanical properties. A series of tests under particular stress ratios $R = 0$ and $R = -0.25$ together with the stress amplitude $\sigma_a = 62.5$MPa are carried out to some specific life fractions N/N_f. The distribution and the evolution of the number of the matrix cracks and the delaminations in the sliced sections of the specimens were counted. The relation between the crack and the delamination density and the reduction of elastic modulus is discussed quantitatively.

R. Pyrz (ed.), IUTAM Symposium on Microstructure-Property Interactions in Composite Materials, 265–276.
© *1995 Kluwer Academic Publishers.*

2. Materials and Test Conditions

The experiments were performed for Graphite/Epoxy (CFRP) laminate tubular specimens shown in Fig. 1. The inside and outside diameters and the gauge length were 15mm, 17mm, and 40mm, respectively. The stacking sequence of the specimens is $[\pm45°]_4$, fabricated from prepreg Toray P3052 (T300/2500) of 0.125mm thick. The volume fraction of the fiber is 59 percent in the prepreg. Table 1 shows the mechanical properties of the specimens employed in this experiment.

The tests were carried out under stress-controlled cyclic tension and compression by use of an electrohydraulic servo-controlled combined tension-compression-torsion machine (SHIMADZU EHF-EB-10/TB-20L). Axial strain of the specimen was measured by use of a clip-on type extensometer MTS-632.11-C20. The change of specimen diameter was also measured during tests. Fatigue tests were performed under constant stress rate of 10MPa/sec (0.02-0.04Hz) at the ambient atmosphere for the stress ratios ranging from $R = (\sigma_{max}/\sigma_{min}) = 0$ to $-\infty$.

Development of internal damage was observed at various stages of the fatigue tests. The tubular specimens were cut at the angle of 90° and 45° with respect to the specimen axis by use of a diamond circular saw (MARUTO MC-100). The surfaces of the sections were observed by a light surface microscope (OLYMPAS BHM-MU).

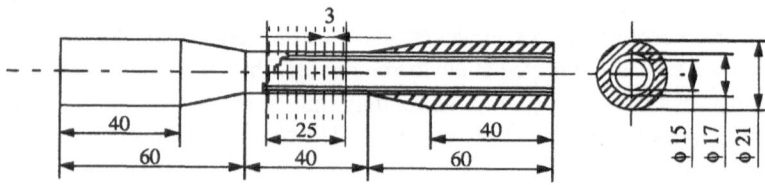

Figure 1. CFRP $[\pm45°]_4$ laminate tubular specimen

TABLE 1. Mechanical properties of CFRP $[\pm45°]_4$ laminate tubular specimen

Young's modulus	16GPa	Shear modulus	30GPa
Tensile yield stress	130MPa	Compressive yield stress	125MPa
Tensile strength	198MPa	Compressive strength	170MPa
Shear strength	420MPa		

3. Inelastic Deformation and Macroscopic Mechanical Behavior in Fatigue Process

The causes and the mechanisms of inelastic deformation will be discussed first in order to elucidate the relationship between the evolution of the internal damage and the resulting inelastic deformation.

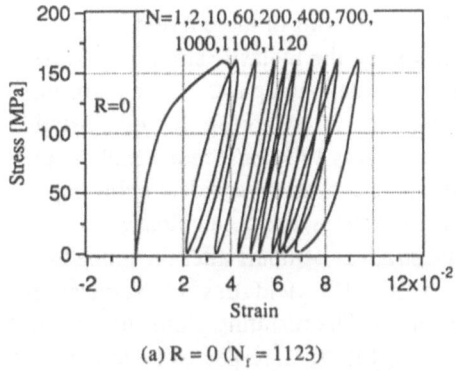

(a) R = 0 (N_f = 1123)

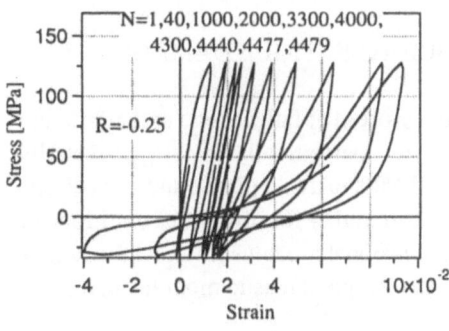

(b) R = -0.25 (N_f = 4480)

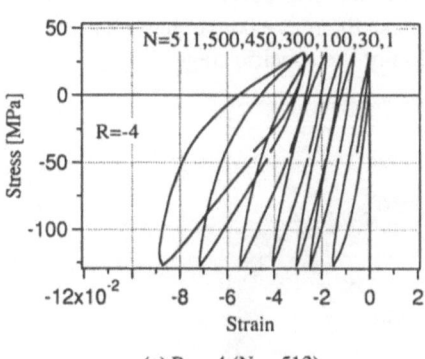

(c) R = -4 (N_f = 512)

Figure 2. Cyclic stress-strain curves under various stress ratios with constant stress

amplitude σ_a = 80MPa

3.1 CYCLIC INELASTIC DEFORMATION IN CROSS-PLY CFRP TUBES

Fatigue tests were performed on the tubular specimens of Fig. 1 under five different stress ratios R = ($\sigma_{max}/\sigma_{min}$) = 0, -0.25, -1, -4 and -∞ for a constant stress amplitude σ_a = 80MPa. The resulting stress-strain hysteresis curves are shown in Fig. 2.

As observed in these figures, hysteresis loops show significant non-linearity, and the shape of the loops changes in the process of fatigue. Though graphite fibers usually show linear elastic behavior under tension up to fracture, the elastic limit of composites for compression in the fibers direction and that for the shear parallel to the fibers are far smaller than that for tension. Thus the causes of the non-linear behavior in Fig. 2 may be attributable to the following five factors:

1) Viscoelastic-plastic deformation of the epoxy resin for the matrix and the inter-lamina adhesive,
2) Development of microscopic cracking in the epoxy resin for the matrix and the inter-lamina adhesive,
3) Delamination induced by the crack growth in the lamina interface,
4) Debonding at fiber-matrix interface and fiber breakage,
5) Reduction of bending rigidity induced by the development of delamination, and the resulting local and global buckling due to compressive loading.

The causes 1), 2) are related to the mechanical property of epoxy resin, and result in strain-hardening during the cyclic viscoelastic-plastic deformation as well as in the strain softening due to the initiation of micro-cracking. The causes 3), 4) and 5), on the other hand, are concerned with the structure of composite materials. The causes 3) is induced mainly by the shear stress between laminae, while 5) is related not only to the mechanical property of composite laminate but also to the geometry of the laminate tube. The salient changes of hysteresis loops in the fatigue process of tubular specimens shown in Fig. 2 can be interpreted in terms of these five causes.

In order to discuss the effects of the above factor 1), the inelastic strain in off-axis composite materials should be separated into the viscoelastic and plastic strains. For this purpose we first examine the plastic property of the material by performing quasi-static tension and compression tests. The stress rate in the tests was 10 MPa/sec identically to the succeeding fatigue tests. Thus the effects of rate dependence was disregarded in this paper.

Specimens were loaded monotonically to a specified stress, and then unloaded to zero. The plastic strain was determined from the residual strain at 24 hours after the unloading, since viscoelastic strain essentially recovers in this period. The yield stress of the materials was specified by the proof stress of 0.2% plastic strain. The resulting yield stresses for tension and compression were σ_{yt} = 130 MPa and σ_{yc} = 125 MPa, respectively, as entered in Table 1.

3.2 REDUCTION OF YOUNG'S MODULUS DUE TO FATIGUE DAMAGE

Fig. 3 shows the reduction of Young's modulus in the process of fatigue tests under different loading conditions. Young's modulus was measured periodically in fatigue tests by performing local unloading in the stress range of 0-20 MPa to avoid plastic and viscoelastic effects. In order to compare the change of Young's modulus for different conditions of loading, the normalized relation between the non-dimensional modulus E/E_0 and the fatigue life fraction N/N_f were used in Fig. 3, where E_0 is the incipient elastic modulus measured before the tests.

Despite some scattering of the results, Fig. 3 shows that the normalized relations between E/E_0 and N/N_f for different R have almost identical evolution. Moreover since the values E/E_0 can be measured accurately at every stage, we can use them as a convenient measure to estimate the damage state and the residual life fraction of the specific stage.

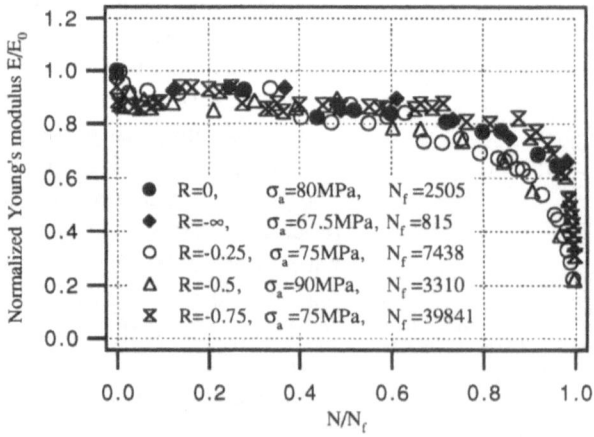

Figure 3. Reduction of Young's modulus due to fatigue

To elucidate the evolution of the internal damage process of the above Figs. 2 and 3, we performed also the micrographical observations on the sections cut out in the direction of 45° with respect to the specimen axis at several stages of life fraction N/N_f. The details of the observations were reported elsewhere[7].

3.3 EFFECTS OF STRESS AMPLITUDE AND STRESS RATIO ON FATIGUE LIFE

Fig. 4 shows the relation between the maximum stress and the number of life cycles with stress ratio as a parameter. It will be observed from this figure that the fatigue lives are largely governed by the maximum stress rather than by the stress amplitude. This may be a different feature observed in metals for which the stress amplitude or strain amplitude is the governing factor of the fatigue life. The deleterious effects on fatigue lives is more significant in the cyclic loading tests with stress reversal R < 0 than the cases of cyclic compression (R = ∞). Namely, the local bulging of thin tubular specimens of R < 0 causes salient inelastic deformation, accelerates the evolution of delaminations, and thus may cause the reduction of the fatigue life. Therefore, fatigue lives under various loading conditions may be reasonably classified into three characteristic damage mechanisms shown schematically by three hatched regions in Fig. 4.

The region A hatched by vertical lines, representing the loading condition of cyclic tension-compression with high stress level, shows the sudden fracture of specimens due to the local disperse defects without significant progressive damage. In the region B of cyclic tension-compression with rather low stress levels, hatched by horizontal lines, the final fracture results from the degradation of composite materials due to accumulation of progressive damage.

In the region C for the stress reversal R < 0, shadowed by small dots, on the other hand, the final fracture is caused by the global buckling of thin tubular specimens resulting from the marked delamination over whole range of stress level. The stress-fatigue life relation for the stress ratio R = -0.25 (marked by Δ in Fig. 4) almost coincides with that of R = -∞. Though the loading condition of R = -0.25 includes a small amount of compressive stress,

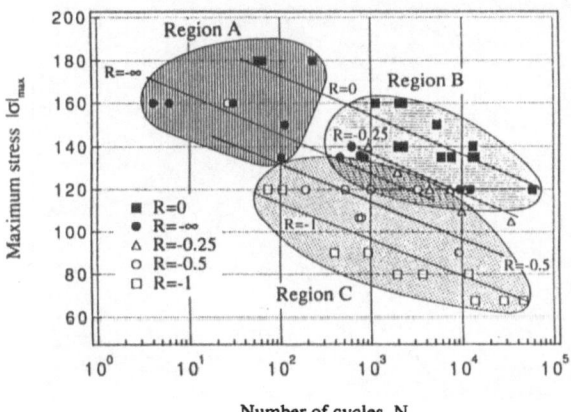

Figure 4. Fatigue limit under axial cyclic loading and the effect of maximum (or minimum) stress

the fatigue life decreases significantly because of the salient inelastic deformation due to the local bulging of tubular specimens.

4. Quantitative Observation of Internal Damage

4.1 OBSERVATION OF INTERNAL DAMAGE AND ITS QUANTIFICATION

In order to elucidate the relation between the microstructural change and the macroscopic mechanical properties discussed so far, more detailed observation and the quantitative evaluation of the evolution of internal damage will be needed. For this purpose, we further observed the development and densities of the matrix cracks and the delamination in the specimens subjected to fatigue at several stages of life fraction N/N_f. The fatigue tests were performed at a stress amplitude $\sigma_a = 62.5\text{MPa}$ and under two levels of stress ratios $R = 0$ and $R = -0.25$. The average fatigue lives under these conditions were $N_f \cong 30,000$ $(R = 0)$ and $46,000$ $(R = -0.25)$, respectively. Micrographical observations were performed at the typical stages of the fatigue life fractions of $N/N_f = 0.03, 0.44, 0.77, 0.93,$ and 0.96 for $R = 0$ and $N/N_f = 0.15, 0.60, 0.80, 0.90$ and 0.96 for $R = -0.25$.

Specimens subjected to the fatigue of specific life fractions N/N_f were sliced perpendicularly to the specimen axis at 8 sections of 3mm interval, as shown in Fig. 1, to examine the distribution of internal damage in the axial direction. The surfaces of the sections were ground and lapped, and were observed by a light micrographic apparatus. Fig. 5 shows a micrograph of the internal damage taken by dark field viewing with reflected light microscope.

4.2 QUANTIFICATION OF INTERNAL DAMAGE

4.2.1 *Average Matrix Crack Length and Matrix Crack Density*

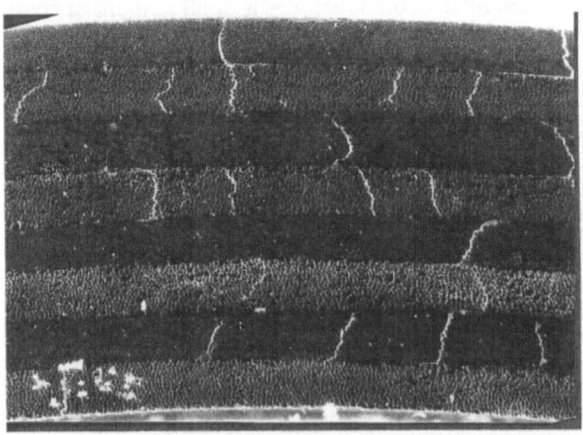

Figure 5. Observation of internal damage $(R = 0, \sigma_a = 62.5\text{MPa}, N/N_f = 0.96)$

For the quantitative evaluation of internal damage, the size and the geometries of the defects as well as their number should be taken into account. As already observed in Fig. 5, the internal damage induced in the present specimens consists mainly of the matrix cracks and the inter-lamina delamination.

The number of defects in the innermost layer (the first layer) and in the lamina-interface between the first and the second layer (D12) were much larger than other regions regardless of the life fractions and stress ratios. Since this may be attributable to the additional deterioration caused in the process of preparing sliced specimens for the micrographical observation, we will exclude the matrix cracks in the first layer together with the delamination at D12 in the later discussion.

Since the matrix cracks have various length and configuration, we will classify them into the following three groups according to their length l projected in the direction perpendicular to the lamina: (a) $0 < l / l_0 < 0.5$, (b) $0.5 \leq l / l_0 < 1$, and (c) $l / l_0 = 1$, where l_0 is the thickness of the lamina. Then, as the quantitative measures of the damage state, we will further define the average matrix crack length l_{AVE} and the matrix crack density D_{MTC} as follows:

$$l_{AVE} = (Na \times 0.25 + Nb \times 0.75 + Nc \times 1) \times l_0 / N_{TOTAL} \tag{1}$$
$$D_{MTC} = (Na \times 0.25 + Nb \times 0.75 + Nc \times 1) \times l_0 / A \tag{2}$$
$$N_{TOTAL} = Na + Nb + Nc \tag{3}$$

Na, Nb and Nc are the number of cracks of the above categories (a), (b) and (c) while N_{TOTAL} is the total number of the cracks (a), (b) and (c). The symbol A in equation (2), on the other hand, is the total cross-sectional area of 8 sliced specimens.

4.2.2 Total Area of Delamination and Delamination Density

According to the micrographical observation, the delaminations are induced mainly by the extension of matrix cracks to the lamina-interface, and then by their development along the interface. In the micrograph, the delaminations are observed as segments of line, and thus we will measure the number and the total length l_{DEL} of the individual delamination. Since the delaminations extend also in the axial direction, we calculate the area of the delamination A_{DEL} by multiplying the observed length of the segments by the thickness 3mm of the sliced specimen: $A_{DEL} = l_{DEL} \times 3$. We will further define the delamination density D_{DEL} by

$$D_{DEL} = A_{DEL} / A_{INT} \tag{4}$$

where A_{INT} denotes the total inter-lamina area of 8 sliced specimens.

4.3 DEVELOPMENT OF MATRIX CRACK

In order to elucidate the development of damage in a whole tubular specimen, we first examined the distribution of the matrix crack density in the sliced specimens at several life fractions N/N_f of the fatigue life. Fig. 6 shows the distribution of the matrix crack density in

272

radial, circumferential and axial directions at the life fraction $N/N_f = 0.96$. The circular section of each slice was divided into 8 portions A, B, C, ···, H in circumferential direction. It will be observed from these figures that, in both cases of $R = 0$ and -0.25, the distribution of matrix crack density is sufficiently uniform in the radial direction. As regards the axial direction, the crack density is rather dominant at the central part of the specimen (slices No. 4~7). This confirms the validity of the test results obtained by tubular specimens.

The tubular specimens are free from edge effects, and can be employed not only to the compressive stress tests but also to those of combined stress. However, the damage states may be non-uniform in radial direction because of the effects of the curvature of the tubes and the resulting radial compressive stress. Thus, Krempl and An[8] have analyzed the stress distribution in CFRP $[\pm45°]_s$ laminate tubes, besides performing the corresponding experiment, and confirmed insignificant effects of the specimen curvature. As observed in Fig. 6, the distribution of matrix crack density for both $R = 0$ and $R = -0.25$ is essentially uniform, and hence the effects of curvature may be disregarded also in the present tests.

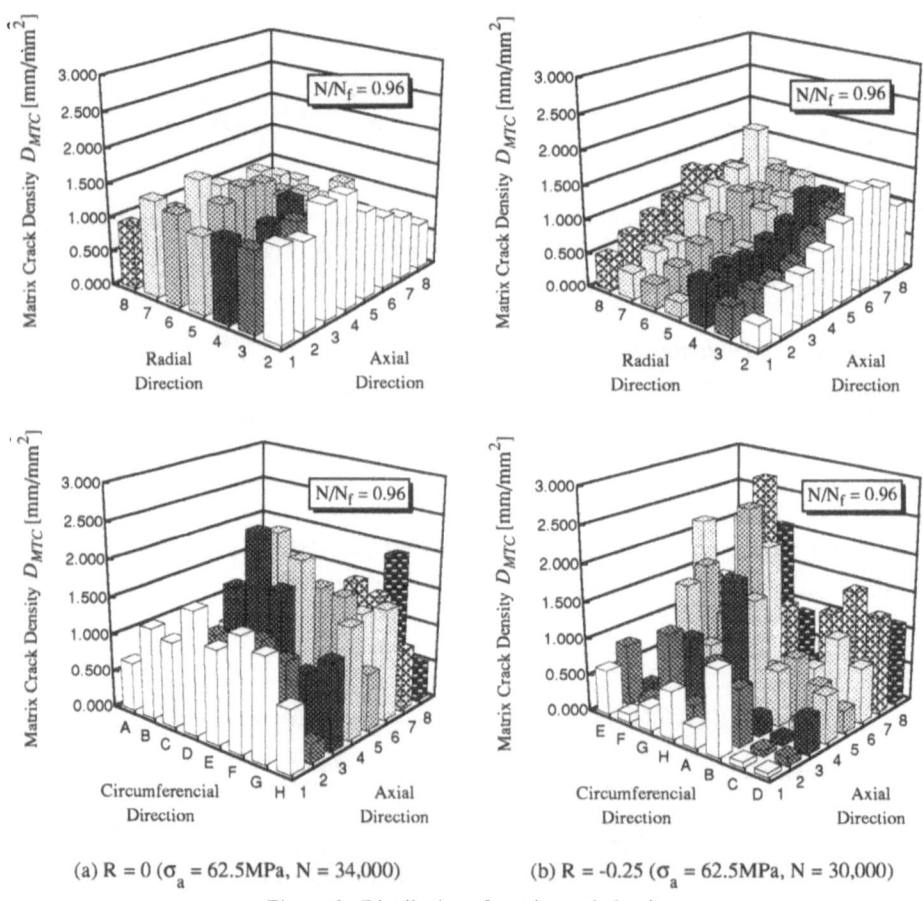

(a) $R = 0$ ($\sigma_a = 62.5$MPa, $N = 34,000$) (b) $R = -0.25$ ($\sigma_a = 62.5$MPa, $N = 30,000$)

Figure 6. Distribution of matrix crack density

Fig. 7 shows the increase of the number of the matrix cracks Na, Nb and Nc classified by their length . As the increase of the life fraction N/N_f, the fraction of the number of the longer cracks Nc increases. The marked increase of matrix cracks in the case of R = -0.25 is concentrated more to the last stage of fracture than the case of R = 0.

Figs. 8 and 9 show the increase of the average matrix crack length l_{AVE}/l_0 and that of matrix crack density D_{MTC} for the cases of R = 0 and -0.25. In the case of R = -0.25, the matrix crack density D_{MTC} increases markedly at $N/N_f \geq 0.8$, and this induces rapid increase of the delamination density (cf. Fig. 12 below). In the case of R = 0, on the other hand, the matrix crack density is sufficiently large at $N/N_f = 0.8$. This may be accounted for by the fact that the maximum stress σ_{max} = 125MPa for R = 0 is larger than σ_{max} = 100MPa for R = -0.25, and this larger value of the maximum stress induces larger number of matrix cracks. However, in the case of R = -0.25, the compressive stress accelerate the growth of the delamination at $N/N_f > 0.8$ (later Fig. 12).

4.4 DEVELOPMENT OF DELAMINATION

Now, let us examine the development of the delamination. Fig. 10 is the distribution of

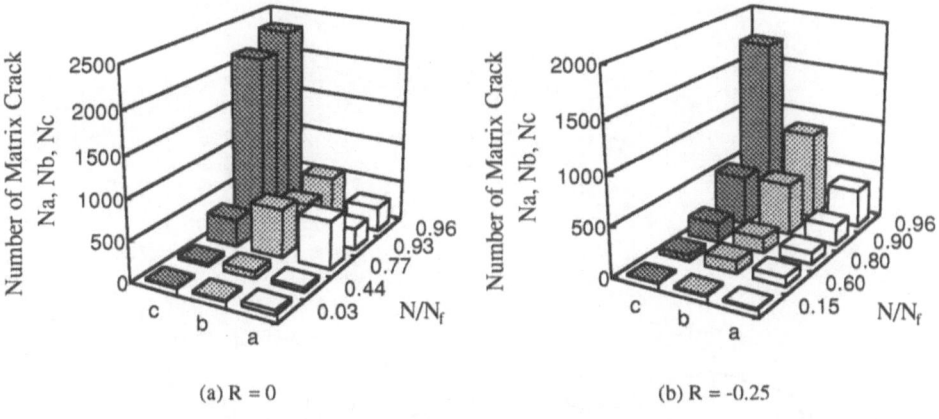

(a) R = 0   (b) R = -0.25

Figure 7. Development of matrix cracks

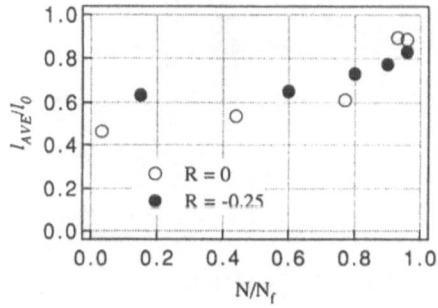

Figure 8. Change of average matrix crack length

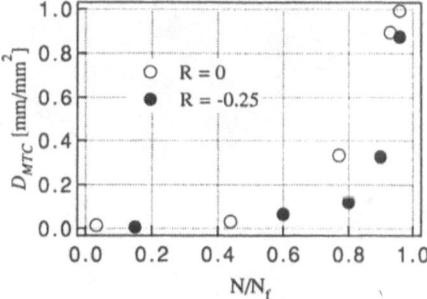

Figure 9. Change of matrix crack density

delamination density in tubular specimens at $N/N_f = 0.96$ for the cases of $R = 0$ and $R = -0.25$. As mentioned already, the delaminations occurring on the interface between the first and the second layers (D12) have been excluded from Fig. 10.

In both cases of $R = 0$ and $R = -0.25$, the locations of the largest delamination density in the axial direction coincide with those of the maximum matrix crack density (i.e., the central part) shown in Fig. 6. However, as regards the distribution in the circumferential direction, the largest delamination density is localized saliently in a few limited regions, and these localized regions locate in the sequence of D→A→F ($R = 0$) and or B→A→H→G ($R = -0.25$). By noting the thickness 3mm of the sliced specimens, these regions are situated along the direction of 45° with respect to the specimen axis. Namely, in view of the mechanism of the delamination which occurs as a result of the extension of matrix cracks to the lamina-interface and their growth along the interface, delamination may develop in the fiber direction of the layer.

As regards the distribution in the radial direction, the largest delamination density is

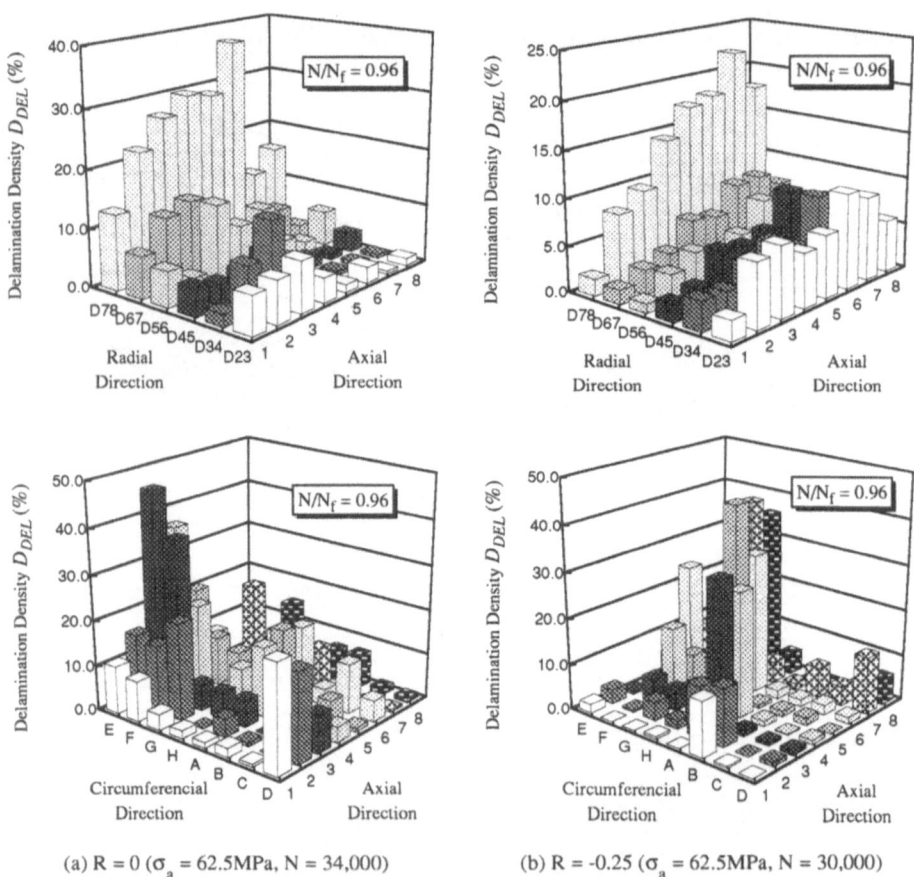

(a) $R = 0$ ($\sigma_a = 62.5$MPa, $N = 34,000$)   (b) $R = -0.25$ ($\sigma_a = 62.5$MPa, $N = 30,000$)

Figure 10. Distribution of delamination density

observed in the outermost interface (D78) in both cases of R = 0 and -0.25 in Fig. 10, and
this situation is observed also at early stages of N/N_f. This is in contrast to the matrix crack
distribution shown in Fig. 6, where the radial distribution is observed rather uniform
throughout the fatigue process. If we note the stress-strain hysteresis curves shown in Fig.
2, local and global buckling may have occurred at the stresses around their minimum.
Then, the outermost lamina will expand radially, and the resulting tensile and shear stresses
in radial direction will facilitate the delamination.

The increase in the number and the density of the delamination is shown in Figs. 11
and 12. In contrast to the development of the matrix crack density of Fig. 9, the delamination
densities of Fig. 12 show almost identical increase both for R = 0 and R = -0.25. Though
maximum stress σ_{max} = 100MPa for R = -0.25 is smaller than σ_{max} = 125MPa for R = 0, the
compressive stress for R = -0.25 may have accelerated the delamination; this may account
for the results of Fig. 12.

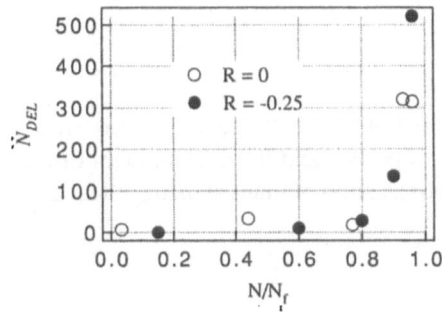

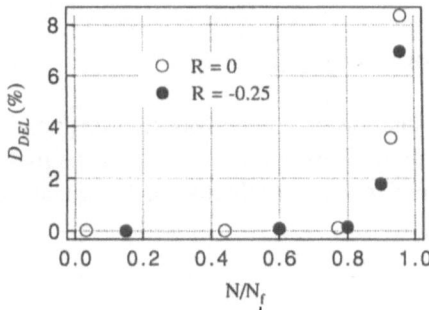

Figure 11. Change of number of delamination   *Figure 12.* Change of delamination density

4.5 REDUCTION OF YOUNG'S MODULUS AND ITS RELATION TO THE DEVELOPMENT OF INTERNAL DAMAGE

Fig. 13 shows the reduction of Young's modulus induced by the fatigue process of the
stress ratios of R = 0 and -0.25 and the stress amplitude of σ_a = 62.5MPa. Young's modulus
was obtained by the procedure of Sec. 3.3, by performing local unloading at specific stages
of N/N_f and by measuring the related change of elastic strain.

Though some decrease in Young's modulus is observed from the onset of experiment,
this may be due to the stabilization of the internal structure of the specimen. However, the
succeeding and accelerating decrease in E/E_0 is mainly due to the significant development
of the matrix cracks and the delaminations.

It should be noted from Fig. 14 that the normalized Young's modulus decreases almost
linearly with the increase of the crack and the delamination densities. The slopes of the
relations in Fig. 14 depends on the stress ratios; i.e., the slope for R = 0 is smaller than that
of R = -0.25. Namely, as observed in Figs. 6 and 10, the distribution of internal damage in
axial direction is less uniform in the case of R = -0.25, and hence more significant decrease

of Young's modulus after $N/N_f = 0.90$ is induced in this case of R.

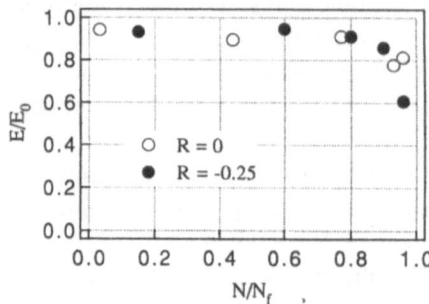

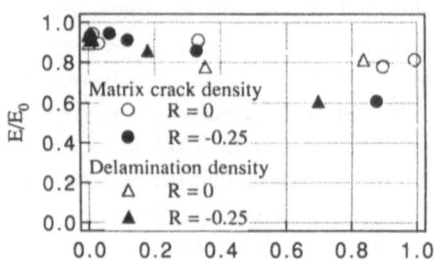

Figure 13. Decrease of normalized Young's modulus

Figure 14. Relation between matrix crack density, delamination density and normalized Young's modulus

5. Conclusions

The cyclic tension-compression tests on graphite/epoxy $[\pm 45°]_4$ laminate tubes are performed. The relation between the change of the inelastic properties and the development of the internal damages was discussed. The quantitative observation of internal damage was also performed.

Acknowledgment

A part of this work was supported by the Ministry of Education of Japan under the Grant-in-Aid for Development of Scientific Research B-1 (No.04555029) for the fiscal years of 1992 through 1994. The authors would like to appreciate their support.

References

[1] Krempl, E. and Niu, T.M. (1982) *J. Composite Materials* 16, 172-187.

[2] Reifsnider, K.L., Schulte, K and Duke, J. C. (1984) *ASTM STP 813*, 136-159.

[3] Daniel, I.M. and Charewicz, A. (1986) *Engineering Fracture Mechanics* 25-6, 793-808.

[4] O' Brien, T.K. (1987) *J. American Helicopter Society* 32-13, 13-18.

[5] Rotem, A. (1989) *J. Composites Technology and Research* 11-2, 59-64.

[6] Voyiadjis, G. Z. (1993) *Damage in Composite Materials*, Elsevier Since Publishers.

[7] Kanagawa, Y., Ishida, T., Murakami, S. and Tanaka, K. (in press) *Sci. Engng Composite Materials*.

[8] Krempl, E. and An, D. (1989) *RPI (Rensselaer Polytechnic Institute) Report* MML 89-1.

INTERACTION BETWEEN A MAIN CRACK AND INCLUSIONS
AT SHEAR STRESS STATE

V.PETROVA
Research Institute of Mathematics,
Voronezh State University
University sq.,1, Voronezh 394693, Russia

Abstract.

The problem of a crack interaction with an arbitrary set of small rigid inclusions under the shear loading has been solved using singular integral equations and a small parameter method. Attention has been paid to analyze the possible contact of the crack edges.

1. Introduction.

The work is concerned with the investigation of interaction between a main crack and the field of small rigid inclusions at shear loading parallel to macrocrack line. Similar problem has been solved by Petrova (1988) under the influence of tensile loading using the small parameter method as given by Romalis and Tamuzh (1984). In this work attention has been paid to analyze the possible contact of macrocrack edges.

This phenomenon was observed for the shear loading of macro- and microcrack system (Tamuzh and Petrova (1993)) and for the heat flux (Tamuzh V.,Petrova V. and Romalis N. (in press)). It was found that depending on the locations and orientations of the small cracks the main crack could be partially or completely closed that means the overlapping of the crack faces. Therefore in order to obtain the correct macrocrack tip stress intensity factor K_I the appearance of the contact zones was taken into consideration.

Early this effect for the shear loading of dissimilar materials with cracks has been revealed in the sixties Commninou and references can be found in paper of Dundurs and Commninou (1979).

Applying the method similar to proposed by Tamuzh and Petrova (1993), the problem of crack — microinclusions interaction under the shear loading is solved with the following assumptions:

R. Pyrz (ed.), IUTAM Symposium on Microstructure-Property Interactions in Composite Materials, 277–287.
© 1995 Kluwer Academic Publishers. Printed in the Netherlands.

1) one opening zone with the unknown length $2c$ is located on the macrocrack;
2) on the closed zones of the crack the smooth contact is assumed, i.e. the friction is not taken into account.

2. Statement of the problem disregarding the contact zones

Let an elastic isotropic plane contain a macrocrack of size $2a_0$ and N microdefects of size $2a_k \ll 2a_0$ (microcracks or small rigid inclusions) (*Figure 1*). It will be assumed that all $a_k = a$. Cartesian coordinates x and y are centered at the midpoint of the main crack such that the crack line lies along the x-axis. The local coordinate systems x_k and y_k are attached to each microdefect positioned at the angle α_k to the x-axis. The macro- and microdefect midpoint coordinates are denoted by z_k^0 ($k=0,1,\ldots,N$).

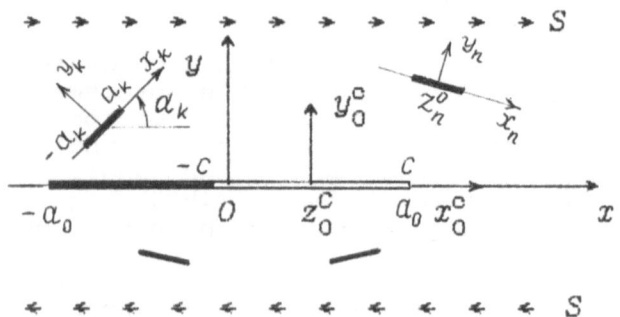

Figure 1. A location scheme of the main crack (with contact zone) and an array of small defects.

The crack edges are supposed to load not self-balanced forces

$$\sigma_n^{\pm} - i\,\tau_n^{\pm} = p_n(t) \pm q_n(t), \quad |t| < a_n \tag{1}$$

Derivatives with respect to displacements

$$2G\,\frac{d}{dt}(u_n^{\pm} + iv_n^{\pm}) = f_n'(t) \pm g_n'(t), \quad |t| < a_n, \tag{2}$$

are assigened to lines of rigid inclusions, where G is the shear modulus. Stresses and rotation are absent at infinity.

Singular integral equations for the system of defects have been derived by Berezhnitskiy et.al.(1983)

$$\sum_{\substack{k=0 \\ k\neq n}}^{N} \int_{-a_k}^{a_k} [Q_k(t,\rho_k^*,\rho_k)K_{nk}(t,x,\rho_n^*,\rho_k^*)+\overline{Q_k(t,\rho_k^*,\rho_k)}L_{nk}(t,x)+$$

$$+t\,\frac{G_n(\rho_n^*,t)}{\bar{T}_k-\bar{X}_n}\,e^{i(\alpha_k-2\alpha_n)}]dt=\varkappa F_n(\rho_n^*,\phi_n)\quad(n=0,1,\ldots,N)\tag{3}$$

Here,

$$Q_k(t,\rho_k^*,\rho_k)=\frac{G_k(\rho_k,t)-G_k(\rho_k^*,t)}{\rho_k-\rho_k^*}$$

and K_{nk} and L_{nk} are regular kernals:

$$K_{nk}(t,x,\rho_n^*,\rho_k^*)=-e\frac{e^{i\alpha_k}}{t}\left[\frac{\rho_n^*}{T_k-X_n}+\frac{\rho_k^*\,e^{-2i\alpha_n}}{\bar{T}_k-\bar{X}_n}\right]\tag{4}$$

$$L_{nk}(t,x)=-e\frac{e^{-i\alpha_k}}{t}\left[\frac{1}{\bar{T}_k-\bar{X}_n}+\frac{T_k-X_n}{(\bar{T}_k-\bar{X}_n)^2}\,e^{-2i\alpha_n}\right]$$

$$X_n=x\,e^{i\alpha_n}+z_n^0,\qquad T_k=t\,e^{i\alpha_k}+z_k^0\tag{5}$$

The notations

$$[G_n(\rho_n^*,t_n),\,F_n(\rho_n^*,t_n)]=\begin{cases}[-2q_n(t_n),\,-2p_n(t_n)];\ \rho_n^*=-1;\\[2mm][2g_n'(t_n),\,2f_n'(t_n)];\ \rho_n^*=\varkappa.\end{cases}\tag{6}$$

are introduced here (Berezhnitskiy et al (1983)). The parameter $\rho_k^*(\rho_k)$ is assigned the following values: $-1(\varkappa)$ for the cracks, and $\varkappa(-1)$ for the inclusions, where $\varkappa=3-4\mu$ for plane deformation, $\varkappa=(3-\mu)/(1+\mu)$ for plane stress state, μ is Poisson's ratio.

If shear stress S are applied at infinity, this problem is reduced to a problem with conditions on the defect boundary. In our case

$$F_n(\rho_n^*,\phi_n)=-2iS\,e^{-2i\alpha_n}+4iG\phi_n\tag{7}$$

where ϕ_n is the rotation angle of the rigid inclusion; $\phi_n=0$ for the cracks. These deflection angles are dertermined from the condition whereby the principal moment of all

forces acting on each inclusion is equal to zero
(Muskhelishvily (1968))

$$M_n(\Phi_n) = \frac{\rho_n^*}{\rho_n - \rho_n^*} \ Re \ \{a_n^2 \int_{-1}^{1} \chi [G_n(\rho_n, \chi)/\rho_n^* + \overline{G_n(\rho_n, \chi)}] d\chi\}$$ (8)

$$M_n(\Phi_n) = 0$$

For the self-balanced forces on the edges

$$G_n(\rho_n^*, t_n) = 0$$

and system (3) is simplified.

Equations (3) can be solved by taking into account the following conditions

$$\int_{-a_k}^{a_k} G_n(\rho_n, t_n) dt = 0$$ (9)

For the macrocrack-microdefects problem we have to single out an equation for the main crack ($n=0$) in system (3). The others may be either for small rigid inclusions or microcracks. Attention has been paid to the first case. The second one has been investigated by Tamuzh and Petrova (1993) and can be used for the check of solution.

3. Consideration of the crack closure

Let us assume that only one open region with the unknown length $2c$ and center coordinate z_0^C will arise on the main crack (Figure 1). Of course, this zone can coincide with whole macrocrack. To solve the problem by taking into account the crack closure it is necessary to formulate new boundary conditions. For the open zone of the main crack and for inclusions the previous boundary conditions (1),(2) are valid. On the closed zones of the macrocrack, the conditions are as follows

$$\tau_0^{\pm} (x,0) = S$$ (10)

$$[v_0] = 0$$

The initial system of equations (3) has to be divided into two sets of real and imaginary parts according to Goldshtein et al (1992)

$$\frac{2\rho_0^*}{\rho_0-\rho_0^*} \int_{-c}^{c} \frac{A_0(\rho_0,t)}{t-x} \, dt + \qquad\qquad |x|< c$$

$$\sum_{k=1}^{N} \frac{1}{\rho_k-\rho_k^*} \int_{-a}^{a} [B_k(\rho_k,t)Re(K_{0k}^C+L_{0k}^C)- A_k(\rho_k,t)Im(K_{0k}^C-L_{0k}^C)]dt=\pi C_0$$

$$\frac{2\rho_n^*}{\rho_n-\rho_n^*} \int_{-a}^{a} \frac{A_n(\rho_n,t)}{t-x} \, dt +$$

$$\frac{1}{\rho_0-\rho_0^*} [\int_{-a_0}^{a_0} B_0(\rho_0,t)Re(K_{no}+L_{no})dt- \int_{-c}^{c} A_0(\rho_0,t)Im(K_{no}^C-L_{no}^C)dt]+$$

$$\sum_{k=n}^{N} \frac{1}{\rho_k-\rho_k^*} \int_{-a}^{a} B_k(\rho_k,t)Re(K_{nk}+L_{nk})-A_k(\rho_k,t)Im(K_{nk}- L_{nk})]dt=\pi C_n$$

$$n=1,2,\ldots,N \qquad\qquad |x|< a \qquad (11)$$

$$\frac{2\rho_0^*}{\rho_0-\rho_0^*} \int_{-c}^{c} \frac{B_0(\rho_0,t)}{t-x} \, dt -$$

$$\qquad\qquad\qquad\qquad |x|< a_0$$

$$\sum_{k=1}^{N} \frac{1}{\rho_k-\rho_k^*} \int_{-a}^{a} [B_k(\rho_k,t)Im(K_{0k}+L_{0k})+ A_k(\rho_k,t)Re(K_{0k}-L_{0k})]dt=\pi D_0$$

$$\frac{2\rho_n^*}{\rho_n-\rho_n^*} \int_{-a}^{a} \frac{B_n(\rho_n,t)}{t-x} \, dt -$$

$$\frac{1}{\rho_0-\rho_0^*} [\int_{-a_0}^{a_0} B_0(\rho_0,t)Im(K_{no}+L_{no})dt+ \int_{-c}^{c} A_0(\rho_0,t)Re(K_{no}^C-L_{no}^C)dt]-$$

$$\sum_{k=n}^{N} \frac{1}{\rho_k-\rho_k^*} \int_{-a}^{a} B_k(\rho_k,t)Im(K_{nk}+L_{nk})+A_k(\rho_k,t)Re(K_{nk}- L_{nk})]dt=\pi D_n$$

$$n=1,2,\ldots,N \qquad\qquad |x|< a \qquad (12)$$

In systems (11),(12) the following notations have been used

$$G_n(\rho_n, t) = B_n(\rho_n, t) + i A_n(\rho_n, t) \tag{13}$$

$$F_n(\rho_n^*, \phi_n) = C_n(\rho_n^*, \phi_n) - i D_n(\rho_n^*, \phi_n) \tag{14}$$

and for macrocrack

$$G_0(\rho_0, t) = 2g_0'(t), \quad F_0(\rho_0^*, t) = -2p_0(t)$$

where

$$g_0'(t) = 2G\frac{\partial}{\partial t}([u_0] + i[v_0]) = \frac{1}{2}(B_0 + i A_0) \tag{15}$$

$[u_0]$, $[v_0]$ are the shear and transverse displacement jumps.

The expressions of kernels K_{ok}^C, L_{ok}^C and K_{no}^C, L_{no}^C can be obtained from (8),(9) replacing the parameters of macrocrack by parameters of the opened zone of macrocrack, i.e. X_0 and T_0 (formula (5) at $n=0$, $k=0$) should be replaced by

$$T_0^C = t + z_0^C, \qquad X_0^C = x + z_0^C.$$

Conditions (9) for (11),(12) are the following

$$\int_{-c}^{c} A_0(\rho_0, t) dt = 0, \quad \int_{-a_n}^{a_n} A_n(\rho_n, t) dt = 0 \quad (n=1,2,\ldots,N)$$

$$\int_{-a_n}^{a_n} B_n(\rho_n, t) \, dt = 0 \qquad\qquad (n=0,1,\ldots,N) \tag{16}$$

Using variables $t = a_k \tau$, $x = a_n \chi$ the systems (11),(12) can be expressed in a dimensionless form and the solution is sought as a sequence of small parameter $\lambda = a/a_0$

$$A_n = \sum_{p=0}^{\infty} A_{np} \lambda^p, \quad B_n = \sum_{p=0}^{\infty} B_{np} \lambda^p \tag{17}$$

The kernels K_{nk} and L_{nk} are expended in the series of λ too. This expressions are cited in Tamuzh et al (1993).

Taking into account the conditions (16) the systems of equations (11), (12) are regularized by Karleman-Vekua method (Muskhelishvili (1968)). Then, inserting series (17) in (11),(12) and equating the expressions at like powers of λ, the recurrent relations are obtained for the subsequent determination of coefficients in (17). We calculated them with an accuracy to λ^2

$$A_{00}(\rho_0, \chi) = 0,$$

$$A_{02}(\rho_0, \chi) = -\frac{\rho_0 - \rho_0^*}{4\rho_0^*} \frac{S}{\sqrt{1-\chi^2}} \sum_{k=1}^{N} \frac{1}{\rho_k^*} [\text{Re}(\frac{2G}{S} \phi_k - J_k) \text{Re}(m_{0k1}^C + n_{0k1}^C) -$$

$$\text{Im}(J_k) \ \text{Im}(m_{0k1}^C(\chi) - n_{0k1}^C(\chi))]$$

$$A_{n0}(\rho_n, \chi) = -\frac{\rho_n - \rho_n^*}{\rho_n^*} \frac{S\chi}{\sqrt{1-\chi^2}} \ \text{Im}(J_n) \qquad\qquad (18)$$

$$B_{00}(\rho_0, \chi) = \frac{\rho_0 - \rho_0^*}{\rho_0^*} \frac{S\chi}{\sqrt{1-\chi^2}}$$

$$B_{02}(\rho_0, \chi) = -\frac{\rho_0 - \rho_0^*}{4\rho_0^*} \frac{S}{\sqrt{1-\chi^2}} \sum_{k=1}^{N} \frac{1}{\rho_k^*} [\text{Re}(\frac{2G}{S} \phi_k - J_k) \text{Im}(m_{0k1} + n_{0k1}) +$$

$$\text{Im}(J_k) \text{Re}(m_{0k1} - n_{0k1}]$$

$$B_{n0}(\rho_n, \chi) = -\frac{\rho_n - \rho_n^*}{\rho_n^*} \frac{S\chi}{\sqrt{1-\chi^2}} \ \text{Re}(\frac{2G}{S} \phi_n - J_n) \qquad\qquad (19)$$

where

$$J_k = -\frac{1}{2}\left[1 + \rho_k^* + \frac{(2e^{-2i\alpha_k}-1)\bar{u}_k}{(\bar{u}_k^2 - 1)^{1/2}} - \frac{\rho_k^* u_k}{(u_k^2 - 1)^{1/2}} - \frac{(\bar{u}_k - u_k)e^{-2i\alpha_k}}{(\bar{u}_k^2 - 1)^{3/2}}\right]$$

$$(20)$$

$$\frac{2G}{S} \phi_k = \text{Re}(J_k)$$

$$m_{0k1}^C(\chi) = \frac{1}{\varepsilon^2} \frac{e^{i\alpha_k}}{i}\left[\frac{\rho_0^* e^{i\alpha_k(\delta_k\chi-1)}}{(\chi-\delta_k)^2(\delta_k^2-1)^{1/2}} + \frac{\rho_k^* e^{-i\alpha_k(\bar{\delta}_k\chi-1)}}{(\chi-\bar{\delta}_k)^2(\bar{\delta}_k^2-1)^{1/2}}\right]$$

$$n^C_{0k1}(\chi) = \frac{e^{-i\alpha_k}}{i\varepsilon^2}\left[\langle u_k - \bar{u}_k\rangle e^{-i\alpha_k}\frac{-\chi^2 + 2\bar{\delta}_k^3\chi - 3\bar{\delta}_k^2 + 2}{\varepsilon(\chi - \bar{\delta}_k)^3(\bar{\delta}_k^2 - 1)^{3/2}} + \right.$$

$$\left. \frac{(e^{-i\alpha_k} - e^{i\alpha_k})(\bar{\delta}_k\chi - 1)}{(\chi - \bar{\delta}_k)^2(\bar{\delta}_k^2 - 1)^{1/2}}\right] \tag{21}$$

$$u_k = z_k^0/a_0, \quad \varepsilon = c/a_0, \quad \delta_k = \frac{u_k - d_0}{\varepsilon}, \quad d_0 = z_0^c/a_0$$

Expressions of m_{0k1}, n_{0k1} looks similar, only $\varepsilon = 1, d_0 = 0$.

4. Stress intensity factors

The stress intensity factors can be obtained from $G_n(\rho_n, t)$ as follows:

$$K_I^{\pm} - iK_{II}^{\pm} = \mp \lim_{\chi \to \mp 1}\sqrt{a_0}\sqrt{1-\chi^2}(B_n(\rho_n, \chi) + iA_n(\rho_n, \chi)) \tag{22}$$

The subscript "+" refer to the right side of the crack tip and "-" — to the left.

The macrocrack stress intensity factor K_{II} is defined by shear displacement jumps u'_0 (15). It was founded that the main crack closed zones don't change the value K_{II}. The macrocrack tip stress intensity factor is defined by gradient displacement jumps v'_{02} (15). So, the K_I contains the unknown macrocrack opening zone $2C$ and it's midpoint coordinate.

Putting the expressions (18)-(21) into equation (22) and separating the real part, there results

$$K_{I0}^{\pm} = \mp S\sqrt{a_0}\,\frac{\lambda^2}{4}\sum_{k=1}^{N}\frac{1}{\rho}*\left\{ Re(\frac{2G}{S}\Phi_k - J_k)\,Re\left[m^C_{0k1}(\pm 1) + n^C_{0k1}(\pm 1)\right]\right.$$

$$\left. -Im(J_k)\,Im\left[m^C_{0k1}(\pm 1) - n^C_{0k1}(\pm 1)\right]\right. \tag{23}$$

$$K_{II\bar{0}}^{\pm} = S\sqrt{a_o}\left\{1\pm\frac{\lambda^2}{4}\sum_{k=1}^{N}\frac{1}{\rho^*}\left\{Re(-\frac{2G}{S}\,\phi_k-J_k)\,Im\left[m_{0k1}(\pm1)+n_{0k1}(\pm1)\right]\right.\right.$$

$$\left.\left. +Im(J_k)\,Re\left[m_{0k1}(\pm1)-n_{0k1}(\pm1)\right]\right.\right. \tag{24}$$

J_{k0}, ϕ_k, m_{0k1}^C, n_{0k1}^C are the formulas (20).

The unknown boundaries of the opening zone are found from the singularity absence in the solution near these boundaries (Goldshtein et al (1992))

$$K_I\,(\pm c) = 0 \tag{25}$$

Practically the calculation of opened zone is provided by iterative process. At first step the region without macro-crack edges overlapping is taken. Then the size C have to be changed until equality (25) is fulfilled with requirement accuracy.

5. Numerical results and conclusion

The domains where inclusions of different orientation cause a full or partial closure of the main crack is defined by K_I (22) and presented in *Figure 2*.

To estimate the effect of crack closure on the stress intensity factor, a numerical analysis is carried out for a single inclusion with coordinates $x=0.$, $y=0.5\,a_o$ and slope angle $\alpha_k=-45^0$, $\lambda=0.1$. The value of K_I^-, disregarding the closed zone on the main crack is $K_I^-/S(a_o)^{1/2} = 0.459\cdot10^{-3}$, but taking into account the closure , we obtain $K_I^-/S(a_o)^{1/2}=0.182\cdot10^{-2}$. On the right side the closure zone expands up to the crack tip and $K_I^+=0$. The half of opened zone is equal to $c= 0.536\,a_o$.

It reveals that absolute value of K_I is rather small, because in the undamaged body $K_I=0$, and the increment of K_I is caused by influence of one inclusion with size $0.1a_o$. But the relative change of K_I due to the phenomenon of crack closure is rather significant.

286

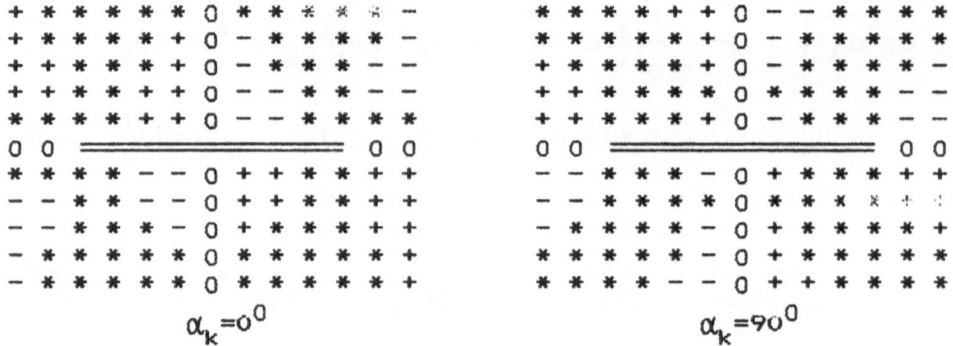

$$\alpha_k = 0^0 \qquad\qquad \alpha_k = 90^0$$

Figure 2. Domains of inclusions cause the full (denoted by "-") or partial ("*") closure of the main crack or opened it (denoted by "+"). 0 - zero macrocrack opening.

Conclusions

It has been revealed that inclusions can cause full or partial closure of the main crack.

The value of K_{II} at the main crack tip can be calculated disregarding the contact zone.

For correct determination of K_I the presence of contact zone should be taken into account.

6. References

Petrova V.(1988) Interaction between a main crack and inclusions of a given orientation, *Mechanics of Composite Materials* ,3 , 288-294.

Romalis N.and Tamuzh V. (1984)Propagation of a main crack in a body with distributed microcracks, *Mechanics of Composite Materials* , V.20,N 2, 35-43 .

Tamuzh V. and Petrova,V.(1993) Crack in damaged body at shear stress state , in V.V.Panasyuk e.l.(eds.), *Fracture mechanics: Successes and problems, Collection of Abstracts, Part* 1, Lviv, pp.106-107

Tamuzh V. and Petrova,V.(1993) Macrocrack in the field of microcracks under shear loading, *Phiziko-Himicheskay Mehanika Materialov*, 3, 147-157. (in Russian)

Tamuzh V., Petrova,V. and Romalis N. (in press) Influence of microcracks on thermal fracture of macrocrack. Account of the crack edge closure.

Dundurs J., Comninou M. (1979) The interface crack in retrospect and prospect, in G.C.Sih, V.P.Tamuzh (eds.), *Fracture of Composite Materials, Proc.of 1-st.USA-USSR Symp.*, pp.93-107.

Berezhnitskii L.T., Panasyuk V.V. and Stashchuk N.G. (1983) *Interaction between Rigid Linear Inclusions and Cracks in a Deformable Body*, Kiev (in Russian)

Muskhelishvili,N.I. (1968) *Singular Integral Equations*, Nauka, Moscow (in Russian).

Goldshtein R., Zhitnicov Y. and Morozova T. (1991) Equilibrium of the cuts system with forming the closing and opening regions on them, *Applied Mathematics and Mechanics* , V.55, N 4, 672-678 (in Russian).

RECENT WORK ON MESOSTRUCTURES AND THE MECHANICS OF FIBER COMPOSITES

M.R. PIGGOTT

Department of Chemical Engineering and Applied Chemistry
University of Toronto, Toronto, Ontario M5S 1A4 Canada

Abstract

Mesostructures are middle size structures, and, in fiber composites, influence many important properties. This paper reviews papers presented at a recent meeting devoted to mesostructures and the mechanics thereof. Rapid developments are being made in the identification and quantification of fibre misalignments, waviness, and packing anomalies. These are shown to affect properties such as stiffness, strength and toughness. In addition, the mesostructure affects warping and cracking in composites. Developments in mechanics go some way towards explaining mesostructure effects in notched strength and other properties. However, much remains to be done. There is a very definite series of links between manufacture, mesostructure and final properties, so that a better knowledge of mesostructures will surely lead to higher quality composites.

1. Introduction

Our models for fibre composites normally consist of straight rod-like fibres, uniformly packed in hexagonal or square arrays. While this is useful for estimating properties which depend principally on the fibre properties, it can be very misleading for matrix and interface dependent properties.

For example, tensile strength and Young's modulus for unidirectional laminates are not very dependent on the detailed structure of the composite. Meanwhile, compressive strength is directly affected by fibre waviness [1] and is also probably influenced by fibre bundling [2]. These are examples of what has been termed *orientation mesostructures* and *packing mesostructures* [3].

Mesostructures are middle size structures. The are larger than the microstructures, which generally refers to the micron level, and involves such fine structures as those observed at the fiber-matrix interface [4,5]. The upper size limit for mesostructures is somewhat less than the macrostructure, e.g. the details of the laminate/honeycomb layup. Thus mesostructures range roughly from 0.03 to 3 mm in size.

R. Pyrz (ed.), IUTAM Symposium on Microstructure-Property Interactions in Composite Materials, 289–300.
© 1995 *Kluwer Academic Publishers.*

Two main classes of mesostructure have been identified [3] and each has two subclasses.

1. Secondary Disorder
1a. Orientation: fibre waviness is an example of this
1b. Packing, e.g. fibre rich areas, resin rich areas and voids

2. Secondary Order
2a. Orientation, e.g. fibre preferred orientations in injection moldings.
2b. Packing, e.g. fibre bundling and fibre end synchronization.

The effect of fibre waviness on compressive strength has already been mentioned. It probably also plays a major role in shear strength [6] and fatigue endurance [7]. Fibre bundling affects toughness [8] as well as compressive strength.

Mesostructures originate in the production process, and are determined by the precise nature of the material constituents used, as well as other factors of production. So far we know little about this, except that the effects can be very subtle: for example in the case of the mold material, graphite molds give straighter fibres than nickel ones, at least with reinforced thermoplastics [9].

There is clearly a pathway involved in this i.e., production processing→mesostructure → mechanics of mesostructures → properties. So mesostructures could provide the key for developing production processes to obtain desired properties. Again there is huge potential for development. Furthermore, quantification of mesostructures could also be used for quality control of the final product.

With this in mind, a two day meeting was called to discuss mesostructures and mechanics in fibre composites. All the participants were invited, and this paper highlights the issues discussed.

2. Identification and Quantification

Most work here appears to have been devoted to the measurement of fibre orientation. Hine et al [10] used sectioning, polishing, and a unique image analysis facility, to obtain three dimensional orientation measurements on long glass fibre reinforced nylon 6,6 injection moldings. Each fibre intersection with the polished plane appears as an ellipse and its ellipticity and orientation were used to evaluate the fibre orientation relative to three orthogonal axes. The molded sheet contained the 1 and 3 axis, and the 2 axis was normal to it. The 3 axis was the molding direction. A typical result, shown in fig. 1, is given in terms of the squares of the cosines of the angles with respect to the three axes, θ_1, θ_2 and θ_3. These data can be used directly to calculate upper and lower bounds for the elastic constants E_{ij}, G_{ij} and ν_{ij}. Comparison with measured values (using ultrasonics) showed that most of the elastic constants fell within the bounds when allowance was made for experimental scatter, and further, that the upper and lower bounds differed by no more than about 30% of the upper bounds.

An alternative method of determining the three dimensional orientation of glass fibres, described by McGrath and Wille [11], could be performed non destructively. It was also

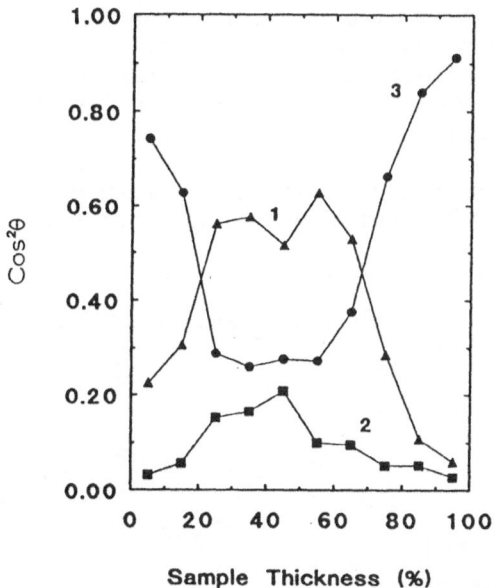

Figure 1.   Fibre orientations close to the injection point in a glass-nylon molding, relative to: 1) the direction in the plane of the molding and perpendicular to the injection direction; 2) the normal to the plane of the molding; and 3) the molding direction [10].

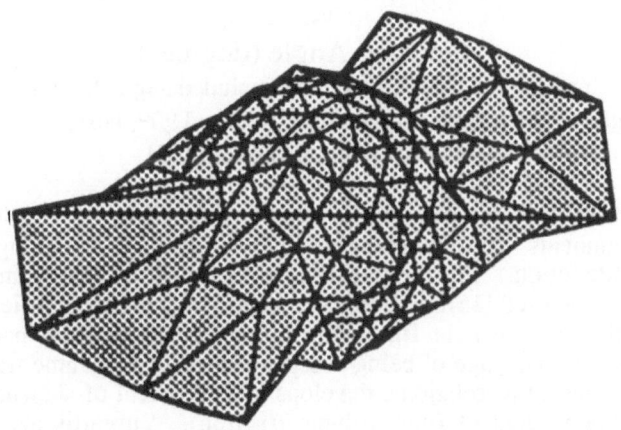

Figure 2.   Fibre orientation distribution diagram visualized as a 3-D histogram: fibres more or less aligned in one direction [11].

292

rapid and economical. However, it needed opaque tracer fibres, and the composite needed to be transparent. (This was effected by matching the fibre and polymer refractive indices.) Carbon fibres were used as tracers, and constituted 0.1 wt % of the composite. 3D histograms, such as that shown in fig. 2, were produced, and these could be rotated or zoomed on the computer monitor.

Frankle [12] described the use of eddy currents to evaluate fibre waviness in carbon-carbon tubes. This method is also non-destructive and provides useful data on resistance to compressive collapse.

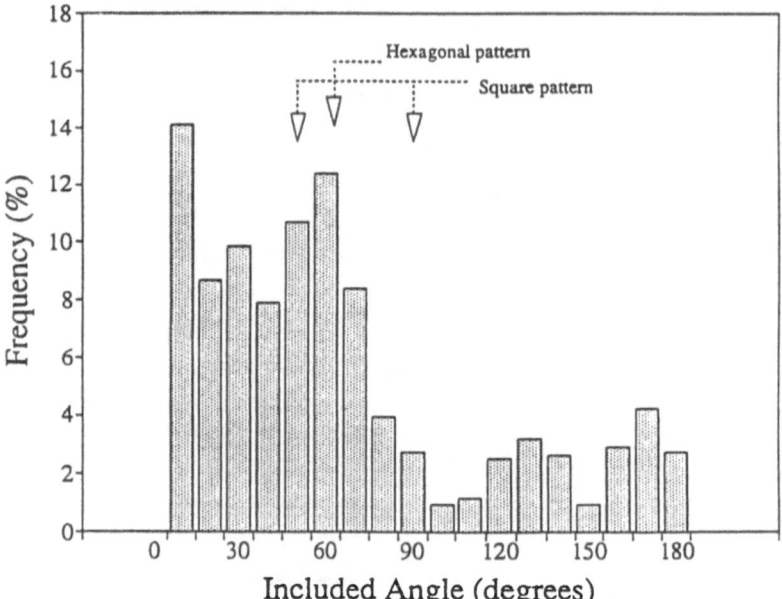

Figure 3.   Fibre packing as given by a typical included angle distribution for nearest neighbours. (Square packing gives 45° and 90°; hexagonal packing gives 60°, as shown.) [13].

Yurgartis [13] quantified packing mesostructures as well as misalignments. (His method for misalignments - described previously [14] - involves sectioning at 5° or 10° to the main fibre direction.) For the packing geometry the Hough Transform image analysis technique was used [15]. Spatial distribution was then quantified in terms of the included angle, as shown in fig. 3 for square and hexagonal packing. This representation has the advantage of being independent of fibre volume fraction. (Pyrz [16], using the method of tesselations, develops a "coefficient of skewness" which is also apparently independent of fibre volume fraction). Yurgartis also described a skeletonization method for the image processing of microcracks. The skeletonized images were then used to determine crack orientation and spacing distributions.

Finally, Ifju et al [17] described the use of a Moire Interferometry method to identify structural details in 3-D woven and 2-D braided materials. In this method the specimen has a fine cross grating attached to it, "carrier fringes" are produced [18] and

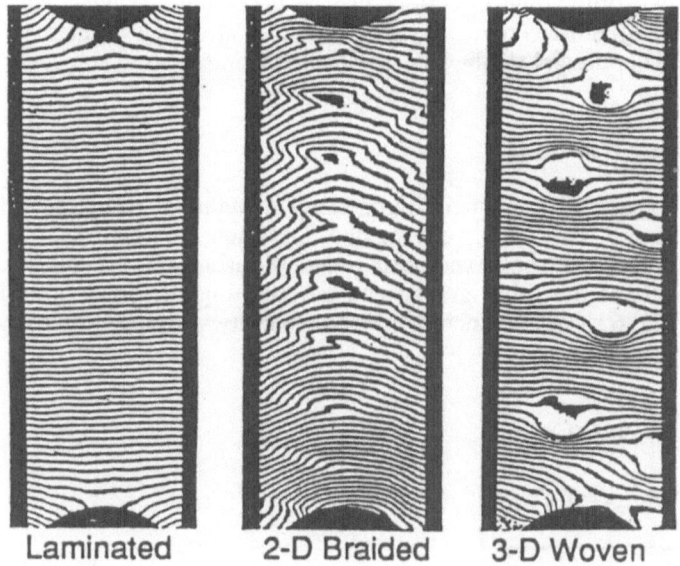

Figure 4.   Displacement field normal to the shear plane for a cross ply laminate, a 2-
D triaxial braid and a 3-D weave. In the last two cases the coarse
structure reveals itself in the strain field (The specimen size was about 21
x 6.5 mm^2) [17].

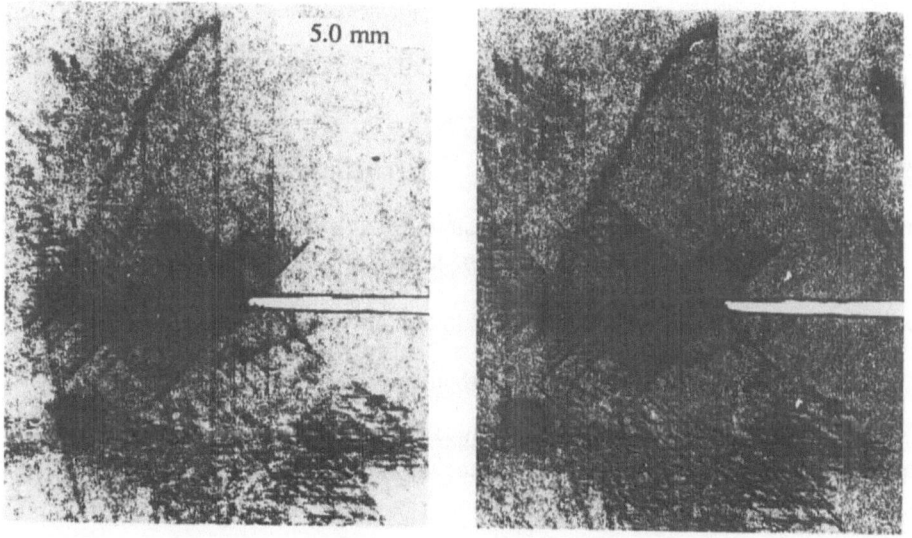

Figure 5.   Microcracking damage in a compact tension carbon-epoxy laminate {0,
-45, 90, +45}$_{2s}$ [19].

294

displacements in two orthogonal directions can be measured. When a specimen is stressed, the internal structure is revealed, as shown in fig. 4, which compares a straight fibre laminate, a 2-D braid and a 3-D weave. These specimens were subject to a shear stress.

3.   Specific Mesostructures

These include microcrack arrays, which can be detected and made visible by using high atomic number "stains" and then x-raying the specimen [19], see fig. 5, and by ultrasonic backscattering. The polar backscattering of ultrasound is particularly effective as an NDE method for detecting cracks and determining their positions in three dimensions [20]. Fig. 6 shows the increase of crack density with increasing strain for a carbon fibre-polymide laminate.

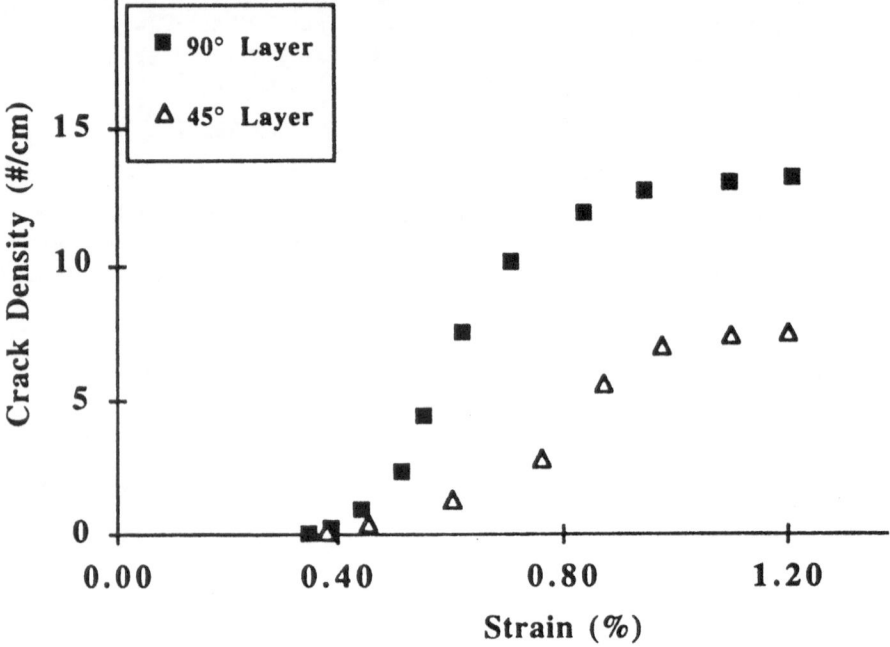

Figure 6.   Increase of crack density with strain for a carbon-polyimide laminate {45, 90, -45, 0}$_{2s}$ tested at -53°C [20].

The x-ray staining technique can also be used to tag individual fibres which may then be used as flow tracers in the molding process [19]. In addition, thermal expansion measurements can be used to detect fibre re-arrangements during molding [21].

Variations in crystallinity in reinforced thermoplastics also constitute a type of mesostructure. These were shown to influence the fracture behaviour of the composite [22].

4.   Property Relationships

Compressive strength: Adams [23] showed that having wavy layers in cross-ply
laminates of carbon-epoxy reduced the compressive strength in a linear fashion, up to
about 35% loss. Thereafter, there appeared to be no further effect, fig. 7. The waviness
was discrete, as shown in fig. 8. Different degrees of waviness were introduced by
causing waviness in more, or fewer layers, while amplitude and wavelengths were held
more or less constant.

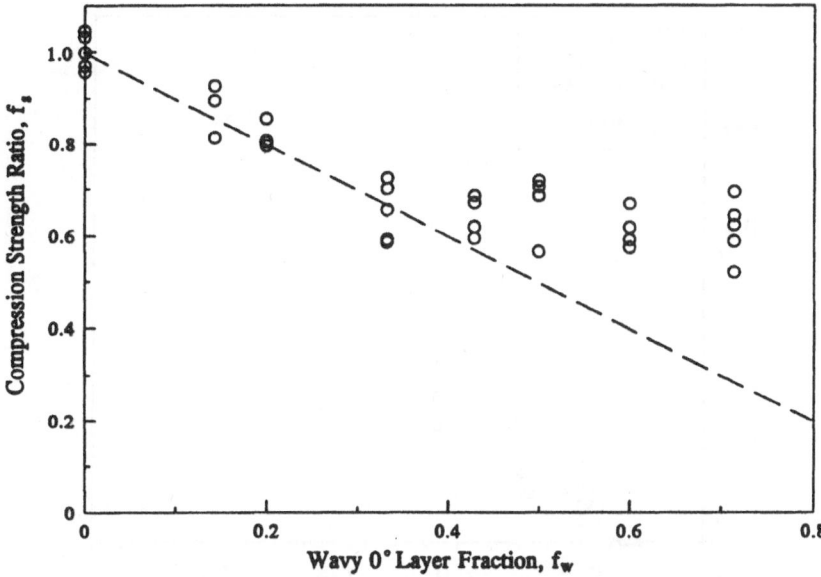

Figure 7.   Compressive strength vs fraction of wavy layers for carbon-epoxy [23].

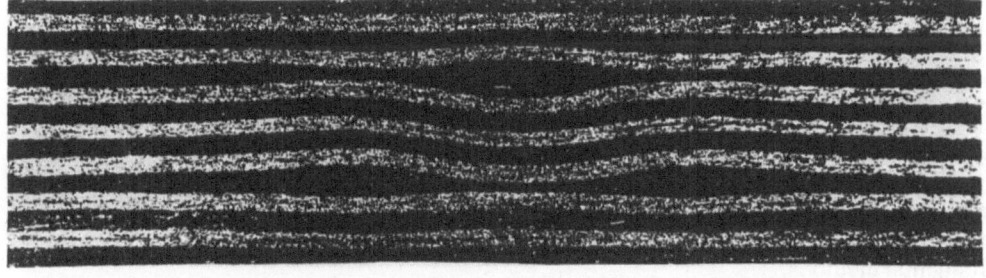

Figure 8   Cross ply carbon epoxy laminate having several wavy layers [23].

Piggott [24] reviewed some previous work on the effect of waviness on compressive
strength and modulus, shear strength, and fatigue endurance, then went on to show that
significant increases in delamination resistance were possible, also, when the fibres are
wavy. Other work on delamination, suggesting that there is, indeed, an effect, is
reported in the next section (see section 5)

In a written submission, Varna et al [25] showed that voids influenced the generation of cracks in glass fabric-vinyl ester resin transfer molded laminates. When the composites were stressed in the transverse direction, two types of crack were produced: large ones which traversed the specimen almost completely, and small ones which arrested at transverse plies. Only the small cracks were influenced by the void content, V_V. Fig. 9 shows the density of small cracks vs strain at $V_V = 0.01$, 0.03 and 0.04. These cracks appeared to toughen the material - the stress-strain curves, for the higher void content materials, fig. 10, showed considerable pseudo ductility.

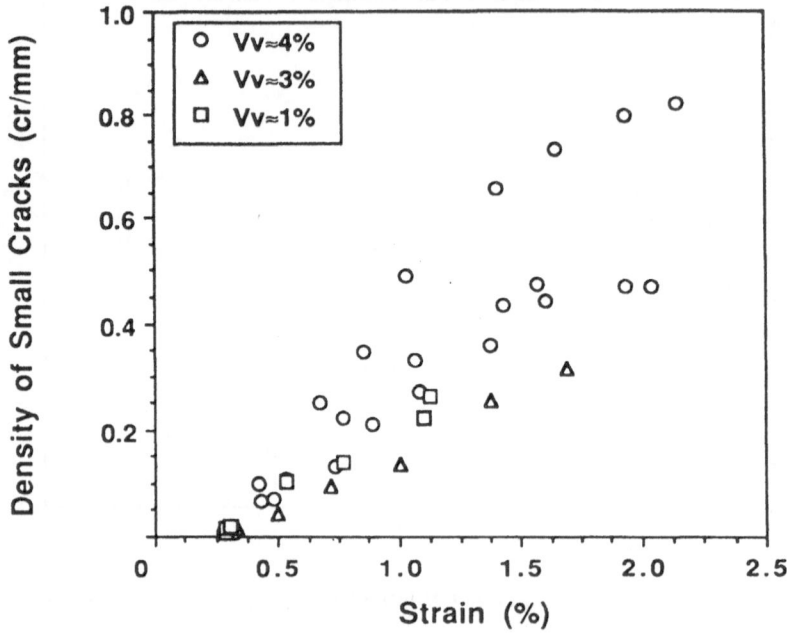

Figure 9.   Density of small cracks vs strain at void contents of 1, 3 and 4% [25].

5.   Manufacture-Mesostructure-Property Relationships

A wide range of manufacturing variables are likely to affect the mesostructure, and through it, the properties. This is inevitable because there are a very large number of types of material, ranging from very short random fibre glass-nylons through sheet molding compounds to highly organized forms such as laminates. Each has its own production methods.

In the random fibre case, Michaeli and Heber were able to link manufacturing processes and warpage in lamp covers made with sheet molding compounds [26]. Sectioning samples gave orientation information. The orientations were used to evaluate the elastic constants. The shrinkage could also be estimated and hence, using the elastic constants, the warpage was predicted. Experimental tests showed good agreement with the predictions.

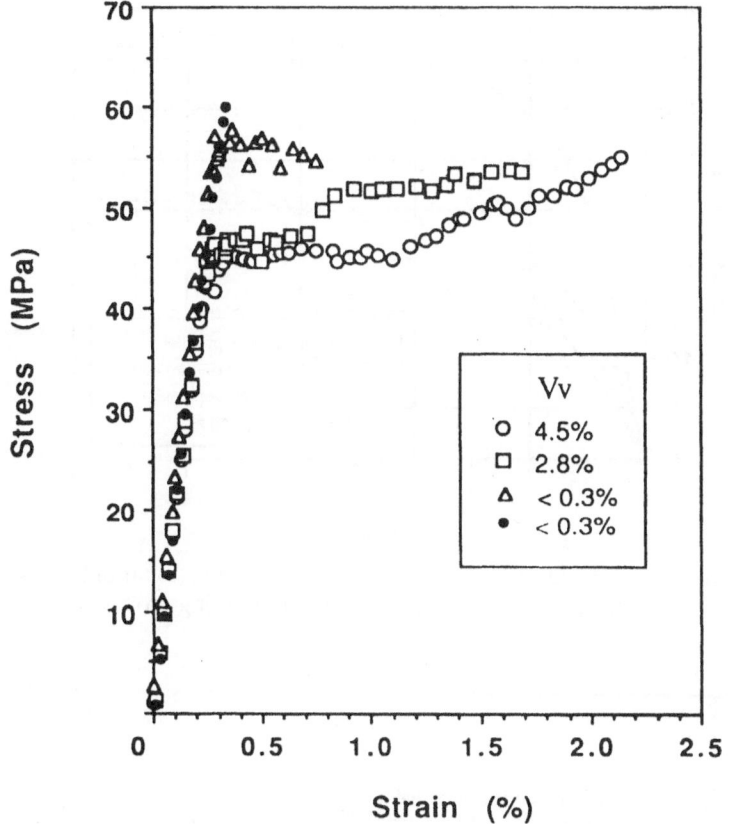

Figure 10.   Stress-strain curves for glass fabric-vinyl ester transfer molded laminates
with different void contents. Stress transverse to main fibre direction
[25].

Ludwig et al [27] were able to orient short glass fibres in polypropylene in injection molding by using the push-pull process. (In this at least two gates, G_1 and G_2 are used. The mold is first filled through G_1 and some solidification occurs. More material is then injected via G_2, so that the molten core is ejected through G_1. This process continues by alternate injection-ejection cycles.)   This made the material highly anisotropic; see fig. 11, and increased the strength parallel to the flow direction by about 65%.

In more highly organized materials, Wang and Wang [28] were able to show that there should be a size effect in 3-dimensionally woven materials. This is because the structure at the corners and edges is different from the centre, and the volume of this material decreases proportionately with increasing size. The effect is influenced by the braiding angle, and is particularly noticeable at 20-30°.

In the case of woven Kevlar-epoxy laminates modifications to the fabric before molding affected the work of delamination. Williams, with Briscoe [29] had earlier shown that

298

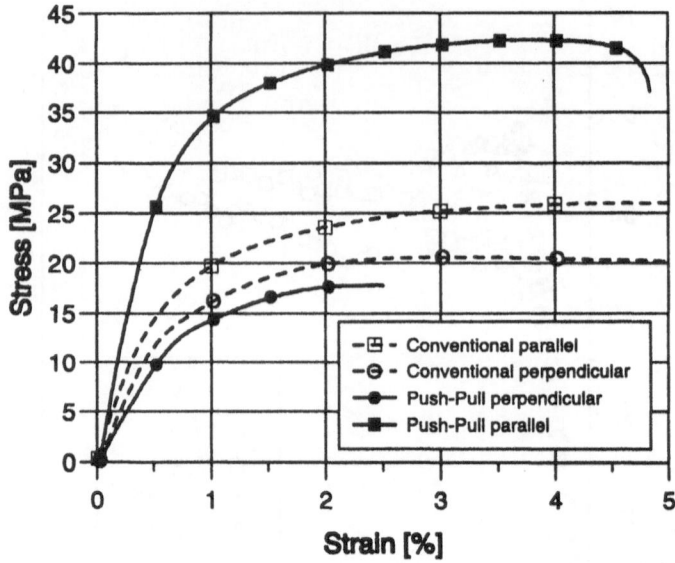

Figure 11.   Stress-strain curves for glass-polypropylene injection moldings showing anisotropy induced by push-pull molding process [27].

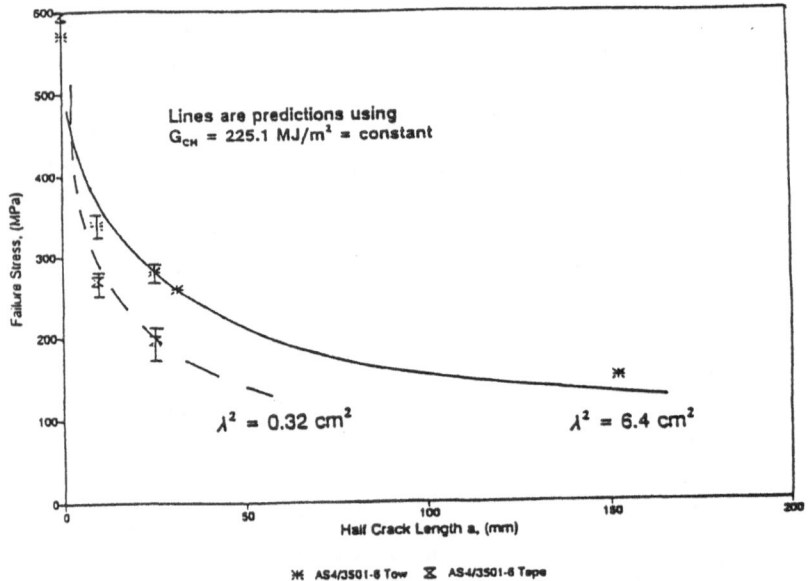

Figure 12.   Predicted vs experimental strengths for notched carbon-epoxy laminates [31].

lubrication of the fibres facilitated their movement in the molding process, so that they became more sinuous. Thus, in a delamination test, they tended to cross the fracture plane more frequently than when dried fibres were used. This gave rise to increased toughness. In the later paper [30] more fibre bridging was produced by abrading the fibres. This more than doubled the fracture resistance.

Cairns et al [31] developed an inhomogeneity parameter and used this to explain the difference in notched tensile strength for samples made by hand lay up, and by an automated tow placement process. Hand lay up gave the most homogeneous structure, and this had a lower strength. The inhomogeneities were defined in terms of a Cosserat length, λ, and fig. 12 compares the predictions using this approach and the experimental results.

6. Conclusions

Mesostructures have been identified as being important features of fibre composites. They affect many properties, and can be influenced by the manufacturing process. Their study is therefore likely to lead to a better control over the quality of fibre composites.

7. References

1. A. Mrse and M.R. Piggott, (1993), Comp. Sci. Tech. 46, 219-27.

2. M.R. Piggott, (1981), J. Mater Sci. 16, 2837-45.

3. M.R. Piggott, (1992), Proc. 37 Int. SAMPE Symp., 738-46.

4. L.T. Drzal, (1983), SAMPE J., 19, (#5), 7-13.

5. V.D. Scott, R.L. Trumper and M. Yang, (1991), Comp. Sci. Tech., 42, 251-74.

6. M.R. Piggott and K. Liu, (1993), Proc. CANCOM '93, 23-32.

7. M.R. Piggott, P.W.K. Lam, (1991), ASTM STP 1110, 686-95.

8. M. Fila, C. Bredin and M.R. Piggott, (1972), J. Mater Sci. 7, 983-8.

9. A. L. Highsmith, J.J. Davies, and K.L.E. Helms, (1992), ASTM STP 1120, 20-36.

10. P.J. Hine, N. Davidson, R.A. Duckett and I.M. Ward, (1994), Proc. 1st Conf. on Mesostructures and Mesomechanics in Fibre Composites, (Ed. M.R. Piggott, University of Toronto, 1994), 1-14.

11. J.J. McGrath and J.M. Wille, ibid, 15-31.

12. R.S. Frankle, (1994), ibid, 56-70.

300

13.   S.W. Yurgartis, *ibid,* 32-55.

14.   S.W. Yurgartis, (1987), Comp. Sci Tech., 30, 279-94.

15.   S.W. Yurgartis and M.N. Purandare, (1991), J. Computer Aided Microscopy, 3, 117-125.

16.   R. Pyrz, (1993), Polymers and Polymer Comp. 1, 283-96.

17.   P.G. Ifju, J.C. Masters and W.C. Jackson, (1994), Proc. 1st Conf. on Mesostructures and Mesomechanics in Fibre Composites, (Ed. M.R. Piggott, University of Toronto, 1994), 71-86.

18.   D. Post, B. Han and P.G. Ifju, *High Sensitivity Moire, Experimental Analysis for Mechanics and Materials* (Springer-Verlag, New York, 1994).

19.   M.T. Kortschot and C.J. Zhang, (1994), Proc. 1st Conf. on Mesostructures and Mesomechanics in Fibre Composites, (Ed. M.R. Piggott, University of Toronto, 1994), 101-12.

20.   K.V. Steiner, R.F. Eduljee, X. Huang and J.W. Gillespie, Jr., *ibid,* 131-44.

21.   T. Vu Khanh and B. Liu, *ibid,* 113-30.

22.   L. Ye, A. Behag and K. Kriedrich, *ibid,* 87-100.

23.   D.O. Adams and S.J. Bell, *ibid,* 159-73.

24.   M.R. Piggott, *ibid,* 145-158.

25.   J. Varna, R.Joffe and L.A. Berglund, *ibid,* 234-51.

26.   W. Michaeli and M. Heber, *ibid,* 210-22.

27.   H-C Ludwig, G. Fischer and H. Becker, *ibid,* 223-32.

28.   Y.Q. Wang and A.S.D. Wang, *ibid,* 174-96.

29.   B.J. Briscoe and D.R. Williams, (1993), Comp. Sci. Tech., 46, 277-86.

30.   M. Kim and D.R. Williams, (1994), Proc. 1st Conf. on Mesostructures and Mesomechanics in Fibre Composites, (Ed. M.R. Piggott, University of Toronto, 1994), 233.

31   D.S. Cairns, L.B. Ilcewicz, T. Walker and P.J. Minquet, *ibid,* 197-209.

STIFFNESS REDUCTION IN COMPOSITE LAMINATES DUE TO TRANSVERSE MATRIX CRACKS

G.N.PRAVEEN and J.N.REDDY
Department of Mechanical Engineering
Texas A&M University
College Station, TX 77843-3123

ABSTRACT: Cross–ply laminates with transverse matrix cracks are studied using the variational approach. An upper bound on the reduced stiffness has been derived, complementary to Hashin's lower bound, for a given density of matrix cracks. Staggerring in crack patterns is also considered. The associated elasticity problem is solved using the *Layerwise Laminate Theory of Reddy*. The effect of crack density on the crack opening displacements and stress transfer is examined.

1. Introduction

Fiber composites are gaining increasing primacy as load bearing components in both small and large scale structures. In fiber matrix composites, matrix cracking is a typical damage mode. Matrix cracks by their very nature, introduce multiple stress concentration points (crack tips) which are potential sites for delamination. The different approaches to modeling transverse cracks are the shear lag model in its one-dimensional and two–dimensional forms, the variational model, the self–consistent schemes, and more recently, the internal state variable model.

Bailey and his coworkers (1977) proposed the shear lag model. In their experiments on cross–ply laminates, cracks in the transverse ply showed a remarkably even spacing. Highsmith and Reifsnider (1982) reported the discovery of the "characteristic damage state", CDS, which refers to the state of a saturation in crack spacing. Fukunaga *et al.* (1984) studied the failure of $[0/90]_s$ and $[90/0]_s$ laminates based on a statistical strength analysis, using an in–plane shear lag model. Flaggs(1985) used a mixed mode strain energy release rate criterion along with an approximate two dimensional shear lag model. Laws and Dvorak (1987) in their analysis accounted for the existence of residual stresses.

The variational model was proposed by Hashin (1985). An approximate state of stress in the vicinity of a transverse crack was obtained by minimizing stress perturbations using the principle of minimum complementary energy. Predictions of this approach agreed well with experimental data of Highsmith and Reifsnider. Nairn (1989), in his analysis included the effect of residual thermal stresses. Varna and Berglund (1991), extended the thermoelastic analysis of $[0_m/90_n]_s$ laminates by taking into account a non–linear stress distribution in the thickness direction of

R. Pyrz (ed.), IUTAM Symposium on Microstructure-Property Interactions in Composite Materials, 301–312.
© 1995 *Kluwer Academic Publishers.*

the $0°$ plies. Herakovich *et al.*, (1988) developed an approximate analytical model to study stiffness reduction of cracked cross–ply laminates. The predictions of this model agreed reasonably well with experiments.

Gudmondson and Östlund (1992) obtained stiffness reduction in cross–ply laminates based on the concept of change in elastic energy due to the appearance of cracks in the plies. Talreja (1986) developed a damage vector concept for matrix cracking and delamination. Allen *et al.*, (1987) proposed a thermomechanical constitutive theory for elastic composites, treating cracks as internal variables.

In Hashin's variational approach, the axial stress perturbations in cracked laminates under tension, are assumed to be independent of the thickness coordinate. The predictions of the theory agree well with experiments; nevertheless, the stresses are not accurate owing to the assumption of the invariance of axial stress perturbations through the thickness. Moreover, the stiffness reduction associated with staggered cracking cannot be very easily handled using Hashin's approach. In what follows an approach similar to that of Hashin will be presented. The approach in this paper is a displacement based approach and governing equations are derived using a concept based on plate theory.

2. Stiffness Reduction Model

In the following strain energy based model for laminates made from linear Hookean elastic material, with cracks, we invoke the principle of superposition. The cracked and uncracked bodies are subjected to the same external boundary conditions. The field variables of the cracked body are obtained by a simple superposition of the field variables in an uncracked body and corresponding perturbations.

Following a method complementary to that of Hashin, by considering the expression for the strain energy of the cracked body in terms of the strain energy of the uncracked body and the perturbation strain energy, one arrives at the following:

$$E^* \leq E^0 - \frac{2U'}{\varepsilon_0^2 V} \tag{1}$$

The above equation yields an upper bound on the effective axial modulus of the cracked laminate. Hashin's complementary energy formulation yields a similar result, for the lower bound, as follows:

$$\frac{1}{E^*} \leq \frac{1}{E^0} + \frac{2U'_c}{\sigma_0^2 V} \tag{2}$$

where, U'_c is the perturbation complementary strain energy and σ_o is the uniform applied stress and ε_o is the uniform applied strain. We now consider a laminate with a number of parallel transverse matrix cracks in the off–axis plies. The results due to Highsmith and Reifsnider (1982) show that the cracks are evenly spaced. In the light of the above fact, we will assume an even distribution of cracks in the transverse plies. With this assumption, the crack density can be defined as the number of cracks per unit length, in a direction that is normal to the surface of the crack. Due to the equispacing of transverse cracks, a repeating pattern is evident. It suffices to solve the perturbation boundary value problem in any one of these repeating patterns. The behavior of the stresses and displacements in the repeating unit is typical of that in the entire laminate. This repeating unit is therefore called a representative volume element, hereinafter referred to as the RVE.

The upper bound on the reduced axial stiffness may be written in terms of the crack spacing, '$2a$', as,

$$2a\varepsilon_0^2 h \Delta E = \int_{S_c} T_i' u_i' dS \qquad (3)$$

where, 'ΔE', is the change in effective axial stiffness. This equation will henceforth be referred to as the *stiffness reduction equation.*

The laminate is assumed to be wide enough in the direction parallel to the cracks, and that a state of plane strain exists corresponding to that direction. To accurately model the load transfer mechanism in the presence of matrix cracks, it is necessary to model the kinematics of deformation in the presence of cracks. The kinematics assumed here are a layerwise representation of the displacement functions. In the residual problem, where the RVE is analysed, the length to thickness ratio of the RVE varies considerably from a high value of 12 to values less than unity, depending on the crack density. It has been reported by Highsmith and Reifsnider (1982) that the saturation crack density in $[0/90_3]_s$ glass laminate is about 0.75 cracks/mm, for a laminate thickness of 1.612 mm. In this case, the length to thickness ratio is around 0.83. Therefore, the RVE tends to behave more often, like a thick plate under the inplane perturbation boundary conditions. The layerwise laminate theory of Reddy is best suited to model displacements in these situations. The theory allows for a descritization of the RVE through its thickness, and yields governing equations in displacements at several planes through the thickness.

The displacement field in an $n-$layered laminated plate composed of orthotropic laminae in the (x, y, z) system, may be written as (Reddy, 1987),

$$u(x, z) = \sum_{j=1}^{n} u^j(x)\Phi^j(z)$$

$$w(x, z) = \sum_{j=1}^{n} w^j(x)\Phi^j(z) \qquad (4)$$

where u^j and w^j are undetermined coefficients and Φ^j are any piecewise continuous functions. The governing equations of the layerwise theory are derived from the principle of virtual displacements, which, in the absence of body forces is written as,

$$\int_{S_T} \hat{T}_i \delta u_i dS = \int_V \sigma_{ij} \delta \varepsilon_{ij} dV \tag{5}$$

Taking the origin of the z coordinate axis to coincide with the lower surface of the plate, the function $\Phi^j(z)$ is defined as follows;

$$\Phi^j(z) = \begin{cases} \frac{(z-z_{j-1})}{(z_j-z_{j-1})} & \text{if } z_{j-1} \leq z \leq z_j \\ \frac{(z_{j+1}-z_j)}{(z_{j+1}-z_j)} & \text{if } z_j \leq z \leq z_{j+1} \\ 0 & \text{otherwise} \end{cases} \tag{6}$$

The Euler–Lagrange equations of the layerwise theory are as follows:

$$N^j_{xx,x} - Q^j_{xz} = 0$$
$$N^j_{xz,x} - Q^j_{zz} = 0 \tag{7}$$

The resultants are evaluated using the modified plane-strain stiffnesses , as follows :

$$N^j_{xx} = \int_{\Delta z_j} \sum_{m=1}^{n} \left[Q^j_{11} u^m_{,x} \Phi^m + Q^j_{13} w^m \Phi^m_{,z} + Q^j_{15} \left(w^m_{,x} \Phi^m + u^m \Phi^m_{,z} \right) \right] \Phi^j dz$$

$$N^j_{xz} = \int_{\Delta z_j} \sum_{m=1}^{n} \left[Q^j_{51} u^m_{,x} \Phi^m + Q^j_{53} w^m \Phi^m_{,z} + Q^j_{55} \left(w^m_{,x} \Phi^m + u^m \Phi^m_{,z} \right) \right] \Phi^j dz$$

$$Q^j_{xz} = \int_{\Delta z_j} \sum_{m=1}^{n} \left[Q^j_{51} u^m_{,x} \Phi^m + Q^j_{53} w^m \Phi^m_{,z} + Q^j_{55} \left(w^m_{,x} \Phi^m + u^m \Phi^m_{,z} \right) \right] \Phi^j_{,z} dz$$

$$Q^j_{zz} = \int_{\Delta z_j} \sum_{m=1}^{n} \left[Q^j_{31} u^m_{,x} \Phi^m + Q^j_{33} w^m \Phi^m_{,z} + Q^j_{35} \left(w^m_{,x} \Phi^m + u^m \Phi^m_{,z} \right) \right] \Phi^j_{,z} dz \tag{8}$$

where $\Delta z_j = [z_{j-1}, z_{j+1}]$ is the interval over which Φ^j is non–zero and the Qs with numerical subscripts are the stiffness terms in the modified plane–strain constitutive equations.

The equations of the layerwise theory form a system of coupled second–order ordinary linear differential equations. These equations may be solved exactly using the matrix operator method. But when the number of equations is large, a finite element solution is a better recourse. In their present form the equations require a simple one–dimensional finite element mesh. The mesh is used on each discretizing plane of the layerwise scheme, to obtain a finite element model of the computational domain. The finite element model is not presented here for the lack of space.

3. Results of the Model

3.1. STIFFNESS REDUCTION

In the present section, the predictions of the stiffness reduction model are presented. The model is validated by comparing the normalized reduced axial stiffness of cross–ply laminates with a general $[0_n/90_m]_s$ layup as a function of crack density, for glass epoxy composite laminate, with other models.

Fig. 1 is a plot of the reduced axial stiffness normalized by the axial stiffness of the uncracked laminate, as a function of the crack density, compared with other models. The crack density is measured in terms of the number of cracks per mm along the longitudinal direction. The results presented herein are for a glass epoxy laminate whose properties are the same as in Hashin (1985).

The upper bound on the reduced stiffness as predicted by the present model shows excellent agreement with the experimental results. The experimental results fall between Hashin's lower bound solution and the present variational upper bound solution. Both the solutions tend towards an asymptotic behavior in the limit of large crack densities. Also plotted are the predictions of the two–dimensional finite element model and the approximate analytical model due to Herakovich et al (1988). It is seen that the latter models underpredict the reduction in stiffness. Fig. 2

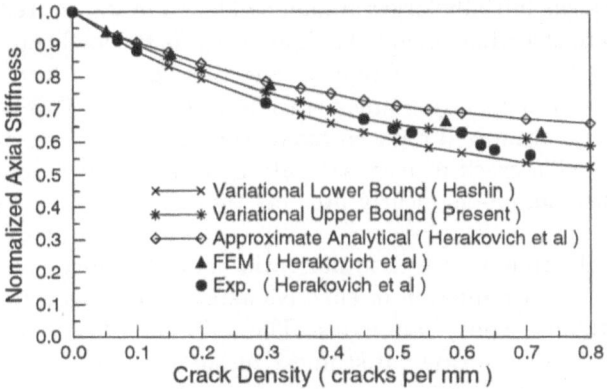

Figure 1: Comparison of various models for $[0/90_3]_s$ glass epoxy laminate.

shows the variation of the reduced axial stiffness normalized by the stiffness of the uncracked laminate, for glass epoxy laminates with increasing ratio of thickness of the cracked plies and the uncracked longitudinal plies. The effect of staggering in the pattern of cracks is also considered. It is seen that as the relative thickness of the cracked transverse plies increases, the reduction of stiffness at any given crack density increases. This is because, a greater amount of load is carried by the transverse plies in laminates with a greater relative thickness of these plies, in the uncracked configuration. Consequently, any cracking in the transverse plies leads to

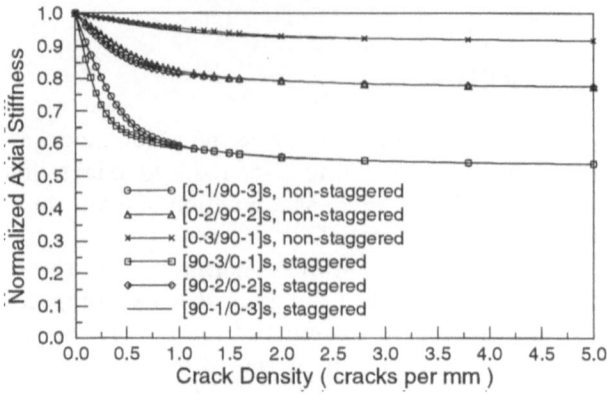

Figure 2: Stiffness reduction v/s crack density for glass epoxy laminates.

a greater loss in load carrying ability of these plies. For the same external strain of the cracked and uncracked laminates, the cracked laminate requires a lesser load, due to the reduction in the effective axial stiffness. The stiffness reduction tends to saturate very quickly after a certain value of the crack density. At low crack densities, the reduction in stiffness is very high with even a slight increase in crack density. This is because, initially larger amounts of load are transferred or "shed" by the transverse plies to the adjacent longitudinal plies. As the load transfer increases, the load carried by the cracked plies reduces and therefore, the load transferred is expected to saturate after a certain crack spacing is achieved. This corresponds to the saturation in the normalized reduced axial stiffness. The crack spacing attains a saturation value. This crack density saturation state is called the characteristic damage state. Thus, the model clearly predicts the existence of the characteristic damage state.

For steel and aluminum with modular ratio equal to unity, as expected, the model predicted a greater reduction in effective axial stiffness as compared to both glass–epoxy and graphite–epoxy laminates. The reduced stiffness was found to be around 25% and 27% of the original stiffness in the case of aluminum and steel respectively. The stiffness reduction curves for both aluminum and steel laminates were found to be almost the same. This implies that the reduction in stiffness is influenced only by the modular ratio and not by the absolute value of the moduli, for a given laminate at any particular crack density.

3.2. CRACK OPENING DISPLACEMENTS

The COD depends on the crack spacing, the layup, the placement of the transverse plies, their relative thickness as compared to the longitudinal plies, the applied strain and the reduction in stiffness for the particular crack spacing. By assuming that the profile of the crack opening is similar to that of a single crack in an infinite isotropic plate under uniform far–field tension, the following normalization was done in conjunction with the stiffness reduction equation:

$$\frac{\pi}{4} \frac{E_2}{\Delta E} \frac{n_c}{n\varepsilon_o} \frac{u'}{a} \approx \sqrt{1 - \frac{z^2}{t^2}} \tag{9}$$

where E_2 is the transverse modulus, ΔE is the reduction in effective axial modulus, n_c is the number of cracked plies, n is the total number of plies, t is the half thickness of the transverse plies and a is the half–crack spacing. The right hand side of the above equation is a function of $\frac{z}{t}$ alone. The crack opening displacement as predicted by the layerwise theory is normalized as in the left hand side of the above equation. The closeness of the normalized layerwise solution of the COD profile to the profile of the ellipse determines the accuracy of the approximation made about the crack opening displacement.

Fig. 3 shows the normalized crack opening displacements and the unit ellipse profile for $[0/90_3]_s$ glass epoxy laminate. It is seen that at crack densities upto the saturation value, the normalized COD profile and the unit ellipse profile are almost identical.

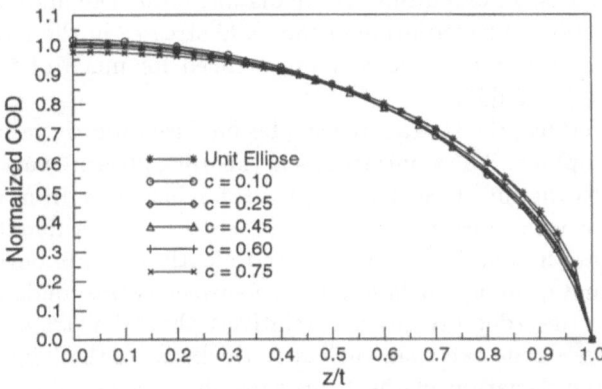

Figure 3: Normalized COD for a $[0/90_3]_s$ glass epoxy laminate.

3.3. STRESS FIELDS IN THE RVE AT 0.1 CRACKS/MM.

Fig. 4 shows the variation of the axial stresses in an RVE of a $[0/90_3]_s$ glass epoxy laminate. The stresses are normalized by the axial stresses in the uncracked laminate at the corresponding locations, through the thickness of the laminate. The thickness coordinate of the discretizing planes are normalized by half the thickness of the transverse plies and is less than, equal to or greater than unity for planes located in the transverse ply group, coincident with the interface or in the longitudinal ply group, respectively. The axial stresses increase almost exponentially from zero

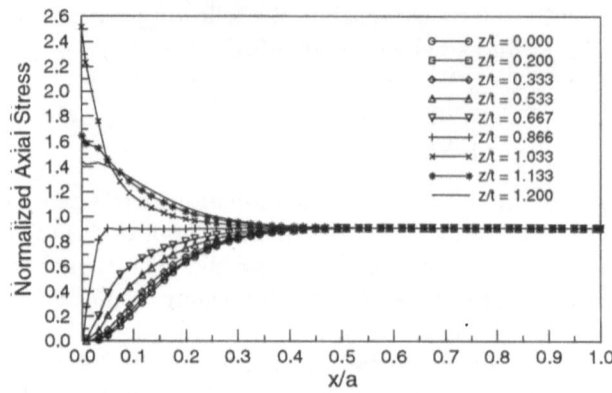

Figure 4: Variation of normalized axial stresses 0.1 cracks/mm.

at the crack face to a far–field value in a distance of about 45% of the half crack spacing. In this region, most of the load is transferred to the adjacent plies. This load transfer occurs primarily at the interface in a region called the shear lag region. The axial stresses increase in magnitude as the distance from the midplane increases in the thickness direction. The variation of the axial stresses in the transverse plies along the thickness coordinate is not very pronounced for much of the transverse ply group, as seen at $\frac{z}{t} = 0.667$.

Closer to the interface, the stresses reach a far field asymptotic value faster than at points near the midplane. The axial stresses in the 0 deg plies are very much higher along planes closer to the interface than at planes farther away from it. The axial stresses reach a peak value midway between the cracks, in the longitudinal plies. The stress concentration zone is pronounced and clearly extends beyond $\frac{z}{t} = 1.2$. In the $[0/90_3]_s$ laminate, the upper laminate free surface corresponds to $\frac{z}{t} = 1.333$. Thus, in cases where the 0 deg ply group is relatively thin, the stress concentration zone or the disturbance zone can span the entire thickness of the longitudinal plies.

Fig. 5 shows the variation of the transverse shear stress along the distance from the crackface to the midpoint of the representative volume element, at different planes through the thickness. The shear stress along each of the planes attains a maximum at a certain distance from the crackface. This behavior is in qualitative agreement with the results of the variational model of Hashin. The shear stresses decay to zero midway between the cracks. A comparison of the shear stress distribution and the axial stress distribution reveals the shear lag region, wherein, there is a high non–zero shear in the transverse plies upto around 45% of the normalized half crack spacing. Thus, the reduction of the axial stresses in the cracked plies over a certain distance from the crack face manifests as an increased transverse shear stress over the same distance in the transverse plies and an increased axial stress over the same distance in the constraining longitudinal plies. This behavior of the stresses defines the extent of the shear lag region. As shown in Fig. 6, the transverse

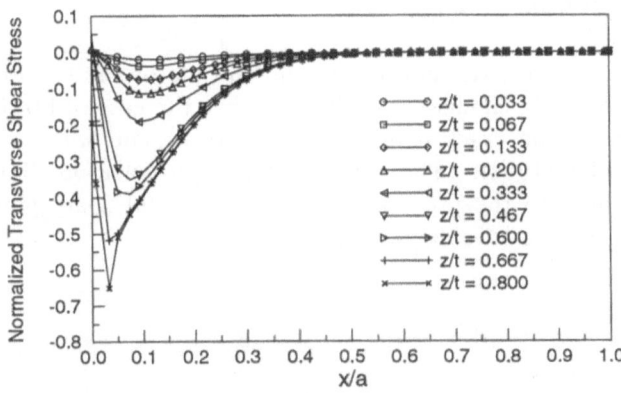

Figure 5: Normalized transverse shear stresses at a crack density of 0.1 cracks/mm.

normal stresses in the transverse plies are compressive near the crack face and reach a maximum at the midplane. The behavior changes from compression to tension, a small distance away from the crack face and attains a positive maximum which is relatively small , and then decays to zero towards the end of the shear lag region. The self–equilibrating nature of the transverse normal stresses is evident. The transverse normal stresses along any plane must be self–equilibrating because the shear lag region is bounded by vertical planes that are shear free and the traction free laminate top surface. In the constraining plies, the transverse normal stresses were again found to be self–equilibrating. The stresses were found to be tensile upto a certain distance from the crack plane. In general, they were of a behavior opposite to that in the transverse plies. These peeling stresses are responsible for the onset of delaminations at the crack tip.

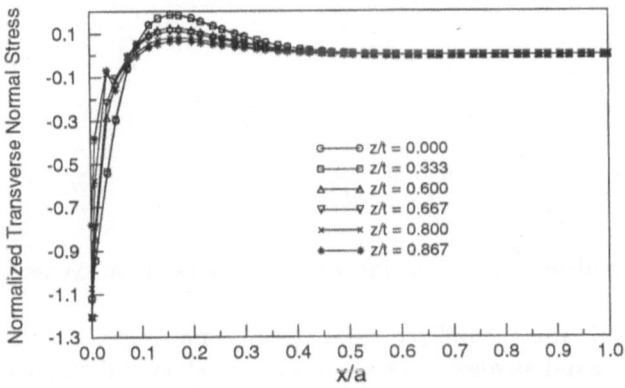

Figure 6: Normalized transverse normal stress at a crack density of 0.1 cracks/mm.

3.4. STRESSES IN THE RVE AT CDS

As the crack density increases, the nature and magnitude of the stresses in the representative volume element changes considerably. For the case of $[0/90_3]_s$ glass epoxy laminates, a saturation crack density of 0.75 cracks/mm was reported by Highsmith and Reifsnider (1982). The stress distribution corresponding to this crack spacing is compared to the previous case. Figs. 7 and 8 are plots of the stresses in the RVE at chracteristic damage state.

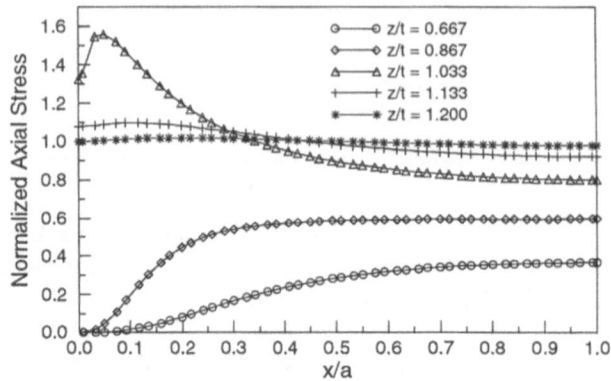

Figure 7: Normalized axial stress at the characteristic damage state.

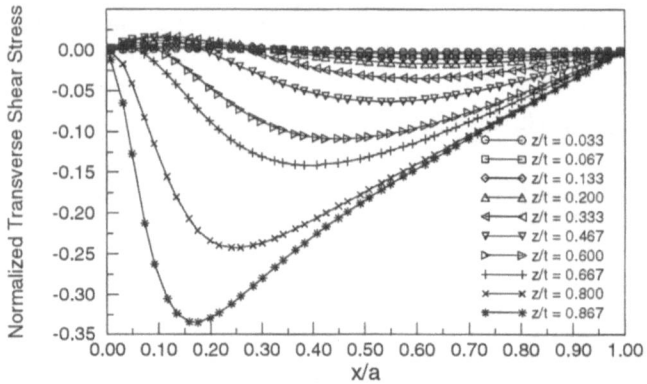

Figure 8: Normalized transverse shear stress at CDS (transverse plies).

As seen in Fig. 7, the normalized axial stress is much lesser than at $c = 0.1$ cracks/mm, and the axial stresses near the midplane show a distinct compressive behavior very close to the crack face. The transverse shear stress shown in Fig. 8 exhibit characteristics that are similar to the case where the crack density was 0.1

cracks/mm. The normalized values of the shear stresses at any plane at c = 0.75 cracks/mm are lesser than at c = 0.1 cracks/mm. This is because the present method is a displacement (strain) based method wherein the laminates with crack density of 0.1 and 0.75 cracks/mm are subjected to the same external strain ε_o. Under these conditions, the load required by the laminate with a higher crack density is very much lesser. The transverse normal stresses were found to attain a large compressive value at the crack face and a tensile value between the cracks. The self–equilibrating nature of the stresses was again noticed. This behavior extended over the entire length between the cracks, suggesting that the shear lag region spans the entire length between the cracks at the saturation crack density.

4. Conclusions

A strain energy based variational method has been employed to derive an upper bound on the effective axial stiffness of laminates with cracked transverse plies. Results of the model agreed well with experiments and the lower bound solution due to Hashin. Laminates with staggered cracks suffered greater loss of stiffness as compared to non–staggered cracking. The profile of the crack opening displacement was found to match that of a single crack in an infinite isotropic elastic solid under uniform far–field tension. A detailed study of the normalized stresses was carried out at two different crack densities including the saturation crack density, to examine the mechanics of load transfer and the mode of load shedding between the cracked transverse plies and the constraining longitudinal plies.

ACKNOWLEDGMENT The support of this research by Sandia National Laboratories is gratefully acknowledged.

5. References

1. Allen, D.H., Harris, C.E. and Groves, S.E. (1987). A thermomechanical constitutive theory for elastic composites with distributed damage - I. Theoretical development. *Int. J. Solids and Structures.* Vol. 23, **9**, 1301–1318.

2. Flaggs, D.L. (1985). Prediction of tensile matrix failure in composite laminates. *Journal of Composite Materials.* **19**, 29–50.

3. Fukunaga, H., Chou, T.W., Peters, P.W.M., and Schulte, K. (1984). Probabilistic Failure Strength Analyses of Graphite/Epoxy Cross–Ply Laminates. *Journal of Composite Materials*, Vol. 18, 339–358.

4. Garrett, K.W. and Bailey, J.E. (1977). Multiple transverse fracture in 90^0 cross–ply laminates of glass fibre-reinforced polyester. *Journal of Materials Science.* **12**, 157–168.

5. Gudmundson, P. and Östlund, S. (1992). First–order analysis of stiffness reduction due to matrix cracking. *Journal of Composite Materials.* Vol. 26, **7**, 1009–1030.

6. Hashin, Z. (1985). Analysis of cracked laminates : A variational approach. *Mechanics of Materials.* **4**, 121–136.

7. Herakovich, C.T., Aboudi, J., Lee, S.W. and Strauss, E.A. (1988). Damage in composite laminates: Effects of transverse cracks. *Mechanics of Materials.* **7**, 91–107.

8. Highsmith, A.L. and Reifsnider, K.L. (1982). Stiffness reduction mechanisms in composite laminates. *Damage in Composite Materials*, ASTM-STP **775**, K.L.Reifsnider, Ed., American Society for Testing and Materials, 103–117.

9. Laws, Norman and Dvorak, George .J. (1988). Progressive transverse cracking in composite laminates. *Journal of Composite Materials.* **22**, 900–916.

10. Nairn, John.A. (1989). The strain energy release rate of composite microcracking: A variational approach. *Journal of Composite Materials.* **23**, 1106–1129.

11. Reddy, J.N. (1987). A generalization of the Two–Dimensional Theories of Laminated Composite Plates. *Communication in Applied Numerical Methods*, **3**, 173–180.

12. Reddy, J.N. and Robbins, D.H. (1994). Structural Theories and Computational Models for Composite Laminates. *Applied Mechanics Reviews.* **47**,(6), Part 1, 147–170.

13. Talreja, Ramesh. (1986). Stiffness properties of composite laminates with matrix cracking and interior delamination. *Engineering Fracture Mechanics.* Vol. 25, **5,6**, 751–762.

14. Varna, J. and Berglund, L., (1991). Multiple Transverse Cracking and Stiffness Reduction in Cross–Ply Laminates. *Journal of Composites Technolgy & Research*, JCTRER. Vol. 13, **2**, Summer 1991, 99–106.

DISCRETE MODEL OF FRACTURE IN DISORDERED TWO-PHASE MATERIALS

R. PYRZ[1] AND B. BOCHENEK[2]

Institute of Mechanical Engineering
Aalborg University, Pontoppidanstræde 101
9220 Aalborg, Denmark [1]

Institute of Mechanics & Machine Design
Cracow University of Technology
Warszawska 24, 31-155 Cracow, Poland[2]

1. Introduction

Improved characterization of present-day composites that are manufactured to obtain optimal property values by microstructure effects necessitates a more thorough knowledge of the microstructural features and their cooperative interaction which determine the physical characteristics of the material. This is particularly true for the description of strongly non-linear phenomena such as fracture. Neglecting geometrical disorder of fillers does not introduce a significant error in the prediction of the elastic and transport phenomena. By contrast, fracture is a highly localized phenomenon, and the local geometrical disorder cannot be neglected. The microfailure threshold is dominated by extreme fluctuations of the stress field, and these local hotspots are strongly influenced by a distribution pattern of inclusions. Continuum material models cannot capture the effect of random local material inhomogeneities on the localization of damage and failure. Also finite element models are not able to reflect the influence of local disorder on overall properties of composites. Moreover, an accurate modelling of a large number of inhomogeneities necessitates the size of the mesh to be sufficiently small compared to the typical size of an inclusion. Similar problems with a simultaneous modelling of microcracks nucleation and growth excludes this technique due to prohibitive costs and time consumption.

In the present study transversely loaded unidirectional composite is discretized into a network of structural entities that are endowed with their own properties which resemble the microstructure on the mesoscopic length scales. The main objective of the investigation is to study the effect of inclusions' distribution on the load-carrying capacity of the material and to relate it to some geometrical descriptors of the microstructure.

R. Pyrz (ed.), IUTAM Symposium on Microstructure-Property Interactions in Composite Materials, 313–326.

2. Statistical Analysis of Patterns

Several attempts have been made to determine parameters that characterize the distribution of fillers, for example: Altan et al (1990), Everett and Chu (1993), Green and Guild (1991), Taya et al (1991). Pyrz (1993a, 1994a) has introduced second order statistics to characterize the distribution patterns of fibres. The statistical descriptors used in the present study are illustrated in Fig. 1.

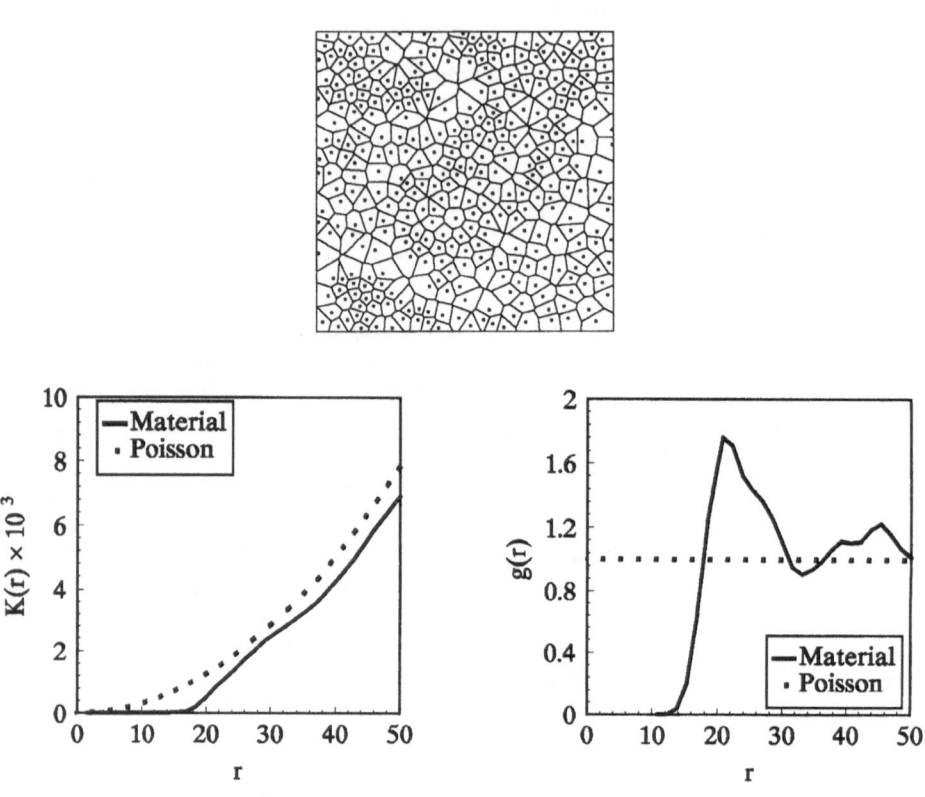

Figure 1. Descriptors of point pattern.

The centre points of fibres, detected from the composite by image analysis technique, serve as nuclei for a construction of the Dirichlet tessellation that uniquely maps the point pattern onto contiguous polygons covering the area of observation, Green and Sibson (1978), Pyrz (1993a). The second order intensity function K(r) is defined as the number of further points expected to lie within a radial distance r of an arbitrary point and divided by the number of points per unit area.

The pair distribution function g(r) is related to the probability g(r)dr of finding an inclusion whose centre lies in an infinitesimal circular region with radius dr about the point r, provided that the coordinate system is located at the centre of a second inclu-

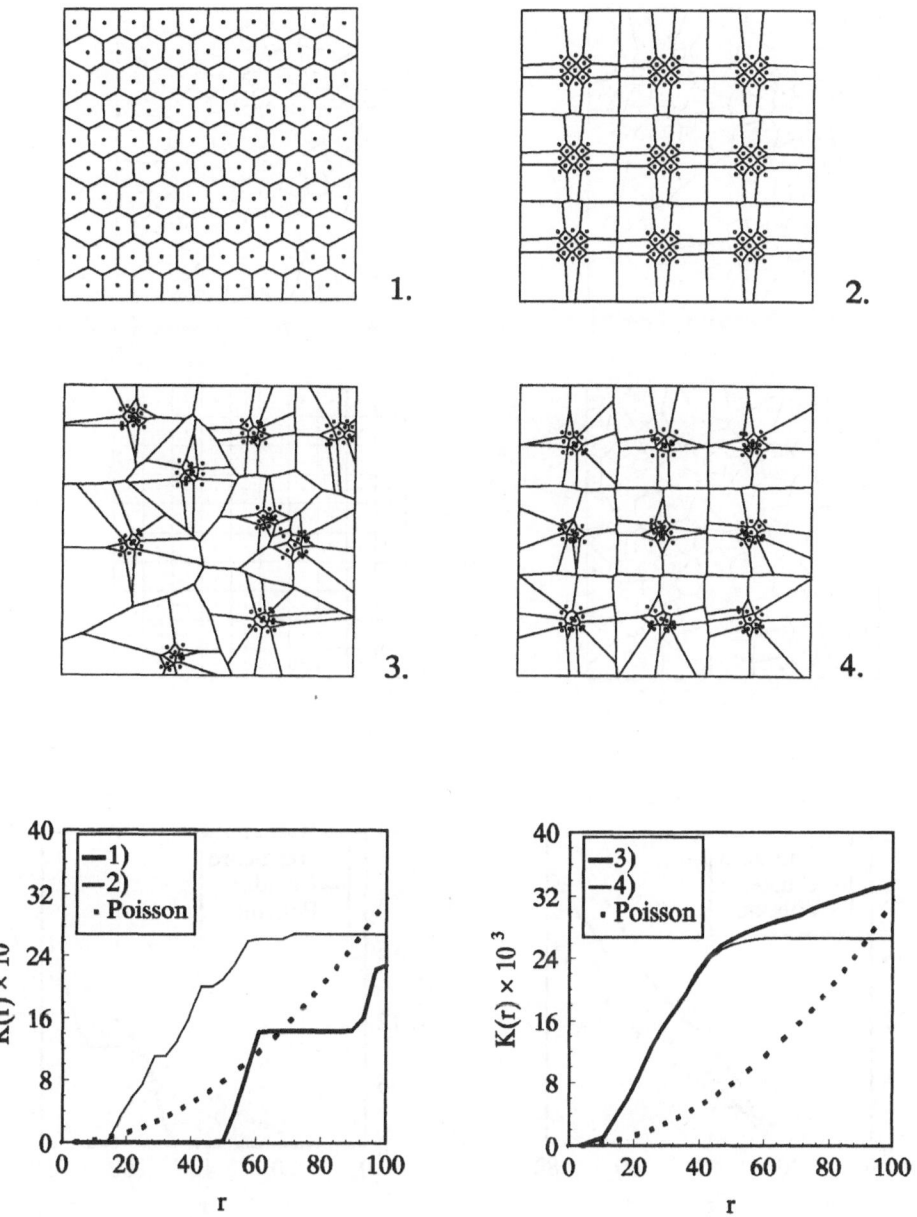

Figure 2. Examples of point patterns and corresponding K(r) functions.

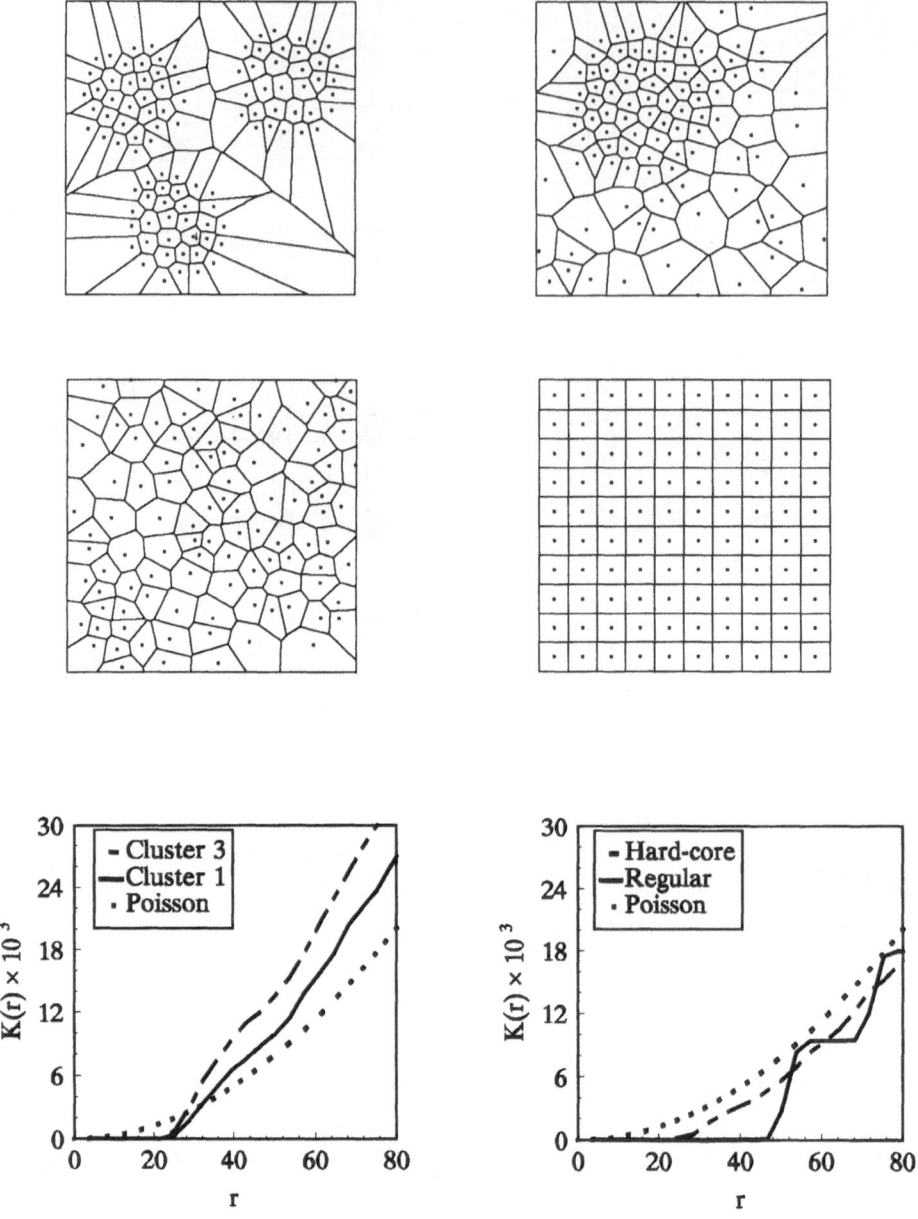

Figure 3. Further examples of point patterns and corresponding K(r) functions.

sion. As shown in Pyrz (1994b) the functions K(r) and g(r) are related

$$g(r) = \frac{1}{2\pi r} \frac{dK(r)}{dr} \tag{1}$$

For the Poisson pattern which is considered as a completely random distribution, Getis and Boots (1978), $K(r) = \pi r^2$ and $g(r) = 1$. The pair distribution function g(r) characterizes the occurrence intensity of inter-inclusion distances. Local maxima indicate the most frequent distances between points; local minima correspond to the least frequent distances in the pattern. Thus the supreme g_{max} of the local maxima determines a characteristic distance between pairs of inclusions. This value of the pair distribution function is, in the sequel, correlated with failure behaviour of patterns. The ability of the K(r) function to discriminate between different point patterns is illustrated in Figs. 2 and 3. The second order intensity function K(r) for more regular patterns always lies below the corresponding function for the Poisson pattern contrary to clustered patterns whose K(r) functions are placed above the K(r) function of the Poisson pattern.

The short- and long-range regularity is detected by a characteristic "stair" shape of the function K(r). Horizontal fragments of the function indicate empty space at corresponding distances either between points of the regular distribution or between clusters which are regularly distributed. The function K(r) for the Poisson pattern serves as a dividing line between clustered patterns and patterns with a certain degree of order.

3. Calculation Model and Simulation Results

The Dirichlet network of polygons is created on transverse sections of unidirectionally reinforced material. It is then possible to identify all neighbours of a given fibre and to define uniquely the zone of influence for each pair of neighbouring fibres as denoted in Fig. 4 by the quadrangle ABCD.

Thus the total area of observation is divided into a set of contiguous quadrangles which embrace the pairwise zones of influence. The stress field within an individual zone of influence and its geometry determine the properties of bonds which replace the continuous material in the discrete model. The pattern of embedded fibres with periodic boundary conditions is subjected to remote unit load in vertical direction. The stress components are calculated for each pairwise zone of influence at four nodes that correspond to the centres of gravity of four quadrangles into which the zone is divided, Fig. 4. The stress calculation is based upon the superposition method that takes into account interactions between neighbouring fibres, Axelsen and Pyrz (this issue). Assuming an equivalence between the energy stored in the matrix material bounded by the pairwise zone of influence and the energy that would have existed if the zone of influence had been replaced by an elastic rod connecting centres of neighbouring fibres, the stiffness of rods may be calculated, and then the microstructure is transformed into a truss-like structure, Pyrz and Bochenek (1994). The model introduces the geometrical disorder through mapping of fibres' positions onto the lattice network and through a

318

quenched disorder in rods' stiffness. The pattern of fibres with the Dirichlet tessellation and corresponding truss network is shown in Fig. 5.

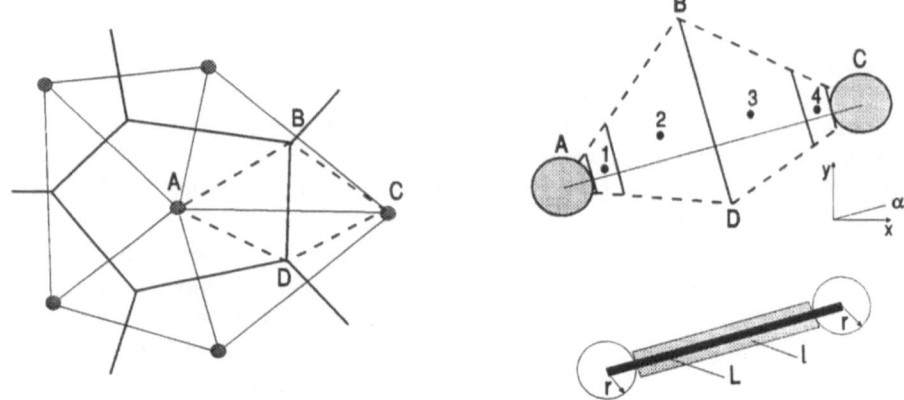

Figure 4. Discretization model.

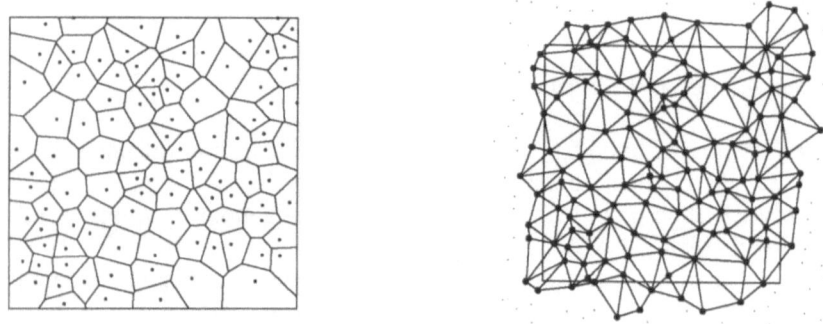

Figure 5. Calculation model.

The network is loaded in displacement mode control until the most overstressed rods reach a predetermined critical strain value. Then the broken rods are removed and the lattice configuration is recalculated to obtain a new equilibrium. The displacement is increased again and this procedure is repeated until the network breaks fully apart. The simulations with the force control mode of loading have been investigated in Pyrz and Bochenek (1994) and Pyrz (1993b).

The force-displacement diagrams for a model composite, having dispersion of fibres as in Fig. 3, are shown in Fig. 6. The triple cluster model starts to soften at smallest maximum force as compared to other distributions. This early stage of softening is caused by a microcracking resulted from intensified stress components due to strong

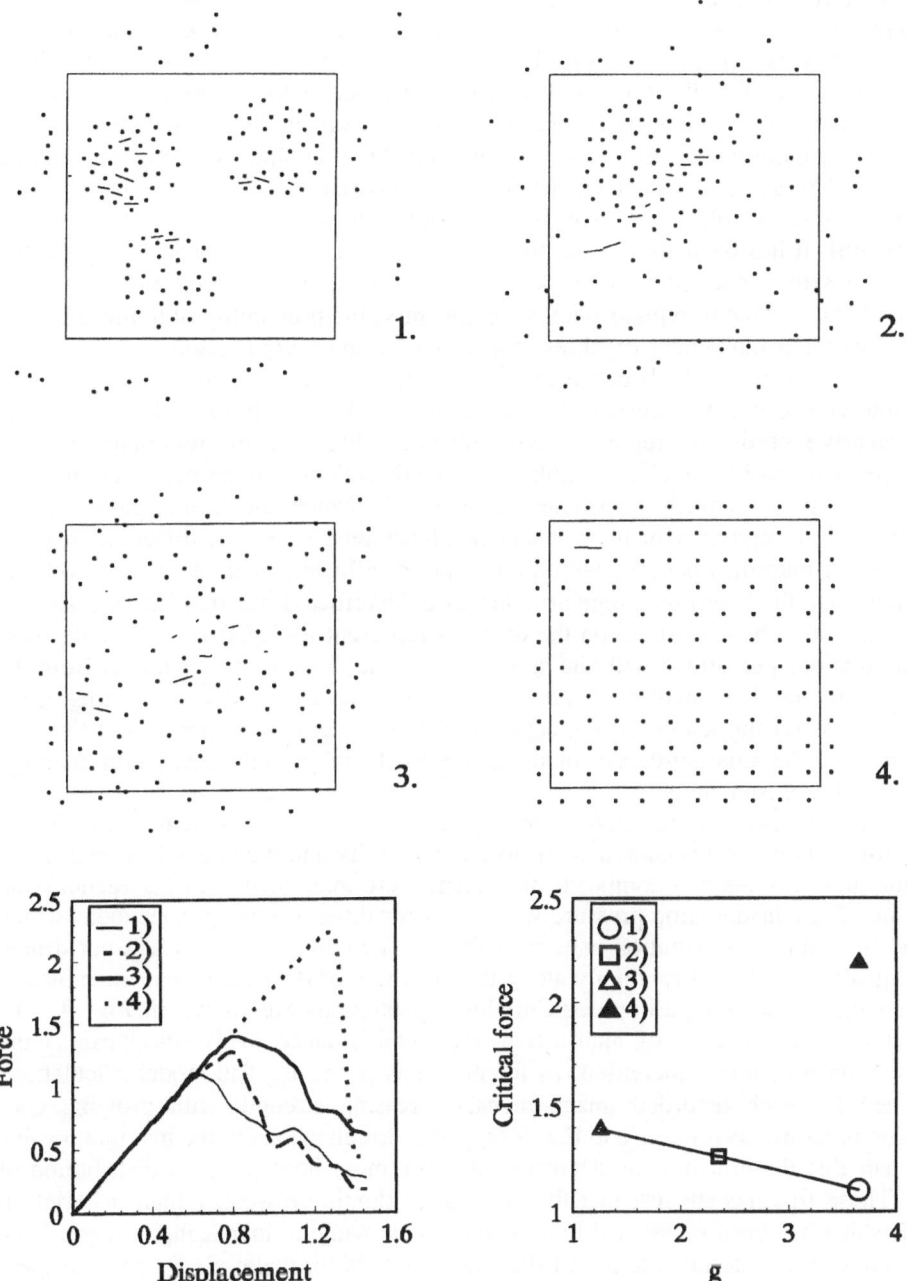

Figure 6. Deformation behaviour of four patterns.

interaction effects between fibres in densely populated clusters. A number and positions of broken rods have been monitored during loading sequences. Each broken rod nucleates a microcrack that has its trace being identified with the length and the position of that side of the Dirichlet polygon which is perpendicular to the broken rod. A susceptibility of clustered patterns to microcracking is seen in Fig. 6, where the triple cluster clearly exhibits largest number of microcracks in comparison to other patterns subjected to the same displacement level. The behaviour of the discrete model fully supports the result of the stress analysis performed with the same distribution of fibres, Pyrz (1994b). It has been shown that the local stress concentration that belongs to the higher values of a spectrum of the local stress variations embraces more fibres in clustered patterns than in regular ones. Consequently, the probability of a micro-failure increases as the arrangement of fibres in a matrix becomes aggregated.

The critical forces which correspond to maximal values attained by models are correlated with respective values of the pair distribution function for each pattern, Fig. 6. It is clearly seen that the regular pattern falls into a distinctly different range than the other three patterns for which the regularity of the dispersion is disturbed or completely lost. This property is further documented in Fig. 7, which shows that patterns with either inter-fibre regularity or inter-cluster regularity tend to occupy different areas on the parameter map than patterns with a pronounced influence of randomness. Both the short-range and the long-range regularity increase the critical force that can be sustained by the material. The error bars on the diagram represent a dispersion of results from eight simulations performed with the hard-core and the random cluster model from the area A and the random-regular cluster model from the area C. The deteriorating effect of randomness on the load-carrying capacity of the model distributions from the area C is apparent. The random-cluster model is more vulnerable to microcracking than the double regular cluster model.

In order to assess if the model predictions are in accordance with a qualitative correlation tendency between a microstructure variability and the strength of real materials, the glass fibre-epoxy composite specimens were manufactured by a vacuum bag technique. The consolidation pressure was varied for three batches of specimens which resulted in different distribution pattern of fibres. The pattern analysis was performed at five quadrants selected randomly along the thickness of the specimens. Examples of recorded images with superimposed Dirichlet tessellations are shown in Fig. 8. The material A exhibits bands of matrix-rich areas which successively disappear as the consolidation pressure is increased, as for materials B and C. The model calculations performed for each recorded image reveal increasing strength with growing consolidation pressure, as reference to Fig. 9 suggests. Simultaneously, the maximum value of the pair distribution function decreases due to a more homogeneous distribution of fibres. These findings suggest that the material A should be weaker than materials B and C, which has been confirmed in experiments showing an increasing strength from 18.2 MPa for the material A to 34.1 MPa and 52.5 MPa for materials B and C, respectively, Pyrz (1993b).

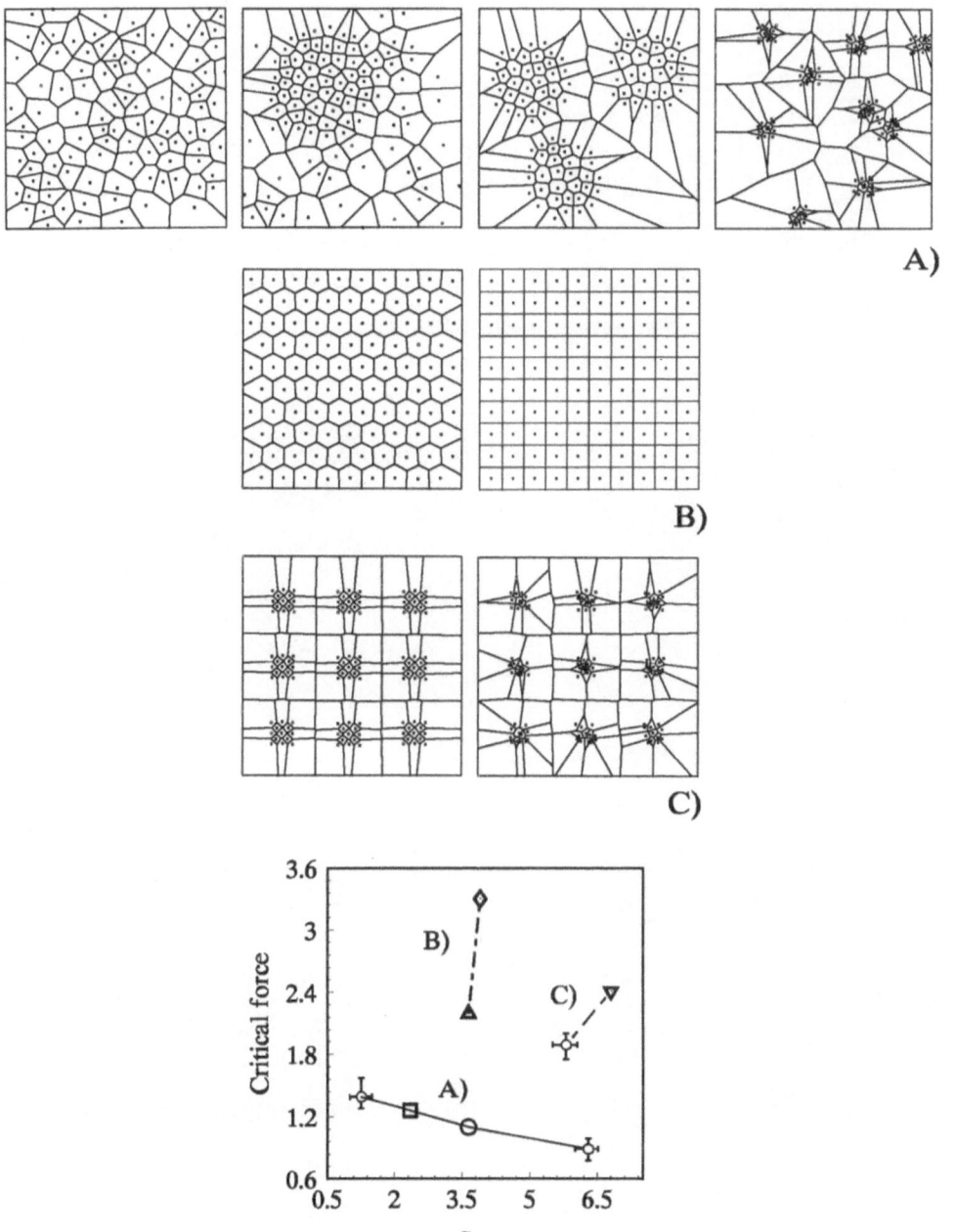

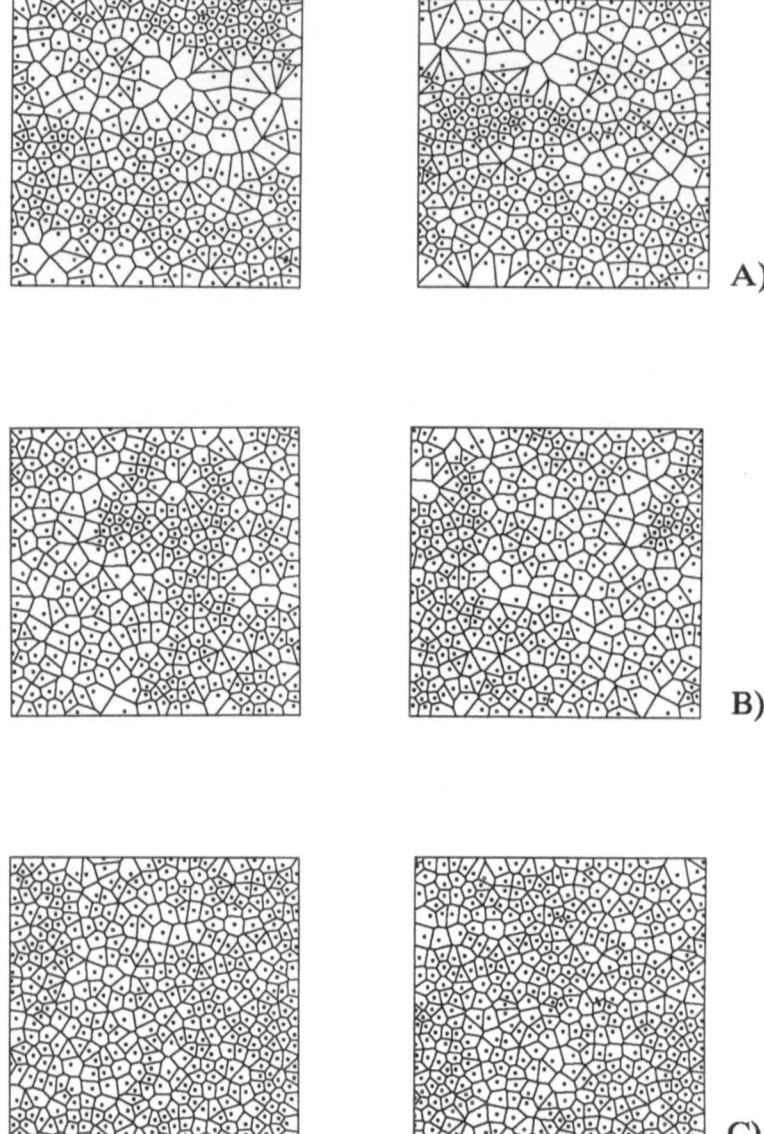

A)

B)

C)

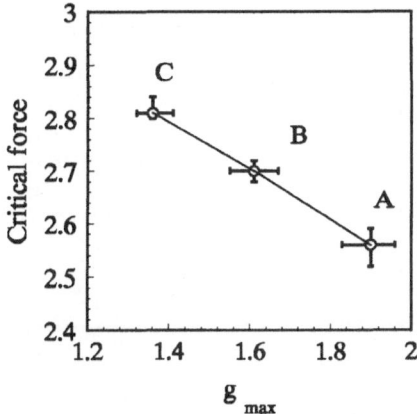

Figure 9. Critical force correlated with the pair distribution function for three materials.

In the network model the medium is discretized in such a way that all network's entities are equivalent and one does not need to make the network finer in regions of higher stress or strain gradients. This fact makes the qualitative analysis of the microcracking process time effortable. Figures 10 and 11 present a sequence of microcracks nucleated in the triple cluster pattern of fibres embedded into the hard-core background. The figures illustrate two cases with pre-existed cracks and one case where the microcracking process starts from the weakest bonds. Obviously, the morphology of the breaking process is different in each case as well as an overall behaviour differs significantly.

The existence of the initial crack enforces the percolating fracture front to propagate along a continuous path with a few short side branches and a limited number of separated microcracks. If the initial crack does not exist the system exhibits damaging behaviour which results in an appearance of a large number of dispersed microcracks. The macrocrack is formed by joining individual microcracks at the late stage of loading. However, the breaking process may also be localized for this case provided that only one most overstressed rod is removed at each step of loading. Then a single macrocrack dominates the failure process in as much similar fashion as for the system with the initial crack present. In order to avoid eventual singularities in the calculation procedure of the truss network the failure of the rods has been associated with a significant decrease of their stiffness rather than a physical removal from the network. Thus the failure behaviour of the model may be further extended if the broken rods are left with some residual strength.

The morphological analysis of the microcracking process and its spatial pattern depending upon an underlying disorder will be presented elsewhere.

324

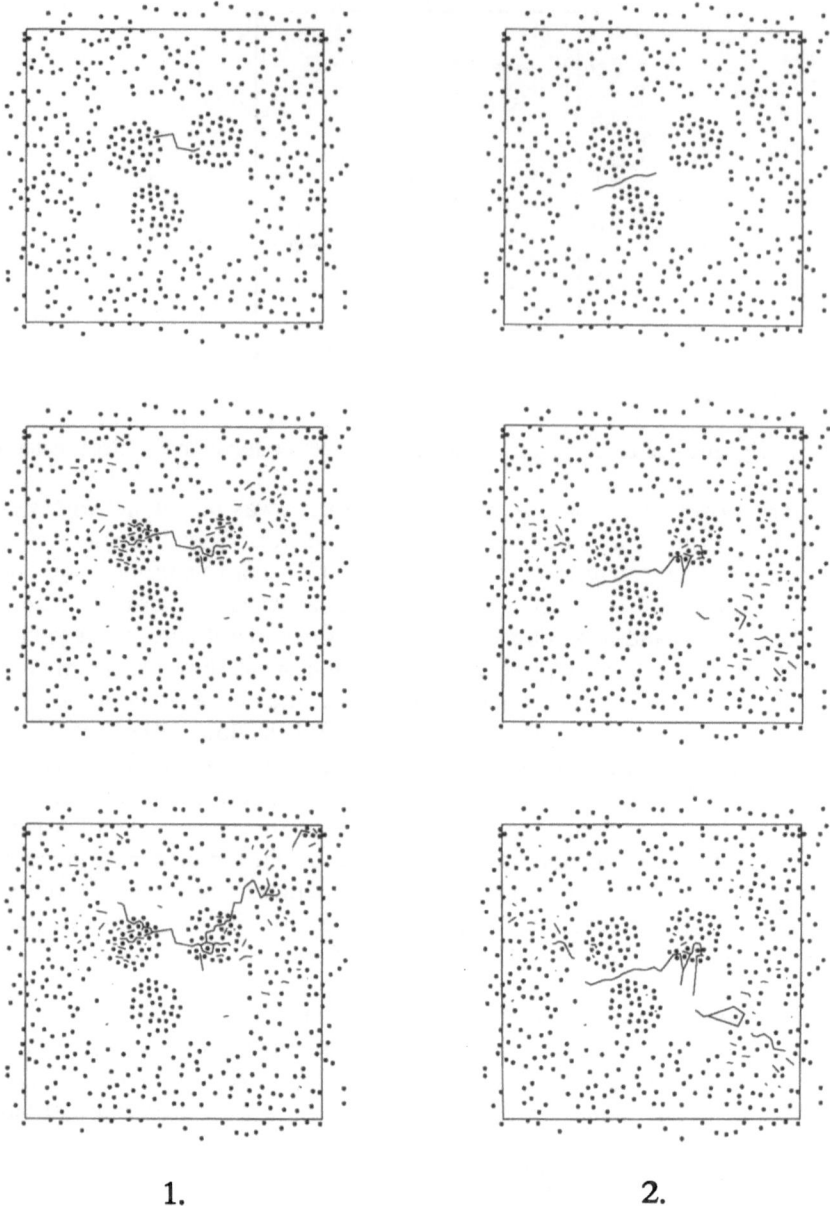

1.   2.

Figure 10. Microcracking sequence with pre-existed cracks.

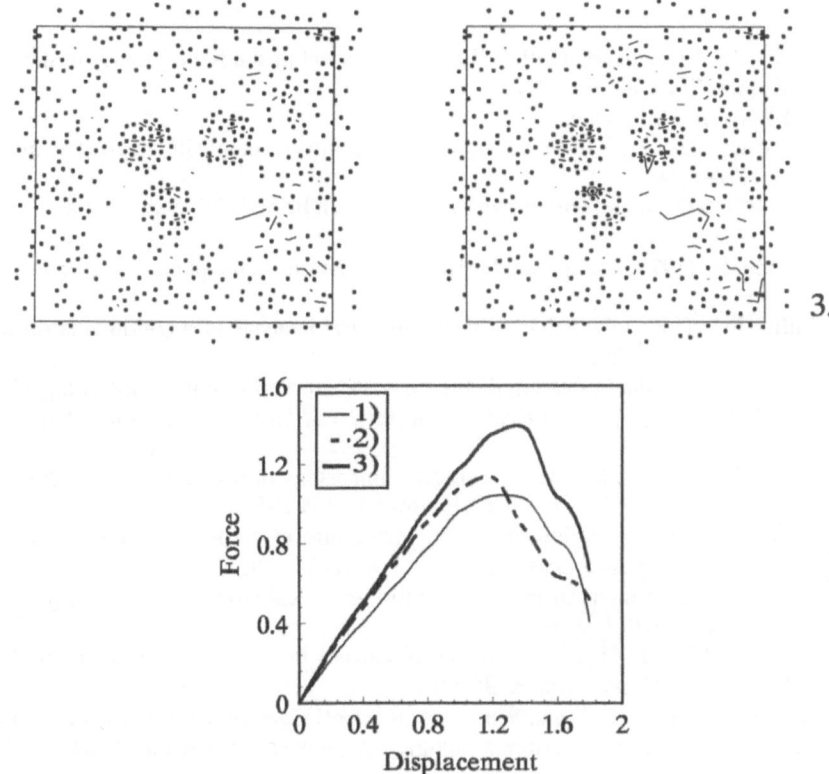

Figure 11. Microcracking sequence for the triple cluster and loading diagram for all three cases.

4. Conclusion

In order to have some insight about fluctuations of the fracture characteristics, which are inherent for all types of materials, the network model is advantageous as it allows to introduce disorder in a natural way and to investigate the trends as a function of disorder parameters. The results from the present investigation strongly suggest that calculations based upon the unit cell concept, which means regularity of the microstructure, overestimate an actual load-carrying capacity of materials. Inherent in the usual continuum approaches are certain concepts like the locality of action and the homogeneity that constrain a microstructure-property connection to establish. Hence considerations of the discreteness of the material from the onset combined with probabilistic aspects of the microstructure and computer simulations seem to point a way for the successful establishment of a more comprehensive failure theory.

5. References

Altan, B., Misirh, Z. and Yorucu, M. (1990) The determination of the homogeneity in multi-phase mixtures, *Z. Metallkde.* **81**, 221-228.

Axelsen, M.S. and Pyrz, R., this issue.

Everett, R.K. and Chu, J.H. (1993) Modelling of non-uniform composite microstructure, *J.Comp. Mat.* **27**, 1128-1144.

Getis, A. and Boots, B. (1978) *Models of Spatial Processes*, Cambridge University Press, Cambridge.

Green, P.J. and Sibson, R. (1978) Computing Dirichlet tessellations in the plane, *Compt. J.* **21**, 168-173.

Green, D. and Guild, F.J. (1991) Quantitative microstructural analysis of a continuous fibre composite, *Composites* **22**, 239-242.

Pyrz, R. (1993a) Stereological quantification of the microstructure morphology for composite materials, in P. Pedersen (ed.), *Optimal Design with Advanced Materials*, Elsevier, Amsterdam, pp. 81-95.

Pyrz, R. (1993b) Interpretation of disorder and fractal dimensions in transversely loaded unidirectional composites, *Polymer & Polymer Composites* **1**, 283-295.

Pyrz, R. (1994a) Quantitative description of the microstructure of composites. Part I: morphology of unidirectional composite systems, *Comp. Sci. Techn.* **50**, 197-208.

Pyrz, R. (1994b) Correlation of microstructure variability and local stress field in two-phase materials, *Mat. Sci. Engng.* **A177**, 253-259.

Pyrz, R. and Bochenek, B. (1994) Statistical model of fracture in materials with disordered microstructure, *Sci. Engng. Comp. Mat.* **3**, 95-110.

Taya, M., Muramatsu, K., Lloyd, D.J. and Watanabe, R. (1991) Determination of distribution patterns of fillers in composites by micromorphological parameters, *JSME Int. J.* **34**, Ser.I, No. 2.

EVOLUTION CONCEPTS FOR MICROSTRUCTURE-PROPERTY INTERACTIONS IN COMPOSITE SYSTEMS

K.L. REIFSNIDER
Materials Response Group, Dept. of Engineering Science and Mechanics
Virginia Polytechnic Institute and State University
Blacksburg, VA 24061 USA

ABSTRACT. Modern durability, damage tolerance, and reliability predictive methodologies rely on an accurate estimate of local stress states and material states in the micro-regions that control remaining strength and life of composite materials systems. Major advances have been made in the determination of local stresses with both closed form and numerical approximate analysis methods. However, the progress in the development of understandings and models of the evolution of <u>material</u> states in the presence of degradation during service has been more modest. These material state changes are particularly important when the material components operate at elevated temperatures, as is the case for many of the ceramic matrix composites in use today, and planned for the future. The present paper addresses this subject for fibrous composite material systems, with a focus on the degradation of stiffness and strength associated with elevated temperature and cyclic load exposure. Special attention is given to the effects of creep and static fatigue on the deformation and remaining strength of ceramic composites. Interpretations of observations will also be discussed using the performance simulation codes that the authors have developed for this purpose. Examples of predictions and observations will be compared to illustrate the concepts that are to be demonstrated. A coherent philosophy for life prediction will be proposed.

1. Fundamentals of Property Evolution due to Damage Accumulation

In the limited space available here, only a brief outline of our fundamental approach will be recorded. We address the problem of remaining strength of a fibrous composite system, and assume that such strength is defined by the (statistical) accumulation of micro-variations in the properties, geometry, and arrangement of the constituents and the interfaces / interphase regions between them. In that context, we set three fundamental premises:

1. Fracture strength is an elastic property, i.e., we can describe fracture as an instability in terms of some potential, U, such that

$$\frac{\delta^2 U}{\delta \epsilon^2} = 0$$

in which ϵ is a measure of system strain, and U is defined over some (carefully identified) domain.

2. Although fracture strength (as we will discuss it) is a scaler, it is a function of the material strength tensor and stress tensor for the instant of definition (analogous to stiffness, conductivity, ... which are collectively defined by a state of material <u>in relation to</u> an applied (tensor) condition).

327

R. Pyrz (ed.), IUTAM Symposium on Microstructure-Property Interactions in Composite Materials, 327–336.
© 1995 *Kluwer Academic Publishers.*

328

3. We construct our definition with the familiar scaler "continuity," ψ, defined as (1 - Fa) (or more generally, some function of Fa) where Fa is a mechanistically or phenomenologically based failure criterion defined in a "critical element" (which defines our domain) for which rupture defines rupture of the system.[1]

To this conceptual foundation, we add the physical observations:

1. (instantaneous tangent stiffness) = $E(\psi)$, i.e., the change in any stiffness component can be defined in terms of ψ.

2. Remaining strength and life can be defined in terms of ψ.

Then we construct a fundamental evolution equation for strength with the following rationale.

1. Helmholtz free energy $f = f(\psi, \epsilon_{ij})$, becomes $f = U\psi$ as in classical Kachanov theory.[1] Hence:

$$df = \sigma_{ij} d\epsilon_{ij} - Q d\psi \quad where \quad Q = -U \qquad (1)$$

where Q is associated with the increment of <u>entropy</u> created by damage, and has the nature of energy released by degradation of the material state. Fa may not be the usual engineering failure criterion for these definitions.

2. The central issue is the kinetic equation. We assume that the kinetics are defined by a specific (damage accumulation) <u>process</u> for a specific fracture mode, and define rates for all such processes of interest.

We start with the most general, common kinetic equation (a power law), such that: $\frac{\delta \psi}{\delta \tau} = A \psi^n$, where τ is a normalized, generalized time variable (monotonic increasing), and n is a material constant.

Generally, $\tau = \frac{t}{\hat{t}}$ where $\hat{t}$ is the characteristic time constant for the process at hand. $\hat{t}$ could be a creep time constant, a creep rupture life, a fatigue life, etc., such that, for example, $\tau = \frac{n}{N}$.

Then:

$$\int_{\psi^o}^{\psi^i} d\psi = A \int_{0}^{t^i} (\psi(\tau))^n d\tau \qquad (2)$$

The left hand side is

$$\psi^i - \psi^o = 1 - Fa^i - 1 + Fa^o \qquad (3)$$

If we set A = 1, and approximate n = 1, then

$$Fa^i = Fa - \int_0^{\tau^i} (1 - Fa(\tau)) \, d\tau \qquad (4)$$

which is the instantaneous value of the failure function for this process.

Then we define a "residual strength" Fr such that

$$Fr = 1 - \int_0^{\tau} (1 - Fa(\tau)) \, d\tau \qquad (5)$$

where all quantities are defined in the critical element and for the process characterized by the characteristic time $\hat{\tau}$. A degenerate special case of equation 5 occurs for $\tau \to \dfrac{n}{N}$; $\hat{\tau} = N$ for which $Fa(\tau) \to \dfrac{S_a}{S_u}$, the ratio of unidirectional applied stress over unidirectional strength, whereupon $Fr = \dfrac{S_r}{S_u}$, and equation 5 integrates to

$$\frac{Sr}{Su} = 1 - (1 - \frac{Sa}{Su}) \frac{n}{N} \qquad (6)$$

a linear degradation of strength from initial to final value. (This form has been suggested by Eisseman.) Equation 6 is also an identity in the sense that it satisfies the end points of the residual strength curve, i.e., it is correct at the limits. In general, however:

$$Fa = Fa\left(\frac{\sigma_{ij}(n)}{X_{ij}(n)}\right) ; \quad N = N(n) \qquad (7)$$

or

$$Fa = Fa\left(\frac{\sigma_{ij}(t)}{X_{ij}(t)}\right) ; \quad N = N(t) \qquad (8)$$

If we claim that the rate equation is explicit in generalized time and recast the basic kinetic law to read $\dfrac{\delta \psi}{\delta \tau} = \psi \, \tau^{j-1}$, we obtain the final kinetic equation in the form

$$Fr = 1 - \int_0^{\tau} (1 - Fa(\tau)) \, (\tau)^{j-1} \, d\tau \qquad (9)$$

which is the form we will use in the present paper, and is essentially the form we first postulated in 1981.[2]

330

2. Applications of the Ideas

To apply this philosophy to the problem at hand, we start by identifying a specific failure mode in the laboratory for the material system and conditions of interest. Our experience has taught us the number of independent failure modes is comparatively small, and that their physical character (tensile failure of fibers, microbuckling, etc.) is relatively general with variations in only the degree of associated details in many cases.[3,4] Careful laboratory identification of a failure mode defines the boundary value problem we are to solve to estimate the fracture strength of the system, i.e., it defines the representative volume for the distributed damage state. At this point, two new concepts are introduced. First, the representative volume is defined by the (eminent) fracture state, rather than by the initial (pristine) condition as is typically done. And second, the representative volume is divided into a "critical element" and subcritical elements, as depicted in Fig. 1.

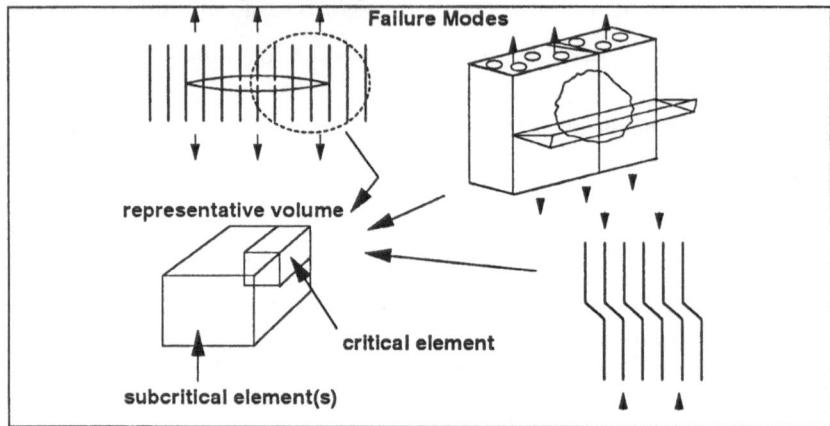

Figure 1 Schematic of critical element concept.

The critical element is the part of the representative volume that defines failure, i.e., when the critical element fails the system (global component) fails. The identity of this physical element must also be determined carefully from experiments. The remainder of the volume in the representative volume is "subcritical," in the sense that failures (such as delamination, cracking, chemical degradation, etc.) may occur without causing failure of the representative volume. In many ("fiber controlled") composites typical of current applications, the fibers are often part of the "critical element," while the matrix behavior is often a major factor in the behavior of the "subcritical elements."

As mentioned in our section on "fundamentals," the failure function, Fa, is defined in, and by, the critical element. The simplest example might be the ratio of the fiber normal stress to the unidirectional strength, if that is determined to be the controlling condition, although the function is generally more complex and should reflect all of the pertinent local information that controls the final fracture event under the applied conditions considered. Also, generally, the "state of the material" in our system is defined by the critical element, and is usually stated in terms of continuum constitutive equations. Damage that develops in the subcritical elements alters the state of stress in the representative volume, and can greatly influence the local conditions in which the critical element operates, but it does not change the state of the material, as we have defined it.

3. Micro-structure - Strength Interactions

The present approach is entirely micro-structure based; all calculations of remaining strength are made at the local level (the ply or fiber/matrix level). In the brief space available, we will discuss two aspects of the interaction, and then present one example of the "operation" of the models we have developed.

The X_{ij} that we defined in equation (8) are, in our model, micromechanical representations of the principle strengths of the fibrous material, i.e., tensile and compressive strength in the fiber direction and transverse to the fiber direction, and the in-plane shear strength. These are typical for laminar materials, but others may be more appropriate. We consider a tensile fiber direction strength model to illustrate our point.

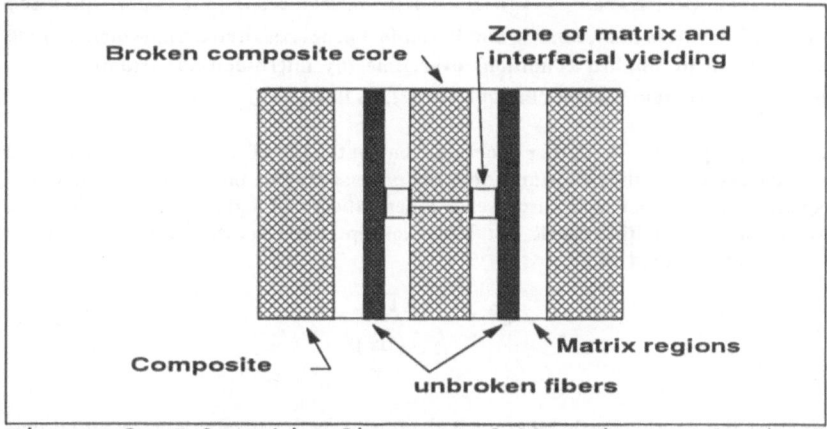

Figure 2   Schematic diagram of the micromechanical tensile strength model.

The authors have shown that it is possible to represent ply-level strength in terms of fiber/interphase/matrix level micromechanics[5,6]. For example, the tensile strength in the direction of the fibers has been represented by a nonlinear boundary value problem, based on a elastic-plastic shear lag formulation in which the number of broken fibers is calculated as the local failure process progresses. The model includes a region of broken fibers, a region of matrix plasticity (or fiber/matrix slipping), and an outer region of elastic behavior. Fig. 2 is an indication of the manner in which the boundary value problem is set. This type of micromechanical representation has several distinct advantages. Since the properties of the fibers, matrix, and interphase regions enter the analysis directly, damage and material state changes can be directly entered into a calculation of the strength to estimate the strength evolution as a function of the behavior of the constituents and the interphases between them. Hence, this presents the opportunity to represent the manner in which material systems come apart in terms of the parameters which define how they are put together. This opens an entirely new vista of opportunity to design materials for specific behavior on the basis of such formulations.

The second fundamental tie to microstructure is associated with the "characteristic time constant" for the process that controls failure in the critical element, as discussed in equation (2). As mentioned earlier, if the process that controls life in the critical element is unidirectional fiber-direction fatigue response at the ply level, then the "characteristic time" may be the unidirectional fatigue life, N, which may be related to applied conditions by an equation such as

$$\frac{\sigma_f}{X_t} = A - B \left(\log \left(N \right) \right)^p \tag{10}$$

where σ_f is the fiber-direction normal stress, X_t is the unidirectional tensile strength, and A,B, and p are material constants. Of course, this equation is applied to the critical elements, i.e., the fibers in this case, so that under long-term loading or environmental conditions both σ_f and X_t are functions of (generalized) time or cycles, reflecting such things as local stress redistribution (which changes σ_f) and changes in the ply strength (which changes X_t). Indeed, the micromechanics model we discussed above for X_t involves fiber, matrix, and interphase properties and arrangements which may be altered by applied conditions over time (by micro-defects, oxidation, etc.). Hence, N is a dynamic variable in equation (2), as is σ_f.

Of course, it is possible that other processes control the life of the critical element, and that a "characteristic time constant" for that process may be more appropriate to use in equation (2). Christensen discusses a kinetic theory of failure in which (using a generalization of Griffith crack instability concepts) the critical applied stress for instability is postulated as

$$\sigma^2 = \frac{2\Gamma}{h \, J(\frac{\rho}{c})} \tag{11}$$

in which ρ is a characteristic length, c is the flaw growth velocity, h is the initial flaw dimension, and J is the material creep function.[7,8] If one assumes a creep function of the form

$$J(t) = \frac{J_o \left(1 + \gamma_1 t^n \right)}{\left(1 + \gamma_2 t^n \right)} \tag{12}$$

and integrates equation 12 (taking only leading terms), one obtains the time to failure under constant (instantaneous) stress as

$$\hat{t} = \frac{1}{(\frac{1}{n} - 1) \, \theta^{\frac{2}{n}}} \tag{13}$$

where the applied stress is normalized by the instantaneous "fast fracture" (intrinsic) strength and n is a material constant. Christensen also discusses a rate equation and associated characteristic time for combined creep and chemical degradation.

Obviously, if the processes that control the failure of the critical element are clearly identified and carefully characterized in the laboratory, one can specify the form of $\hat{t}$. The important point in the present context is that the resulting expression depends on applied conditions and material constants; when applied to the critical element, those conditions and material parameters will, in general, be functions of time, cycles, or history of the component being modelled.

Before presenting an example calculation, we remind ourselves that rate equations (as functions of applied stress, temperature, and environmental conditions) must also be determined for such things as matrix cracking and creep or other stress relaxation mechanisms (to modify the modulus at the local level) as well as rate equations that represent such things as (diffusion or chemical rate controlled) strength degradation of constituents or interphase regions (to modify the inputs to our X_{ij} correctly).

4. An example

The present philosophy has been applied to several ceramic composite systems, including SiC/SiC, Aluminum based oxide/oxide systems, Nicalon / CAS, Nicalon / LAS, and numerous carbon reinforced polymer systems.[9-12] Only a few representative results will be presented here. A comprehensive description of the details of the simulations, especially the manner in which the rate equations for creep, creep rupture, oxidation, and matrix cracking were obtained, is beyond the scope of this paper, but is of critical importance to the success of the effort. Some of those details are not yet published, but will appear shortly. To that extent, these results should be viewed as illustrative and preliminary.

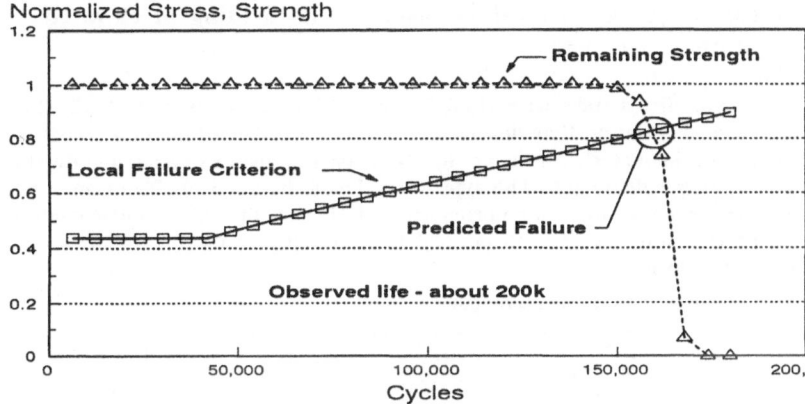

Figure 3 Predicted remaining strength and predicted and observed life for a Ni/CAS laminate at 1000 °C.

Figure 3 shows an example of the fatigue performance of a cross ply Nicalon / CAS laminate at about 1000 °C. (The data are normalized to protect the proprietary nature of the material.) The degradation of the laminate is a strong combination of micro-damage due to the cyclic mechanical loading, and oxidative degradation of the carbon coating on the fibers and the fibers themselves. The degradation of the fundamental strength of the material reduces the denominators of the failure criterion used, and the relaxation of stress in the matrix by matrix cracking increases the stress in the fibers and, thereby, increases the numerator of the failure function. The net result is a steady rise of the local failure criterion throughout the test, and an acceleration of the failure event (compared to the room temperature result, for example, which has a life of over one million cycles at this load level). When the accumulation of damage reaches a certain point, the remaining strength decreases quickly, and the specimen fails, at a predicted life which is quite similar to the observed value of about 200k cycles. This "sudden death" type of failure is quite common in these materials under these conditions.

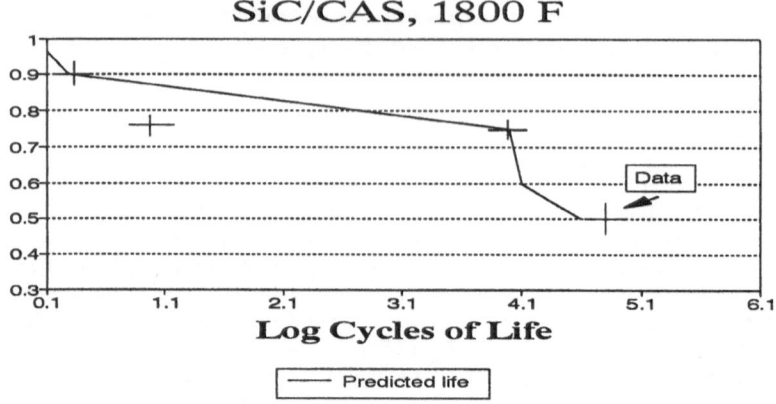

Figure 4 Predicted and observed creep-fatigue lives of Ni/CAS specimens subjected to tension-tension cyclic loading at elevated temperature.

Figure 4 shows predicted and observed life for several loading levels at 1000 °C. It is seen that at low stress levels, time-dependent oxidation is predicted (and observed) to control the life of the material, while at high cyclic load levels the degradation due to mechanical loading is dominant. This type of combined (interactive) effect cannot be gotten from linear models nor from phenomenological curve fits. Mechanistic models based on the constituent-level behavior and physics, such as the present one, are necessary for this task.

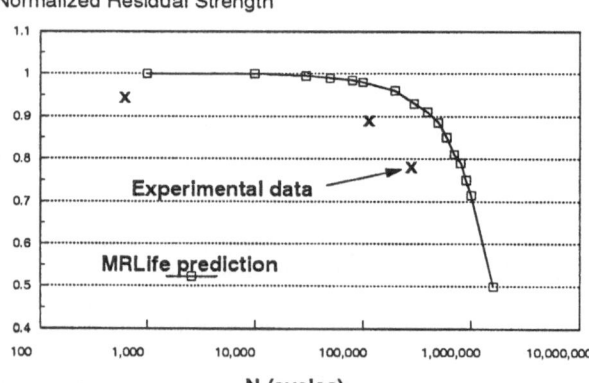

Figure 5 Remaining strength predictions and observations for a Quasi-isotropic center notched coupon fatigue loaded with R=-1.

Numerous predictions for polymer composite have also been successfully made. Figure 5 shows an example of predictions of remaining strength as a function of cycles for a fully reversed loading of an AS-4/PEEK(APC-2) center-notched coupon, compared to experimental data. The predictions were made based on data for unidirectional fatigue of that material under tensile loading, only. Rate equations for matrix cracking were estimated on the basis of past experience. The predictions were made with MRLife8 for this complex laminate, geometry, and loading. The agreement is seen to be within about 10 percent, which is typical of our experience for the prediction of remaining strength.

5. Closure

The present paper suggests that the durability and damage tolerance of composite systems can be modeled in terms of micromechanical representations of strength which make it possible to represent changes in global composite properties and performance in terms of local changes in the constituents, the interfaces and interphases between constituents, and the local geometry (as influenced by defect development). In this way, micromechanical modeling can be used to represent microstructure-property interactions and evolution, as the properties of the constituents and local geometry change. Since these constituent changes or local geometry changes can be measured directly and independently, the approach provides a sound and systematic method of representing, and in some cases predicting, the evolution of strength and life at the global level. Examples of applications of the approach have also been shown, with good success in a performance simulation of the remaining strength and life of composite laminates (even at elevated temperatures) with limited input data. This approach holds the promise of providing the community with a systematic method of relating the details of how composite systems are made to the manner in which they perform under long-term exposure to complex combinations of mechanical, thermal, and other environmental combinations. Although the approach is new in many respects, many applications over the last ten years have provided a sound basis for progress, and a good motivation for continued effort in this direction. Continuing work is especially focused on situations in which complex combinations of changes in material state and stress state dominate the remaining strength and life of complex material systems, especially high temperature polymer and ceramic based composites.

6. Acknowledgements

The authors gratefully acknowledge the support of this research effort by the National Science Foundation - Science and Technology Center for High Performance Polymeric Adhesives and Composites at Virginia Tech, contract no. DMR-8809714, the Virginia Institute for Material Systems, the Babcock and Wilcox Corporation, and the support of NASA Langley Research Center under contract no. NAG1-1-1608.

7. References

[1] Kachonov, L.M., "Introduction to Continuum Damage Mechanics," Martinus Nijhoff Pub., Boston, 1986.
[2] Reifsnider, K.L., and Stinchcomb, W.W., "Cumulative Damage Model for Advanced Composite Materials," Phase II Final Report to Air Force Materials Laboratory, Wright Patterson AFB, October 1983 (see also General Dynamics Report Number FZM-7149, contract no. F33615-81-C-5049, Fort Worth, TX, 1983)
[3] Reifsnider, K.L., Editor, 1982. Damage in Composite Materials, Philadelphia: ASTM STP 775, American Society for Testing and Materials.

336

[4] Reifsnider, K.L., Editor, 1991. <u>Fatigue of Composite Materials</u>, London: Elsevier Science Publishers.

[5] Gao, Z. and Reifsnider, K.L., 1991. "Micromechanics of Tensile Strength in Composite Systems," in <u>Proc. Fourth Symp. on Composite Materials: Fatigue and Fracture</u>, Indianapolis, IN: ASTM, (in press).

[6] Xu, Y. and Reifsnider, K.L., 1992. "Micromechanical Modeling of Composite Compressive Strength," submitted to <u>J. Composite Materials.</u>

[7] Christensen, R.M., "Lifetime Predictions for Polymers and Composites under Constant Load," Journal of Rheology, 25(5), (1981) pp. 517-528.

[8] Christensen, R.M., 1979. <u>Mechanics of Composite Materials</u>, New York: John Wiley & Sons.

[9] Reifsnider K.L. and Stinchcomb, W.W., 1986. "A Critical Element Model of the Residual Strength and Life of Fatigue-loaded Composite Coupons, in <u>Composite Materials:Fatigue and Fracture</u>, ASTM STP 907, H.T. Hahn, Ed., Philadelphia: Am. Soc. for Testing & Materials, pp.298-313.

[10] Reifsnider, K.L., 1991. "Performance Simulation of Polymer-based Composite Systems, in <u>Durability of Polymer-Based Composite Systems for Structural Applications</u>, A.H. Cardon & G. Verchery, Eds., New York: Elsevier Applied Science, pp.3-26.

[11] Reifsnider, K.L., 1992. "Use of Mechanistic Life Prediction Methods for the Design of Damage Tolerant Composite Material Systems," in <u>ASTM STP 1157</u>, M.R. Mitchell & O. Buck Eds., Philadelphia: American Society for Testing and Materials, pp.205-223.

[12] Reifsnider K.L. and Gao, Z., 1991. "Micromechanical Concepts for the Estimation of Property Evolution and Remaining Life," in <u>Proc. Intl. Conf. on Spacecraft Structures and Mechanical Testing</u>, Noordwijk, the Netherlands, 24-26 April 1991: (ESA SP-321, Oct. 1991) pp.653-657.

OFF-AXIS FATIGUE LIFE PREDICTION USING MICROSTRESS ANALYSIS

G. SHEN*, G. GLINKA* and A. PLUMTREE*
* Department of Mechanical Engineering, University of Waterloo
Waterloo, Ontario, Canada

1. Introduction

Microscopic observations have revealed [1] that fatigue cracks in unidirectional off-axis polymer-based composites initiate and grow within the matrix along a plane parallel to the fibre direction. The fatigue process is controlled by the microstresses on these critical planes as well as the properties of the matrix. Also, due to the fibre-matrix interaction a complex stress state occurs in the matrix resulting in a multiaxial stress state on the critical planes. A micromechanical stress analysis is conducted in the present study on unidirectional graphite/epoxy, boron/aluminum and glass/epoxy composites to calculate the local stress/strain fields which serve as a basis for determining a multiaxial stress/strain fatigue parameter used to correlate the fatigue life data and to derive a generalized fatigue stress/strain-life curve. This curve is then used to predict the fatigue lives of the particular composite tested under a variety of loading conditions.

2. Micromechanical Analysis

The present paper includes the micromechanics of deformation in long fibre composites loaded at an angle to the fibre and involves the finite element analysis of fibres arranged in a square unit cell. The fibres may be packed in an edge (0° square) or diagonal (45° square) sequence, as shown in Figs. 1 and 2. Considering transverse monotonic loading, they represent the stiffest and most compliant uniaxial tensile directions respectively.

It has been shown that the elastic response of a composite loaded off-axially is sensitive to the spatial distribution of large concentrations of long fibres especially when the matrix is ductile [2,3]. Under such conditions the monotonic flow stress at a fixed plastic transverse strain was found to be the highest for edge packing (0° square) and lowest for diagonal packing (45° square), with hexagonal packing and random distributions lying between the two. Hence, the edge and diagonal arrangements may be regarded as representing upper and lower bounds.

The first step in the microstress analysis establishes the relationship between the macrostresses S_{ij} in the lamina and the point to point microstresses σ_{ij} in the matrix and fibre. The macrostresses are usually used as input data in further analyses since they are linearly related to the applied load. Because of the periodic structure, adjacent basic blocks containing one fibre match exactly without any overlapping. Therefore, rectangular blocks must remain rectangular under deformation. This constraint allows the basic deformation to be modelled by only five deformation states [4].

Finite element analysis based on either square packing arrangement is carried out for each deformation state. Superposition of the calculated stress components allows the corresponding stress tensors S_{ij} and σ_{ij} to be determined.

337

R. Pyrz (ed.), IUTAM Symposium on Microstructure-Property Interactions in Composite Materials, 337–347.
© 1995 *Kluwer Academic Publishers.*

338

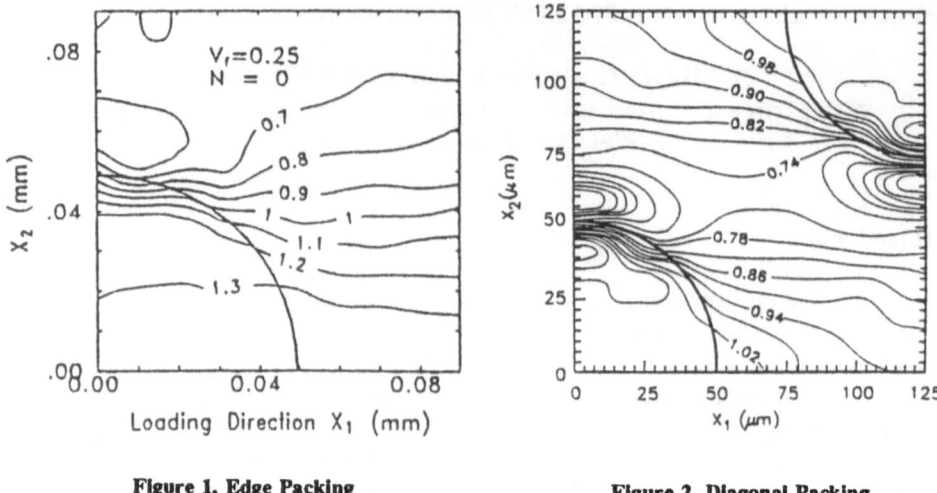

Figure 1, Edge Packing   Figure 2, Diagonal Packing

Figures 1 and 2. Normalized max. principal stress contours for 6061 Al/0.25B under transverse loading in X_1 direction.

Following this procedure, the microstress fields for the two square unit cells were calculated for 0.25 volume fraction boron fibres in 6061 aluminum alloy loaded in the transverse direction. The normalized maximum principal stress contours for edge packing (0° square) are shown in Fig. 1. In this figure, X_1 represents the loading direction. The highest stress concentration of 1.35 is located in the matrix where $|X_2|$ tends to zero. The maximum principal stress decreases with increase in the value of $|X_2|$. It can be seen that the stress falls below the nominal value when $|X_2| > r_f$. The microstresses σ_{ij} were normalized with respect to the uniformly applied macrostress S_{11}. Fig. 2 shows the normalized principal stress contours for diagonal packing (45° square). In this case, the highest stress concentration is 1.04.

Using the maximum and minimum values of the microstress range, the stress fields at different stages of the fatigue life may be determined. Because of the fatigue process, the coefficients in the microstress analysis vary due to the application of the constitutive equation containing a damage variable which reflects the deterioration of material properties with the number of applied cycles. Accordingly, the stress state must be evaluated for different stages of fatigue life. Considering the microstress distribution for a 0° square unit cell representing the boron reinforced composite, the stress concentration of 1.35 adjacent to the fibre is established immediately on the initial load cycle. During fatigue, nucleation and coalescence of microvoids occurring in the matrix greatly reduce the local stiffness resulting in a more uniform stress distribution. The stress concentration decreases to 1.24 after 5×10^5 cycles ($N=0.67N_f$). On the other hand, because of the high strength of boron, damage in the fibre is relatively small and its stiffness remains essentially unchanged.

3. Critical Plane Model

Critical plane models proposed by Findley [5], McDiarmid [6], Brown and Miller [7] and Lohr and Ellison [8], as well as others, have been based upon a physical interpretation of the fatigue process whereby cracks were observed to form and grow on particular planes, known as critical planes. Also the strain-life approach and uniaxial strain fatigue failure criteria were found to be relatively successful in predicting the fatigue life of notched components in plane stress or plane strain. Therefore, most of the proposed critical plane criteria are given in the form of expressions involving a combination of shear and tensile strain amplitudes. One of the parameters which is often used in correlating fatigue data obtained under various multiaxial stress states is that proposed by Brown and Miller [7], given by:

$$\tilde{\gamma}^* = \tilde{\gamma}_{13} + k\tilde{\varepsilon}_{11} \tag{1}$$

This parameter $\tilde{\gamma}^*$ has also been found to be successful in correlating fatigue data of composite materials tested under multiaxial cyclic loading [9] and is related to the fatigue life (N_f) by:

$$\log \tilde{\gamma}^* = A + m \log 2N_f \tag{2}$$

where coefficient A and exponent m are determined from experimental data.

Equation (1) is often criticized for the lack of formal correctness from the continuum mechanics viewpoint. This is due to the difficulties concerning the interpretation of Eqn. (1), being an algebraic sum of the normal and shear strain amplitudes acting in the critical plane. Also, these amplitudes are weighted with the help of an experimental parameter, k, claimed to be constant for a given material. However, Fatemi and Kurath [10] have shown that parameter k varies with fatigue life. In addition, it has been suggested [11] that in the case of multiaxial fatigue, life estimates based on strain amplitudes are unsatisfactory since the stresses also contribute to cyclic damage. These are highly dependent upon the strain path, especially for non-proportional loading.

To account for the effects of stress and strain, a strain energy density relation for the critical plane, which is analogous to Eqn. (1), can be formulated:

$$W^* = \tilde{\gamma}_{13} \, \tilde{\sigma}_{13} + \tilde{\varepsilon}_{11} \, \tilde{\sigma}_{11} \tag{3}$$

The parameter W^* represents the sum of the strain energy density and the complementary strain energy density contributed by the stress and strain components on the critical plane. It can be shown that for a non-zero shear stress range, σ_{13} Eqn. (3) can be re-arranged in a similar form to that of Eqn. (1).

$$W^*/\tilde{\sigma}_{13} = \tilde{\gamma}_{13} + (\tilde{\sigma}_{11}/\tilde{\sigma}_{13}) \, \tilde{\varepsilon}_{11} \tag{4}$$

The ratio of $\tilde{\sigma}_{11}/\tilde{\sigma}_{13}$, corresponding to parameter k in Eqn. (1), is not constant. Also, it is not clear how Eqn. (4) should be interpreted in the case of $\tilde{\sigma}_{13} \to 0$. Therefore, the energy relation given by Eqn. (3) seems to be more appropriate as it is non-singular for both uniaxial and multiaxial loading. In addition, it is acceptable from the point of view of the formalism of continuum mechanics because energy components $\tilde{\sigma}_{13} \, \tilde{\gamma}_{13}$ and $\tilde{\sigma}_{11} \, \tilde{\varepsilon}_{11}$

are scalars and can be added algebraically. The novelty of the parameter W^* lies in the fact that it represents a fraction of the overall strain energy density contributed only by the stresses and strains on the critical plane. In other words, it can be interpreted as the flux of strain energy density associated with the direction defined by the normal to the critical plane.

4. Validation - Smooth Specimens

4.1 Boron-Aluminum

The fatigue data of a 6061-0 aluminum alloy reinforced with 0.25 volume fraction 100 μm diameter boron fibres were adapted from work carried out by Toth [12]. The elastic moduli of the fibre and matrix were 410 GPa and 69 GPa and the Poisson's ratios were 0.2 and 0.345 respectively. The specimens were 100 mm long, 6.35 mm wide and 1.65 mm thick with a gauge length of 25.4 mm. All the specimens were tested under constant amplitude loading with a stress ratio of R = 0.2.

From the fracture surface photograph taken by Toth [12] of a 45° unidirectional plate, it was apparent that the fracture occurred in the matrix parallel to the fibre axis. Therefore, the present matrix-controlled fatigue model could be used for these specimens.

Two sets of data were obtained for unidirectional specimens tested under axial tension-tension loading with fibre to load axis angles (α) of 20° and 45°. The third set of data was obtained from cross-ply specimens tested under cyclic tension-tension with a fibre/load angle of ±45°. The original data (S_{max}-N_f) of Toth [12] showed that the fatigue life was strongly influenced by the load/fibre angle α.

The local shear strain amplitude γ_{13} and the normal strain amplitude ε_{11} at the highest stress concentration site were calculated using linear elastic analysis. The calculation was carried out for each applied load. Using γ_{13} and ε_{11}, the fatigue control parameter γ^* was determined by Eqn. (1) and then correlated with the fatigue lives by Eqn. (2). The best fit was found using k=0 since the same fatigue lives were observed for the ±45° laminate and 45° laminae, indicating that the normal strain had a negligible effect. From laminate stress analysis, it is known that these two lay-ups have the same shear strain and different normal strain. Consequently only k=0 can result in the same fatigue control parameter γ^* for the same N_f.

The experimental data and the correlation are shown in Fig. 3. It is seen that the data for both square packing sequences follow Eqn. (2) and have the same slope. For edge packing A = -2.76, m = -0.017 and for diagonal packing A = -2.86, m = -0.017.

Nakamura and Suresh [3] have shown that monotonic transverse tensile simulations for 0.46 volume fraction boron fibres randomly distributed in a 6061 aluminum alloy matrix give results close to those for edge packing rather than diagonal packing. Since it is apparent from the present work that the predicted and experimental results are in very good agreement regardless of the packing sequence, then the edge packing arrangement will be employed for the remainder of this work.

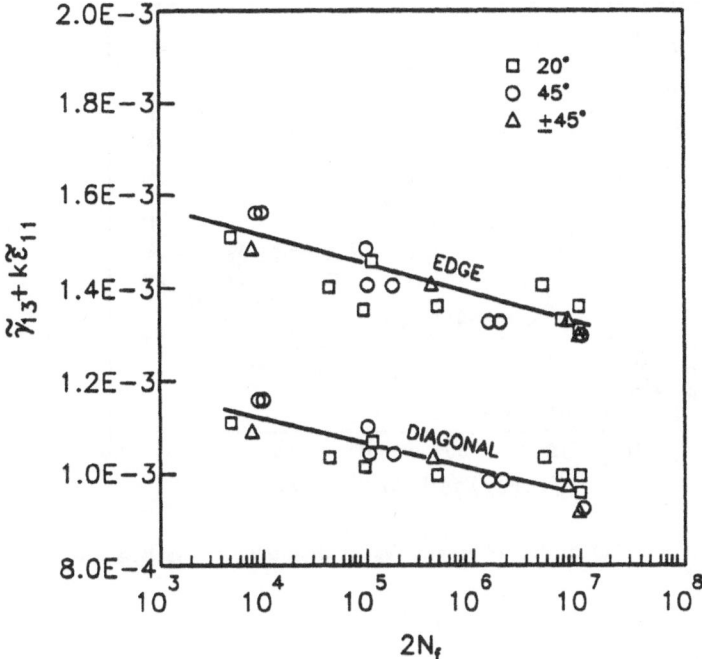

Figure 3. Fatigue lives of 6061 Al/0.25β for edge and diagonal packing as a function of the multiaxial parameter. Angles between fibre direction and loading axis are indicated.

4.2 Graphite/Epoxy

Experimental data[1] for an epoxy based composite reinforced with 0.66 volume fraction graphite fibres were also analyzed. The results obtained for the load/fibre angle of 10° were used to derive the generalized multiaxial strain-life relation in the form of Eqn.(2). The best fit was found when k = 2.4 and A = -1.198, m = -0.069. This equation was then used to predict the fatigue lives of specimens tested at angles of 20°, 30°, 45°, and 60°. The parameter ($\tilde{\gamma}_{13}$ + $k\tilde{\varepsilon}_{11}$) satisfactorily described all the experimental data and allowed the effect of the load/fibre angle to be predicted.

When the loading angle is 90° the shear strain amplitude $\tilde{\gamma}_{13}$ tends to zero and the shear strain based critical plane model overestimates the fatigue life[13]. With this in mind, the fatigue life curves for loading angles of 60° to 90° were estimated by linearly extrapolating the data for the angles of 45° to 60° in accordance with previous experimental work[2]. Also, since the fibre properties dominate the fatigue life when α = 0, the life can be provided by axial fatigue tests. Combining the results of zero angle tests with those predicted for the 10° loading angle a full range of fatigue life curves from 0 to 90° loading angles can be obtained. These are given in Fig. 4.

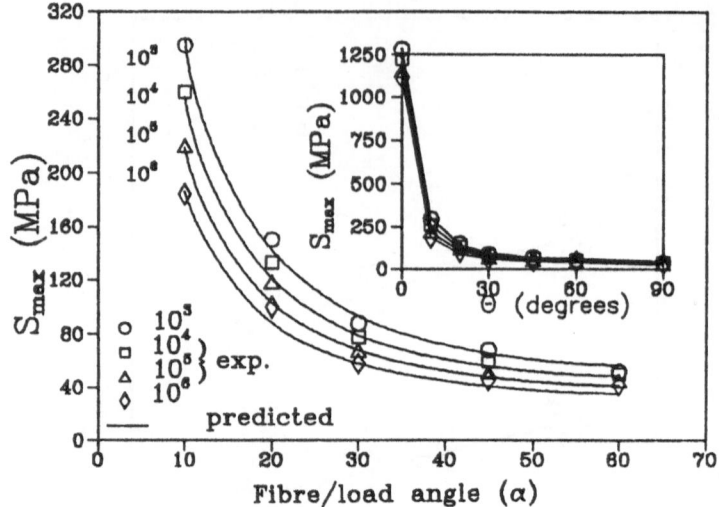

Figure 4. Effect of load/fibre on fatigue lives of graphite/epoxy composite; comparison of experimental [1] and predicted results.

4.3 Glass/Epoxy

Experimental data (R = 0.1) obtained for an epoxy based composite reinforced with 0.6 volume fraction E-glass fibres was taken from reference [14]. There were six sets of data corresponding to different angles of the load to fibre axis, namely 5°, 10°, 15°, 20°, 30° and 60°. The set of data obtained for the load/fibre angle of 10° was used to derive the generalized multiaxial strain-life relation in the form given by Eqn. (2). The best fit was found when k=1 and A = -1.395, m = -0.101.

This relation was then used to predict the fatigue lives of specimens tested at the angles of 5°, 15°, 20°, 30°, and 60°. Comparison of the predicted and experimental results indicated that the parameter $(\tilde{\gamma}_{13} + k\tilde{\varepsilon}_{11})$ accurately described all the experimental data and that the effect of the load/fibre angle can be predicted. When the data were plotted in the form of $lg\tilde{\gamma}^* - lgN_f$ all the experimental results collapsed into one master curve.

5. Validation - Notched Specimens

5.1 Prediction of Fatigue Crack Initiation Sites

The validity of $\tilde{\gamma}^*$ as a fatigue parameter can be illustrated by its ability to predict the location of crack initiation sites in a notched composite plane subjected to constant amplitude cyclic tensile loading. The experimental results were taken from work

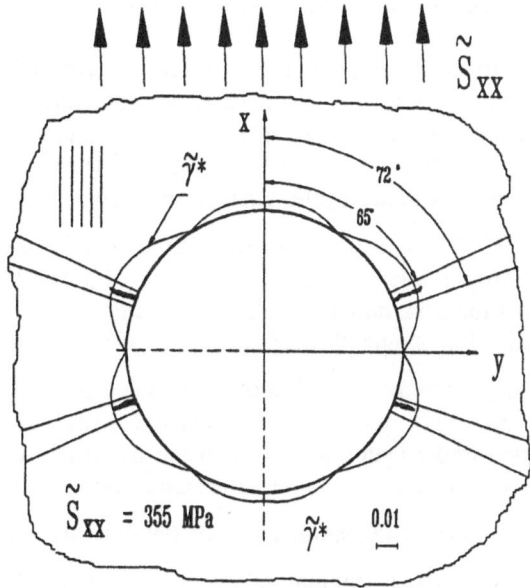

Figure 5. Distribution of fatigue parameter $\tilde{\gamma}^*$ around hole in unidirectional SCS-6/Ti 15-3 composite plate.

of Newaz and Majumdar[15] who conducted fatigue tests $(R = 0.1, \tilde{S}_{xx} = 355MPa)$ on unidirectional eight-ply silicon carbide fibre reinforced titanium (SCS-6/Ti 15-3) plates with central circular holes 9.53 mm in diameter. The specimens were of rectangular shape 152.4 mm long by 19 mm wide and 1.55 mm thick. The angle between the loading direction and the fibre axis was zero and after cycling it was observed that four cracks initiated at the hole circumference, at angles $\theta = \pm(65 \rightarrow 72°)$ and $\theta = \pm(155 \rightarrow 162°)$. These cracks were symmetric with respect to the x and y axes, as shown in Fig. 5.

The circumferential stress ($S_{\theta\theta}$) around the edge of the circular hole was calculated using the analyses of Shen et al [9]. The highest stress concentration ($K_t = 3.42$) occurred at $\theta = \pm90°$ which did not correspond to the fatigue crack initiation sites. It is apparent that the normal stress $S_{\theta\theta}$ was not the fatigue controlling parameter. Subsequently, the macrostress components $S_{\theta\theta}$ around the edge of the hole were used as the input data to calculate the microstress components σ_{ij}.

Substitution of the calculated shear $\tilde{\gamma}_{13}$ and normal $\tilde{\varepsilon}_{11}$ strain amplitudes into

Eqn. (1), enabled the fatigue parameter $\tilde{\gamma}^*$ around the edge of the hole to be determined.

The distribution of parameter $\tilde{\gamma}^*$, calculated for a macrostress amplitude of

$\tilde{S}_{xx} = 355MPa$, is shown in Fig. 5. The locations of the maximum $\tilde{\gamma}^*$ (1.3%) coincide with the observed crack initiation sites, indicating that the parameter γ^* can be used to determine the location of fatigue crack initiation sites in notched composite components.

5.2 Prediction of Fatigue Life Under Cyclic Tension and Torsion

It is now possible to consider predicting the fatigue lives of notched tubes subjected to a variety of loading modes such as tension, torsion, and a combination of both using material data obtained from simple experiments associated with one loading mode, i.e. cyclic tension. Therefore, the first step was to determine whether the experimental fatigue data obtained under cyclic tension and torsion may be unified by using the fatigue parameter $\tilde{\gamma}^*$. The tension and torsion fatigue data for a graphite/epoxy ([±45°], lay up) tube containing a hole were taken from Ref. [16]. The specimens were in the form of a laminate tube 254 mm long, 25.4 mm diameter and 0.6 mm wall thickness. All tubes contained a single through thickness circular hole of 4.8 mm diameter. The distribution of the parameter $\tilde{\gamma}^*$ around the edge of the hole in the -45° laminae under cyclic tension with the maximum tension stress of 108MPa was determined and also obtained under cyclic torsion with the maximum torsion stress of 110MPa. Both series of tests were carried out under the same stress ratio of R = 0.1. The distribution of parameter $\tilde{\gamma}^*$ in the +45° laminae can be obtained by a 180° rotation of that determined for the -45° ply. The maximum values of parameter $\tilde{\gamma}^*$ were found at angles $\theta = \pm90°$ in the -45° laminae subjected to cyclic tensile loading. However, in the case of torsion loading four maxima located at angles $\theta = +123°, +147°, -33°$ and $-57°$ were found in the -45° laminae. When the maximum values of parameter $\tilde{\gamma}^*$ at these locations are plotted against the experimental fatigue lives it is apparent that the $\tilde{\gamma}^*$ parameter is quite capable of normalizing the fatigue lives obtained under entirely different loading modes allowing the relationship to be expressed by Eqn. (2). The corresponding coefficient A(=1.216) and exponent m (= - 0.058) were determined by the least squares method using the experimental tensile and torsion data [16]. They agree very closely with those obtained under combined cyclic axial tension and internal pressure given in Ref. [16] which result in the expression

$$\tilde{\gamma}^* = 1.241(2N_f)^{-0.053} \tag{5}$$

Eqn. (5) was subsequently used to predict fatigue lives under combined cyclic tension and torsion loading. For these conditions, the $\tilde{\gamma}^*$ parameter around the circumference of the hole was first calculated to determine the locations of the fatigue crack initiation sites due to the different tension to torsion load ratios. The initiation sites were assumed to coincide with the sites of maximum $\tilde{\gamma}^*$.

The generalized $\tilde{\gamma}^* - N_f$ curve (Eqn. (8)) was used to determine the fatigue lives of three sets of tubes tested under combined cyclic tension and torsion with tension: torsion load ratios of 1:05, 1:1 and 1:2 [16]. The fatigue lives were predicted using Eqn. (5) and the maximum values of $\tilde{\gamma}^*$ found on the notch circumference. Comparison of the experimental and calculated fatigue lives is given in Fig. 6. Good agreement between the predicted and experimental fatigue lives is apparent for the whole range of load combinations.

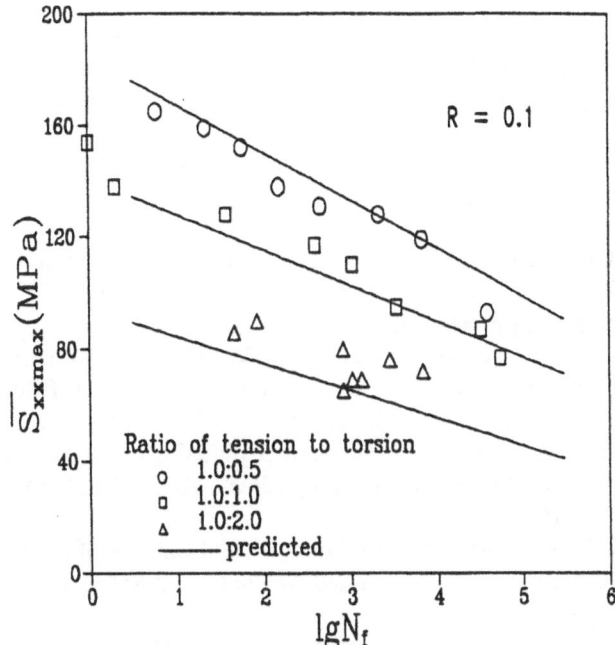

Figure 6. Effect of combined tension and torsion on fatigue life of notched graphite/epoxy tubes.

6. Conclusions

It has been shown that microstress/strain analysis in conjunction with the multiaxial fatigue parameter can be applied to successfully predict the off-axis fatigue lives of long fibre composites. The multiaxial fatigue parameter is expressed as a combination of the shear and normal strains on the critical plane parallel to the fibres. This parameter enables a generalized strain-life relationship to be determined using limited experimental data. Once the generalized relationship is known the life of the composite cycled under any combination of loads and load-fibre angle may be predicted.

The fatigue process of the composites considered in this work was controlled mainly by the local shear strain on the critical plane and the contribution of the normal strain was secondary. By applying the multiaxial fatigue parameter very good agreement was achieved between the experimental and predicted lives of smooth specimens of graphite/epoxy, glass/epoxy and boron/aluminum.

Considering the fatigue life of a notched silicon carbide/titanium composite the location of the crack initiation sites was predicted using the local fatigue parameter incorporating the multiaxial alternating strains on the critical plane parallel to the fibre axis. The application of this parameter also enables the generalized strain-life relationship to be determined from a limited amount of experimental data, which can then be used to predict the life of notched graphite/epoxy composite components subjected to complex multiaxial loading conditions.

346

7. Acknowledgements

The authors would like to thank the Natural Science and Engineering Research Council for financial support and Marlene Dolson for typing the manuscript.

8. References

[1]   J. Awerbuch and H.T. Hahn, "Off-Axis Fatigue of Graphite/Epoxy Composite", Fatigue of Fibrous Composite Materials, ed. J.B. Wheeler, STP 723, ASTM, Philadelphia, 1981, pp. 243-273.

[2]   Brokenbrough, J.R., Suresh, S. and Wienecke, H.A., "Deformation of Metal-Matrix Composites with Continuous Fibers: Geometrical Effects of Fiber Distribution and Shape", Acta Metall. Mater. vol. 39, 1991, pp. 735-754.

[3]   Nakamura, T. and Suresh, S., "Effects of Residual Stresses and Fiber Packing on Deformation of Metal-Matrix Composites", Acta Metall. Mater., vol. 41, 1993, pp. 1665-1681.

[4]   T.H. Lin, D. Salinas and Y.M. Ito, "Initial Yield Surface of a Unidirectionally Reinforced Composite", J. of Applied Mechanics, ASME, 1972, vol. 39, pp. 321-326.

[5]   Findley, W.M., "Theory for the Effect of Mean Stress on Fatigue of Metals under Combined Torsion and Axial Load or Bending", Journal of Engineering for Industry, vol. 81, No. 4, 1959, pp. 301-306.

[6]   McDiarmid, D.L., "A General Criterion of Fatigue Failure under Multiaxial Stress", in Proceedings, 2nd Int. Conf. on Pressure Vessel Technology, vol. II-61, 1973, pp. 851-862.

[7]   Brown, M.W. and Miller, K.J., "A Theory for Fatigue under Multiaxial Stress-Strain Conditions", in Proceedings of the Institution of Mechanical Engineers, vol. 187, 1978, pp. 745-755.

[8]   Lohr, R.D. and Ellison, E.G., "A Simple Theory for Low Cycle Multiaxial Fatigue", Fatigue of Engineering Materials and Structures, vol. 3, No. 1, 1980, pp. 1-17.

[9]   Shen, G., Glinka, G. and Plumtree, A., "Fatigue Life Prediction of Notched Composite Components", Fatigue & Fracture of Engineering Materials & Structures, vo. 17, pp. 77-91, 1994.

[10]   Fatemi, A. and Kurath, P., "Multiaxial Fatigue Life Predictions under the Influence of Mean-Stresses", Journal of Engineering Materials and Technology", ASME, vol. 110, No. 3, 1988, pp. 380-388.

[11]   Bannantine, J.A. and Socie, D.F., "A Multiaxial Fatigue Life Estimation Technique", ASTM, STP 1122, eds. Mitchell, M.R. and Landgraf, R.W., American Society for Testing and Materials, Philadelphia, 1992, pp. 249-275.

[12]   Toth, I.J., "An Exploratory Investigation of the Time Dependent Mechanical Behaviour of Composite Materials", Technical Report AFML-TR-69-9, Air Force Materials Laboratory, Ohio, 1969.

[13]   Socie, D.F., "Multiaxial Fatigue Damage Assessment", Low Cycle Fatigue and Elasto-Plastic Behaviour of Materials, ed. K.T. Rie, Elsevier Applied Science, London and New York, 1987, pp. 465-472.

[14]   Hashin, Z., and Rotem, A., "A Fatigue Failure Criterion for Fibre Reinforced Materials", J. Composite Materials, vol. 7, 1973, pp. 448-462.

[15]   Newaz, G.M. and Majumdar, B.S., "Crack Initiation Around Holes in a Unidirectional MMC Under Fatigue Loading". Engineering Fracture Mechanics, vol. 42, pp. 697-711.

[16]   Francis, D.H., Walrath, D.E., Sims, D.F. and Weed, D.N., "Biaxial Fatigue Loading of Notched Composites." NASA Report No. Cr-145198, 1977.

MICROSTRUCTURE CHARACTERIZATION AND FE – MODELING OF PLASTIC FLOW IN A DUPLEX STEEL

T. SIEGMUND AND F.D. FISCHER
Christian Doppler Laboratory for Micromechanics
of Materials and Institute for Mechanics
Montanuniversität Leoben, Austria

AND

E.A. WERNER
Institute for Metal Physics
Montanuniversität Leoben, Austria

1. Introduction

To calculate the plastic flow properties of two-phase materials, a large number of micromechanical models based on continuum mechanics principles has been established throughout the past years. Self consistent methods [1-3], rules of mixture (ROM) [4-9], non-linear generalizations of the classical Hashin-Shtrikman bounds [10-12] and the finite-element (FE) method [6, 13-19] have been used for this purpose. All these approaches aim to predict the properties of two-phase materials from the given properties of their constituent phases. Quantitative information from the microstructure, however, is incorporated seldomly in the models, with the exception of those with specific reference to matrix-inclusion-type microstructures. Up to now, only little attention was paid to microstructures different from these special types due to the lack of stereological parameters to characterize general two-phase microstructures quantitatively, and so even finite element based micromechanical modeling methods have been restricted to more or less special two-phase microstructures [21]. It is the aim of this report to present a micromechanical model applicable to general two-phase microstructures. In this course, a stereological parameter is introduced, which fully quantifies the geometrical continuity (GC) of the constituent phases. GC is a quantity of eminent importance, when dealing with various physical prop-

349

R. Pyrz (ed.), IUTAM Symposium on Microstructure-Property Interactions in Composite Materials, 349–360.
© 1995 *Kluwer Academic Publishers.*

erties of coarse grained two-phase alloys possessing interpenetrating phases [20].

The proposed micromechanical model is used to simulate and to analyze uniaxial tensile tests with respect to the overall flow stress, as well as the local deformation behavior within the two phases. The numerical results are compared with the predictions from a new rule of mixture-type equation that combines the quantitative characterization of the microstructure with the results from the non-linear generalization of the Hashin-Shtrikman bounds.

2. The Micromechanical Model

The essence of the micromechanical model is the unit cell approach combined with finite element calculations. The model, which is similar to the one used in [16], has been described in detail in [20] and will be reviewed briefly here. During the production of the investigated duplex stainless steel, hot forging of the material results in a pronounced uniaxially anisotropic microstructure formed of elongated domains of the two phases ferrite and austenite arranged parallel to the axis of the forged rod. Therefore, the real microstructure can be approximated by a model microstructure that is generated as periodic arrangement of identical unit cells on a rectangular mesh. The unit cell consists of hexagonal prisms resembling the phase domains. One octant of a unit cell is given in fig. 1. The whole unit cell (dimensions $2a$, $2b$, $2c$) is obtained by starting from this octant, and repeatedly reflecting it about the coordinate planes. This geometry is implemented into the FE code ABAQUS using generalized plane strain elements. Properties calculated for this unit cell are characteristic for the overall microstructure. For the simulation of uniaxial tensile tests, boundary conditions are imposed on the unit cell octant so that the deformed microstructure can be constructed in the same way as the undeformed one. This is achieved by forcing the coordinate planes to remain plane and fixed in their initial positions. All other bounding planes remain plane and parallel to their initial setup. Uniaxial loading is achieved by prescribing a displacement in the e.g. x_1-direction, only. The unknowns to be determined from the FE calculations are the principle normal stress σ_1 acting on the planes normal to x_1, and the displacements U_2 and U_3.

The two phases ferrite (α) and austenite (γ) of the duplex steel are assumed as isotropic elastic-perfectly plastic, with yield stresses σ_y^α and σ_y^γ, respectively. The von Mises yield criterion is used to calculate the onset of plastic deformation. In the calculations, the flow stress ratio $\sigma_y^\gamma/\sigma_y^\alpha$ is varied between 1 and 10. For the isotropic elastic properties the values $E^\alpha = E^\gamma = 200\,000\text{MPa}$ and $\nu^\alpha = \nu^\gamma = 0.3$ are used.

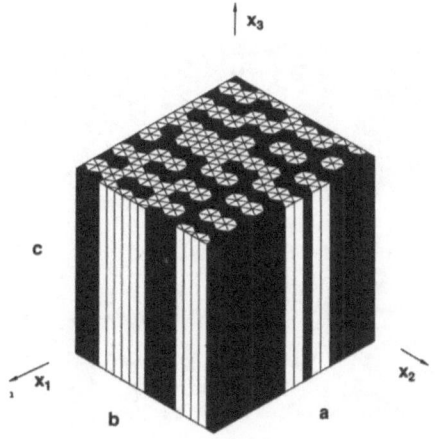

Figure 1. One octant of the unit cell used in the FE calculations.

3. Microstructures

In a number of previous investigations it has been reported that the GC of the phases in the two-phase microstructure significantly influences the properties of the composite [1, 4, 6, 18, 22-26]. In coarse two-phase alloys both phases can be inclusion and/or matrix phase. A few examples of such microstructures are depicted in fig. 2. These plane two-phase microstructures are generated by random placement of the two phases ferrite (black) and austenite (white) on an array of 108 hexagons (resembling the grains) per unit cell octant according to a prescribed volume (area) fraction. A second set of similar microstructures can be gained immediately by color inversion. On increasing the volume fraction of the second phase (γ) a microstructural transition from initially isolated γ-regions in an α-matrix to isolated α-regions in a γ-matrix occurs. This transition takes place in a range of volume fractions where both phases can be continuous and can form interconnected networks [27, 28].

Many different microstructures are possible at fixed volume fractions. Figure 3 shows three topologically different realizations for $V_V^\alpha = V_V^\gamma = 0.5$ (α-matrix with γ-inclusions, both phases interwoven and γ-matrix with α-inclusions). The parameter *fraction of clusters r* is introduced as a new concept to describe quantitatively the geometrical continuity of phase regions. Its determination starts with counting the numbers of mutually enclosed grain clusters of each phase, N^α and N^γ, as suggested in [27]. An enclosed grain cluster consists only of grains of the considered phase, and is completely surrounded by grains of the other phase. From the numbers N^α and

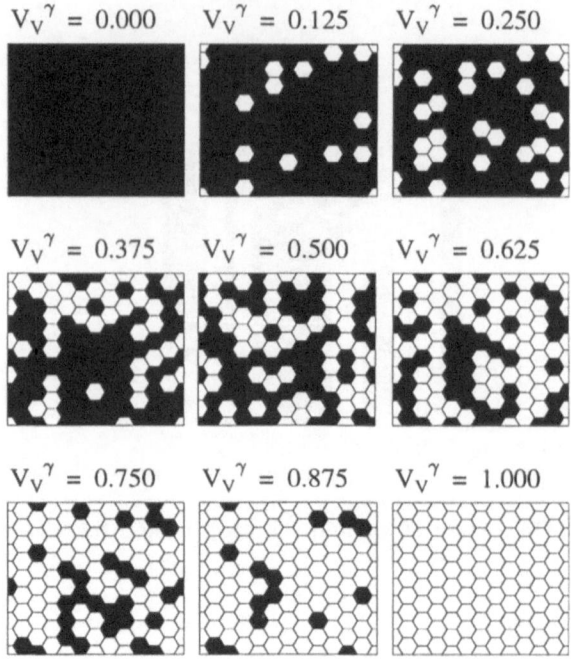

Figure 2. Examples of unit cell microstructures. Ferrite is the dark phase.

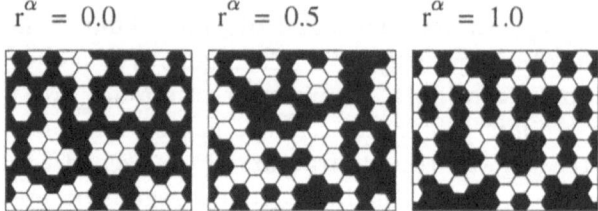

Figure 3. Three different realizations of the volume fraction $V_V^\gamma = 0.5$.

N^γ, the fraction of clusters, r^α and r^γ, now can be defined as proposed in [20, 23]:

$$r^\alpha = \frac{N^\alpha}{N^\alpha + N^\gamma}, \qquad r^\gamma = 1 - r^\alpha = \frac{N^\gamma}{N^\alpha + N^\gamma}. \tag{1}$$

In fig. 4 the parameter fraction of clusters is given as a function of V_V^γ for plane microstructures similar to the unit cell octants shown in Fig. 2. Microstructures with α-inclusions in a γ-matrix are characterized by $r^{\alpha} \to 1$, $r^{\gamma} \to 0$. For microstructures with γ-inclusions in an α-matrix one obtains $r^{\alpha} \to 0$, $r^{\gamma} \to 1$. At intermediate volume fractions and for random placements of the constituents the parameters are approximately $r^{\alpha} \sim r^{\gamma} \sim 0.5$. These microstructural geometries with both phases equally interwoven shall be termed *duplex* microstructures. As can be seen from the values on top of each microstructure in fig. 3, the r-parameter is quite appropriate to destinguish between microstructures of constant volume fractions. All microstructures used in the FE calculations are isotropic with respect to the geometrical arrangement of the two phases. Since both phases are modeled as isotropic continua, the duplex microstructures can be assumed as transverse isotropic in their properties.

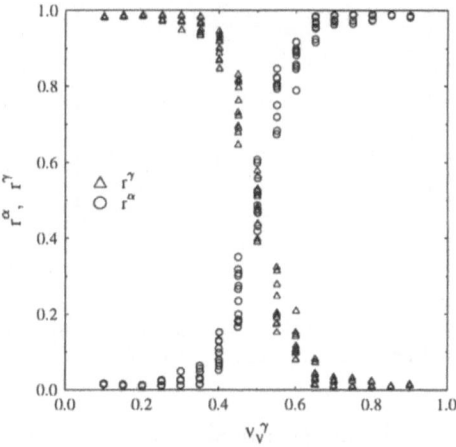

Figure 4. The parameter fraction of clusters, r^{α} and r^{γ}, plotted over the volume fraction V_V^{γ} of the γ-phase for microstructures similar to those of fig. 2.

4. The Flow Stress of the Duplex Steel

The plastic flow stress of the duplex steel is investigated for elastic idealplastic constituents. This simple model system is chosen, because analytical reference solutions are available from the results by Bao et al. [1], Ponte Castañeda et al. [10] and Suquet [11, 12].

4.1. THE LIMIT FLOW STRESS AS FUNCTION OF THE VOLUME FRACTION

For the microstructures introduced in fig. 2, tensile tests with a controlled boundary displacement are simulated in longitudinal (x_3) as well as in the transverse (x_1, x_2) directions of the unit cell. The flow stress ratio of

354

the phases is put to $\sigma_y^\gamma/\sigma_y^\alpha = 5$ in these calculations. Results of simulated tensile tests in the x_1-direction are given in fig. 5 for the microstructures with $V_V^\gamma = 0.125, 0.5, 0.875$, together with those for the single phases. The stress-strain curves of the composite materials exhibit a nonlinear transition from the elastic range to the limit state. The dependence of the limit flow stress σ_L^d reached at high strains on the volume fraction of the constituents will now be studied in detail.

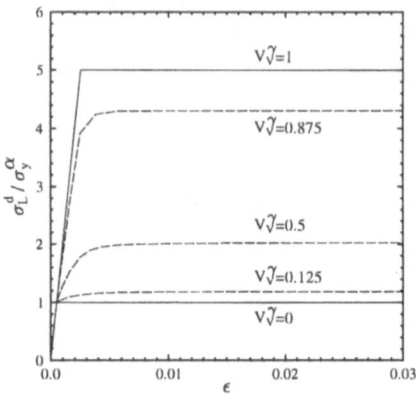

Figure 5.   Result of tensile test simulations for microstructures with various volume fractions of the γ-phase. Loading is done in the x_1-direction.

The limit flow stress of the composite is, in a first approximation, bounded by the Voigt-upper bound $\sigma_L^{d,\text{Voigt}} = \sigma_y^\alpha(1 + V_V^\gamma(\sigma_y^\gamma/\sigma_y^\alpha - 1))$ and the Reuss-lower bound $\sigma_L^{d,\text{Reuss}} = \sigma_y^\alpha$. The limit flow stress for loading in the x_3-direction as predicted by the micromechanical model is in all cases very well described by the upper bound $\sigma_L^{d,\text{Voigt}}$ (the linear ROM). The following considerations focus, therefore, on the transverse mechanical properties. In [10-12] improved bounds resp. estimates for the bounds for the limit flow stress of a transversely isotropic two-phase composite are given in form of a nonlinear generalization of the linear Hashin-Shtrikman bounds. If the yield strengths of the constituents fulfill the condition $\sigma_y^\alpha/\sigma_y^\gamma \leq \frac{1}{2}\sqrt{1 + V_V^\gamma}$, an estimate for the lower bound for the limit flow stress of the unidirection-ally anisotropic composite is given by $\sigma_L^{d,L} = \sigma_y^\alpha\sqrt{1 + V_V^\gamma}$. This estimate is close to that for the limit flow stress of a particle composite consisting of the same constituents ($\sigma_L^{d,L} = \sigma_y^\alpha\sqrt{1 + \frac{3}{2}V_V^\gamma}$, if $\sigma_y^\alpha/\sigma_y^\gamma \leq \frac{2}{5}\sqrt{1 + \frac{3}{2}V_V^\gamma}$). In [10] no analytic expression for the upper nonlinear Hashin-Shtrikman bound for the unidirectional composite has been stated explicitly. Therefore, the upper Hashin-Shtrikman bound for the particle composite is used instead. This upper bound, $\sigma_L^{d,U}$, which is certainly only an approximation for the upper bound for the limit flow stress of the unidirectional composite, is given by [10, 11]

$$\sigma_L^{d,U} = \sigma_y^\gamma \left[1 + V_V^\alpha \sqrt{\frac{3}{A}} \left(\sqrt{\left(\frac{\sigma_y^\alpha}{\sigma_y^\gamma}\right)^2 - \frac{2V_V^\gamma}{A}} - \sqrt{\frac{3}{A}} \right) \right], \quad A = 3 + 2V_V^\gamma.$$

(2)

Using this approximation seems to be justified, since the elastic moduli predicted by the linear Hashin-Shtrikman bounds for particle and unidirectionally reinforced composites are similar, too [29].

In fig. 6 the results of the computer calculations for $\sigma_L^d/\sigma_y^\alpha$ are depicted as a function of the volume fraction of the hard phase V_V^γ. The circles denote the mean values of the individual FE runs on the two microstructural realizations for each volume fraction. All microstructures are tested in the x_1- and x_2-directions. Additionally, the bounds and estimates for the limit flow stress are depicted.

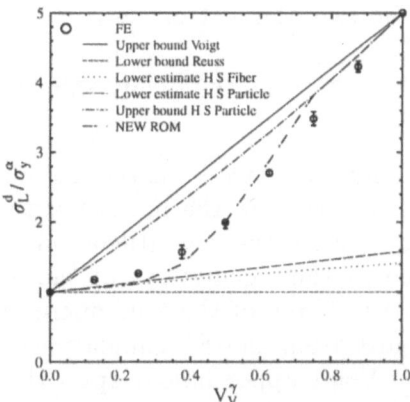

Figure 6. Dependence of $\sigma_L^d/\sigma_y^\alpha$ on the volume fraction V_V^γ of the γ-phase, compared with analytical results (H S denotes Hashin-Shtrikman).

Several observations can be made from fig. 6. The increase in limit flow stress obtained by the FE-model is small at low volume fractions of the γ-phase, and a considerable increase can be achieved only for $V_V^\gamma > 0.3$. Comparing these results with the detailed informations on the microstructures (figs. 2 and 4), it becomes evident that the observed behavior results from the change in the geometrical arrangement of the phases occurring at intermediate volume fractions. The lower estimates are a good approximation for the computed limit flow stress at low volume fractions ($r^\alpha \to 0, r^\gamma \to 1$), whereas at high volume fractions of the hard phase ($r^\gamma \to 0, r^\alpha \to 1$) the upper bounds are close to the computed values. At intermediate volume fractions, the regime of the interwoven *duplex*-type microstructures ($r^\alpha \approx r^\gamma$), the computed values of the limit flow stress deviate considerably from both the lower estimates and the upper bounds, and a continuous transition of the FE-results from the lower estimates to the upper bound solutions can be observed. A combination of the microstructural information given by the parameter fraction of clusters and the estimates and bounds for the limit flow stress can describe the FE-results quite well. For that purpose a ROM-type equation is suggested in the form

$$\sigma_L^d = r^\gamma \sigma_L^{d,L} + r^\alpha \sigma_L^{d,U} . \tag{3}$$

At low or high volume fractions of the γ-phase, this equation coincides with the limiting solutions (see fig. 6). Furthermore, it predicts an S-shaped transition of σ_L^d, a tendency also followed by the FE-results. The current result could be improved further by inserting better estimates and bounds into the ROM. The S-shaped curve is also observed experimentally in tensile tests on ferritic-martensitic two-phase steels [24].

4.2. THE DEPENDENCE OF THE LIMIT FLOW STRESS ON THE MICROSTRUCTURE

The dependence of the limit flow stress on the geometrical arrangement of the phases austenite and ferrite in the microstructure can be explored also for equal volume fractions of the constituents ($V_V^\alpha = V_V^\gamma = 0.5$). The yield strength ratios used in the calculations of the limit flow stress are $\sigma_y^\gamma / \sigma_y^\alpha = 1, 2, 3, 4, 5$ and 10. Three of the four microstructures tested are shown in fig. 3. The results from the FE-simulations are summarized in fig. 7. For $\sigma_y^\gamma / \sigma_y^\alpha < 2$ the Voigt upper bound approximates the limit flow stress of the composite, which is rather insensitive to the type of microstructure present in the material. For high yield strength ratios, however, the limit flow stress strongly depends on the details of the microstructural arrangement. In microstructures with $r^\alpha = 0$ (*i.e.* hard γ inclusions in a weak matrix) the presence of the second phase does not lead to an increase in the limit flow stress. The properties of the inverse inclusion-matrix type microstructure with $r^\alpha = 1$ (*i.e.* weak inclusions in a hard matrix) are completely different. Irrespectively of the yield strength ratio, the hard phase now contributes to the limit flow stress. This finally leads to much higher limit flow stresses for this type of microstructure.

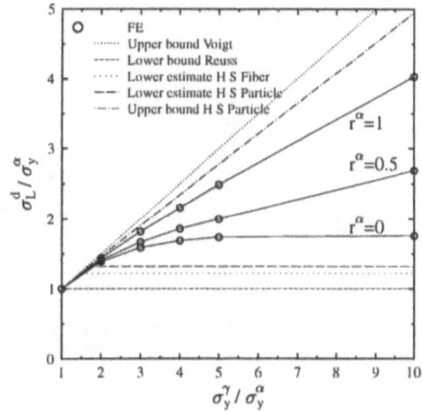

Figure 7. The dependence of the limit flow stress on the microstructure for equal volume fractions of the phases. FE calculations compared with analytical results.

Bao *et al.* [1] apply a three-phase self-consistent model to particle reinforced composites and inclusion-matrix type microstructures and obtain similar results with respect to the effect of interchanging the particle and matrix phases. The current model, however, allows to go further, and makes possible the investigation of duplex-type microstructures. The mechanical properties of these materials are distinctly different from materials possessing an inclusion-matrix type microstructure. For $r^\alpha = 0.5$, inversion of the phases does not influence the mechanical properties, and a single value of limit flow stress can be assigned to these microstructures. The computed results fall between the limit flow stresses obtained for the two inclusion-matrix type microstructures and are approximately equidistant from these limits. The microstructures with $r^\alpha = 0$ and $r^\alpha = 1$ are close to the estimates for the lower Hashin-Shtrikman bounds and the upper Hashin-Shtrikman bounds, respectively.

5. Influence of the Microstructure on the Local Deformations

FE-based micromechanical modeling gives insight also into the local material behavior. Contour plots of the equivalent plastic strain (PEEQ) provide a convenient overview of the influence of the geometrical arrangement of the phases on the local deformation behavior. Figures 8 and 9 are contour plots as obtained from FE-simulations on the microstructures of fig. 3.

The composite is strained to a deformation of $\epsilon = 0.02$ in the x_1-direction of the unit cell assuming $\sigma_y^\gamma / \sigma_y^\alpha = 5$. Figure 8a shows PEEQ for the microstructure with $r^\alpha = 0$, *i.e.* a microstructure with hard austenitic

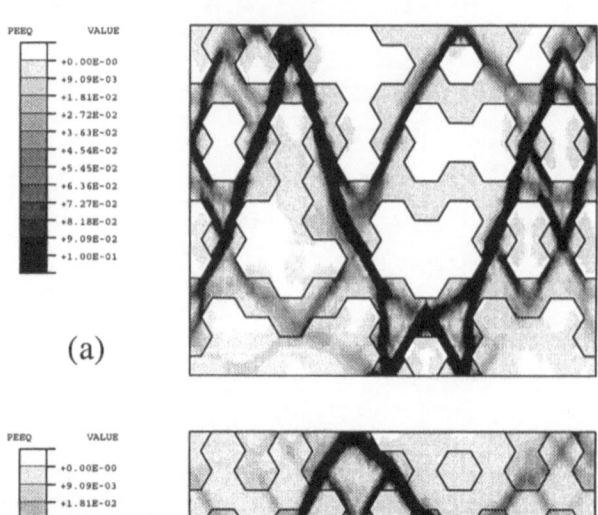

(a)

(b)

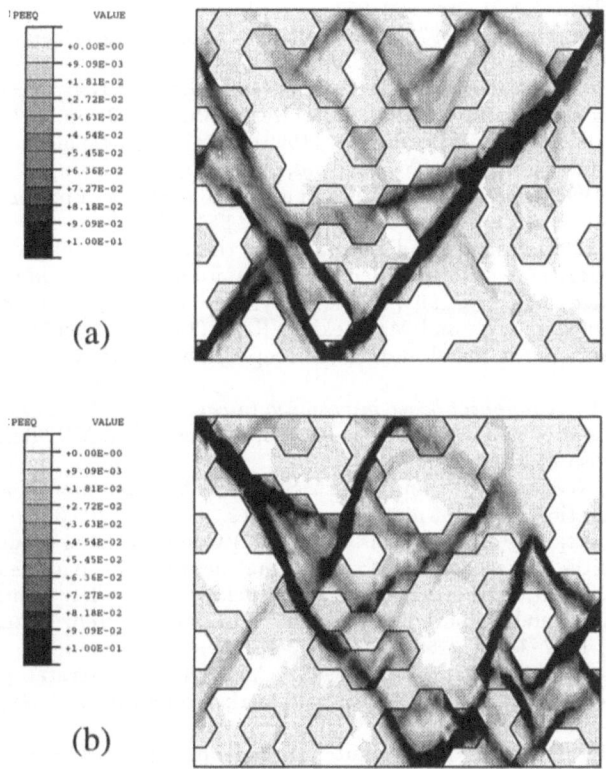

(a)

(b)

Figure 9. Contour plots of the plastic equivalent strain (PEEQ) at an overall total strain $\epsilon_{11} = 0.02$ for the two duplex-type microstructures ($r^\alpha = 0.5$). Microstructure (a) is phase-inverse to (b). The maximum local strain is 0.1.

References

1. Bao, G., Hutchinson, J.W. and McMeeking, R.M. (1991) The flow stress of dual-phase non-hardening solids, *Mech Mater*, **12**, pp. 85–94.
2. Berveiller, M. and Zaoui, A. (1985) Modeling of the plastic behavior of inhomogeneous media, *J Engng Mater Technol*, **84**, pp. 1–4.
3. Weng, G.J. (1990) The overall elastoplastic stress-strain relations of dual-phase materials, *J Mech Phys Solids*, **38**, pp. 419–441.
4. Cho, K. and Gurland, J. (1988) The law of mixtures applied to the plastic deformation of two-phase alloys of coarse microstructures, *Metall Trans*, **A19**, pp. 2027–2040.
5. Durand, L. (1987) Models of tensile behavior of two-phase alloys from their components, *Mat Sci Technol*, **3**, pp. 105–109.
6. Fischmeister, H.F. and Karlsson, B. (1977) Plastizitätseigenschaften grob-zweiphasiger Werkstoffe, *Z Metallkd*, **68**, pp. 311–327.
7. Koo, J.Y., Young, M.J. and Thomas, G. (1980) On the law of mixtures in dual-phase

steels, *Metall Trans*, **A11**, pp. 852–854.

8. Lee, H.C. and Gurland, J. (1978) Hardness and deformation of cemented tungsten carbide, *Mater Sci Engng*, **33**, pp. 125–133.

9. Poech, M.H. and Fischmeister, H.F. (1992) Deformation of two-phase materials: A model based on strain compatibility, *Acta Metall Mater*, **40**, pp. 487–494.

10. Ponte Castañeda, P. and DeBotton, G. (1992) On the homogenized yield strength of two-phase composites, *Proc R Soc Lond*, **A438**, pp. 419–431.

11. Suquet, P.M. (1993) Overall potentials and extremal surfaces of power law or ideally plastic composites, *J Mech Phys Solids*, **41**, pp. 981–1002.

12. Suquet, P.M. (1993) Bounds and estimates for the overall properties of nonlinear composites, in *MECAMAT 93 Int Seminar on micromechanics of materials*, Moret-sur Loing, France, 6-8 July, 1993, Editions Eyrolles, Paris, pp. 361–382.

13. Bao, G., Hutchinson, J.W. and McMeeking, R.M. (1991) Particle reinforcement of ductile matrices against plastic flow and creep, *Acta Metall Mater*, **39**, pp. 1871–1882.

14. Böhm, H.J. (1991) *Computer based micromechanical investigations of the thermomechanical behaviour of metal matrix composites*, Fortschr.-Ber. VDI Reihe 18 Nr.101, VDI-Verlag, Düsseldorf.

15. Christman, T., Needleman, A. and Suresh, S. (1989) An experimental and numerical study of deformation in metal-matrix composites, *Acta Metall*, **37**, pp. 3029–3050.

16. McHugh, P.E., Asaro, R.J. and Shih, C.F. (1993) Computational modeling of metal matrix composites materials, part I: Isothermal deformation patterns in ideal microstructures, part II: Isothermal stress-strain behavior, part III: Comparisons with phenomenological models, part IV: Thermal deformations, *Acta Metall Mater*, **41**, pp. 1461–1476, 1477–1488, 1489–1499, 1500-1510.

17. Needleman, A. and Tvergaard, V. (1993) Comparison of crystal plasticity and isotropic hardening predictions for metal-matrix composites, *J Appl Mech*, **60**, pp. 70–76.

18. Poech, M.H., Fischmeister, H.F., Kaute, D. and Spiegler, R. (1993) FE-modelling of the deformation behavior of WC-Co alloys, *Comp Mater Sci*, **1**, pp. 213–224.

19. Weissenbek, E. and Rammerstorfer, F.G. (1993) Influence of the fiber arrangement on the mechanical and thermomechanical behavior of short fiber reinforced MMCs, *Acta Metall Mater*, **41**, pp. 2833–2843.

20. Werner, E.A., Siegmund, T., Weinhandl, H. and Fischer F.D. (1994) Properties of random polycrystalline two-phase materials, *Appl Mech Rev*, **47**, pp. S231–S240.

21. Baskes, M.I., Hoagland, R.G. and Needleman, A. (1992) Summary report: computational issues in the mechanical behavior of metals and intermetallics, *Mater Sci Engng*, **A159**, pp. 1–34.

22. Gurland, J. (1979) A structural approach to the yield strength of two-phase alloys with coarse microstructures, *Mater Sci Engng*, **40**, pp. 59–71.

23. Siegmund, T., Werner, E.A. and Fischer, F.D. (1993) Irreversible deformation of a stainless duplex steel under thermal cycling, *Mater Sci Engng*, **A169**, pp. 125–134.

24. Uggowitzer, P. and Stüwe, H.P. (1982) Plastizität von ferritisch-martensitischen Zweiphasenstählen, *Z Metallkd*, **73**, pp. 277–285.

25. Werner, E.A. and Stüwe, H.P. (1984/85) Phase boundaries as obstacles to dislocation motion, *Mater Sci Engng*, **68**, pp. 175–182.

26. Bornert, M., Herve, E., Stolz, C. and Zaoui, A. (1994) Self-consistent approaches and strain heterogeneities in two-phase materials, *Appl Mech Rev*, **47**, pp. S66–S76.

27. DeHoff, R.T. (1968) Curvature and the topological properties of interconnected phases, in *Quantitative Microscopy*, R.T. DeHoff and F.N. Rhines (eds), McGraw-Hill Book Company, New York, pp. 291–325.

28. Poech, M.H. (1992) Mechanische Eigenschaften von Martensit-Austenit-Modellegierungen (Gefügeanordnung und Bruchzähigkeit), *Z Metallkd*, **83**, pp. 379–385.

29. Hashin, Z. (1983) Analysis of composite materials – A survey, *J Appl Mech*, **50**, pp. 481–505.

PROPERTIES OF TWO–DIMENSIONAL MATERIALS CONTAINING INCLUSIONS OF VARIOUS SHAPES

M. F. THORPE, J. CHEN
Department of Physics & Astronomy
and I. M. JASIUK
Department of Materials Science & Mechanics
Michigan State University, East Lansing, MI 48824

ABSTRACT. We discuss the effect of inclusion shape on the dielectric and elastic properties of two–dimensional materials. We show how the dielectric constant is determined by the dipole moment of the inclusion, conveniently found from the far field. Conformal mapping techniques are useful in two–dimensions to obtain the changes in the bulk properties in the special cases when the inclusions are holes or perfectly conducting or rigid. Our results are interpreted in terms of the reciprocity theorem for dielectrics and in terms of a similar, recently proved, theorem for elastic materials. We illustrate these techniques for polygonal inclusions and compare our results with those for circles and ellipses.

1. Introduction

In dielectrics and in elastic materials, it is possible to obtain exact closed form solutions when a single elliptical inclusion is embedded in a matrix. This is because the appropriate quantity (electric field for dielectrics and stress for elastic materials) is uniform inside the inclusion. Using these solutions, it is possible to find the change in the bulk properties to first order in the number of inclusions. This assumes that the inclusions are far enough apart that they are non–interacting. These solutions have been invaluable over the years as model systems and as the basis for constructing effective medium theories. However it would be useful to add some other shapes to our lexicon for comparison which is the purpose of the present paper. In the next section we show how conformal mapping techniques can be used to map a polygon onto a circle. This mapping can be used to solve the problem of a hole (or perfectly conducting inclusion) in a dielectric. If a second complex function is also introduced, then the solution for a polygonal inclusion (hole or rigid) can be found for elastic materials but only as an infinite series. In section 3, we present exact results for polygonal holes (and perfectly conducting inclusions) for dielectrics. We show how the reciprocity theorem limits the form for a general inclusion. We compare these results to those for circles and ellipses, to better understand the importance of the inclusion shape in determining the physical properties. In section 4, we show how the conformal mapping technique can be set up as a rapidly converging series

361

R. Pyrz (ed.), IUTAM Symposium on Microstructure-Property Interactions in Composite Materials, 361–373.

for polygons (holes and rigid inclusions) and obtain the first few terms for these hypotrochoids. We show how these results are compatible with the recently proved CLM theorem. Here also we compare the results to those for circles and ellipses.

2. Conformal Mapping

Conformal mapping is a useful technique in two-dimensions to map a complex shape onto a simple shape like a circle. For simple boundary conditions (Dirichlet or Neumann), a solution to the problem can then be obtained in some cases. The only boundary conditions that survive the mapping are for holes and perfectly conducting inclusions in the dielectric case, and holes and perfectly rigid inclusions in the elastic case. Despite these limitations, these mappings provide us with solutions for additional shapes.

The conformal mapping of an n-sided polygon (n-gon) in the z-plane onto the unit circle in the w-plane can be found using the Schwarz–Christoffel transformation for straight lines as (Thorpe, 1992)

$$z = \int_c^w \left(1 - \frac{1}{w^2}\right)^{2/n} dw \tag{1}$$

where c is some constant. The ratio of the area A of the n-gon to that of the unit circle is given by

$$A = \frac{4\pi^2 n}{\tan\left(\frac{\pi}{n}\right)} \frac{\Gamma^2\left(\frac{2}{n}\right)}{\Gamma^4\left(\frac{1}{n}\right)} \tag{2}$$

For the case of dielectrics, this mapping can be carried through in its entirety, to yield closed form solutions for the effective conductivity when the material contains just a few inclusions (Thorpe, 1992). For elastic media, the problem is more complex, and closed form solutions are not possible. It is therefore necessary to expand (1) as

$$z = w\left(1 + \sum_{r=1}^{\infty} a_r \omega^{-nr}\right)$$

$$= w\left(1 + \frac{2}{n(n-1)w^n} + \frac{(n-2)}{n^2(n-1)w^{2n}} + \frac{(n-2)(2n-2)}{3n^3(3n-1)w^{3n}} \cdots\right) \tag{3}$$

and use as many terms as possible (Jasiuk, Chen and Thorpe, 1994). The area of the complete n-gon is given by (2), and the area of the rounded n-gon is given by the series using (3) as

$$A = \pi \left[1 - n \sum_{r=1}^{\infty} r |a_r|^2 \right]$$

(4)

$$= \pi \left[1 - \frac{4}{n^2(n-1)} - \frac{(n-2)^2}{n^4(2n-1)} - \frac{(n-2)^2(2n-2)^2}{9n^6(3n-1)} - \ldots \right]$$

For example the first 3 terms approximating the triangle are shown in Fig. 1. The respective areas are $7\pi/9 = 2.4435...$, $314\pi/405 = 2.4357...$ and $31\pi/40 = 2.4347...$ for one, two and three non–trivial terms in (4) for $n = 3$. As the number of terms approaches infinity, the area is given by (2) which for $n = 3$ is 2.4343...

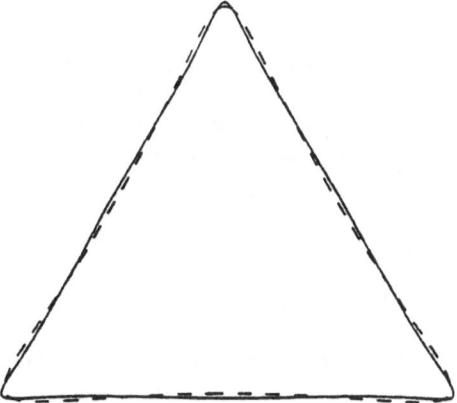

Figure 1.   Showing the shape corresponding to the conformal mapping of an equilateral triangle from a circle given in (1) and (3), using two terms (dashed line) and three terms (solid line).

3. Conductivity

The dielectric problem is the easiest of this general class of problems. By dielectric problem, we refer to any system with a potential φ that obeys Laplace's equation $\nabla^2 \varphi = 0$ in the interior. This applies to thermal conductivity, dielectric media and electrical conduction. We will use the language of the latter here. Consider a large system containing a few polygonal holes as shown in Fig. 2, with a uniform electric field applied across the sample. We will consider the dilute limit of this problem, when the perturbations to the current flow produced by the inclusions are independent of one another. This is not so in the sketch in Fig. 2, but it is convenient to sketch the more concentrated case.

Figure 2.   Showing a conducting sheet containing randomly located and oriented triangular inclusions.

The material is initially isotropic and characterized by a single parameter, the conductivity σ_0. The conductivity of a two–dimensional sheet containing a vanishingly small area fraction of inclusions, can be written as

$$\frac{\sigma}{\sigma_0} = 1 - \alpha f \tag{5}$$

where f is the area fraction of inclusions. The coefficient α is obtained from the far field perturbation caused by the current flow around a single inclusion (Thorpe, 1992). For polygonal holes, the result is independent of the orientation of the hole, as the polygonal hole acts as an equivalent circular hole in the far field. This can also be seen by symmetry arguments. A second rank tensor, in a symmetry group that contains an n–fold axis (where n is any integer greater than or equal to 3) is isotropic.

3.1. RECIPROCITY THEOREM

The reciprocity theorem in two dimensions (Keller, 1964) states that if a two phase medium, with conductivity σ has components with conductivities σ_1 and σ_2 respectively, then if the geometry is kept fixed and the two components interchanged

$$\sigma(\sigma_1, \sigma_2)\sigma(\sigma_2, \sigma_1) = \sigma_1\sigma_2. \tag{6}$$

This is useful in the present context as it relates the case when the inclusions are holes, to that where the inclusions are perfectly conducting. The reciprocity theorem is proved by using a rotation by $\pi/2$ to relate the two cases. This rotation interchanges the equipotentials and the electric field lines. If (5) is the result for holes, then if the holes are replaced by perfect conductors, we obtain

$$\frac{\sigma}{\sigma_0} = 1 + \alpha f \tag{7}$$

with the only difference being the sign change.

3.2. ELLIPTICAL INCLUSIONS

A complete solution is available for elliptical inclusions (Hetherington and Thorpe, 1992). If the host material σ_0 contains an area fraction f of elliptical inclusions with conductivity σ_1, then the conductivity of the composite σ is given by

$$\frac{\sigma}{\sigma_0} = 1 + \frac{1}{2}\left[\frac{1}{a\sigma_1 + b\sigma_0} + \frac{1}{b\sigma_1 + a\sigma_0}\right](\sigma_1 - \sigma_0)(a+b)f \tag{8}$$

to first order in the area fraction f. Here a, b are the semi–major, minor axes of the ellipse. To obtain this formula we have performed an average over the orientations of the ellipses to produce an isotropic material. In the limit for holes when $\sigma_1 = 0$, we find that

$$\alpha = 1 + \frac{1}{2}\left(\frac{a}{b} + \frac{b}{a}\right). \tag{9}$$

From the discussion earlier, this also gives the α for the perfectly conducting elliptical inclusion. Thus α achieves its minimum value of 2 for the circular hole, and is larger for the ellipse, eventually approaching ∞ in the needle limit.

3.3. POLYGONS

The solution for an n–gon is possible in the dielectric case, because only the induced dipole moment of the inclusion is required. Because the mapping (1) does not have a dipole part (w^{-1} term), the induced dipole is the same as that of the circle into which it is conformally mapped. Thus we only need to consider the change in area of the inclusion as it is mapped, and the general result is

$$\alpha = \frac{2\pi}{A} \tag{10}$$

where the area A that maps onto the unit circle is given by (2). Thus we find that

$$\alpha = \frac{\tan\left(\frac{\pi}{n}\right)\,\Gamma^4\left(\frac{1}{n}\right)}{2\pi n\,\Gamma^2\left(\frac{2}{n}\right)} \tag{11}$$

for a regular n–gon. The results are tabulated in table 1.

It can be seen that the corners in the n–gons increase the value of α, hence depressing the conductivity from the value that it would have for holes with the same area fraction. For perfectly conducting inclusions, the effect is in the opposite sense. Thus the n–gons are *more* effective in scattering the current from its direct path across the material. A similar depression of α was seen for ellipses in (9), although in that case the *corners* are rounded of course.

TABLE 1. Showing the exact results (11) for the coefficient α in the conductivity expression (5) for n–gons (Thorpe, 1992).

Shape	n	α
triangle	3	2.5811...
square	4	2.1884...
pentagon	5	2.0878...
hexagon	6	2.0454...
heptagon	7	2.0298...
octagon	8	2.0197...
.	.	.
.	.	.
circle	∞	2

If only the first few terms in the conformal mapping (3) are retained, successive approximations to the n-gon are obtained. These are probably best regarded not as approximations, but rather as exact solutions for the hypotrochoids shown in Fig. 1. As for the complete series, the result can be expressed in terms of the area via (10). The areas are given in (4), so that the results for α for the triangular shape keeping one, two and three non–trivial terms in (4) are $18/7 = 2.5714...$, $405/157 = 2.5796...$, and $80/31 = 2.5806...$, which are converging rapidly to the result $\alpha = 2.5811...$ for the triangle. Thus while the sharp corners have a dramatic influence on the local electric field near the corners of the n–gon (the electric field becomes divergent), the singularity is weak and has hardly any effect on global quantities like the conductivity (Hetherington and Thorpe, 1992).

Similar conformal mappings can be used to solve for other shapes made up of straight edges, rectangles, diamonds, stars etc. We have concentrated on the regular polygons here as these do not reduce the overall symmetry of the material from isotropic. For a general inclusion, where σ_1 is neither zero nor infinity, numerical methods must be used. For example Hetherington and Thorpe (1992) have used a boundary element approach via an integral equation to find α for this more general case. The exact results above serve as useful checks on the accuracy of the numerical procedures.

4. Elasticity

The situation, as always, is considerably more complex for analogous problems in elastic materials. While the conformal mappings described in the previous sections are applicable, a second auxiliary complex function is also required (Savin, 1961). This complicates the problem considerably and means that only a finite number of terms in the series (3) can be treated (Jasiuk, Chen and Thorpe, 1994). Nevertheless from our experience with the dielectric case, we expect this to be a rapidly convergent sequence, and indeed this is the case.

The details of this work have been published elsewhere (Jasiuk, Chen and Thorpe, 1994; Jasiuk, 1994).and we summarize our results here with more terms in some of the series than previously published. The algebra was done using MAPLE (1981–1992); an algebraic manipulation program.

In general two quantities are required to characterize an isotropic elastic sheet, and we choose the two–dimensional Young's modulus E and two–dimensional Poisson's ratio v. All other quantities of interest can be derived from the Young's modulus and Poisson's ratio. In particular, the planar bulk modulus K and shear modulus G are related through

$$K = \frac{E}{2(1-v)}$$

$$G = \frac{E}{2(1+v)} \, .$$

(12)

The other useful relations are

$$\frac{4}{E} = \frac{1}{K} + \frac{1}{G}$$

$$v = \frac{K-G}{K+G} \, .$$

(13)

As with dielectrics, we will restrict our attention to the extreme cases of holes and rigid inclusions, as analytic progress is possible here. The results in the dilute limit for *rigid inclusions* can be written

$$\frac{E}{E_0} = 1 + \alpha f$$

$$v = v_0 + \beta f$$

(14)

where in general α and β both *depend upon the Poisson's ratio* of the host material v_0. In all cases the subscript zero refers to matrix quantities. We will see that indeed the general form (14) holds for rigid inclusions, with both α and β depending on the host Poisson's ratio v_0, but there is a significant simplification for holes.

We note that with n–fold symmetry axis, the elastic tensor, which is a fourth rank tensor, is isotropic except for square symmetry ($n = 4$). Thus for $n = 4$, it is necessary to do an average over two orientations of the inclusions which differ by $\pi/4$. The dielectric tensor is second rank and is therefore isotropic for all n as noted earlier, without the need for any orientational averaging

4.1. CLM THEOREM

The results that we obtain must be consistent with the recently proved CLM theorem (Cherkaev, Lurie and Milton, 1992). This is somewhat analogous to the reciprocity theorem for dielectrics and puts restrictions upon the form of the results. A significant difference is that the CLM theorem does not relate holes and rigid inclusions (as the reciprocity theorem could for dielectrics). The CLM theorem is of little use for rigid inclusions, but for holes tells us that the two–dimensional Young's modulus must be independent of the Poisson's ratio for the host, and also restricts the form of the Poisson's ratio, allowing the system in this dilute limit to be described by just two numbers.

4.2. HOLES

Using the CLM theorem, the results for *holes* can be written as

$$\frac{E}{E_0} = 1 - \alpha f$$

(15)

$$v = v_0 - \alpha\left(v^* - v_0\right)f$$

where the result just depends on the two *pure* numbers α and v^*, which are *independent* of the Poisson's ratio of the host v_0. This is an important simplification and means that we can present the results in tabular form as is done in Tables 2 and 3 for α and v^* respectively.

Table 2.   Showing the result for the coefficient α in the expression for the Young's modulus in (15) for polygonal holes. The number of terms in the series is from the conformal mapping (3). This table contains more extensive results than were given in Jasiuk, Chen and Thorpe (1994).

Shape	n	Number of terms in series							
		1	2	3	4	5	6	7	8
triangle	3	4.1429	4.1897	4.2019	4.2069	4.2096	4.2112	4.2122	4.2129
square	4	3.4260	3.4580	3.4672	3.4711	3.4732	3.4745	3.4753	3.4758
pentagon	5	3.2092	3.2285	3.2342	3.2366	3.2380	3.2387	3.2393	3.2396
hexagon	6	3.1186	3.1306	3.1342	3.1357	3.1366	3.1371	3.1374	3.1376
.	.	.	.	.	.	.	.	.	.
.	.	.	.	.	.	.	.	.	.
circle	∞	3	3	3	3	3	3	3	3

It can be seen that the series converge about as rapidly as similar series for the dielectric, although in the elastic case we are unable to find the limit. It can be seen that the sharpening of the corners has little effect on the parameter α, as we might have anticipated using our experience with holes in dielectrics. Similar studies have been done recently by Kachanov, Tsukrov and Shafiro (1994).

Results for v^* are shown in Table 3, which also exhibit rapid convergence properties. In the case of both α and v^* it can be seen how the circle limit is approached rapidly as the number of sides n increases.

Table 3. Showing the result for the coefficient v^* in the expression for the Poisson's ratio in (15) for polygonal holes. The number of terms in the series is from the conformal mapping (3). This table contains more extensive results than were given in Jasiuk, Chen and Thorpe, (1994).

Shape	n	Number of terms in series							
		1	2	3	4	5	6	7	8
triangle	3	.24138	.23323	.23117	.23035	.22995	.22971	.22956	.22946
square	4	.31008	.30743	.30671	.30642	.30627	.30618	.30612	.30608
pentagon	5	.32485	.32383	.32354	.32343	.32336	.32333	.32331	.32329
hexagon	6	.32952	.32906	.32893	.32888	.32885	.32884	.32883	.32882
.	.	.	.	.	.	.	.	.	.
.	.	.	.	.	.	.	.	.	.
circle	∞	.33333	.33333	.33333	.33333	.33333	.33333	.33333	.33333

4.3. ELLIPTICAL INCLUSIONS

It is interesting to compare the results in Tables 2 and 3 with similar results for elliptical inclusions, as we did for dielectrics. This table contains more extensive results than were given in Jasiuk, Chen and Thorpe, (1994) For elliptical holes (averaged over all orientations), we have

$$\alpha = \frac{1}{v^*} = 1 + \frac{a}{b} + \frac{b}{a} \tag{16}$$

which shows that the rounded corners of the ellipse increase α and decrease v^* as also happens to the hypotrochoids and the polygons. It is interesting to compare (16) with the very similar result (9) for elliptical holes in dielectrics. For the ellipses we have the identity $\alpha v^* \equiv 1$ (for all aspect ratios a/b), and although this is true to better than 5%; it is not exactly obeyed for the hypotrochoids or the regular n–gons. The one exception is the first non–trivial term for the triangle, as can be determined from Tables 2 and 3.

As we noted before, there is no simplification in the form of the results for rigid inclusions. As an example of the complexity, we give the results for α and β for rigid ellipses (Thorpe and Sen , 1985).

$$\alpha = \frac{1-v_0}{3-v_0} + \frac{1}{4}(2+a/b+b/a)\left[\frac{3-2v_0+3v_0^2}{(1+v_0)(3-v_0)} + \frac{1+v_0}{1+a/b+b/a-v_0}\right]$$

$$\beta = \left(1-v_0^2\right)\left(\frac{1}{3-v_0} + \frac{1}{4}(2+a/b+b/a)\left[\frac{1-3v_0}{(1+v_0)(3-v_0)} - \frac{1}{1+a/b+b/a-v_0}\right]\right)$$

(17)

These expressions are not particularly illuminating, except to illustrate the complexity of the behavior of rigid inclusions even in this simple, exactly soluble, geometry.

4.4. RIGID INCLUSIONS

The case of rigid elastic inclusions is considerably more complex than the other situations we have examined in this paper. There are no simplifications that are available to restrict the form of (14) and α and β both depend upon the Poisson's ratio of the host material v_0. This case has recently been examined by Jasiuk (1994), using series methods similar to those described in the previous section. Again a second complex function is required in addition to the conformal mapping (1). The results are complicated to present and we just give a flavor here. In Fig. 3, we show the results graphically for the first three non–trivial terms in the conformal mapping (3) for triangular inclusions. It can again be seen that the convergence is as good as in other cases studied in this paper.

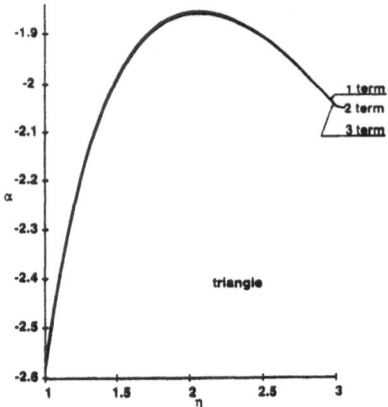

Figure 3. The parameter α defined in (14) for elastic materials with rigid triangular inclusions, approximated by one, two and three non–trivial terms in the conformal mapping (3), is plotted against $\eta = (3-v_0)/(1+v_0)$ (Jasiuk, 1994).

With this result, we can use only the first non–trivial term in (3) to compare different n–gons as is done in Fig. 4. It can be seen that there is rather more shape dependence than in some of the other cases studied in this paper.

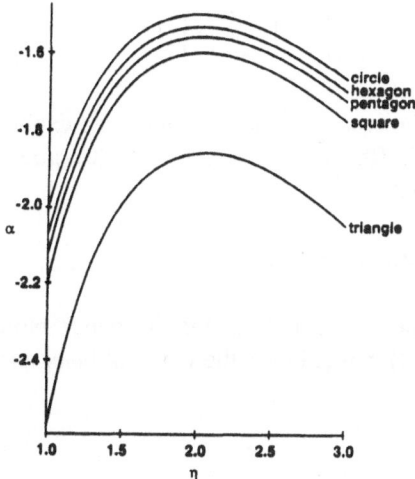

Figure 4. The parameter α in (14) is plotted against $\eta = (3 - v_0)/(1 + v_0)$ for elastic materials with rigid inclusions approximated by the first non–trivial term in (3) (Jasiuk, 1994).

The case of rigid inclusions has also been considered by Zimmerman (1986). He used the same conformal mapping techniques and concentrated the compressibility which is directly related to the area bulk modulus K. The work of Zimmerman leads to a useful bridge between the dielectric and elastic problems via

$$\alpha_{elastic} = 2\alpha_{dielectric} - 1 \tag{18}$$

where we have included subscripts to avoid confusion. The α for the dielectric and elastic problems are given by (5) and (14) respectively. This is a very useful relationship, but unfortunately is only *approximate* as a second integral in the expression for one of the stress functions was neglected [equation (3)] by Zimmerman (1986), which leads to small numerical error. Nevertheless (18) can be used as a useful approximation for any shape with isotropic symmetry. We note that it is exactly obeyed for elliptical inclusions [see equations (9) and (16)]. The first three non–trivial terms in the conformal mapping (3) for the triangle, Zimmerman's approximation (18) gives $29/7 = 4.1429...$, $653/157 = 4.1592...$ and $129/31 = 4.1629$ for $\alpha_{elastic}$ (see section 3.3), whereas the correct answers are $4.1429...$, $4.1897...$ and 4.2019 [from Table 2], and only the first term is in exact correspondence.

From the work in this paper, we are tempted to believe that (18) may be true in general as an inequality

$$\alpha_{elastic} \geq 2\alpha_{dielectric} - 1. \tag{19}$$

Recently a rigorous inequality has been derived by Gibiansky and Torquato (1993) that relates the elastic and dielectric properties of 2d materials containing holes. By combining equations (13) and (14) in Gibiansky and Torquato [and ignoring the terms in $\sigma_2/(\kappa_2 + \mu_2)$] we find

$$\left(\frac{E_0}{E}-1\right) \geq \frac{3}{2}\left(\frac{\sigma_0}{\sigma}-1\right) \tag{20}$$

which is an unusual bound as the area fraction is not involved explicitly. The bound (20) is most useful in the dilute limit [the left hand side of (20) goes negative well before percolation] where (20) becomes

$$\alpha_{elastic} \geq \tfrac{3}{2}\alpha_{dielectric}. \tag{21}$$

which of course is obeyed by the results presented in this paper. Note that (21) is a weaker bound than that proposed in (18), but (21) has the virtue of being proven!

5. Summary

We have examined the effect of the shape of the inclusion in the dilute limit for both dielectric and elastic media, for both holes and perfectly conducting (rigid) inclusions. We find that there are significant shape dependencies, especially for n-gons with small n. However it is shown that the sharp corners have little effect on the bulk properties of the composites. We have also discussed how the reciprocity theorem and the CLM theorem can help in simplifying the results in some cases. This paper gives a brief review of some of our recent work. More details can be found in the references cited, especially Jasiuk (1994), Jasiuk, Chen and Thorpe (1994), Thorpe (1992) and Thorpe and Jasiuk (1992).

6. Acknowledgments

We should like to thank B. Djordjević, J. H. Hetherington and S. Torquato for useful discussions and the Research Excellence Fund of the State of Michigan and the NSF (grant numbers DMR–9024955 and MSS–9402285) for financial support.

7. References

Cherkaev, A., Lurie, K. and Milton, G. W. (1992) Invariant properties of the stress in plane elasticity and equivalence classes of composites, *Proc. Roy. Soc. A* **438**, 519–529.

Day, A. R., Synder, K. A., Garboczi, E. J. and Thorpe, M. F. (1992) The elastic moduli of a sheet containing circular holes, *J. Mech. Phys. Solids* **40**, 1031–1051.

Gibiansky, L. V. and Torquato, S. (1993) Link between the conductivity and elastic moduli of compoisite materials, *Phys. Rev. Letts.* **71**, 2927–2930.

Jasiuk, I. M., (1994) Cavities vis–a–vis rigid inclusions: elastic moduli of materials with polygonal inclusions, *Int. J. Solids Structures* (in press)..

Jasiuk, I., Chen, J. and Thorpe, M. F. (1994) Elastic moduli of two dimensional materials with polygonal and elliptical holes, *Applied Mechanics Reviews* **47** (1, Part 2), S18–S28.

Hetherington, J. H. and Thorpe, M. F. (1992) The conductivity of a sheet containing inclusions with sharp corners, *Proc. Roy. Soc. A* **438**, 591–604.

Kachanov, M., Tsukrov, I. and Shafiro, B. (1994) Effective moduli of a solid with holes and cavities of various shapes, *Applied Mechanics Reviews* **47** (1, Part 2), S151–S174.

Keller, J. B. (1964) A theorem on the conductivity of a composite medium, *J. Math Phys.* **5**, 548–549.

Maple V Release 2. Waterloo Maple Software, Copyright (c) 1981-1992 by the University of Waterloo

Savin, G. N. (1961) *Stress concentration around holes*, Kreiger Publishing Co., Malabar, Florida.

Sen, P. N. and Thorpe, M. F. (1985) Elastic moduli of two–dimensional composite continua with elliptical inclusions, *J. Accoust. Soc. Am.* **77**, 1674–1680.

Thorpe, M. F. (1992) The conductivity of a sheet containing a few polygonal holes and/or superconducting inclusions, *Proc. Roy. Soc. A* **437**, 221–227.

Thorpe, M. F. and Jasiuk, I. M. (1992) New results in the theory of elasticity for two–dimensional composites, *Proc. Roy. Soc. A* **438**, 531–544.

Zimmerman, R. W. (1986) Compressibility of two dimensional cavities of various shapes, *J. Appl. Mech.* **53**, 500–504.

REINFORCEMENT DISTRIBUTION EFFECTS ON FAILURE IN PARTICULATE REINFORCED METALS

Viggo Tvergaard
Department of Solid Mechanics
Technical University of Denmark, Lyngby, Denmark

ABSTRACT – For an aluminium alloy reinforced by SiC particulates a micromechanical study of damage development is carried out numerically. A unit cell model analysis is employed to model the elastic–plastic deformations of the composite, taking into account failure by particulate fracture or by decohesion in the reinforcement–matrix interface. An estimate of the effect of reinforcement distribution is obtained by comparing the behaviour predicted for periodic arrays of either transversely aligned or transversely staggered particulates. Interfacial failure is modelled in terms of a cohesive zone formulation that accounts for decohesion by any combination of normal and tangential separation, while particulate fracture is represented by a critical value of the average tensile stress on a cross–section.

1. Introduction

SiC–particulate reinforcement is used to improve the stiffness and the tensile strength of aluminium alloys, but the reinforcement also results in poor ductility and low fracture toughness due to early void formation by debonding of the matrix–particulate interface or by particle fracture (McDanels, 1985; Zok et al., 1988; Lagace and Lloyd, 1989; Derby and Mummery, 1993). A discussion of current applications of metal matrix composites has been given recently by Koczak et al. (1993). Numerical analyses for a characteristic volume element representative of a composite material allow for accurate modeling of the stress and strain fields, including local stress peaks at sharp particulate edges. Therefore such analyses can be used to obtain a parametric understanding of the effect of material parameters such as shape, distribution and volume fraction of reinforcement, strength of interface and particulates, and matrix yield stress and strain hardening.

For metals reinforced by short fibres a number of numerical studies have focused on the behaviour of perfectly bonded composites (Christman et al., 1989a,b; Teply and Dvorak, 1988; Tvergaard, 1990; Bao et al., 1991). The onset of debonding at flat fibre ends has been analysed by Nutt and Needleman (1987), and debonding leading to fibre pull–out has been analysed by Tvergaard (1990,1991), taking into account also the effect of thermal contraction mismatch during cooling from the processing temperature. In two more recent papers Tvergaard (1993,1994) has analysed the effect of fibre breakage as well as fibre–matrix debonding in whisker reinforced metal, considering both transversely staggered and transversely aligned whiskers. It was found that for a realistic range of material parameters the model predictions reproduce experimentally observed material behaviour involving both types of failure (e.g. see Zok et al., 1988; Derby and Mummery, 1993). Quite recently Finot et al. (1993) have analysed failure of materials with transversely aligned particles, assuming that brittle reinforcement fracture initiates from an initial penny–shaped crack inside the particles. An interesting result of this study is that a critical value of the average tensile stress on a reinforce-

375

R. Pyrz (ed.), IUTAM Symposium on Microstructure-Property Interactions in Composite Materials, 375–385.
© 1995 *Kluwer Academic Publishers.*

ment cross–section gives a good approximation of the prediction resulting from the assumption of an initial crack in the reinforcement.

In addition to failure by fibre breakage or fibre–matrix debonding a number of discontinuously reinforced aluminium matrix composites show ductile failure by the nucleation, growth, and coalescence of voids within the matrix (e.g. see Christman et al., 1989b; Needleman et al., 1993). This type of failure mechanism has been studied numerically by Llorca et al. (1991), using a porous ductile material model to represent progressive failure development in the matrix material. A discussion of failure in metal matrix composites by fibre breakage, decohesion of the fibre–matrix interface, or failure within the matrix material alone has been given by Needleman et al. (1993), including both experimental observations and results of numerical model simulations.

In the present paper the effect of different reinforcement distributions is studied for particulate reinforced metal. The distribution of particulates in a real material is more or less random, and the relative location of neighbouring particulates may have a significant influence on the onset of failure by debonding or breakage. Such distribution effects may be conveniently studied in the context of plane strain models (e.g. see Christman et al., 1989b; Brockenbrough et al., 1991), but in the case of short fibre or particulate reinforced metals this gives a rather poor approximation of the actual three dimensional stress fields around reinforcements. Since these stress fields are important in studies of damage, the numerical model would have to be full three dimensional, e.g. extending studies such as those of Levy and Papazian (1990), Hom (1992), or Sørensen et al. (1994) to also account for failure; or an axisymmetric model can be used, which allows for a much finer mesh to resolve failure development. The present investigation is based on axisymmetric models, with particulates modeled as short cylinders with aspect ratio one. Analyses for periodic arrays of particulates, which are taken to be either transversely aligned or transversely staggered, are here used to get some insight in the effect of local reinforcement distributions. Assuming that each of these idealised distributions is relevant to some region of a real material, the distribution giving first failure would represent the most critical situation. The possibility of ductile failure within the matrix material alone will not be studied here.

2. Problem Formulation and Numerical Model

The particulate reinforced material is modelled in terms of axisymmetric unit cell–models, as shown in Fig. 1, where particulates are represented as short circular cylindrical fibres with aspect ratio one. For transversely aligned fibres (Fig. 1b) the axisymmetric approximation of the periodic deformation pattern requires that the circular cylindrical unit–cell remains circular cylindrical throughout the deformation history (detailed boundary conditions are given by Tvergaard, 1982; 1994; Nutt and Needleman, 1987; Christman et al., 1989a). For transversely staggered fibres (Fig. 1c) the same unit–cell is used (Fig. 1a), with different boundary conditions, as has been specified by Tvergaard (1990a; 1993). In both cases the boundary conditions are specified such that the average logarithmic strain in the axial direction, ε_1 , is the prescribed quantity, and the ratio of the average true stresses σ_1 and σ_2 in the axial and transverse directions remains fixed,

$$\sigma_2 = \varrho\sigma_1 \tag{2.1}$$

The initial radius and length of the unit–cell are denoted r_c and ℓ_c , respectively, and the particulate (or fibre) geometry is specified by the initial half length ℓ_f and radius r_f. Thus, the fibre volume fraction f is

$$f = \frac{r_f^2\ell_f}{r_c^2\ell_c} \tag{2.2}$$

The initial fibre aspect ratio α_f and cell aspect ratio α_c , respectively, are specified by

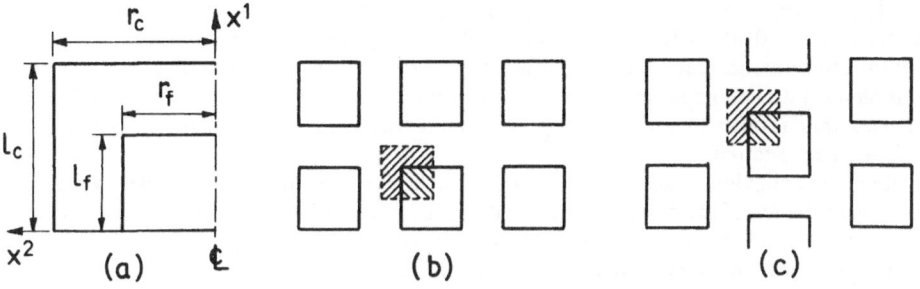

Figure 1. Periodic arrays of cylindrical particulates with parallel axes. (a) Axisymmetric cell–model. (b) Transversely aligned particulates. (c) Transversely staggered particulates.

$$\alpha_f = \ell_f/r_f \quad , \quad \alpha_c = \ell_c/r_c \tag{2.3}$$

In all calculations to be reported here the values of these aspect ratios are taken to be $\alpha_f = \alpha_c = 1$. It is noted that the cell aspect ratio α_c gives a measure of the average fibre spacings in the axial and transverse directions.

The matrix material deformations are taken to be described by J_2 flow theory, with isotropic hardening. For the analysis, a convected coordinate, Lagrangian formulation of the field equations is used, in which g_{ij} and G_{ij} are metric tensors in the reference configuration and the current configuration, respectively, with determinants g and G. A cylindrical reference coordinate system is used, and the displacement components on the reference base vectors are denoted u^i. The Lagrangian strain tensor is given by

$$\eta_{ij} = \frac{1}{2}\left(u_{i,j} + u_{j,i} + u^k_{,i}u_{k,j}\right) \tag{2.4}$$

where $(\ )_{,j}$ denotes the covariant derivative in the reference frame. The contravariant components τ^{ij} of the Kirchhoff stress tensor on the current base vectors are related to the components of the Cauchy stress tensor σ^{ij} by

$$\tau^{ij} = \sqrt{G/g}\ \sigma^{ij} \tag{2.5}$$

The finite strain generalization of J_2 flow theory, in which the incremental stress–strain relationship is of the form $\dot{\tau}^{ij} = L^{ijkl}\dot{\eta}_{kl}$, is described in more detail elsewhere (e.g. see Tvergaard, 1990). The uniaxial stress–strain behaviour is represented by

$$\varepsilon = \begin{cases} \dfrac{\sigma}{E} & \text{for } \sigma \leq \sigma_y \\[2ex] \dfrac{\sigma_y}{E}\left(\dfrac{\sigma}{\sigma_y}\right)^n & \text{for } \sigma > \sigma_y \end{cases} \tag{2.6}$$

where σ_y is the uniaxial yield stress, n is the strain hardening exponent, and E is Young's modulus.

The particles are approximated as rigid, to simplify the debonding analysis. The elastic modulus of SiC is much higher than that of aluminium $\left(E_f \simeq 5.7\ E_{A\ell}\right)$, and it has been found for perfectly bonded fibres (Tvergaard, 1990a) that predictions for elastic or rigid fibres differ rather little when plastic yielding has occurred. Breakage is taken to occur when the average tensile stress in the middle of the particle (at $x^1 = 0$) reaches a critical value σ_f. When the peak fibre stress σ_f is reached, the two halves of the reinforcement are allowed to separate, and the axial force on the fibre cross–section, $\sigma_f \pi r_f^2$, is stepped down to zero in a few subsequent increments. During this force reduction the average axial strain is taken to be constant, $\dot{\varepsilon}_I = 0$, rather than specifying a traction separation law for the fibre cross–section.

The debonding behaviour is specified in terms of a cohesive zone model proposed by Tvergaard (1990b) as an extension of the model of Needleman (1987). A set of interface constitutive relations give the dependence of the normal and tangential tractions T_n and T_t on the corresponding components u_n and u_t of the displacement difference across the interface. The model is chosen such that in pure normal separation $(u_t \equiv 0)$ it coincides with that of Needleman (1987).

A non–dimensional parameter λ is defined as

$$\lambda = \sqrt{\left(\frac{u_n}{\delta_n}\right)^2 + \left(\frac{u_t}{\delta_t}\right)^2} \tag{2.7}$$

and a function $F(\lambda)$ is chosen as

$$F(\lambda) = \frac{27}{4}\sigma_{max}\left(1 - 2\lambda + \lambda^2\right), \quad \text{for} \quad 0 \le \lambda \le 1 \tag{2.8}$$

Then, as long as λ is monotonically increasing, the interface tractions are taken to be given by the expressions

$$T_n = \frac{u_n}{\delta_n}F(\lambda), \quad T_t = \alpha\frac{u_t}{\delta_t}F(\lambda) \tag{2.9}$$

It is seen that in pure normal separation $(u_t \equiv 0)$ the maximum traction is σ_{max}, total separation occurs at $u_n = \delta_n$, and the work of separation per unit interface area is $9\sigma_{max}\delta_n/16$. In pure tangential separation $(u_n \equiv 0)$ the maximum traction is $\alpha\sigma_{max}$, total separation occurs at $u_t = \delta_t$, and the work of separation per unit interface area is $9\alpha\sigma_{max}\delta_t/16$. The values of the four parameters δ_n, δ_t, σ_{max} and α have to be chosen such that the maximum traction and work of separation in different situations are well approximated. The incremental expressions for (2.9), needed in the numerical solution, are specified by Tvergaard (1990b), together with expressions for interface unloading during reversed loading and expressions defining friction between fibre and matrix during fibre pull–out.

Numerical solutions are obtained by a finite element approximation, using a linear incremental method based on the incremental principle of virtual work. The elements used are quadrilaterals, each built up of four triangular axisymmetric linear displacement elements. The meshes used for the computations are analogous to those applied for short fibre composites by Tvergaard (1993, 1994), using a small initial blunting at the centre of the fibre, to be able to resolve the large strains that develop at the tip of the penny shaped crack resulting from fibre fracture.

The principle of virtual work for the cell–model is

$$\int_V \tau^{ij}\delta\eta_{ij}dV + \int_{S_I} (T_n\delta u_n + T_t\delta u_t)dS = \int_S T^i\delta u_i dS \tag{2.10}$$

where V and S denote the reference volume and surface of the cell, S_I denotes the internal surface, at which debonding or frictional sliding may take place, and T^i are the specified nominal surface tractions. The normal displacement difference u_n and the tangential displacement difference u_t across the interface are expressed in terms of the axial displacement of the rigid fibre and the displacements u^i on the matrix side of the interface. The boundary conditions with a fixed principal stress ratio (2.1) are implemented using a Rayleigh–Ritz finite element method (Tvergaard, 1976), and this procedure is also used to control the numerical stability during debonding.

3. Results

For the aluminium matrix the material parameters are chosen equal to those used for an aluminium alloy 2124–SiC whisker–reinforced material tested by Christman *et al.* (1989a). Thus, the uniaxial stress–strain curve is approximated by the power law (2.6) with $\sigma_y/E = 0.005$ and $n = 7.66$ (i.e. $\sigma_y = 0.3$ GPa and $E = 60$ GPa), and with Poisson's ratio $\nu = 0.3$. The fibre and cell aspect ratios are taken to be $\alpha_f = \alpha_c = 1$, and in most of the studies the fibre volume fraction $f = 0.20$ is considered.

For SiC fibres the failure strain is often around 0.01 (Teply and Dvorak, 1988; Lloyd, 1989; Koczak *et al.*, 1989), somewhat larger for whiskers and somewhat smaller for reinforcements with larger diameter, and the SiC elastic modulus is about 5.7 times that for aluminium. A failure strain of 0.01 corresponds to a value 0.057 of $\sigma_f/E_{A\ell}$, or to $\sigma_f/\sigma_y = 11.4$ for the present material parameters. Since particulates have often larger diameters, it is chosen here to focus on the values 10, 5 and 2.5, respectively, of the ratio σ_f/σ_y. Also in the description of decohesion at the particle–matrix interface different values of the peak stress σ_{max} are considered, and the values of the remaining parameters in the model (2.7)–(2.9) are here taken to be $\delta_n = \delta_t = 0.02 r_f$ and $\alpha = 1$.

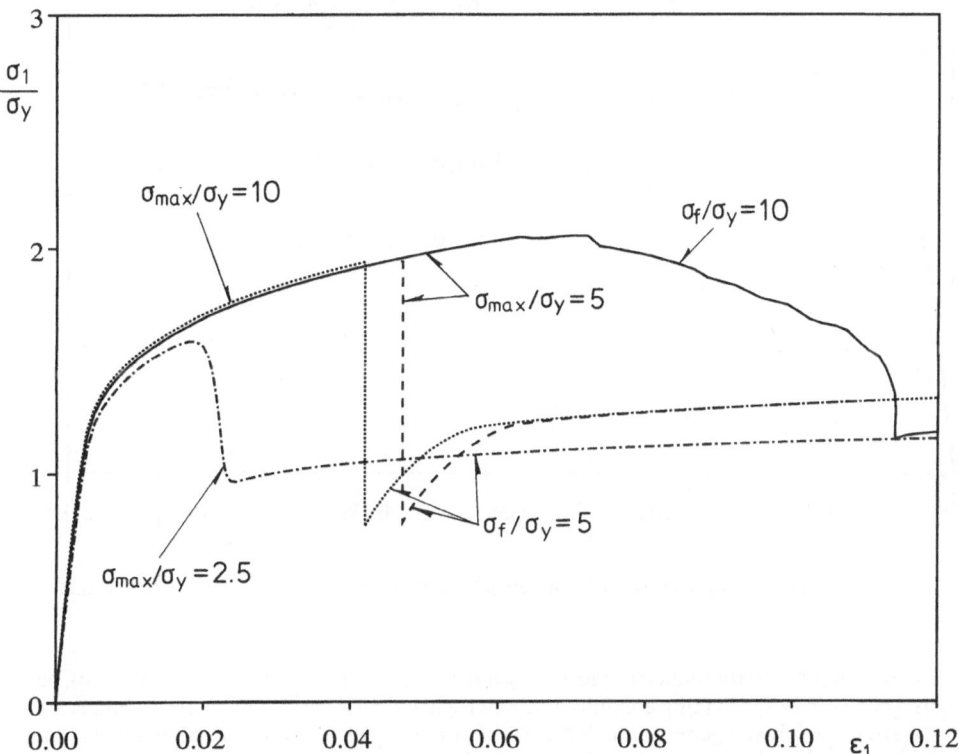

Figure 2. Stress–strain curves predicted for transversely aligned particulates, $\varrho = 0$ and $f = 0.20$.

380

Fig. 2 shows computed overall stress–strain curves for materials with transversely aligned particulates. The materials are subjected to uniaxial tension, $\varrho = \sigma_2/\sigma_1 = 0$, and have $f = 0.20$. For $\sigma_f/\sigma_y = 10$ and $\sigma_{max}/\sigma_y = 5$ no fibre breakage is predicted, but debonding at the sharp particle edge starts at $\varepsilon_1 \simeq 0.061$ and the flat end of the particle is completely debonded at $\varepsilon_1 \simeq 0.117$. The other three curves in Fig. 2 assume a lower fibre strength, $\sigma_f/\sigma_y = 5$. For $\sigma_{max}/\sigma_y = 5$ this results in particle breakage at $\varepsilon_1 \simeq 0.047$, prior to the onset of debonding found on the solid curve in Fig. 2, and practically no debonding occurs in this case as the crack through the particle centre opens up. For $\sigma_{max}/\sigma_y = 10$ this particle breakage occurs a little earlier, as a result of the reduced compliance of the interface and less tendency towards beginning interfacial failure. On the other hand, for $\sigma_{max}/\sigma_y = 2.5$ failure by debonding at the flat particle end takes over again, and no breakage is predicted.

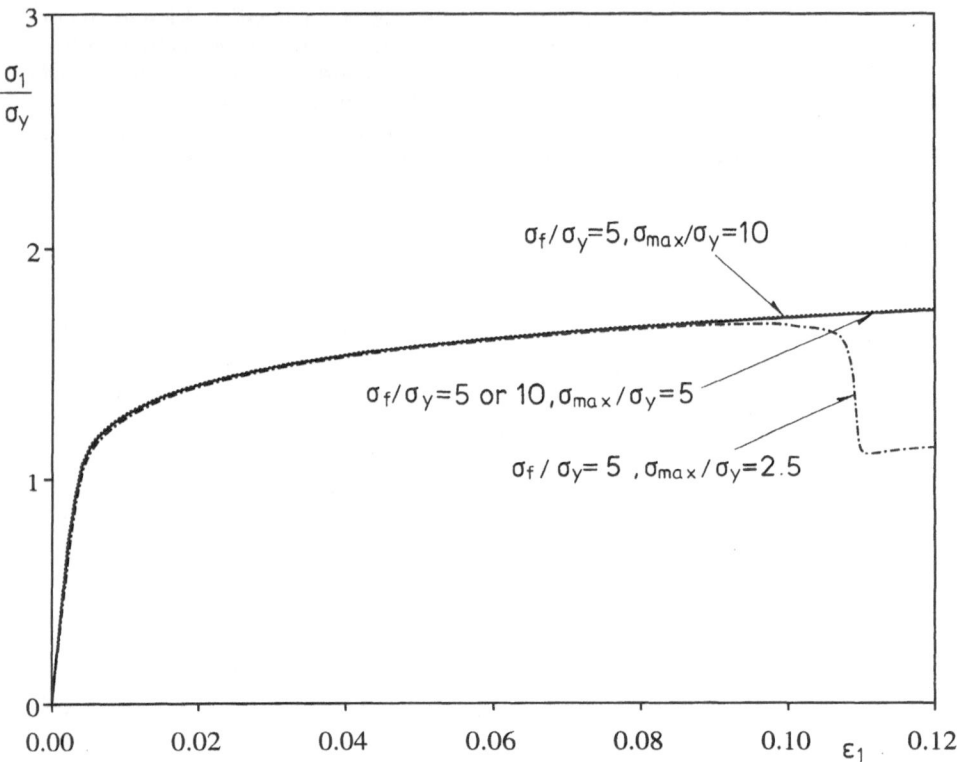

Figure 3. Stress–strain curves predicted for transversely staggered particulates, $\varrho = 0$ and $f = 0.20$.

Fig. 3 shows predictions for a metal reinforced by transversely staggered particulates (Fig. 1c), but for material parameters and loading conditions otherwise identical to those considered in Fig. 2. It is seen that the transversely staggered particulates give significantly lower overall stress levels and thus also a much reduced tendency for material failure. In fact, three of the cases analysed show practically identical stress–strain curves with no failure at all in the range considered. In the fourth case, for

$\sigma_{max}/\sigma_y = 2.5$ and $\sigma_f/\sigma_y = 5$, debonding at the sharp particle edge starts rather late, at $\varepsilon_1 \simeq 0.095$, and full debonding at the flat endface has occurred at $\varepsilon_1 = 0.110$.

It is noted that the relatively high stress levels for transversely aligned particulates are a result of the highly constrained plastic flow that develops for this reinforcement distribution, as has also been found for whisker reinforced metals (Christman *et al.*, 1989a,b; Tvergaard, 1994). The generally high stress level in the material promotes failure by particle fracture as well as by particle matrix decohesion. The transversely staggered array of particulates gives less constraint on plastic flow and thus generally lower stress levels in the material; but a cross–section through particulate centres contains only half as many particulates in the transversely staggered array, and this should tend to increase the stress levels inside particulates and thus give earlier breakage. For a whisker reinforced metal it has been shown (Tvergaard, 1994) that the latter mechanism dominates, so that failure by fibre breakage occurs at a smaller strain and a smaller average tensile stress with transversely staggered fibres. The same trend is not found for the particulate reinforced materials illustrated by Figs. 2 and 3. Here, the transversely staggered particulates give no breakage at all in the range analysed, and the only failure event predicted in Fig. 3 occurs at a much larger strain and a somewhat larger average tensile stress than that predicted in Fig. 2 for the corresponding material with transversely aligned particulates.

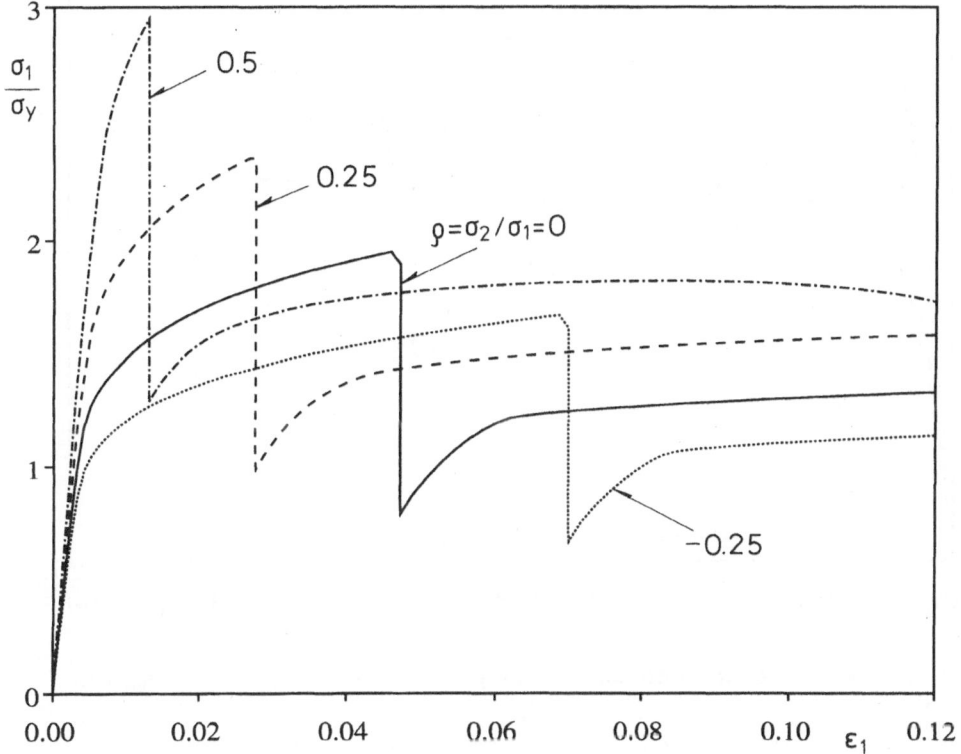

Figure 4. Stress–strain curves predicted for transversely aligned particulates, $\sigma_f/\sigma_y = 5$, $\sigma_{max}/\sigma_y = 5$ and $f = 0.20$.

382

In Figs. 4 and 5 the effect of stress triaxiality, as measured by $\varrho = \sigma_2/\sigma_1$, is compared for the two different types of particulate distributions. The material parameters are as those in the previous two figures, with $f = 0.20$ and $\sigma_f/\sigma_y = \sigma_{max}/\sigma_y = 5$, and thus the solid curves in Figs. 4 and 5 are identical to curves also shown in Figs. 2 and 3, respectively. It is seen that increased stress triaxiality gives rise to much higher tensile stress levels, which promotes earlier onset of failure, as is clearly illustrated in Fig. 4, where particle fracture is the cause of failure in all four cases. For the corresponding cases with transversely staggered particulates Fig. 5 shows no occurrence of failure, but it is noted that for $\varrho = 0.5$ failure by particle fracture has been predicted outside the range illustrated in the figure, at $\varepsilon_1 \simeq 0.127$ and $\sigma_1/\sigma_y \simeq 3.54$. Thus, also here the trend is opposite to that found for short fibre reinforced metals (Tvergaard, 1994), since for $\varrho = 0.5$ transversely staggered particulates give much later failure at a higher average tensile stress. In a real material, where the particulates are more or less randomly distributed, it is thus more likely that failure will initiate in regions where the particulates are transversely aligned than in regions with transversely staggered particulates.

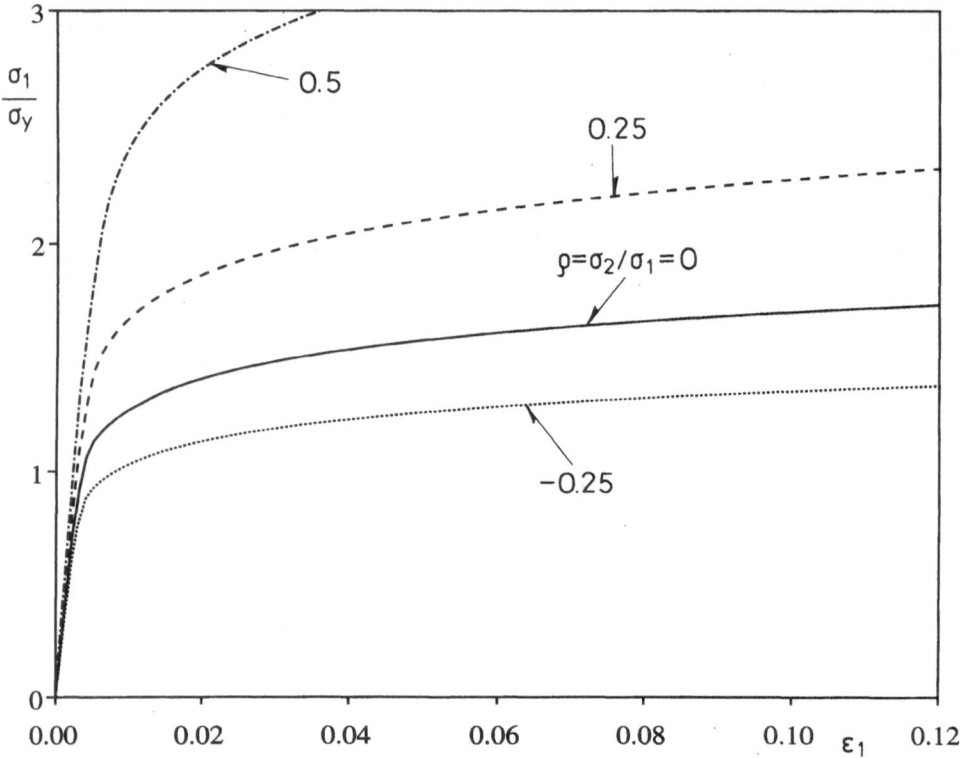

Figure 5. Stress–strain curves predicted for transversely staggered particulates, $\sigma_f/\sigma_y = 5$, $\sigma_{max}/\sigma_y = 5$ and $f = 0.20$.

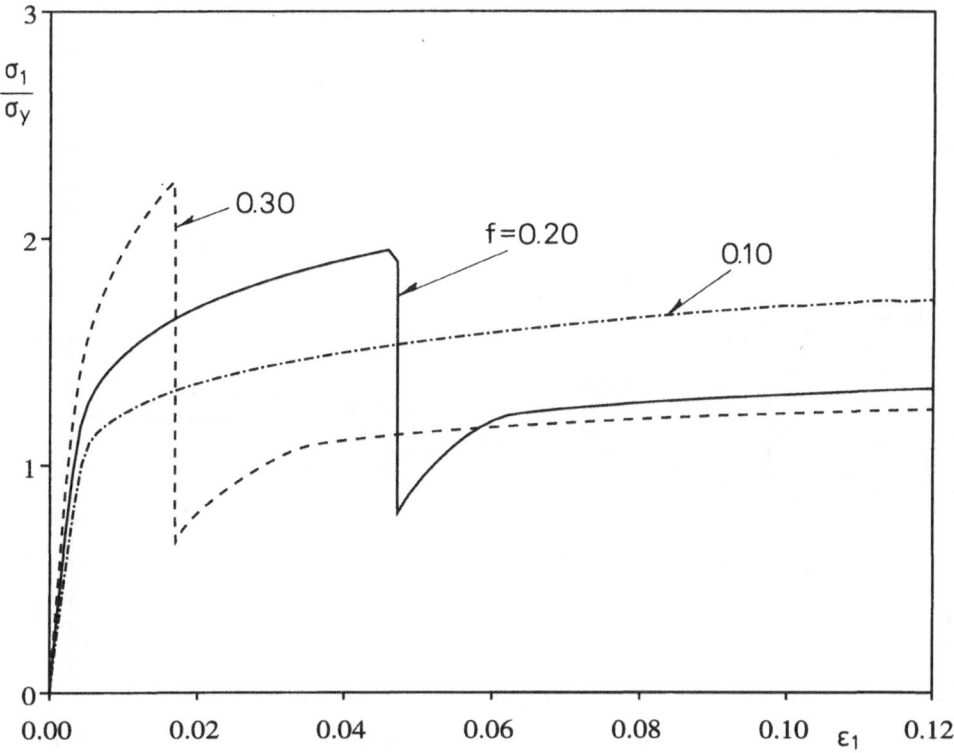

Figure 6. Stress–strain curves predicted for transversely aligned particulates, $\varrho = 0$, $\sigma_f/\sigma_y = 5$ and $\sigma_{max}/\sigma_y = 5$.

The effect of differences in the volume fraction of particulates is illustrated in Fig. 6 for a material with $\sigma_{max}/\sigma_y = \sigma_f/\sigma_y = 5$ and with transversely aligned reinforcement, subject to uniaxial tension, $\varrho = 0$. Increasing the value of f from 0.20 to 0.30 gives significantly higher overall stresses, as expected, and this results in failure by particle fracture at a much smaller strain. For the smaller volume fraction, $f = 0.10$, the stress level in the particulates is so much reduced that no breakage occurs in the range analysed; but in this case a region of decohesion near the sharp particulate edges starts to develop at $\varepsilon_1 \simeq 0.098$, rather slowly since at $\varepsilon_1 = 0.148$ the flat end face is not yet fully debonded. The same three sets of material parameters have also been analysed for transversely staggered particulates, but here the stress levels are so much lower than those in Fig. 6 that neither breakage nor debonding are predicted in the range analysed.

Results for transversely staggered particulates with $\sigma_f/\sigma_y = 2.5$ and $\sigma_{max}/\sigma_y = 5$ are shown in Fig. 7, for $\varrho = 0$. Here, due to the smaller particulate strength, breakage is predicted in all three cases. Qualitatively, the behaviour is quite in agreement with that in Fig. 6, showing that higher particulate volume fraction gives rise to more rapid growth of the average tensile stress, which leads to breakage at a smaller strain.

384

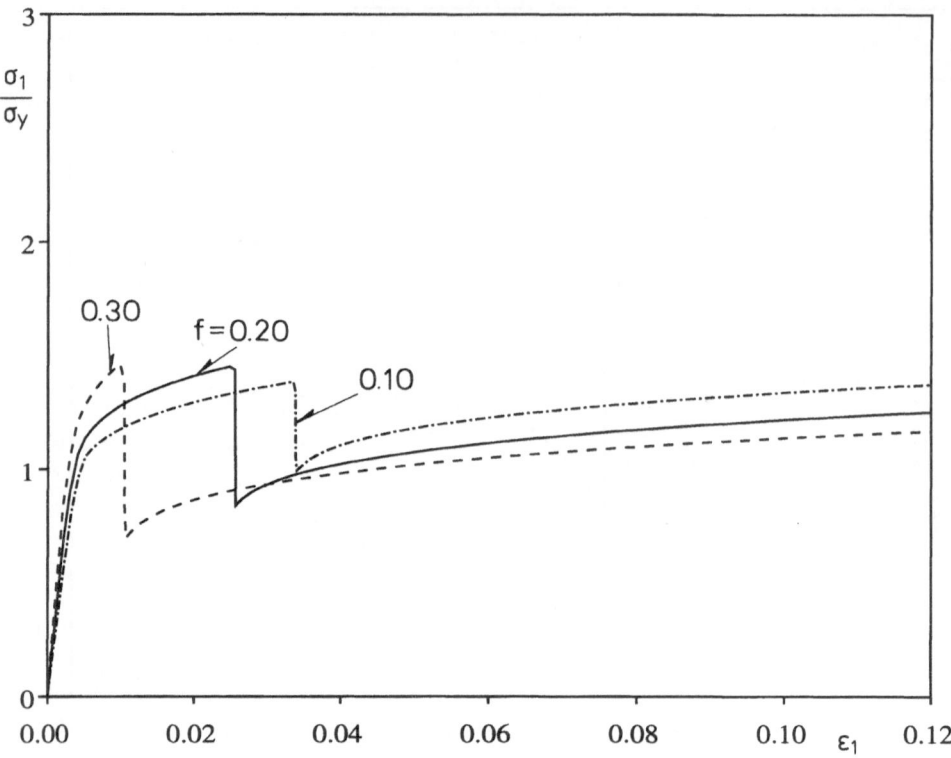

Figure 7. Stress–strain curves predicted for transversely staggered particulates, $\varrho = 0$, $\sigma_f/\sigma_y = 2.5$ and $\sigma_{max}/\sigma_y = 5$.

References

Bao, G., Hutchinson, J.W. and McMeeking, R.M. (1991) Particle reinforcement of ductile matrices against plastic flow and creep, *Acta Metall. Mater.* **39**, 1871–1882.

Brockenbrough, J.R., Suresh, S. and Wienecke, H.A. (1991) Deformations of metal–matrix composites with continuous fibres: Geometrical effects of fibre distribution and shape, *Acta Metall. Mater.* **39**, 735.

Christman, T., Needleman, A., Nutt, S. and Suresh, S. (1989a) On microstructural evolution and micromechanical modelling of deformation of a whisker–reinforced metal–matrix composite, *Mater. Sci. Engng.* **A107**, 49–61.

Christman, T., Needleman, A. and Suresh, S. (1989b) An experimental and numerical study of deformation in metal–ceramic composites, *Acta Metall.* **37,** 3029–3050.

Derby, B. and Mummery, P.M. (1993) Fracture behaviour, in S. Suresh, A. Mortensen and A. Needleman (eds.), *Fundamentals of Metal Matrix Composites*, Butterworth–Heinemann, Boston, pp. 251–268.

Finot, M., Shen, Y.–L., Needleman, A. and Suresh, S. (1993) Micromechanical modelling of rein-forcement fracture in particulate–reinforced metal–matrix composites, Brown University, Division of Engineering, Report.

Hom, C.L. (1992) Three–dimensional finite element analysis of plastic deformation in a whisker–re-inforced metal matrix composite, *J. Mech. Phys. Solids* **40**, 991–1008.

Koczak, M.J., Prewo, K., Mortensen, A., Fishman, S., Barsoum, M. and Gottschall, R. (1989) Inor-ganic composite materials in Japan: status and trends, Scientific Monograph ONR/AFOSR/ARO.

Koczak, M.J., Khatri, S.C., Allison, J.E. and Bader, M.G. (1993) Metal–matrix composites for ground vehicle, aerospace, and industrial applications, in S. Suresh, A. Mortensen and A. Needle-man (eds.), *Fundamentals of Metal Matrix Composites*, Butterworth–Heinemann, Boston, pp. 297–326.

Lagacé, H. and Lloyd, D.J. (1989) Microstructural analysis of Al–SiC composites, *Canadian Metall. Q.* **28**, 145–152.

Levy, A. and Papazian, J.M. (1990) Tensile properties of short fiber–reinforced SiC/Al composites: Part II. Finite–element analysis, *Metall. Trans.* **21A**, 411–420.

Llorca, J., Needleman, A. and Suresh, S. (1991) An analysis of the effects of matrix void growth on deformation and ductility in metal–ceramic composites, *Acta Metall. Mater* **39**, 2317–2335.

Lloyd, D.J. (1989) Metal matrix composites – an overview, Alcan Int. Ltd., Kingston Lab., Ontario.

McDanels, D.L. (1985) Analysis of stress–strain, fracture, and ductility behaviour of aluminum matrix composites containing discontinuous silicon carbide reinforcement, *Metall. Trans.* **16A** 1105–1115.

Needleman, A. (1987) A continuum model for void nucleation by inclusion debonding, *J. Appl. Mech.* **54**, 525–531.

Needleman, A., Nutt, S.R., Suresh, S. and Tvergaard, V. (1993) Matrix, reinforcement and interfa-cial failure, in S. Suresh, A. Mortensen and A. Needleman (eds.), *Fundamentals of Metal Matrix Composites*, Butterworth–Heinemann, Boston, pp. 233–250.

Nutt, S.R. and Needleman, A. (1987) Void nucleation at fiber ends in Al–SiC composites, *Scripta Metallurgica* **21**, 705–710.

Sørensen, N.J., Suresh, S., Tvergaard, V. and Needleman, A. (1994) Effects of reinforcement orientation on the tensile response of metal matrix composites, Risø National Laboratory, Materi-als Department, Report.

Teply, J.L. and Dvorak, G.J. (1988) Bounds on overall instantaneous properties of elastic–plastic composites, *J. Mech. Phys. Solids* **36**, 29–58.

Tvergaard, V. (1976) Effect of thickness inhomogeneities in internally pressurized elastic–plastic spherical shells, *J. Mech. Phys. Solids* **24**, 291–304.

Tvergaard, V. (1990a) Analysis of tensile properties for a whisker–reinforced metal–matrix compos-ite, *Acta Metall. Mater.* **38**, 185–194.

Tvergaard, V. (1990b) Effect of fibre debonding in a whisker–reinforced metal, *Mater. Sci. Engng.* **A125**, 203–213.

Tvergaard, V. (1991) Effect of thermally induced residual stresses on the failure of a whisker–rein-forced metal, *Mech. Mater.* **11**, 149–161.

Tvergaard, V. (1993) Model studies of fibre breakage and debonding in a metal reinforced by short fibres, *J. Mech. Phys. Solids* **41**, 1309–1326.

Tvergaard, V. (1994) Fibre debonding and breakage in a whisker reinforced metal, *Danish Center for Appl. Math. and Mech.*, Report No. 479.

Zok, F., Embury, J.D., Ashby,M.F. and Richmond,O. (1988) The influence of pressure on damage evolution and fracture in metal–matrix composites, in S.I. Andersen *et al.* (eds.), *Mechanical and Physical Behaviour of Metallic and Ceramic Composites*, Risø Nat. Lab., Denmark, pp. 517–526.

ON THE MICROMODELLING OF DYNAMIC RESPONSE FOR THERMOELASTIC PERIODIC COMPOSITES

CZ. WOŹNIAK
Institute of Fundamental Technological Research
Świętokrzyska 21, PL 00-049 Warszawa, Poland

M. WOŹNIAK (Ms.)
Technical University of Łódź, Department of Geotechnical and Structure Engineering
Al. Politechniki 6, PL 93-590 Łódź, Poland

1. Introduction

As it is known, the asymptotic homogenization methods for micro-periodic composites, leading to the effective modulus theories, neglect inertial aspects of microstructural features related to the size of constituents (cf. [1,2] and the references therein). The main aim of this contribution is to propose a new approach to the formulation of macro--models for micro-periodic thermoelastic composite materials. This approach takes into account a length-scale effect on a dynamic response of a composite and is simple enough to be applied in analysis of engineering problems and for quasi-stationary processes reduces to the special effective modulus theory, [3,4]. Theories of this type for elastic composite materials and structures were discussed in [5-7] and are termed *refined macro-theories*. In this paper governing equations of the refined macro--thermoelastodynamics are formulated on the basis of heuristic hypotheses concerning the expected form of disturbances in displacement and temperature fields, caused by the micro-inhomogeneity of a composite. At the same time a special form of macro--modelling approximations is used. The resulting equations are obtained without any

R. Pyrz (ed.), IUTAM Symposium on Microstructure-Property Interactions in Composite Materials, 387–395.
© *1995 Kluwer Academic Publishers.*

reference to a boundary value problem on the representative volume element, that is required in asymptotic homogenization approaches, [1,2]. The general considerations are illustrated by the simple example the aim of which is to compare results of the refined macro-theory and those of the effective modulus theory. It is shown that the microstructure length scale effects, described by the proposed macro-theory, play an essential role in investigations of the non-stationary behaviour of the composites.

The analysis will be carried out in the framework of the linear thermo-elastodynamics under assumption of the perfect bonding between constituents of the composite. The considerations are restricted to micro-periodic bodies, i.e., the maximum length dimension of the representative volume element is sufficiently small compared to the minimum characteristic length dimension of the body.

Denotations. The region in the reference space, occupied by the undeformed composite body, will be denoted by Ω. By $\mathbf{x} \equiv (x_1, x_2, x_3)$ and τ we denote points of Ω and a time coordinate, respectively, and x_1, x_2, x_3 are Cartesian orthogonal coordinates in the reference space. Subscripts i, j, k, l related to these coordinates run over the sequence 1, 2, 3. Superscripts a, b, run over 1,...,n being related to a certain micro-discretization of the representative volume element $V = (0, l_1) \times (0, l_2) \times (0, l_3)$ of the periodic composite structure. The summation convention holds both for i, j, k, l and a, b. For any V-periodic integrable function $f(\mathbf{x})$ we introduce the averaging operator

$$\langle f \rangle \equiv \frac{1}{l_1 l_2 l_3} \int_V f(\mathbf{x}) dv \quad ,$$

where $dv = dx_1 dx_2 dx_3$. The area element of the boundary $\partial\Omega$ will be denoted by da. The remaining basic denotations will be given at the beginning of the subsequent section.

2. Analysis

2.1. FOUNDATIONS

Foundations of the proposed approach are based on the governing equations of thermo-
-elastodynamics. Denoting by u_i, θ, s_{ij}, q_i, ρ, ε, b_i, α, s_i, q displacements, temperature,
stresses, heat fluxes, mass density, specific energy, body forces, heat supply, boundary
tractions and boundary heat supply, respectively, we shall postulate the principle of
balance of momentum and that of balance of energy in the following weak form

$$\frac{d}{d\tau}\int_\Omega \rho(\mathbf{x})\ddot{u}_i(\mathbf{x},\tau)\delta u_i(\mathbf{x})dv = \oint_{\partial\Omega} s_i(\mathbf{x},\tau)\delta u_i(\mathbf{x})da - \int_\Omega s_{ij}(\mathbf{x},\tau)\delta u_{i,j}(\mathbf{x})dv +$$

$$+\int_\Omega \rho(\mathbf{x})b_i\delta u_i(\mathbf{x})dv \; , \tag{1}$$

$$\frac{d}{d\tau}\int_\Omega \varepsilon(\mathbf{x},\tau)\delta\theta(\mathbf{x})dv = \oint_{\partial\Omega} q(\mathbf{x},\tau)\delta\theta(\mathbf{x})da - \int_\Omega q_i(\mathbf{x},\tau)\delta\theta_{,i}(\mathbf{x})dv +$$

$$+\int_\Omega [\alpha(\mathbf{x},\tau)+s_{ij}(\mathbf{x},\tau)\dot{u}_{i,j}(\mathbf{x},\tau)]\delta\theta(\mathbf{x})dv \; ,$$

where δu_i, $\delta\theta$ are sufficiently regular test functions. The constitutive equations will be
assumed in the linearized form

$$\varepsilon(\mathbf{x},\tau) = \frac{1}{2}C_{ijkl}(\mathbf{x})\,u_{(i,j)}(\mathbf{x},\tau)\,u_{(k,l)}(\mathbf{x},\tau) + B_{ij}(\mathbf{x})\,u_{(i,j)}(\mathbf{x},\tau)\,\theta(\mathbf{x},\tau) +$$

$$+\frac{1}{2}c(\mathbf{x})\,\theta^2(\mathbf{x},\tau) \; , \tag{2}$$

$$s_{ij}(\mathbf{x},\tau) = C_{ijkl}(\mathbf{x})\,u_{(i,j)}(\mathbf{x},\tau) + B_{ij}(\mathbf{x})\,\theta(\mathbf{x},\tau) \; , \quad q_i(\mathbf{x},\tau) = K_{ij}(\mathbf{x})\,\theta_{,j}(\mathbf{x},\tau) \; ,$$

where the elastic modulae $C_{ijkl}(\cdot)$, the thermal conductivity modulae $K_{ij}(\cdot)$, the thermal
expansion modulae $B_{ij}(\cdot)$ and the specific heat $c(\cdot)$ are V-periodic functions. Eqs (2)
have to satisfy the known dissipation condition.

If Eqs (1) hold for arbitrary regular test functions δu_i, $\delta\theta$ then Eqs (1), (2) are equivalent to the well known equations of the linear thermo-elastodynamics, which for composite materials have to be considered together with the continuity conditions $[\sigma_{ij}]n_j = 0$, $[h_i] n_i = 0$ on the interfaces between constituents. However, due to the highly oscillating (V-periodic) form of Eqs (2), the aforementioned equations do not constitute the proper analytical tool for investigations of composite bodies. In order to formulate macro-modelling assumptions leading from Eqs (1), (2) to the equations of the refined thermo-elastodynamics we have to introduce certain preliminary concepts.

2.2. PRELIMINARY CONCEPTS

The first preliminary concept we are to introduce is related to the expected form of disturbances in displacement and temperature fields caused by the inhomogeneity of the medium. To this end we shall assume that from the qualitative viewpoint these disturbances can be described by a sequence of n linear independent functions $h^a(x)$, $x \in R^3$, which are V-periodic, continuous, have piece wise continuous first derivatives $h^a_{,i}$ suffering jump discontinuities across the interfaces between constituents and satisfy conditions: $<h^a> = 0$, $<h^a_{,i}> = 0$, $<\rho h^a> = <\alpha h^a> = <ch^a> = <B_{ij}h^a> = 0$. Moreover, we assume that $h^a(x) \in \mathcal{O}(l)$, where $l \equiv \max l_i$ is the microstructure length parameter, and that the values $h^a_{,i}(x)$ of the derivatives of h^a are independent of l. Functions $h^a(\cdot)$ are called *micro-shape functions*; their specification depends on the material structure of the representative volume element V of the periodic composite and can be also related to a certain discretization procedure of V; for particulars the reader is referred to [3-5].

Let λ be a small macro-accuracy parameter related to the calculations of a certain real-valued function $F(\cdot)$ defined on Ω (F can also depend on the time coordinate τ). Function F will be called V-macro function if for every $x,y \in \Omega$ such that $x-y \in V$ condition $|F(x) - F(y)| < \lambda$ holds. If the condition of this form also holds for all derivatives of F then F will be referred to as the *regular V-macro function*. For any integrable V-periodic function $f(\cdot)$, micro-shape function $h^a(\cdot)$ and regular V-macro function $F(\cdot)$, we obtain

$$\int_{\Omega} f(x)F(x)dv = <f> \int_{\Omega} F(x)dv + \mathcal{O}(\lambda) \ , \tag{3}$$

$$(h^a(x)F(x))_{,i} = h^a_{,i}(x)F(x) + \mathcal{O}(\lambda) \ ,$$

The concept of a regular V-macro function is strictly related to the macroscopic description of the behaviour of a composite in which the oscillations of functions within every single periodicity cell can be neglected, [5].

2.3. MACRO-MODELLING ASSUMPTIONS

The passage from Eqs (1), (2) of micromechanics to the proposed macro-model of a composite will be based on the following modelling assumptions.

Micro-Macro Localization Hypothesis. The displacement and temperature fields in the micro-periodic body can be expected in the form

$$u_i(x,\tau) = U_i(x,\tau) + h^a(x)V^a_i(x,\tau) \ , \tag{4}$$

$$\theta(x,\tau) = \Theta(x,\tau) + h^a(x)\Phi^a(x,\tau) \ , \qquad x \in \Omega \ ,$$

where $U_i(\cdot,\tau)$, $V^a_i(\cdot,\tau)$, $\Theta(\cdot,\tau)$, $\Phi^a(\cdot,\tau)$ are arbitrary regular V-macro functions and $h^a(\cdot)$ are micro-shape functions, postulated in every problem under consideration.

Fields U_i and Θ will be called macro-displacement and macro-temperature field, respectively. Fields V^a_i, Φ^a are referred to as correctors and describe, from the quantitative point of view, the possible disturbances in displacements and a temperature caused by the micro-periodic inhomogeneous structure of a composite.

Macro-Balance Assumption. The balance equations (1) are assumed to hold for $\delta u_i = \delta U_i + h^a\delta V^a_i$, $\delta\theta = \delta\Theta + h^a\delta\Phi^a$, where δU_i, δV^a_i, $\delta\Theta$, $\delta\Phi^a$ are arbitrary linear independent regular V-macro functions.

Macro-Modelling Approximation. In the balance equations (1) terms $\mathcal{O}(\lambda)$ in integrals over Ω and terms $\mathcal{O}(l)$ in integrals over $\partial\Omega$ can be neglected.

3. Refined theory

Substituting the right-hand sides of Eqs (2) into Eqs (1) and using the aforementioned macro-modelling assumptions (we apply formulae (3)), after some manipulations and introducing the following V-macro fields

$$S_{ij} = <C_{ijkl}> U_{k,l} + <C_{ijkl}h^a{}_{,k}> V_l^a + <B_{ij}> \Theta + \underline{<B_{ij}h^b> \Phi^b} \ ,$$

$$H_i^a = <C_{ijkl}h^a{}_{,j}> U_{k,l} + <C_{ijkl}h^a{}_{,j}h^b{}_{,l}> V_k^b + <B_{ij}h^a{}_{,j}> \Theta +$$

$$+ <B_{ij}h^a{}_{,j}h^b> \Phi^b \ , \tag{5}$$

$$Q_i = <K_{ij}> \Theta_{,j} + <K_{ij}h^b{}_{,j}> \Phi^b \ ,$$

$$G^a = <K_{ij}h^a{}_{,j}> \Theta_{,i} + <K_{ij}h^a{}_{,i}h^b{}_{,j}> \Phi^b \ ,$$

we obtain

$$<\rho> \ddot{U} - S_{ij,j} = <\rho> b_i \ ,$$

$$\underline{<\rho h^a h^b> \ddot{V}_i^b} + H_i^a = 0 \ , \tag{6}$$

$$<c> \dot{\Theta} - Q_{i,i} + <B_{ij}> \dot{U}_{i,j} + \underline{<B_{ij}h^b{}_{,j}> \dot{V}_i^b} = <\alpha> \ ,$$

$$\underline{<ch^a h^b> \dot{\Phi}^b} + \underline{<B_{ij}h^b{}_{,j}h^a> \dot{V}_i^b} + G^a = 0 \ ,$$

and $S_{ij}n_j = s_i$, $Q_i n_i = q$ on $\partial\Omega$. Substituting the right-hand sides of Eqs (5) into Eqs (6) we arrive at the system of 4+4n equations for macro-displacements U_i, macro--temperature Θ and correctors V_i^a, Φ^a. These equations have constant coefficients and hence, represent a certain macro-model of the periodic body under consideration. The underlined constants in Eqs (5), (6) depend on the microstructure length parameter l and describe the effect of the microstructure on the behaviour of the composite. Hence, Eqs (5), (6) represent the refined macro-thermoelastodynamics of composite materials and will be called macro-constitutive equations and local macro-balance equations, respectively. It has to be emphasized that the equations for correctors (the second and the last from Eqs(6)) are ordinary differential equations and hence, the correctors play a role of certain internal dynamical variables, i.e., they do not enter boundary conditions. For a homogeneous body from Eqs (5), (6) we obtain $V_i^a = 0$, $\Phi^a = 0$, provided that initial values of V_i^a, $\dot{V}_i^a$ and Φ^a are equal to zero. Hence, we see that correctors describe the effect of inhomogeneity on the macro-behaviour of the body.

4. Effective Modulus Theory

Scaling the microstructure down in Eqs (5), (6) by means of $l \rightarrow 0$, we arrive at a certain asymptotic theory; in this case the underlined terms are equal to zero and we arrive at conditions $H_i^a = 0$, $G^a = 0$ representing the systems of linear algebraic equations for the correctors V_i^a, Φ^a. Hence, eliminating correctors from Eqs (5), (6) we obtain equations of a certain special effective modulus theory, given by

$$<\rho> \ddot{U}_i - S_{ij,j} = <\rho> b_i \ , \qquad c^{eff} \dot{\Theta} - Q_{i,i} + B_{ij}^{eff} \dot{U}_{i,j} = <\alpha> \ , \tag{7}$$
$$S_{ij} = C_{ijkl}^{eff} U_{k,l} + B_{ij}^{eff} \Theta \ , \qquad Q_i = K_{ij}^{eff} \Theta_{,j} \ .$$

The constant coefficients in Eqs (7) are termed effective modulae and defined by:

$$C_{ijkl}^{eff} \equiv <C_{ijkl}> - <C_{ijmn} h^a,_m> D_{np}^{ab} <C_{klrp} h^b,_r> \ ,$$
$$B_{ij}^{eff} \equiv <B_{ij}> - <B_{kl} h^a,_l> D_{kp}^{ab} <C_{prij} h^b,_r> \ ,$$
$$K_{ij}^{eff} \equiv <K_{ij}> - <K_{ik} h^a,_k> D^{ab} <K_{jl} h^b,_l> \ ,$$
$$c^{eff} \equiv <c> - <B_{ij} h^a,_j> D_{ik}^{ab} <B_{kl} h^b,_l> \ ,$$

where D_{ik}^{ab} and D^{ab} represent linear transformations inverse to those given by $<C_{ijkl}h^a,_j h^b,_l>$ and $<K_{ij}h^a,_i h^b,_j>$, respectively. The aforementioned results have been derived independently in [3], without any reference to the refined theory.

5. Example of Application

Let us consider a laminated body made of two orthotropic constituents. In this case we introduce one micro-shape function $h(x_1)$ (periodic in a direction x_1 normal to the lamina interfaces, cf. [1]), denoting by V_k, Φ the pertinent correctors related to $h(x_1)$. For the sake of simplicity let us neglect the body forces b_i and heat supply α. The aim of this example is to show a difference between results obtained from the refined theory and those derived from the effective modulus theory. To this end we shall consider the homogeneous boundary conditions for the macro-displacements and macro-temperature: $U_i = 0$, $Q_i n_i = 0$ on $\partial\Omega$, homogeneous initial conditions for the macro-displacements and correctors: $U_i = 0$, $\dot{U}_i = 0$, $V_i = 0$, $\dot{V}_i = 0$, $\Phi = 0$ at $\tau = 0$ and the initial condition for the macro-temperature in the form: $\Theta = \Theta_0$ at $\tau = 0$, $\Theta_0 = \text{const.}$ Then in the framework of the refined theory we obtain $U_i = 0$, $V_2 = V_3 = \Phi = 0$ for every $\tau > 0$, $x \in \Omega$, and

$$\Theta = \Theta_0 K (1 - \frac{<B_{11}h,_1>^2}{<c><C_{1111}(h,_1)^2>} \cos\omega\tau) , \tag{8}$$

$$V_1 = -\Theta_0 K \frac{<B_{11}h,_1>}{<C_{1111}(h,_1)^2>} (1 - \cos\omega\tau) ,$$

where

$$K \equiv \frac{<c><C_{1111}(h,_1)^2>}{<c><C_{1111}(h,_1)^2> - <B_{11}h,_1>^2} , \quad \omega^2 \equiv \frac{<c><C_{1111}(h,_1)^2> - <B_{11}h,_1>^2}{<\rho h^2><c>} .$$

At the same time, the effective modulus theory yields the constant values of Θ and V_1 for every $\tau \geq 0$:

$$\Theta = \Theta_0 \quad , \quad V_1 = -\Theta_0 \frac{< B_{11} h_{,1} >}{< C_{1111}(h_{,1})^2 >} \quad . \tag{9}$$

From Eqs (8) it follows that the inhomogeneity of the medium and the coupling between temperature and deformations produce highly oscillating character of the macro-temperature field; this fact is not described by the effective modulus theory leading to (9). Thus we conclude that in investigations of non-stationary processes in thermo-elastic composites, the refined macro-elastodynamics has to be used instead of the effective modulus theory.

Acknowledgment. This research work was supported by KBN, Warsaw, under grant 3 3310 92 03.

6. References

1. Bensoussan, A., Lions, J. L., Papanicolaou, G. (1980), *Asymptotic Analysis of Periodic Structures*, North-Holland Publ. Comp., Amsterdam.

2. Bachvalov, N. S., Panasenko, G. P. (1984), *Process Averaging in Periodic Media* [in Russian], Nauka, Moskva.

3. Matysiak, S., Woźniak, Cz. (1987) Micromorphic effects in a modelling of periodic multilayered elastic composites, *Int. J. of Engng. Sci.* **25**, 549-559.

4. Woźniak, Cz. (1989) On the modelling of thermo-unelastic periodic composites, *Acta Mechanica* **80**, 81-94.

5. Woźniak, Cz. (1993) Refined macro-dynamics of periodic structures, *Arch. Mech.* **45**, 295-304.

6. Konieczny, S., Woźniak, Cz. Woźniak, M. (1993) A note on dynamic modelling of periodic composites, *Arch. Mech.* **45**, 779-783.

7. Woźnak, Cz. (1993) Micro-macro dynamics of periodic material structures, in T. Moan et al. (eds.), *Structural Dynamics*, Proceedings of Eurodyn '93, A. A. Balkema, Rotterdam, 573-575.

MICROMECHANICAL MODELING OF EFFECTIVE ELECTRO-THERMO-ELASTIC PROPERTIES OF TWO-PHASE PIEZOELECTRIC COMPOSITES

N. YU

Department of Engineering Science and Mechanics
University of Tennessee, Knoxville, TN 37996-2030, U. S. A.

ABSTRACT. A periodic microstructural model is developed to estimate the coupled electro-thermo-elastic response of piezoelectric composites. The model includes vital microstructural parameters, such as the constituent properties and shapes, and provides analytic estimates of the effective electro-thermo-elastic moduli of two-phase piezoelectric composites.

1. Introduction

The coupling of electric, thermal, and elastic responses within piezoelectric composites has attracted extensive attention recently in the light of the development of smart materials. Some of the existing works focus on the derivation of universal relations among the effective electro-thermo-elastic properties of piezoelectric composites [1-8], and some others study the local fields and effective behavior of piezoelectric composites, using various micromechanics models [1, 7-19]. It is noted that most of the existing micromechanics models, which were originally developed for structural composites in purely mechanical applications (see [20] for a comprehensive account), consider representative volume elements of statistically homogeneous piezoelectric composites with random microstructure. On the other hand, as a result of fabrication procedures, some of the piezoelectric composites may exhibit periodic microstructure [21, 22]. In the present work, a general model based on periodic microstructure is developed to estimate

R. Pyrz (ed.), IUTAM Symposium on Microstructure-Property Interactions in Composite Materials, 397–405.
© *1995 Kluwer Academic Publishers.*

the effective electro-thermo-elastic properties and to study the microstructure-property relationships in two-phase piezoelectric composites.

2. The Periodic Microstructural Model

In the present model, the composites are considered to be infinite piezoelectric solids containing periodically distributed piezoelectric inhomogeneities. In view of the periodicity, the composite can be regarded as a collection of unit cells; see Figure 1 for the example of a particulate composite with periodically distributed ellipsoidal particles. Thus, the coupled electro-thermo-elastic response of the unit cell is representative of that of the infinite composite, under applied homogeneous electric, thermal, and mechanical loading.

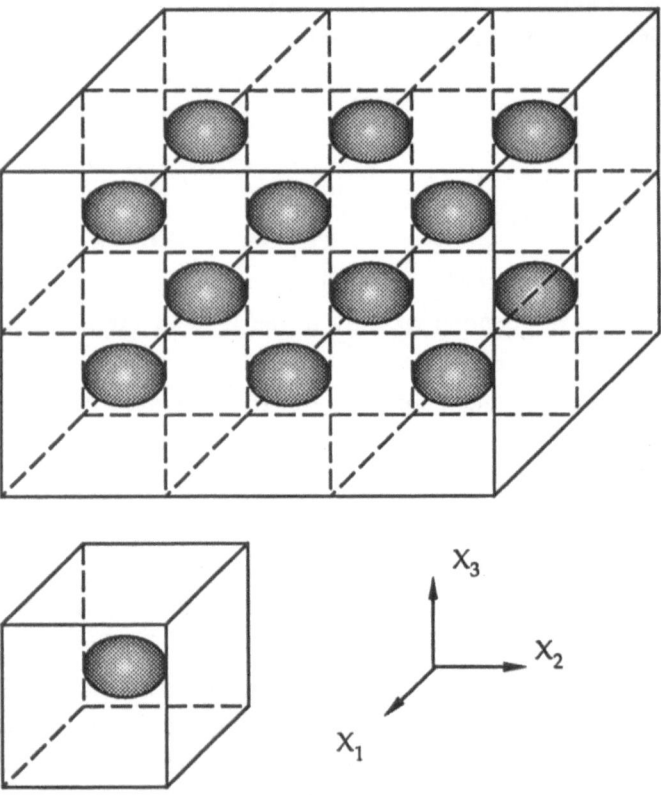

Figure 1. Composite with periodically distributed ellipsoidal particles and its unit cell.

Consider a parallelepiped unit cell, U, consisting of a continuous piezoelectric matrix, M, and a different piezoelectric phase (i.e., an inhomogeneity), Ω, of arbitrary shape. Let the unit cell (or equivalently, the infinite composite with periodic microstructure) be subjected to linear displacements $u(x) = x \cdot \varepsilon^\circ$ and a linear electric potential $\phi(x) = -x \cdot E^\circ$, along with a uniform temperature change $\theta(x) = \theta^\circ$ on its boundary surface, where ε° and E° are the homogeneous strains and electric fields, respectively, and x is the position vector. The superscript o indicates that the corresponding tensors are homogeneous. The average strains, electric fields and temperature change over U are therefore, ε°, E°, and θ°, respectively. However, the presence of periodically distributed inhomogeneities (fibers, whiskers, particles, etc.) induces a periodic perturbation field, of which the average over U vanishes. For example, the local strain field can be expressed by

$$\varepsilon(x) = \varepsilon^\circ + \varepsilon^P(x) = \varepsilon^\circ + \sum_{\zeta \neq 0}^{\pm\infty} \hat{\varepsilon}(\zeta) \exp(i\zeta \cdot x), \tag{1}$$

where the superscript p stands for the periodic perturbation; ζ is a vector with components

$$\zeta_k = \frac{\pi n_k}{a_k} \quad (k=1,2,3; \text{ k not summed}); \tag{2}$$

n_k are integers; a_k are the dimensions of the unit cell; and $i = \sqrt{-1}$. The term associated with $\zeta = 0$ (i.e., the average strains over U, ε°) is excluded from the Fourier series. The Fourier coefficients $\hat{\varepsilon}(\zeta)$ are defined by

$$\hat{\varepsilon}(\zeta) = \frac{1}{V_U} \int_U \varepsilon(x) \exp(-i\zeta \cdot x) \, dV_x \tag{3}$$

with V_U being the volume of the unit cell. Similar expressions exist for the stresses $\sigma(x)$, electric fields $E(x)$, temperature change $\theta(x)$, and electric displacements $D(x)$. The effective electro-thermo-elastic moduli of the piezoelectric composite are defined by

$$\langle\sigma(x)\rangle_U = \overline{C} : \varepsilon^\circ - \overline{e}^T \cdot E^\circ - \overline{\lambda}\theta^\circ, \tag{4}$$

$$\langle D(x) \rangle_U = \bar{e}:\varepsilon^\circ + \bar{\kappa} \cdot E^\circ - \bar{p}\, \theta^\circ, \tag{5}$$

where the angle brackets with the subscript U mean the volume average over U. The superscript T stands for the transpose of the corresponding third-order tensor. $\bar{C}, \bar{e}, \bar{\kappa}, \bar{\lambda},$ and $\bar{p}$ are the effective stiffness, piezoelectric, dielectric, thermal stress, and pyroelectric moduli of the piezoelectric composites, respectively.

3. Homogenization

To determine the effective electro-elastic moduli of piezoelectric composites, let the unit cell be subjected to electro-mechanical loading of $u(x) = x \cdot \varepsilon^\circ$ and $\phi(x) = -x \cdot E^\circ$ on its boundary surface. The effective thermal stress and pyroelectric moduli of the composite are to be determined separately; see Section 4. Instead of dealing with the composite unit cell, one can homogenize the original unit cell by the matrix phase [23]. A suitable amount of transformation stresses, $\sigma^*(x)$, and transformation electric displacements, $D^*(x)$, which are strain-free and electric field-free, respectively, must be introduced in Ω of the homogenized unit cell to satisfy the following constraint conditions:

$$C^\Omega : \langle \varepsilon(x) \rangle_\Omega - e^{\Omega T} \cdot \langle E(x) \rangle_\Omega = C^M : \langle \varepsilon(x) \rangle_\Omega - e^{MT} : \langle E(x) \rangle_\Omega + \langle \sigma^*(x) \rangle_\Omega, \tag{6}$$

$$e^\Omega : \langle \varepsilon(x) \rangle_\Omega + \kappa^\Omega \cdot \langle E(x) \rangle_\Omega = e^M : \langle \varepsilon(x) \rangle_\Omega + \kappa^M \cdot \langle E(x) \rangle_\Omega + \langle D^*(x) \rangle_\Omega, \tag{7}$$

which require the electric, thermal, and elastic fields in Ω to be preserved in an *average* sense after the homogenization. The quantities with the superscripts and subscripts $\Omega(M)$ represent the material properties of and the volume averages over $\Omega(M)$, respectively. To solve (6) and (7) for $\langle \sigma^*(x) \rangle_\Omega$ and $\langle D^*(x) \rangle_\Omega$, one must determine the relationship between transformation fields, $\sigma^*(x)$ and $D^*(x)$, and their resulting perturbation fields, $\varepsilon^P(x)$, and $E^P(x)$. This inclusion problem is solved as follows.

Consider a homogeneous unit cell with stiffness, piezoelectric, and dielectric moduli, $C^M, e^M,$ and κ^M. Furthermore, transformation stresses, $\sigma^*(x)$, and transformation electric displacements, $D^*(x)$, which are also periodic functions of x, are prescribed in Ω. The constitutive relations for such a homogeneous piezoelectric unit cell are given by

$$\sigma(x) = C^M : \varepsilon(x) - e^{MT} \cdot E(x) + \sigma^*(x), \tag{8}$$

$$D(x) = e^M : \varepsilon(x) + \kappa^M \cdot E(x) + D^*(x), \tag{9}$$

where the strains $\varepsilon(x)$ and electric fields $E(x)$ are defined by

$$\varepsilon(x) = \frac{1}{2} \left\{ \nabla \otimes u(x) + [\nabla \otimes u(x)]^T \right\}, \tag{10}$$

$$E(x) = - \nabla \phi(x) \tag{11}$$

with $u(x)$ and $\phi(x)$ being the elastic displacements and electric potential, respectively. The equilibrium and Gauss' law read

$$\nabla \cdot \sigma(x) = 0, \tag{12}$$

$$\nabla \cdot D(x) = 0. \tag{13}$$

Substitution of (8) - (11) into (12) and (13), with the help of Fourier series expansion for all the electro-elastic fields, yields

$$\hat{u}(\zeta) = i \, (\zeta \cdot C^M \cdot \zeta)^{-1} \cdot \left[(\zeta \cdot \hat{\sigma}^*)(\zeta \cdot \kappa^M \cdot \zeta) + (\zeta \cdot e^{MT} \cdot \zeta)(\zeta \cdot \hat{D}^*) \right] / \Delta, \tag{14}$$

$$\hat{\phi}(\zeta) = i \left[(e^{MT} \cdot \zeta) : \hat{\Gamma}(\zeta) : \hat{\sigma}^* - \zeta \cdot \hat{D}^* \right] / \Delta, \tag{15}$$

where $\hat{u}(\zeta)$ and $\hat{\phi}(\zeta)$, defined in a manner similar to (3), are the Fourier coefficients of displacements $u(x)$ and electric potential $\phi(x)$, respectively. Δ and $\hat{\Gamma}(\zeta)$ are defined by

$$\Delta = (e^{MT} \cdot \zeta) : \hat{\Gamma}(\zeta) : (e^{MT} \cdot \zeta) + \zeta \cdot \kappa^M \cdot \zeta, \tag{16}$$

$$\hat{\Gamma}(\zeta) = sym \left[\zeta \otimes (\zeta \cdot C^M \cdot \zeta)^{-1} \otimes \zeta \right] \tag{17}$$

with sym representing the symmetric part of the corresponding fourth-order tensor. Consequently, the strains and electric fields are given by (10) and (11), whereas the stresses and electric displacements are obtained by (8) and (9), respectively. When the transformation stresses and electric displacements are uniform in Ω, i.e.,

$$\sigma^*(x) = \bar{\sigma}^* \text{ and } D^*(x) = \bar{D}^* \quad x \in \Omega, \tag{18}$$

where $\bar{\sigma}^*$ and $\bar{D}^*$ are constant second- and first-order tensors, respectively, the average perturbation strains and electric fields are given by

$$\langle \varepsilon^P(x) \rangle_\Omega = \sum_{\zeta \neq 0}^{\pm\infty} \hat{\varepsilon}^P(\zeta) \, g(\zeta), \tag{19}$$

$$\langle E^P(x) \rangle_\Omega = \sum_{\zeta \neq 0}^{\pm\infty} \hat{E}^P(\zeta) \, g(\zeta), \tag{20}$$

where

$$\hat{\varepsilon}^P(\zeta) = -fg(-\zeta)\hat{\Gamma}(\zeta) \cdot [(\zeta \cdot \kappa^M \cdot \zeta)\bar{\sigma}^* + (\zeta \cdot \bar{D}^*)(e^{MT} \cdot \zeta)]/\Delta, \tag{21}$$

$$\hat{E}^P(\zeta) = fg(-\zeta)\{(e^{MT} \cdot \zeta) : \hat{T}(\zeta) : \bar{\sigma}^* - \zeta \cdot \bar{D}^*\}\zeta/\Delta, \tag{22}$$

f is the inhomogeneity volume fraction, and the geometry of inhomogeneity is accounted for in

$$g(\zeta) = \frac{1}{V_\Omega} \int_\Omega \exp(i\zeta \cdot x) \, dV_x \tag{23}$$

with V_Ω being the volume of Ω. Explicit expressions of $g(\zeta)$ for a broad class of inhomogeneity shape have been reported in [24]. For example, the $g(\zeta)$ for an ellipsoid is given by

$$g = 3(\sin\eta - \eta\cos\eta)/\eta^3, \tag{24}$$

where

$$\eta = \pi\sqrt{(n_1 b_1/a_1)^2 + (n_2 b_2/a_2)^2 + (n_3 b_3/a_3)^2} \tag{25}$$

with b_1, b_2, and b_3 being the principal radii of the ellipsoid, whereas the $g(\zeta)$ for an elliptic cylinder with principal radii b_1 and b_2, and length b_3, is expressed by

$$g = \frac{\sin\left(\pi n_3 b_3/a_3\right)}{\left(\pi n_3 b_3/a_3\right)} \frac{2J_1\left(\pi\sqrt{\left(n_1 b_1/a_1\right)^2+\left(n_2 b_2/a_2\right)^2}\right)}{\left(\pi\sqrt{\left(n_1 b_1/a_1\right)^2+\left(n_2 b_2/a_2\right)^2}\right)}. \tag{26}$$

As can be seen, the shape and size of the inhomogeneity, as well as the material properties of the matrix and inhomogeneity, are explicitly included in the present formulation.

4. Effective Electro-Thermo-Elastic Moduli

Now, turning to (6) and (7), for simplicity, one assumes the transformation stresses and transformation electric displacements to be constant within Ω. Therefore, $<\sigma^*(x)>_\Omega = \bar{\sigma}^*$ and $<D^*(x)>_\Omega = \bar{D}^*$. By substituting (19) and (20) into (6) and (7), one can readily solve the appropriate $\bar{\sigma}^*$ and $\bar{D}^*$, which satisfy the constraint conditions (6) and (7), i.e., the electro-elastic fields are preserved after the homogenization. The effective electro-elastic moduli are then determined by

$$\bar{C}:\varepsilon^\circ - \bar{e}^T \cdot E^\circ = C^M:\varepsilon^\circ - e^{MT}\cdot E^\circ + f\,\bar{\sigma}^*, \tag{27}$$

$$\bar{e}:\varepsilon^\circ + \bar{\kappa}\cdot E^\circ = e^M:\varepsilon^\circ + \kappa^M\cdot E^\circ + f\,\bar{D}^*. \tag{28}$$

The effective thermal stress and pyroelectric moduli, $\bar{\lambda}$ and $\bar{p}$, can be expressed in terms of the effective electro-elastic moduli, $\bar{C}$, $\bar{e}$, and $\bar{\kappa}$, and the thermal stress and pyroelectric moduli of the matrix and inhomogeneity [3, 7, 8, 15].

Acknowledgment

This work was supported by the University of Tennessee, Knoxville, under the 1994-95 UTK Professional Development Award Program for Tenure-Track Faculty. The author would like to thank Ms. Donna Strohl for her skillful typing.

References

1. Benveniste, Y. and Dvorak, G. J. (1992) Uniform fields and universal relations in piezoelectric composites, *J. Mech. Phys. Solids* **40**, 6, 1295-1312.
2. Schulgasser, K. (1992) Relationships between the effective properties of transversely isotropic piezoelectric composites, *J. Mech. Phys. Solids* **40**, 2, 473-479.
3. Dunn, M. L. (1993) Exact relations between the thermoelectrcoelastic moduli of heterogeneous materials, *Proc. R. Soc. Lond.* **A441**, 549-557.
4. Chen, T. (1993) Piezoelectric properties of multiphase fibrous composites: Some theoretical results, *J. Mech. Phys. Solids* **41**, 11, 1781-1794.
5. Nan, C. (1993) Comment on "Relationships between the effective properties of transversely isotropic piezoelectric composites", *J. Mech. Phys. Solids* **41**, 9, 1567-1570.
6. Benveniste, Y. (1993) Exact results in the micromechanics of fibrous piezoelectric composites exhibiting pyroelectricity, *Proc. R. Soc. Lond.* **A441**, 59-81.
7. Benveniste, Y. (1993) Universal relations in piezoelectric composites with eigenstress and polarization fields, Part I: Binary media - Local fields and effective behavior, *J. Appl. Mech.* **60**, 265-269.
8. Benveniste, Y. (1993) Universal relations in piezoelectric composites with eigenstress and polarization fields, Part II: Multiphase media - Effective behavior, *J. Appl. Mech.* **60**, 270-275.
9. Deeg, W. F. (1980) The analysis of dislocation, crack, and inclusion problems in piezoelectric solids, Ph. D. Dissertation, Standford University.
10. Grekov, A. A., Kramarov, S. O., and Kuprienko, A. A. (1987) Anomalous behavior of the two-phase lamellar piezoelectric texture, *Ferroelectrics* **76**, 1-2, 43-48.
11. Grekov, A. A., Kramarov, S. O., and Kuprienko, A. A. (1989) Effective properties of a transversely isotropic piezoelectric composite with cylindrical inclusions, *Mech. Compos. Mater.* **25**, 54-60.
12. Wang, B. (1992) Three-dimensional analysis of an ellipsoidal inclusion in a piezoelectric material, *Int. J. Solids Structures* **29**, 3, 293-308.
13. Benveniste, Y. (1992) The determination of the elastic and electric fields in a piezoelectric inhomogeneity, *J. Appl. Phys.* **72**, 3, 1086-1095.

14. Dunn, M. L. and Taya, M. (1993) Micromechanics predictions of the effective electroelastic moduli of piezoelectric composites, *Int. J. Solids Structures* **30**, 2, 161-175.

15. Dunn, M. L. (1993) Micromechanics of coupled electroelastic composites: Effective thermal expansion and pyroelectric coefficients, *J. Appl. Phys.* **73**, 10, 5131-5140.

16. Chen, T. (1993) An invariant treatment of interfacial discontinuities in piezoelectric media, *Int. J. Engr. Science* **31**, 7, 1061-1072.

17. Dunn, M. L. and Taya, M. (1993) An analysis of piezoelectric composite materials containing ellipsoidal inhomogeneities, *Proc. R. Soc. Lond.* **A443**, 265-287.

18. Dunn, M. L. and Taya, M. (1994) Electroelastic field concentrations in and around inhomogeneities in piezoelectric solids, *J. Appl. Mech.* **61**, 2, 474-475.

19. Dunn, M. L. (1994) Electroelastic Green's functions for transversely isotropic piezoelectric media and their application to the solution of inclusion and inhomogeneity problems, *Int. J. Engr. Science* **32**, 119-131.

20. Nemat-Nasser, S. and Hori, M. (1993) *Micromechanics: Overall properties of heterogeneous materials*, Elsevier, Amsterdam, New York, North Holland.

21. Klicker, K. A., Biggers, J. V. and Newnham, R. E. (1981) Composites of PZT and epoxy for hydrostatic transducer applications, *J. Am. Cer. Soc.* **64**, 5-9.

22. Savakus, H. P., Klicker, K. A. and Newnham, R. E. (1981) PZT-Epoxy piezoelectric transducers: A simplified fabrication procedure, *Mat. Res. Bull.* **16**, 677-680.

23. Eshelby, J. D. (1957) The determination of the elastic field of an ellipsoidal inclusion, and related problems, *Proc. R. Soc. London* **A241**, 376-396.

24. Nemat-Nasser, S., Yu, N. and Hori, M. (1993) Bounds and estimates of overall moduli of composites with periodic microstructure, *Mech. Mater.* **15**, 163-181.

MICROSTRUCTURE EVOLUTION
IN IDEALLY PLASTIC COMPOSITES

M. ZAIDMAN and P. PONTE CASTAÑEDA
Department of Mechanical Engineering
and Applied Mechanics
University of Pennsylvania
Philadelphia, PA 19104, U. S. A.

1. Introduction

In this paper, we present a constitutive model for two-phase rigid/ideally plastic composites which accounts, approximately, for finite changes in the microstructure during a given loading program. To this end, variational estimates are computed for the instantaneous effective yield functions of composites with particulate microstructures consisting of aligned ellipsoidal inclusions of one phase dispersed in a matrix of a second phase. Then, the problem of finding appropriate state variables to characterize the evolution of the microstructure is addressed. It is argued that, under triaxial loading conditions, the aligned ellipsoidal inclusions deform into aligned ellipsoidal inclusions of different size and shape. This suggests that the volume fractions of the phases and the aspect ratios of the inclusions are the appropriate variables to characterize the state of the microstructure. Evolution laws, relating the change in volume and shape of a typical ellipsoidal inclusion to the current value of the average strain-rates in the inclusion, which in turn may be related to the average strain-rate in the composite, are developed. Estimates for the average strain rate in the inclusion are obtained from the works of Hashin and Shtrikman (1963) and Willis (1977) for linear composites, and the results are extended to the rigid/ideally plastic composites by means of the variational principles of Ponte Castañeda (1991). The resulting constitutive model takes the form of effective

R. Pyrz (ed.), IUTAM Symposium on Microstructure-Property Interactions in Composite Materials, 407–418.
© 1995 Kluwer Academic Publishers.

stress/strain-rate relations complemented by ordinary differential equations for the evolution of the volume fraction and aspect ratios of the inclusions.

An important result, in the context of the proposed constitutive model, is that composites with ideally plastic phases, because of microstructure evolution, do not exhibit *effective* ideally plastic behavior, but may actually display hardening, or even softening behavior, depending on the specific loading conditions. The introduction of microstructure evolution is also shown to affect the predictions for strain localization in composite materials. Section 2 presents the model for linearly viscous composites with non-dilute concentrations of aligned ellipsoidal inclusions. Section 3 deals with the extension of the model to incompressible rigid/perfectly plastic composites. Section 4 is concerned with the application of the constitutive model to plane-strain deformations of initially isotropic composites. The final section includes some discussion of the main findings.

2 Microstructure evolution for linearly viscous composites

In this section, we address the problem of determining appropriate state variables to characterize the evolution of the microstructure in linearly viscous composite materials. The class of microstructures that is considered corresponds to aligned, self-similar ellipsoidal inclusions, with aspect ratios $w_1 = l_3/l_1$ and $w_2 = l_3/l_2$ (where l_1, l_2, l_3 are the dimensions of the typical ellipsoid along the x_1, x_2, x_3 axes), distributed in a continuous matrix phase according to "ellipsoidal" two-point correlation functions (Willis, 1977).

2.1. EFFECTIVE VISCOSITY AND STRAIN CONCENTRATION TENSORS

If we denote the proportions and viscosity tensors of the inclusion and matrix phases by $c^{(1)}$, $c^{(2)}$ and $\mathbf{L}^{(1)}$, $\mathbf{L}^{(2)}$, respectively, the effective viscous compliance tensor $\tilde{\mathbf{M}} = \tilde{\mathbf{L}}^{-1}$, may be expressed in the form

$$\tilde{\mathbf{M}} = \left\{ \mathbf{I} + c^{(1)} \left[\left(\mathbf{M}^{(1)} \mathbf{L}^{(2)} - \mathbf{I} \right)^{-1} + c^{(2)} \left(\mathbf{I} - \mathbf{S}^{(2)} \right) \right]^{-1} \right\} \mathbf{M}^{(2)}, \qquad (2.1)$$

where $\mathbf{S}^{(2)}$ denotes the Eshelby (1957) tensor associated with an ellipsoidal inclusion, with the given aspect ratios w_1 and w_2, embedded in a matrix of phase 2.

This effective viscous compliance tensor serves to relate the average Eulerian strain rate $\overline{\mathbf{D}}$ to the the average Cauchy stress $\overline{\sigma}$ via

$$\overline{\mathbf{D}} = \tilde{\mathbf{M}}\overline{\sigma}. \tag{2.2}$$

The average strain-rate in the inclusions $\mathbf{D}^{(1)}$, which we will need later in this section, is given, in terms of the strain-rate concentration tensor (Hill, 1965) $\mathbf{A}^{(1)}$, such that

$$\mathbf{A}^{(1)} = \left[\mathbf{I} + c^{(2)}\mathbf{S}^{(2)}\left(\mathbf{M}^{(2)}\mathbf{L}^{(1)} - \mathbf{I}\right)\right]^{-1}, \tag{2.3}$$

by

$$\mathbf{D}^{(1)} = \mathbf{A}^{(1)}\overline{\mathbf{D}}. \tag{2.4}$$

2.2. EVOLUTION LAWS FOR COMPOSITES SUBJECTED TO TRIAXIAL LOADS

We assume that under a loading process involving uniform triaxial loading, where the loading axes are aligned with the principal axes of the ellipsoids, initially ellipsoidal inclusions remain, on the average, ellipsoidal, with possibly different size and shape (for details, in the context of porous materials, refer to Ponte Castañeda and Zaidman, 1994). This assumption justifies the use of relations (2.1) as instantaneous stress/strain-rate relations. Then, given relations (2.1), the appropriate state variables, which serve to describe the microstructure of the composite, become the aspect ratios w_1, w_2 and the volume fractions $c^{(1)}$, $c^{(2)}$.

By invoking mass conservation of the phases, it may be shown that the volume fractions of the phases satisfy the kinematical relations

$$\dot{c}^{(1)} = -\dot{c}^{(2)} = c^{(1)}c^{(2)}\left(\text{tr}\mathbf{D}^{(1)} - \text{tr}\mathbf{D}^{(2)}\right), \tag{2.5}$$

where $\mathbf{D}^{(2)}$ is the average strain rate in the matrix phase. In addition, evolution equations for the aspect ratios of the inclusions may be obtained from the relations

$$\dot{w}_1 = \left(D_{33}^{(1)} - D_{11}^{(1)}\right)w_1 \quad \text{and} \quad \dot{w}_2 = \left(D_{33}^{(1)} - D_{22}^{(1)}\right)w_2. \tag{2.6}$$

Or, in terms of the strain-rate concentration tensor $\mathbf{A}^{(1)}$ of (2.3), we have that

$$\dot{w}_1 = w_1\left(A_{33ij}^{(1)} - A_{11ij}^{(1)}\right)\overline{D}_{ij} \quad \text{and} \quad \dot{w}_2 = w_2\left(A_{33ij}^{(1)} - A_{22ij}^{(1)}\right)\overline{D}_{ij}. \tag{2.7}$$

The resulting *effective* constitutive model is thus given by equations (2.1) and (2.2), complemented by the evolution equations (2.5) and (2.7) for the state variables $c^{(1)}$, $c^{(2)}$ and w_1, w_2. In other words, equations (2.1), (2.2), (2.5) and (2.7) can be solved for the evolution of $c^{(1)}$, $c^{(2)}$, w_1, w_2, and $\bar{\sigma}$ (if $\mathbf{v} = \bar{\mathbf{D}}\mathbf{x}$ is prescribed on the boundary), or $\bar{\mathbf{D}}$ (if $\sigma\mathbf{n} = \bar{\sigma}\mathbf{n}$ is prescribed on the boundary).

Before proceeding with the extension of the model for nonlinear composites, we specialize the model for the linearly viscous composites to materials with *incompressible* phases. In this case, due to incompressibility, the proportions of the inclusions and the matrix, $c^{(1)}$ and $c^{(2)}$, respectively, remain fixed during a deformation process, so that relations (2.5) become trivially satisfied. We also note that effective moduli tensor in this case may be written in terms of the shear moduli of the inclusions and the matrix $\mu^{(1)}$, $\mu^{(2)}$, respectively, in the form

$$\tilde{\mathbf{M}} = \frac{1}{3\mu^{(2)}}\,\tilde{\mathbf{m}}(y; w_1, w_2),\tag{2.8}$$

where $y = \mu^{(2)}/\mu^{(1)}$. Similarly, the strain-rate concentration tensor may be written in the form

$$\mathbf{A}^{(1)} = \mathbf{A}^{(1)}(y; w_1, w_2).\tag{2.9}$$

3 Microstructure evolution for composites with rigid ideally plastic phases

In this section, we consider two-phase composites made of rigid/perfectly plastic, homogeneous, isotropic and incompressible phases, with ellipsoidal symmetry. The local constitutive behavior of the rigid-plastic materials is defined by the relation

$$\mathbf{D}(\mathbf{x}) = \dot{\lambda}(\mathbf{x})\frac{\partial\Phi}{\partial\sigma}(\mathbf{x},\sigma),\tag{3.1}$$

where σ and $\mathbf{D}$ denote the Cauchy stress and Eulerian strain-rate tensors, respectively and Φ is the local yield function. The effective behavior of the composite is then given by (Hill, 1967; Suquet, 1983)

$$\bar{\mathbf{D}} = \dot{\Lambda}\frac{\partial\tilde{\Phi}}{\partial\bar{\sigma}}(\bar{\sigma}),\tag{3.2}$$

where $\tilde{\Phi}(\overline{\sigma}) = 0$ defines the effective yield surface of the composite, and $\dot{\Lambda}$ is a plastic loading parameter that is to be determined later in this section.

3.1. EFFECTIVE YIELD SURFACES

Using the variational procedure of Ponte Castañeda (1991), Ponte Castañeda and Zaidman (1995) derived estimates for the effective yield surfaces of the above family of composite materials. The yield functions may be written in terms of a one-dimensional optimization problem in the form (see also Suquet, 1993)

$$\tilde{\Phi}(\overline{\sigma}) = \max_{y \geq 0} \left\{ \left[c^{(2)} + c^{(1)} \left(\frac{\sigma_y^{(1)}}{\sigma_y^{(2)}} \right)^2 y \right]^{-1} \overline{\sigma} \cdot \left[\tilde{m}(y)\overline{\sigma} \right] - \left(\sigma_y^{(2)} \right)^2 \right\} \tag{3.3}$$

These yield surfaces are valid for general loading conditions and for arbitrary values of the phase concentrations and aspect ratios of the inclusions.

In Figure 1, we show examples of the resulting yield surfaces for fiber-reinforced materials. These yield surfaces were determined from expressions (2.1), (2.8) and (3.3) by prescribing one of the stress components and then solving for a second component, while the rest of the components was set equal to zero. The fibers have elliptical cross sections and are aligned in x_2 direction ($w_2 = 0$). The volume fractions are taken to be $c^{(1)} = c^{(2)} = 0.5$, and the contrast ratio is $\sigma_y^{(2)} / \sigma_y^{(1)} = 0.5$, so that the weaker material occupies the matrix phase. For illustrative purposes, yield surfaces are plotted for several aspect ratios, namely, $w_1 = 100, 1, 0.5, 0.01$. We observe the existence of weak and strong modes corresponding to flat and curved sectors, respectively, on the yield surfaces. These different domains are separated by areas of high curvature (*i.e.*, corners). Flat sectors on the effective yield surfaces of rigid/perfectly plastic composites with the weaker material in the matrix phase were predicted by the work of Ponte Castañeda and deBotton (1992) for the special cases of laminated and fiber-reinforced composites with circular cross sections. However, Ponte Castañeda and Zaidman (1995) found that the effective yield surfaces of composites with particulate microstructures (with discontinuous reinforcement) are always smooth. We mention that bi-modal yield

412

surfaces for fiber-reinforced composites have been observed experimentally, for example, by Dvorak *et al.* (1988). It was on this basis that Hashin (1980) and Dvorak and Bahei-El-Din (1987) proposed empirically based models for fiber-reinforced composites.

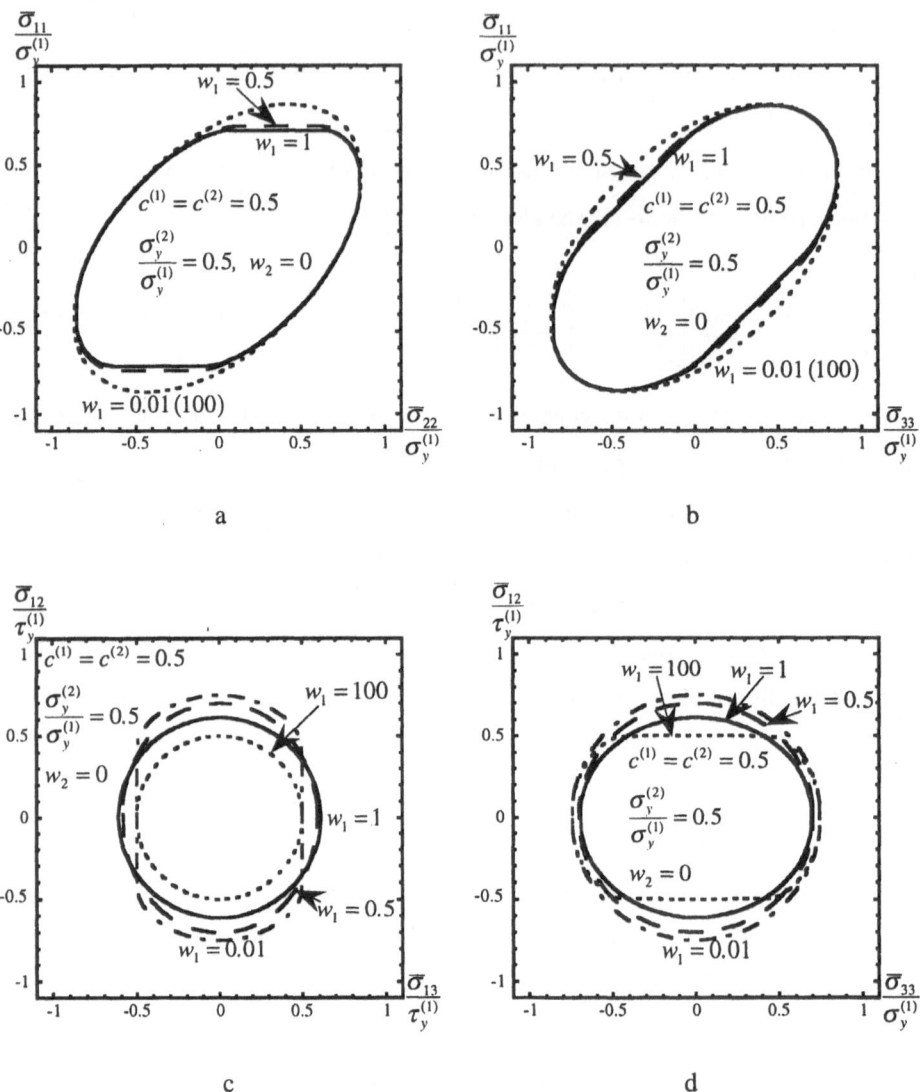

Figure 1 – Different cross sections of the yield surfaces of materials reinforced by fibers aligned in the x_2 direction, with elliptical cross section with aspect ratio w_1 and yield strength ratio $\sigma_y^{(2)}/\sigma_y^{(1)} = 0.5$.

3.2. EVOLUTION LAWS FOR COMPOSITES SUBJECTED TO TRIAXIAL LOADING

It may be demonstrated (see deBotton and Ponte Castañeda, 1993) that the effective stress/strain rate relations for a nonlinear composite may be expressed in the form

$$\overline{\mathbf{D}} = \tilde{\mathbf{M}}\left(\hat{\mu}^{(1)}, \hat{\mu}^{(2)}\right)\overline{\sigma},$$ (3.4)

where $\hat{\mu}^{(1)} = \hat{\mu}^{(1)}(\overline{\sigma})$ and $\hat{\mu}^{(2)} = \hat{\mu}^{(2)}(\overline{\sigma})$ are the optimal values of $\mu^{(1)}$ and $\mu^{(2)}$ from the appropriate optimization problem in the context of the variational principle of Ponte Castañeda (1991). For the special case of rigid/ideally plastic composites, the variational principle gives the optimization problem (3.3) for the variable $y = \mu^{(2)}/\mu^{(1)}$, whose optimal solution $\hat{y} = \hat{\mu}^{(2)}/\hat{\mu}^{(1)}$ may be seen to depend on the applied load. Note that, in spite the similarity between (2.1) and (3.4), the latter defines a non-linear constitutive relation due to the dependence of $\hat{\mu}^{(1)}$ and $\hat{\mu}^{(2)}$ (or $\hat{y}$) on the applied loads.

Being able to express the non-linear stress/strain-rate relations in a form similar to that of a linear constitutive relation suggests also being able to express the average strain rate in the inclusions of a nonlinear composite in terms of an expression similar to (2.4), but with $\mathbf{A}^{(1)}$, from equations (2.4) and (2.9), expressed in terms of $\hat{y}$ instead of y.

Recalling that the phases are incompressible, and therefore the volume fractions of the two phases remain unchanged during the deformation process, it follows from the above observations that evolution laws may be derived for the aspect ratios, w_1 and w_2, of the aligned ellipsoidal inclusions in the composite, from relations (2.6), to obtain

$$\dot{w}_1 = w_1 \left[A^{(1)}_{33ij}(\hat{y}) - A^{(1)}_{11ij}(\hat{y}) \right] \overline{D}_{ij} \quad \text{and} \quad \dot{w}_2 = w_2 \left[A^{(1)}_{33ij}(\hat{y}) - A^{(1)}_{22ij}(\hat{y}) \right] \overline{D}_{ij}.$$ (3.5)

Finally, the plastic loading factor $\dot{\Lambda}$ in relation (3.2) is obtained from the consistency condition, applied to the effective yield function (3.3), which requires that

$$\dot{\tilde{\Phi}}\left(\overline{\sigma}; w_1, w_2\right) = \frac{\partial \tilde{\Phi}}{\partial \overline{\sigma}_{ij}}\dot{\overline{\sigma}}_{ij} + \frac{\partial \tilde{\Phi}}{\partial w_1}\dot{w}_1 + \frac{\partial \tilde{\Phi}}{\partial w_2}\dot{w}_2 = 0,$$ (3.6)

where, we have used the fact that, for triaxial loading conditions, the principal axes of the applied stress $\overline{\sigma}$ do not rotate, and therefore we do not need to distinguish between the objective Jaumann and standard time derivatives. The final result for the effective stress/strain-rate relation may be written in the form

$$\overline{D}_{ij} = \frac{1}{H} \frac{\partial \tilde{\Phi}}{\partial \overline{\sigma}_{ij}} \frac{\partial \tilde{\Phi}}{\partial \overline{\sigma}_{kl}} \dot{\overline{\sigma}}_{kl},$$ (3.7)

where H is the effective rate of hardening, given by

$$H = -\left[\left(w_1 \frac{\partial \tilde{\Phi}}{\partial w_1}\right)\left[A_{33ij}^{(1)}(\hat{y}) - A_{11ij}^{(1)}(\hat{y})\right] \frac{\partial \tilde{\Phi}}{\partial \overline{\sigma}_{ij}} + \left(w_2 \frac{\partial \tilde{\Phi}}{\partial w_2}\right)\left[A_{33ij}^{(1)}(\hat{y}) - A_{22ij}^{(1)}(\hat{y})\right] \frac{\partial \tilde{\Phi}}{\partial \overline{\sigma}_{ij}}\right].$$

Note that even though the constituent phases exhibit perfectly plastic behavior, the effective hardening rate H is not zero in general due to the evolution of the microstructure.

In conclusion, the constitutive behavior of the rigid/ideally plastic composites, under finite deformations, is defined by the effective yield function (3.3), the stress/strain-rate relations (3.7), and the evolution equations (3.5).

4 Plane strain loading conditions for an initially isotropic composite

In this section, we illustrate the application of the above constitutive model by considering the finite deformation of a two-phase rigid/perfectly plastic composite under plane strain conditions. This problem is analogous to that considered by Howard and Brierley (1976) (see also Bilby et al., 1975) for linearly viscous composites with dilute concentrations of inclusions. Thus, we consider triaxial loads with their principal axes aligned with the axes of material symmetry, such that

$$\overline{\sigma}_{11} = 0, \quad \overline{D}_{22} = 0, \quad \overline{\sigma}_{12} = \overline{\sigma}_{13} = \overline{\sigma}_{23} = 0 \text{ and } \overline{D}_{33} = \xi,$$ (4.1)

where ξ is prescribed and taken to be small enough to justify omission of inertial effects.

Equations (3.3) and (3.7) may then be solved for the non-zero stresses. The solutions may be written, in terms of the optimal $\hat{y}$, as

$$\overline{\sigma}_{33}^2 = \frac{c^{(1)}\hat{y}z^2 + c^{(2)}}{\tilde{m}_{3333}(\hat{y})} \sigma_y^{(2)2},$$ (4.2)

$$\bar{\sigma}_{22} = -\frac{\tilde{m}_{2233}(\hat{y})}{\tilde{m}_{2222}(\hat{y})}\,\bar{\sigma}_{33},$$ (4.3)

where

$$z = \frac{\sigma_y^{(1)}}{\sigma_y^{(2)}}.$$

In order to obtain the uniaxial stress/strain curve, equations (4.2) and (4.3) are to be solved together with the differential equations for the aspect ratios (3.5). Solutions with a compressive strain-rate in the axial direction ($\bar{D}_{33} < 0$) for a composite material with volume fractions $c^{(1)} = c^{(2)} = 0.5$, initial aspect ratios $w_1 = w_2 = 1.0$ and contrast ratios $\sigma_y^{(2)}/\sigma_y^{(1)} = 0.5, 0.6, 0.7$ are plotted in Figure 2. Recall that the inclusion and matrix phases are denoted by 1 and 2, respectively, so that with the above choices of contrast ratios, the matrix is always weaker than the inclusions. The plots depict the evolution of the non-zero stresses $\bar{\sigma}_{22}$, $\bar{\sigma}_{33}$, the aspect ratios w_1, w_2 and the hardening-rate coefficient H, all as functions of the average axial strain $\bar{\varepsilon}_{33}$.

It is observed, from Figure 2c, that the aspect ratios of the inclusions decrease under compression, faster in the unconstrained direction x_1 than in the direction in which the applied strains are constrained to vanish (the x_2 direction). Thus, the initially spherical inclusions become shorter in the x_3 direction and longer in x_1 direction. Looking at Figures 2a and b, we observe that the material hardens initially in both the x_2 and x_3 directions, but as the deformation progresses, the material eventually softens in x_2 direction. The overall response, however, is of hardening, as may be observed in Figure 2d, where the hardening rate coefficient H is plotted and found to be always positive.

It is also observed that if the inclusions are stiffer than the matrix, there are values of yield strength ratios for which all the deformation takes place in the matrix. Such a case is described by the plots corresponding to $\sigma_y^{(2)}/\sigma_y^{(1)} = 0.5$. The inclusions in this case remain spherical and do not deform under the prescribed boundary conditions.

The important thing to notice in the above results is that the composite material does not behave in general as a perfectly-plastic material even though the phases themselves are made of rigid/perfectly-plastic constituents.

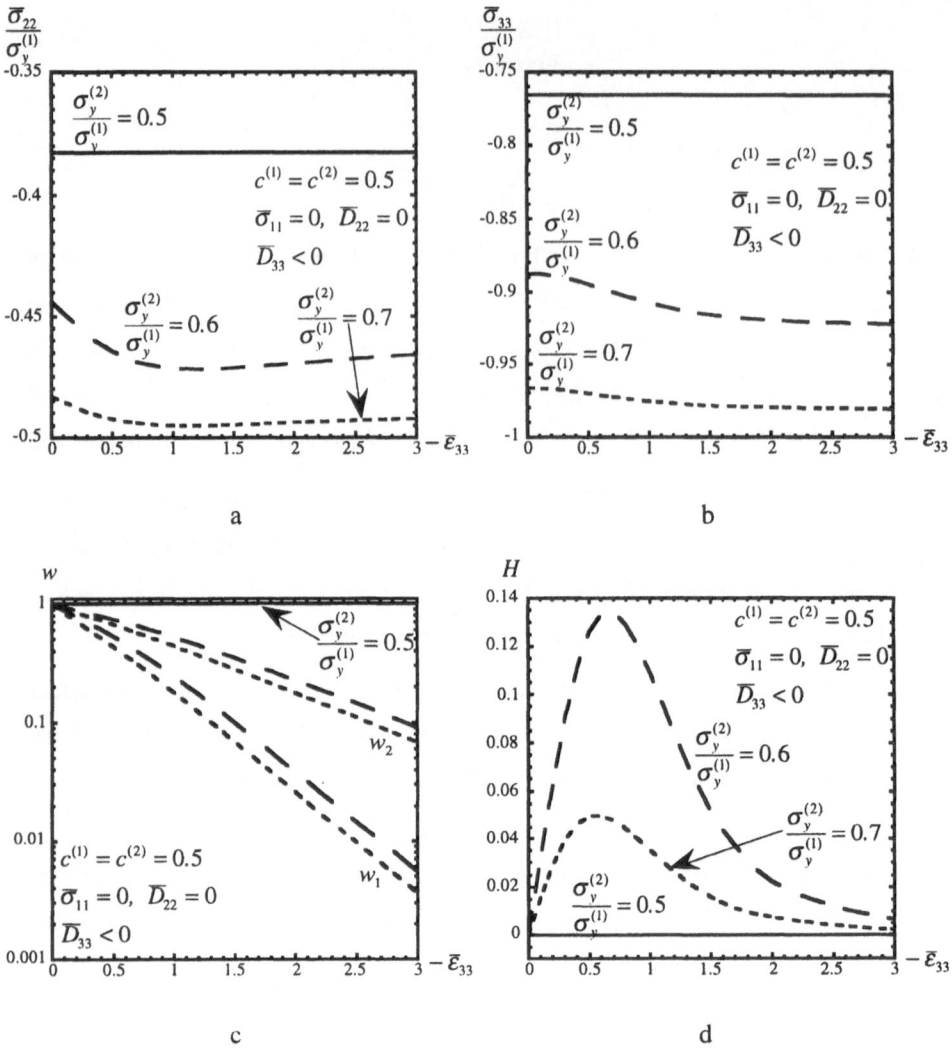

Figure 2 – Evolution of the stresses $\bar{\sigma}_{22}$, $\bar{\sigma}_{33}$, aspect ratios w_1, w_2 and hardening rate coefficient H, of an initially isotropic composite material, under plane strain conditions $\bar{\sigma}_{11} = 0$, $\bar{D}_{22} = 0$, $\bar{D}_{33} < 0$.

Results were also computed for materials with inclusions weaker than the matrix (see Zaidman, 1994). The findings are similar to those presented in this section except that the possibility of situations where the inclusions remain spherical is not available.

We finally remark that the hardening/softening induced by the change in the shape of the inclusions might be crucial in stability predictions for composite materials. For example, for plane-strain loading conditions, the possibility of localization may be identified with the vanishing of the effective hardening rate H (Rice, 1977). This condition, which was not satisfied during the process described in Figure 2, was found to be satisfied, in some cases, for initially anisotropic microstructures (Zaidman, 1994).

5. Concluding remarks

In this document, we have presented a constitutive model for two-phase rigid/ideally plastic composites, which is capable of accounting for the evolution of the microstructure under finite deformations conditions. The model assumes that the microstructure is of the particulate type with aligned, self-similar ellipsoidal inclusions distributed in a continuous matrix with *ellipsoidal symmetry* (Willis, 1977).

The model includes instantaneous effective stress/strain-rate relations, as well as evolution equations for the aspect ratios of the inclusions, which serve to characterize the overall anisotropy of the composite. The stress/strain-rate relations are expressed in terms of effective yield functions which are defined by simple one-dimensional optimization problems involving the Hashin-Shtrikman estimates of Willis (1977) for the effective viscosity tensor of linearly viscous composites with the same type of microstructures as the nonlinear composites. These effective yield functions are valid for any combination of loading conditions, aspect ratios, volume fractions and yield strengths. Some sample yield surfaces, for materials reinforced by fibers with elliptical cross section, have been computed and are presented in the body of the paper.

The evolution equations for the aspect ratios of the inclusions were derived for triaxial loading conditions, for which the principal axes of the applied loads are aligned with the axes of the ellipsoidal inclusions. The instantaneous stress/strain-rate relations and these evolution equations were solved simultaneously to obtain effective stress/strain curves for a sample composite subjected to plane strain conditions. These relations show that the effective behavior of the composite is qualitatively different from that of its

418

constituent phases, due to the evolution of the microstructure. Thus, it was found that the effective response of composites with perfectly plastic phases may exhibit hardening, or softening, depending rather strongly on the specific loading conditions and the initial shape of the inclusions. The transition from hardening to softening behavior, which may signal the onset of instabilities, may be crucial in situations where the stability of the deformation process is important, as in forming processes, for example.

Acknowledgments

This work was begun with the support of NSF grant DDM-90-12230 under subcontract No. 8009-69210-X with the Johns Hopkins University, and completed with the support of NSF grant MSS-92-02513. Additional support was provided by the NSF/MRL at the University of Pennsylvania under grant No. DMR-91-20668 through use of its computational facilities.

References

Bilby, B.A., Eshelby, J.D. and Kundu, A.K. (1975) Change in shape of a viscous ellipsoidal region embedded in a slowly deforming matrix having a different viscosity. *Techtonophysics* **28**, 265-274.

deBotton, G. and Ponte Castañeda, P. (1993) Elastoplastic constitutive relations for fiber-reinforced solids. *Int. J. Solids Structures* **30**, 1865-1890.

Dvorak, G.J. and Bahei-El-Din, Y.A. (1987) A bimodal plasticity theory of fibrous composite materials. *Acta Mech.* **69**, 219-241.

Dvorak, G.J. and Bahei-El-Din, Y.A., Macheret, Y. and Liu, C.H. (1988) An experimental study of the elastic-plastic behavior of a fibrous composite. *J. Mech. Phys. Solids* **26**, 655-687.

Eshelby, J.D. (1957) The determination of the elastic field of an ellipsoidal inclusion, and related problems. *Proc. R. Soc. Lond. A* **241**, 376-396.

Hashin, Z. (1980) Failure criteria for unidirectional fiber composites. A.S.M.E. *J. appl. Mech.* **47**, 329-334.

Hashin, Z. and Shtrikman, S. (1963) A variational approach to the theory of the elastic behavior of multiphase materials. *J. Mech. Phys. Solids* **11**, 127-140.

Hill, R. (1965) Continuum micromechanics of elastoplastic polycrystals. *J. Mech. Phys. Solids* **13**, 89-101.

Hill, R. (1967) The essential structure of constitutive laws for metal composites and polycrystals. *J. Mech. Phys. Solids* **15**, 79-95.

Howard, I.C. and Brierley, P. (1976) On the finite deformation of an inhomogeneity in a viscous liquid. *Int. J. Engng. Sci.* **14**, 1151-1159.

Ponte Castañeda, P. (1991) The effective mechanical properties of nonlinear isotropic solids. *J. Mech. Phys. Solids* **39**, 45-71.

Ponte Castañeda, P. and deBotton, G. (1992). On the homogenized yield strength of two-phase composites. *Proc. R. Soc. Lond. A* **438**, 419-431.

Ponte Castañeda, P. and Zaidman, M. (1994) Constitutive models for porous materials with evolving microstructure. *J. Mech. Phys. Solids* **42**, 1459-1497.

Ponte Castañeda, P. and Zaidman, M. (1995) On the finite deformation of nonlinear composite materials. Part I — Instantaneous effective potentials. Submitted for publication.

Rice, J.R. (1976) The localization of plastic deformation. In *Proceedings of the 14th International Congress of Theoretical and Applied Mechanics*, edited by W.T. Koiter, North-Holland, 207-220.

Suquet, P. (1983) Analyse limite et homogénéisation. *C. R. Acad. Sci. Paris II* **295**, 1355-1358.

Suquet, P. (1993) Overall potentials and extremal yield surfaces of power law or ideally plastic composites. *J. Mech. Phys. Solids* **41**, 981-1002.

Willis, J.R. (1977) Bounds and self-consistent estimates for the overall moduli of anisotropic composites. *J. Mech. Phys. Solids* **25**, 185-202.

Zaidman, M. and Ponte Castañeda, P. (1995) On the finite deformation of nonlinear composite materials. Part II — Evolution of the microstructure. Submitted for publication.

Zaidman, M. (1994) Constitutive models for composite materials with evolving microstructures. Ph.D. Thesis, Department of Mechanical Engineering and Applied Mechanics, University of Pennsylvania.

MICROMECHANICAL MODELLING BASED ON MORPHOLOGICAL ANALYSIS ; APPLICATION TO VISCOELASTICITY

A. ZAOUI*, Y. ROUGIER*‡ and C. STOLZ*
* *Laboratoire de Mécanique des Solides, Ecole Polytechnique, CNRS*
F 91128 Palaiseau Cedex
‡ *Laboratoire de Mécanique, ENSTA, Batterie de l'Yvette, Chemin de la Hunière, F 91120 Palaiseau Cedex*

Abstract

We first show the advantage of dealing with finite composite patterns instead of points in order to take morphological features of inhomogeneous materials into account in the derivation of their overall behaviour. The method allows the derivation of new bounds, tighter than the classical ones, for the Composite Spheres Assemblage and a new definition as well as an extension of the Generalized Self-Consistent Scheme in the case of linear elasticity. A comparison of the classical and the generalized self-consistent predictions is then performed in the case of non ageing linear viscoelasticity: the relaxation spectra of a two-phase isotropic material with Maxwellian incompressible constituents exhibit strong differences according to the choice of the model, what reflects the influence of the phase connectedness. The case of nonlinear viscoelasticity is then addressed and a new formulation, applying both to the classical and the generalized self-consistent schemes, is proposed and illustrated.

1. Introduction

In order to take better into account morphological characteristics of heterogeneous materials in view of the prediction of their overall mechanical behaviour as well as of the local response to a prescribed loading, it is proposed to combine a deterministic description of small, but finite, well-chosen "composite patterns" and a statistical representation of their spatial distribution: such a procedure can, in many cases, express essential morphological features, such as the connectedness of one phase or its dispersion as inclusions in another one, much more directly and easily than by using, as classical, point correlation functions of higher and higher orders.

In the case of linear elasticity, this approach has been applied to the Hashin Composite Spheres (or Cylinders) Assemblage (C.S.A. or C.C.A.) whose overall shear moduli can be bounded more tightly than classically. In addition, when combined with Kröner's theory of perfect disorder, the method leads naturally to the so-called "three-phase model" or "generalized self-consistent scheme" (G.S.C.S.) as the self-consistent treatment of a perfectly disordered C.S.A. (or C.C.A. for fiber reinforced composites). The principle of a computer-aided extension of this scheme to arbitrary morphological patterns is then suggested.

419

R. Pyrz (ed.), IUTAM Symposium on Microstructure-Property Interactions in Composite Materials, 419–430.
© 1995 *Kluwer Academic Publishers.*

This analysis shows clearly that the classical (C.S.C.S.) and the generalized (G.S.C.S.) self-consistent schemes are two basic models adequate for two extreme cases of morphological situations (polycrystal or composite-type morphology). Therefore, it is interesting to investigate further the comparison of their predictions, in view of analysing the influence of the morphology on the overall properties. Attention will be focussed here on viscoelasticity. Linear non ageing viscoelasticity is first considered. The overall relaxation spectra of two-phase isotropic materials whose constituents obey a Maxwell-type behaviour are compared: while both models lead to continuous overall spectra instead of discrete ones for the constituents, the connectedness of one phase according to the G.S.C.S. is responsible for the splitting of the spectrum into two continuous parts, which does not occur when the phases are morphologically intricated.

Finally the case of nonlinear viscoelasticity is addressed. This problem is much more complex and former attempts to propose a (classical) self-consistent formulation have failed to save the viscoelastic nature of the interphase accommodation. So, a new formulation is developed which combines a Hill-type linearisation procedure along the loading path and the use of the Laplace-Carson transform technique, what results in a symbolic elastic problem with eigenstrains. This procedure can be used in principle for any homogenization technique, including the G.S.C.S. one; it is illustrated in the case of the C.S.C.S. only, for a two-phase blend responding to a prescribed shear relaxation.

2. "Pattern" versus "point" approaches

Let us first recall briefly the basis of the proposed "pattern approach" [1] in order to stress its main conclusions. We consider that the given multiphase elastic material has already been decomposed, after some preliminary microstructural analysis aiming at extracting the most important morphological features, into several "morphological phases" (λ) constituted with N_λ identical composite representative domains $D_{\lambda i}$ centered at $\underline{X}_{\lambda i}$ and some residual content (λ_0). For any field quantity $f(\underline{y})$, with $\underline{y} = \underline{x} + \underline{X}_{\lambda i} \in D_{\lambda i}$, we can define two types of average values, namely:

$$f_\lambda^M(\underline{x}) = \frac{1}{N_\lambda} \sum_{i=1}^{N_\lambda} f(\underline{x} + \underline{X}_{\lambda i})$$

$$f_\lambda^m = \frac{1}{V_{\lambda_0}} \int_{V_{\lambda_0}} f(\underline{y}) d\omega \tag{1}$$

Using a non uniform polarization stress field $\mathbf{p}_\lambda(\underline{y})$ which takes equal values $\mathbf{P}_\lambda(\underline{x})$ at homologous points $\underline{y} = \underline{x} + \underline{X}_{\lambda i} \in D_{\lambda i}$ and constant ones $\mathbf{p}_{\lambda_0}$ in V_{λ_0} and the Green operator $\Gamma^\circ(\underline{x},\underline{x}')$ for an infinite body with homogeneous moduli $\mathbf{C}^\circ$ and homogeneous strain conditions $\mathbf{E}$ at infinity, the Hashin-Shtrikman functional $HS^\circ(\mathbf{E},\mathbf{p})$ is:

$$2HS^\circ(\mathbf{E},\mathbf{p}) = \mathbf{E}:\mathbf{C}^\circ:\mathbf{E} + \sum_\lambda \frac{N_\lambda}{V} \int_{D_\lambda} \mathbf{P}_\lambda:(\mathbf{E} + \varepsilon_\lambda^M - \mathbf{H}_\lambda:\mathbf{P}_\lambda) d\omega' + \dots$$

$$\dots + \sum_\lambda \frac{V_{\lambda_0}}{V} \mathbf{p}_{\lambda_0}:(\mathbf{E} + \varepsilon_\lambda^m - \mathbf{H}_\lambda^m:\mathbf{p}_{\lambda_0}) \tag{2}$$

with :

$$H(\underline{y}) = (c(\underline{y}) - C^\circ)^{-1}$$

$$\varepsilon_\lambda^M(\underline{x}) = \varepsilon_0 - \sum_\mu \{ N_\mu \int_{D_\mu} \Gamma^{\circ MM}_{\lambda\mu}(\underline{x},\underline{x}'):P_\mu(\underline{x}')d\omega' + V_{\mu_0}\Gamma^{\circ Mm}_{\lambda\mu}(\underline{x}):p_{\mu_0} \}$$

$$\varepsilon_\lambda^m = \varepsilon_0 - \sum_\mu \{ N_\mu \int_{D_\mu} \Gamma^{\circ mM}_{\lambda\mu}(\underline{x}'):P_\mu(\underline{x}')d\omega' + V_{\mu_0}\Gamma^{\circ mm}_{\lambda\mu}:p_{\mu_0} \} \qquad (3)$$

$$E = \sum_\lambda c_\lambda (\frac{1-\alpha_\lambda}{D_\lambda} \int_{D_\lambda} \varepsilon_\lambda^M d\omega + \alpha_\lambda \varepsilon_\lambda^m)$$

Here $c(\underline{y})$ are the local elastic moduli, ε_0 is an auxiliary strain tensor which is determined by the last eqn (3), c_λ is the volume fraction of the "phase" (λ), α_λ is the relative volume fraction of (λ_0) with respect to (λ); the first upper index of Γ° refers to averaging over $\underline{x}$ and the second one over $\underline{x}'$. The last step aims at optimizing the choice of the polarization stress fields $P_\lambda(\underline{x})$ and p_{λ_0} by making $HS^\circ(E,p)$ stationary with respect to $P_\lambda(\underline{x})$ and p_{λ_0}, which lead to the equations :

$$P_\lambda(\underline{x}) = H_\lambda^{-1}(\underline{x}):\varepsilon_\lambda^M(\underline{x}) \qquad\qquad p_{\lambda_0} = H_\lambda^{m^{-1}}:\varepsilon_\lambda^m \qquad (4)$$

Integration of (4) into eqns (3) shows that they rule (composite) inclusions problems through the Green operators $\Gamma^{\circ MM}_{\lambda\mu}(\underline{x},\underline{x}'), \Gamma^{\circ Mm}_{\lambda\mu}(\underline{x}), \Gamma^{\circ mM}_{\lambda\mu}(\underline{x}')$ and $\Gamma^{\circ mm}_{\lambda\mu}$ which have still to be determined, according to the available informations on the phase spatial distribution. Finally, optimal bounding is obtained by choosing the reference medium moduli C° so as to make H_λ and H_λ^m (just) positive or negative everywhere.

In the case of an isotropic distribution of the phases (λ) and of the domains $D_{\lambda i}$ within (λ), the problem can be solved completely: it reduces to several problems of composite spheres in an infinite homogeneous matrix. Similar conclusions hold for cylindrical patterns and transverse isotropy [2].

The main consequences of this approach may be summed up as follows:

- when the composite domains $D_{\lambda i}$ reduce to points and the morphological phases (λ) to the mechanical ones, the classical Hashin-Shtrikman's theory and results are recovered, as expected.

- when this analysis is applied to an isotropic C.S.A. or to a transversely isotropic C.C.A., new Hashin-Shtrikman-type bounds are found [3] for the overall moduli, through the resolution of one composite sphere/matrix problem: they are tighter than the classical Hashin's [4] and Hashin-Rosen's [5] ones for those moduli which are not exactly determined (shear or transverse shear moduli).

- when Kröner's [6] iterative procedure, related to his theory of graded and perfect disorder, consisting in softening or strengthening gradually the reference medium in order to converge towards the (classical) self-consistent estimate, is applied to the foregoing situation (i.e. new bounds for the C.S.A. or the C.C.A.), the limit estimate coincides with Christensen-Lo's [7] "three-phase model" one. This result is an indication that the three-phase model could apply exactly to a "perfectly disordered" isotropic C.S.A. or transversely isotropic C.C.A.: it gives a foundation to the denomination of "Generalized Self-Consistent Scheme" (G.S.C.S.) which is usually attributed to this model.

- according to the proposed approach, patterns may be arbitrary, provided that numerical (e.g. F.E.M.) calculations of the involved inclusion/matrix problems be performed; the three-phase model could then be generalized further by using arbitrary composite patterns embedded in the homogeneous equivalent medium. Such a "Pattern-based Self-Consistent Scheme" is now in progress [8].

- for the time being, one can consider the C.S.C.S. and the G.S.C.S. as two basic models for the investigation of the influence of morphology on the mechanical behaviour of inhomogeneous media, since they correspond to two extreme cases of morphology: the first one is well-suited to a "polycrystal-type morphology" (each phase domain is surrounded by many others distributed in a random fashion, so that no phase plays any prominent morphological role) and the second one to a "composite-type morphology" (one phase is geometrically continuous and acts as a "matrix" whereas the other ones are distributed as inclusions in it). So, significant conclusions may be drawn concerning the influence of the connectedness of one of the constituent phases of a multiphase material on its overall behaviour by comparing the predictions of these two models for a given loading path, as it is intented to be performed in the following.

- within this framework, efforts can be made towards the extension of these two basic schemes to non elastic constitutive behaviour. Several significant results have already been obtained for elastoplasticity [9, 10] and viscoplasticity [11] and will not be reported here. Alternatively, attention will be focussed on the case of viscoelasticity.

3. Application to linear viscoelasticity

A basic phenomenon of the viscoelasticity of inhomogeneous materials is the so-called "long-range memory effect" related to the delayed mechanical interactions between the constituents. In the linear case, it is well-known [12] that this effect is responsible, for instance, for the fact that the overall behaviour of an aggregate of Maxwellian constituents is no more Maxwellian. If the local constitutive equations have the form

$$\dot{\varepsilon} = a:\sigma + b:\dot{\sigma} \tag{5}$$

the global ones exhibit an additional integral term whose kernel J expresses this effect:

$$\dot{E} = A^{hom}:\Sigma + B^{hom}:\dot{\Sigma} + \int_0^t J(t-s):\dot{\Sigma}(s)ds \tag{6}$$

From that, it looks interesting to investigate the sensitivity of this phenomenon to morphology by comparing the C.S.C.S. and the G.S.C.S. estimates of this J function.

For sake of simplicity and tractability in view of the obtention of closed form results, let us consider a two-phase isotropic material whose constituents obey an isotropic incompressible Maxwellian behaviour, according to the equations :

$$\dot{e} = a_i s + b_i \dot{s} \qquad i = 1,2 \tag{7}$$

where e and s are the local strain and stress deviators and a_i and b_i material constants of the constituent (i). Through the Laplace transform, (7) can be written:

$$s^L(p) = 2p\mu_i^L(p)e^L(p) , \qquad \mu_i^L(p) = \frac{1}{2b_i(p+1/T_i)} \tag{8}$$

where p is the complex variable, $f^L(p)$ the Laplace transform of $f(t)$, $T_i = b_i/a_i$ the relaxation time of phase (i) and $\mu_i(t)$ its shear relaxation function. We are looking for the overall relaxation function $\mu^{eff}(t)$, according to either the C.S.C.S. or the G.S.C.S., with one phase, phase (2) say, included in the other. A convenient representation for our purpose is the spectral one, which allows to write a relaxation function $\mu(t)$ as:

$$\mu(t) = \int_0^\infty g(\tau) e^{-t/\tau} d\tau \qquad (9)$$

where $g(\tau)$ defines the relaxation spectrum of the material (which reduces to a single line for a Maxwell body). A direct comparison of these spectra for the two models [13] is able to illustrate the influence of morphology (especially of the phase connectedness).

The correspondence principle states that the transform $\mu^{eff\,L}(p)$ is linked to $\mu_1^L(p)$ and $\mu_2^L(p)$ through the same equation that the one which links μ^{eff}, μ_1 and μ_2 in the elastic problem, namely:

$$L(\beta,c)X^2 + 2M(\beta,c)X + N(\beta,c) = 0 \qquad (10)$$

where $X = \mu^{eff}/\mu_1$, $\beta = \mu_2/\mu_1$ and c is the volume fraction of phase (2). For the C.S.C.S., we have:

$$L = 1, \quad M = \frac{2-5c}{6}\beta + \frac{5c-3}{6}, \quad N = -\frac{2\beta}{3} \qquad (11)$$

and for the G.S.C.S.:

$$L(\beta,c) = 4[3(\beta-1)x^3 - \eta_3](\eta_1 x^7 - 2\eta_2) - 126\eta_2(\beta-1)x^3(1-x^2)^2$$

$$M(\beta,c) = 3\eta_1(\beta-1)x^{10} + 4\eta_1\eta_3 x^7 - \frac{129}{4}\eta_2(\beta-1)x^3 + ...$$

$$... + \frac{3}{4}\eta_2\eta_3 + 126\eta_2(\beta-1)x^3(1-x^2)^2 \qquad (12)$$

$$N(\beta,c) = -[\frac{9}{2}(\beta-1)x^3 + \eta_3](\eta_1 x^7 + \frac{19}{2}\eta_2) - 126\eta_2(\beta-1)x^3(1-x^2)^2,$$

$$x = c^{1/3}, \quad \eta_1 = 19(\beta-1), \quad \eta_2 = \frac{19\beta+16}{2}, \quad \eta_3 = \frac{3}{2}(2\beta+3).$$

The same equations (10) to (12) hold for $X^L(p) = \mu^{eff\,L}(p)/\mu_1^L(p)$ instead of X by replacing β by $\beta^L(p)$ defined by:

$$\beta^L(p) = \frac{\mu_2^L(p)}{\mu_1^L(p)} = k\frac{p+1/T_1}{p+1/T_2} \quad \text{with} \quad k = b_1/b_2 \qquad (13)$$

The solution $\mu^{eff\,L}(p)$ can be written as the sum of two terms, $f_1^L(p)$ and $f_2^L(p)$ say:

$$f_1^L(p) = -\frac{M(\beta^L(p),c)}{2b_1(p+1/T_1)L(\beta^L(p),c)}$$

$$f_2^L(p) = \frac{[M^2(\beta^L(p),c) - L(\beta^L(p),c)N(\beta^L(p),c)]^{1/2}}{2b_1(p+1/T_1)L(\beta^L(p),c)} \qquad (14)$$

- the inverse of the first term is the sum of two (for the C.S.C.S.) or three (for the G.S.C.S.) decreasing exponential functions : they are responsible for two or three discrete lines in the relaxation spectrum respectively, at times T_1 and T_2 for the C.S.C.S. and T_1, θ_1 and θ_2 for the G.S.C.S., with θ_1 and θ_2 intermediate between T_1 and T_2; nevertheless, their intensity can be negative according to the c value.

- this second term can be rewritten, after some reductions, as:

$$f_2^{LC}(p) = k_c \frac{P_c(p)^{1/2}}{(p + 1/T_1)(p + 1/T_2)} \quad \text{with } P_c(p) = (p + 1/\tau_1)(p + 1/\tau_2)$$

$$\text{and} \quad k_c = \frac{\sqrt{(2 - 5c)^2 k^2 + 2(6 - 5c)(5c + 1)k + (5c - 3)^2}}{12b_1} \tag{15}$$

where τ_1 and τ_2 are intermediate times between T_1 and T_2, for the C.S.C.S.; for the G.S.C.S., this second term is:

$$f_2^{LG}(p) = k_g \frac{P_g(p)^{1/2}}{(p + 1/T_1)(p + 1/\theta_1)(p + 1/\theta_2)} \quad \text{with } k_g \geq 0$$

$$\text{and} \quad P_g(p) = (p + 1/\tau'_1)(p + 1/\tau'_2)(p + 1/\tau'_3)(p + 1/\tau'_4) \tag{16}$$

where τ'_1 and τ'_2 (τ'_3 and τ'_4 resp.) are intermediate times between T_1 and θ_1 (θ_1 and θ_2 resp.) . In both cases, the Laplace inverse $f_2(t)$ can then be derived through the formula:

$$f_2(t) = \frac{1}{2i\pi} \int_\Delta f_2^L(p) e^{pt} dp \tag{17}$$

where the vertical axis Δ has to leave on its left all the critical points of $f_2^L(p)$. By using the theorem of residues, Jordan's lemma and adequate cuts on the real negative axis, one finds (with $T_1 < T_2$ for the discussion) for the C.S.C.S:

$$f_2^C(t) = \pm \frac{k_c}{\pi} \{ -\int_{-1/\tau_1}^{-1/\tau_2} \frac{\sqrt{-P_c(x)}}{(x + 1/T_1)(x + 1/T_2)} e^{tx} dx - \dots$$

$$\dots - \pi \frac{\sqrt{P_c(-1/T_1)}}{1/T_2 - 1/T_1} e^{-t/T_1} + \pi \frac{\sqrt{P_c(-1/T_2)}}{1/T_1 - 1/T_2} e^{-t/T_2} \} \tag{18}$$

and for the G.S.C.S.:

$$f_2^G(t) = \pm \frac{k_g}{\pi} \{ \int_{-1/\tau_1}^{-1/\tau_2} \frac{\sqrt{-P_g(x)}}{(x + 1/T_1)(x + 1/\theta_1)(x + 1/\theta_2)} e^{tx} dx - \dots$$

$$\dots - \int_{-1/\tau_3}^{-1/\tau_4} \frac{\sqrt{-P_g(x)}}{(x + 1/T_1)(x + 1/\theta_1)(x + 1/\theta_2)} e^{tx} dx - \dots$$

$$\dots - \frac{\pi}{(1/\theta_2 - 1/\theta_1)} [\frac{\sqrt{P_g(-1/\theta_1)}}{(1/T_1 - 1/\theta_1)} e^{-t/\theta_1} - \frac{\sqrt{P_g(-1/\theta_2)}}{(1/T_1 - 1/\theta_2)} e^{-t/\theta_2}] \} \tag{19}$$

Through the change of variable $\tau = -1/x$, the integrals in (18) and (19) are put in the form (9), which corresponds to continuous spectra, whereas the exponential functions

outside the integrals are associated to discrete lines. Addition to $f_1{}^C(t)$ for the C.S.C.S. or to $f_1{}^G(t)$ for the G.S.C.S. lead to the following final results:
- the shear relaxation spectrum according to the C.S.C.S. is composed of one continuous spectrum lying between τ_1 and τ_2 (with $T_1 < \tau_1 < \tau_2 < T_2$) with the intensity:

$$g_c(\tau) = \frac{k_c T_1 T_2}{\pi \sqrt{\tau_1 \tau_2}} \frac{\sqrt{(\tau - \tau_1)(\tau_2 - \tau)}}{\tau(\tau - T_1)(T_2 - \tau)}, \quad \tau \in [\tau_1, \tau_2] \tag{20}$$

and of additional lines at $\tau = T_1$ for $c \le 3/5$ (with the intensity $(3-5c)/6b_1$) and at $\tau = T_2$ for $c \ge 2/5$ (with the intensity $(5c-2)/6b_2$); thus, such a spectrum line can be present or not according to the volume fraction of the phases;
- the relaxation spectrum according to the G.S.C.S. is composed of a continuous part which is split up into two spectra lying between τ'_1 and τ'_2 for the former and τ'_3 and τ'_4 for the latter, with the intensity;

$$g_g(\tau) = \frac{\in(\tau) k_g T_1 \theta_1 \theta_2}{\pi \sqrt{\tau'_1 \tau'_2 \tau'_3 \tau'_4}} \frac{\sqrt{(\tau - \tau'_1)(\tau - \tau'_2)(\tau - \tau'_3)(\tau'_4 - \tau)}}{\tau(\tau - T_1)(\theta_1 - \tau)(\theta_2 - \tau)}$$

$$\tau \in [\tau'_1, \tau'_2] \cup [\tau'_3, \tau'_4] \tag{21}$$

with $\quad \in(\tau) = 1$ if $\tau \in [\tau'_1, \tau'_2]$ and $\in(\tau) = -1$ if $\tau \in [\tau'_3, \tau'_4]$

and of additional lines at $\tau = T_1$ in a fixed range of volume fractions and, according to the value of c, T_1, T_2, k and b_1, at $\tau = \theta_1$ or θ_2.

We can see on Figure 1 an illustrative example of these results comparing the relaxation spectra as predicted, for fixed values of the material parameters a_i, b_i, and the volume fraction c, by the C.S.C.S. (a) and the G.S.C.S. (b) respectively. Several features can be emphasized:
- obviously, both models predict a non-Maxwellian overal behaviour since the spectra do not reduce to a single spectrum line; this main result must be kept in mind in view of the treatment of the nonlinear case. In addition to some discrete lines, the continuous (bounded) spectrum expresses the delayed mechanical interactions between the phase domains which are responsible for the "long-range memory effect".
- morphology strongly affects the continuous spectrum shape: whereas the "symmetrical" morphology associated with the C.S.C.S. modelling leads to a unique spectrum whose symmetry is only disturbed by the differences of material parameters and volume fractions, the unsymmetrical morphology inherent to the G.S.C.S. modelling is responsible for a splitting of the spectrum into two well-separated parts and for a spectrum shape which is clearly dependent on the prominent role played by the continuous phase (the "matrix" phase).

This property would benefit in a direct experimental investigation in order to check whether such a "morphological signature" of a connected phase can be observed: some evidences of a temporary broadening of the relaxation spectrum when crossing the glass transition temperature could be related to this phenomenon.

Note that the foregoing analysis could be extended to various loading paths (such as creep, applied sinusoidal stress or strain...) or to more general constitutive equations than Maxwell's ones without too much efforts [14]: the main conclusions would be essentially the same.

426

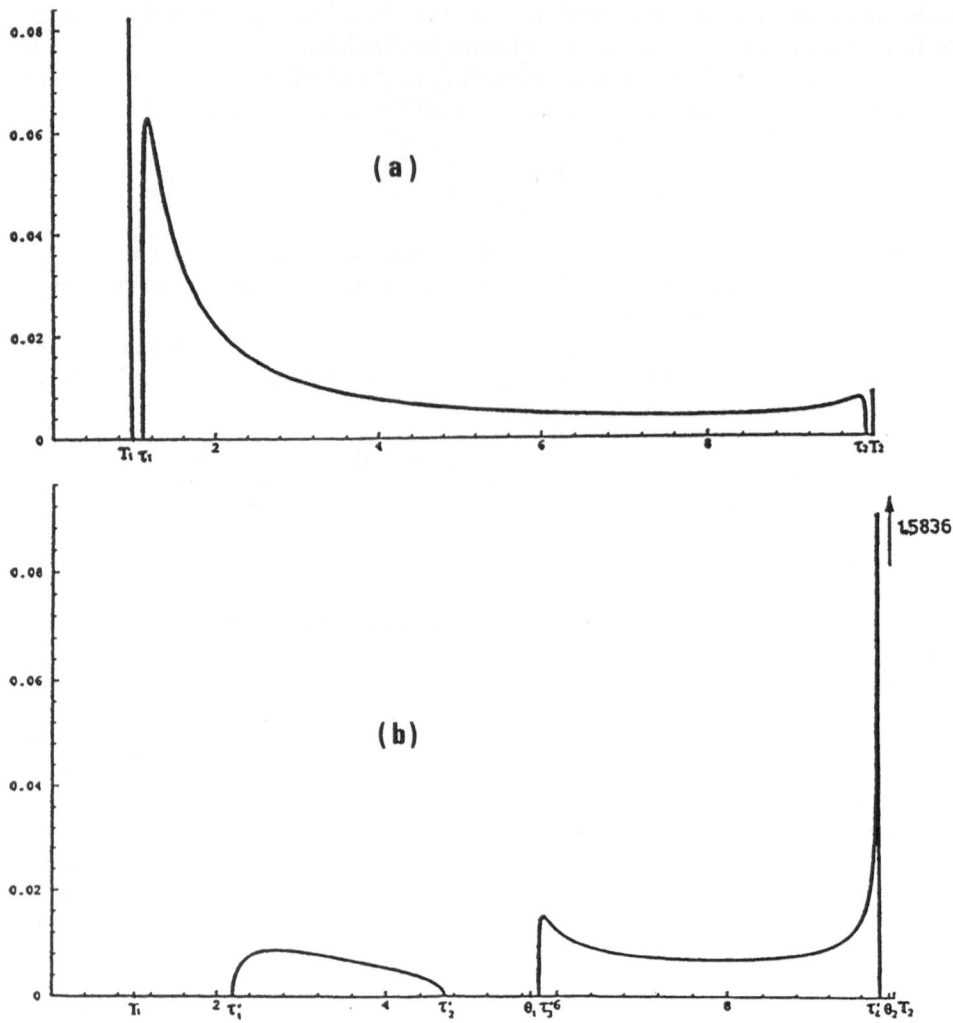

Figure 1: Comparison of the relaxation spectra derived by the C.S.C.S. (a) and the G.S.C.S. (b) for c = 0.5, $a_1 = a_2 = 1$, $T_1 = 1$ and $T_2 = 10$

4. Nonlinear viscoelasticity

4.1. GENERAL FRAME

The problem of nonlinearity in the context of self-consistent modelling is now addressed in order to extend the foregoing analyses to more various situations. This is an open problem even in the case of the classical self-consistent scheme, despite several attempts to solve it during the last decade [15, 16, 17]. The main difficulty to deal with is concerned with the coupling between elasticity and viscosity which is responsible for the simultaneous occurence in the constitutive equations of derivatives of different orders of the mechanical variables σ, ε... So, the classical Hill's (for elastoplasticity [18]) or Hutchinson's (for viscoplasticity [19]) linearisation procedure which allows to save the convenient use of Green techniques for the resolution of the concentration problem (i.e. the derivation of the relations between the local and the global variables) cannot be applied anymore. In addition, nonlinearity prevents to use the Laplace transform facilities which can overcome the basic difficulty in the case of linear viscoelasticity [20]. Here we propose a new formulation [21] which allows to derive a nonlinear viscoelastic (or elastoviscoplastic) scheme from any model developed in the elastic case. Application will concern the C.S.C.S. but the G.S.C.S. or other models could be considered as well.

Let us consider first these simple local nonlinear viscoelastic equations:

$$\dot{\varepsilon} = s{:}\dot{\sigma} + g(\sigma) \qquad (22)$$

where s are the elastic compliances. Starting from time t = 0, we are supposed to have determined the local and overall responses to the load history up to $t = t_n$ and we are looking for the solution for the next infinitesimal time interval $[t_n, t_n + dt]$. Thus we can consider that, for the prediction of the response of the material on $[t_n, t_n + dt]$, the actual equations (22) are approximated by:

$$\dot{\varepsilon}(t) = s{:}\dot{\sigma}(t) + \mathbf{m}_n{:}\sigma(t) + \dot{\varepsilon}_n^0(t, t_n),$$

$$\dot{\varepsilon}_n^0(t, t_n) = g[\sigma(t_n)] - \mathbf{m}_n{:}\sigma(t_n) + ...$$

$$... + \{g[\sigma(t)] - g[\sigma(t_n)] - \mathbf{m}_n{:}[\sigma(t) - \sigma(t_n)]\}[1 - H(t - t_n)], \qquad (23)$$

$$\mathbf{m}_n = \frac{\partial \mathbf{g}}{\partial \sigma}[\sigma(t_n)].$$

It is clear on these equations that the considered behaviour is a Maxwellian one with the *eigenstrain* rate $\dot{\varepsilon}_n^0(t, t_n)$. The crucial point is here: such a strain is actually an eigenstrain because its variation is completely known *a priori* and does not depend on the external loading which is applied beyond t_n; its time derivative is constant beyond t_n and variable but known on $[0, t_n]$. So, such a strain generalizes Eshelby's definition of a "stress-free strain" and can be dealt with in the following such as a (variable) eigenstrain. This would be wrong for the viscous strain ε^v whose time-derivative $g(\sigma)$ is stress history-dependent: considering ε^v as an eigenstrain [15,16,17] leads to an elastic treatment which conflicts with the viscoelastic nature of the interphase strain accommodation and the "long range memory effect [22].

The following steps are straightforward : as it is usual for linear non ageing viscoelasticity, the use of the Laplace-Carson transform defined by:

$$f*(p) = p\int_0^\infty f(t)e^{-pt}dt \qquad (24)$$

which can apply to $\varepsilon_n^0(t, t_n)$ as well, allows to convert the problem into a (symbolical) elastic one with eigenstrains. From the strain (stress resp.) elastic concentration tensor **A** (**B** resp.), the transformed viscoelastic equivalent $\mathbf{A}*(p)$ ($\mathbf{B}*(p)$ resp.) is derived according to the correspondence principle and the whole set of equations has the classical form:

$$\varepsilon* = s*:\sigma* + \varepsilon_n^0 * \qquad \text{or} \qquad \sigma* = c*:(\varepsilon* - \varepsilon_n^0*),$$

$$s*(p) = s + \frac{m_n}{p}, \qquad c*(p) = s*^{-1}(p),$$

$$E* = S^{hom}*:\Sigma* + E_n^0 * \qquad \text{or} \qquad \Sigma* = C^{hom}*:(E* - E_n^0*), \qquad (25)$$

$$C^{hom}*(p) = <c*(p):A*(p)> \qquad \text{or} \qquad S^{hom}*(p) = <s*(p):B*(p)>,$$

$$B*(p) = c*(p):A*(p):<c*(p):A*(p)>^{-1},$$

$$E_n^0*(p) = <B*^T(p):\varepsilon_n^0*(p)>.$$

Due to the nonlinearity, an additional relation between the local and global variables is needed, dependent on the chosen model, in order to determine the mechanical state of each phase. The last step consists in the Laplace-Carson inversion: in general, it has to be performed numerically according to adequate techniques such as the collocation ones.

Note that more general constitutive equations than (22) can be used with the same general procedure [21], such as the following ones:

$$\dot{\varepsilon} = s:\dot{\sigma} + g(\sigma,\alpha),$$
$$\dot{\alpha} = l(\sigma,\alpha):\dot{\sigma} + h(\sigma,\alpha) \qquad (26)$$

but that no coupling with time-independent plasticity is permitted.

4.2. APPLICATION TO THE C.S.C.S.

Application to the C.S.C.S. can be made straightforwardly by specifying the $A*(p)$ tensor in (25) and the additional concentration relation. We have, as classical:

$$A_r^{SC}*(p) = [I + P_r^{SC}*(p):\delta c_r^{SC}*(p)]^{-1}:<[I + P^{SC}*(p):\delta c^{SC}*(p)]^{-1}>^{-1},$$

$$\delta c_r^{SC}*(p) = c_r*(p) - C^{SC}*(p), \qquad (27)$$

$$P_r^{SC}*(p) = \int_{\Omega_r} \Gamma^{SC}*(p,\underline{x}_r,\underline{x}_r')d\omega_r',$$

$$\varepsilon_r*(p) = [I + P_r^{SC}*(p):\delta c_r^{SC}*(p)]^{-1}:\{\varepsilon_0*(p)+\dots$$

$$\dots + P_r^{SC}*(p):[c_r*(p):\varepsilon_{nr}^0*(p) - C^{SC}*(p):E_n^0*(p)]\}, \qquad (28)$$

$$<\varepsilon*(p)> = E*(p).$$

where Ω_r is the ellipsoidal inclusion for phase (r), superindex ^{SC} stands for "self-consistent"and $\Gamma*(p)$ is the Laplace-transform of the viscoelastic Green operator associated to the elastic one by the correspondence principle. If all the ellipsoids have the same aspect ratios and orientation, P^{SC} is independent on the phase and the "normalisation term" $<[I + P^{SC}*(p):\delta c^{SC}*(p)]^{-1}>$ reduces to unity.

The tractability of this method has been checked on a quite simple case [14], namely a shear relaxation test performed on the two-phase material already studied before in the linear case, but now with the constitutive local equations:

$$\dot{e} = a\sigma_{equ}^{m-1}s + b\dot{s} \qquad (29)$$

where σ_{equ} is the von Mises equivalent stress; in this particular case, the problem reduces to a scalar one. Let the uniaxial applied strain be $E(t) = E_0 H(t)$: we are looking for the global stress response $S(t)$. The details of the numerical treatment are reported elsewhere [23]: suffice it to say here that the closed form solution of the linear case reported hereabove can be used at each step for the Laplace transform inversion but that collocation and FFT techniques are also used. Typical results are reported on Figure 2 which shows clearly both the nonlinear nature of the response and the variation of the overall eigenstrain $E_n^0(t, t_n)$ at each step .

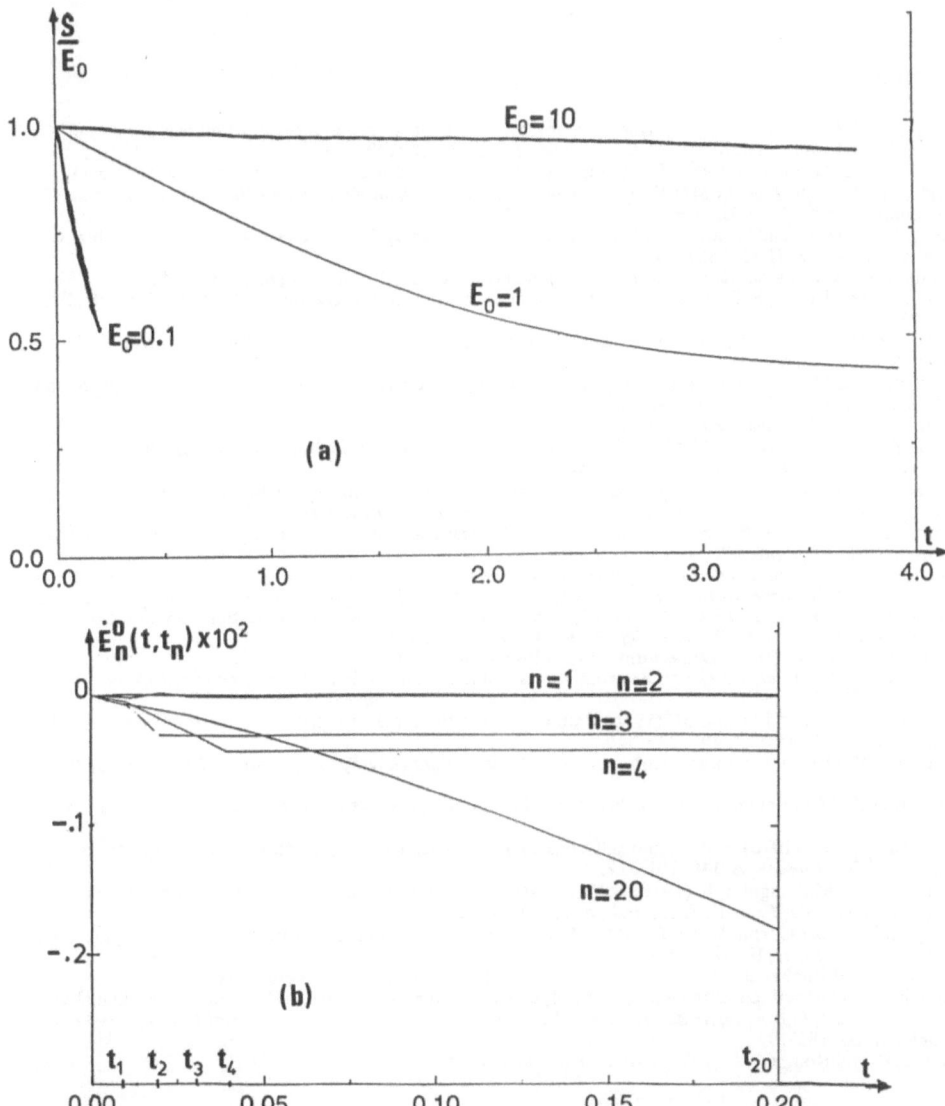

Figure 2: (a) normalized shear stress relaxation for a two-phase material according to the C.S.C.S. (m = 2, c = 0.5, b1 = b2 = 1, a1 = 1, a2 = 5); (b) time variation of the overall eigenstrain for different steps t_n.

430

Further developments concerned with other models, including the G.S.C.S., and more general loading and material situations are now in progress.

Acknowledgements

Part of this work (section 4) was funded by EdF-SEPTEN for nuclear applications. The valuable assistance of Dr P. Navidi (LMS) for developing the numerical treatment of &4.1 is gratefully acknowledged.

References

[1] Stolz, C. and Zaoui, A. (1991) Analyse morphologique et approches variationnelles du comportement d'un milieu élastique hétérogène, *C. R. Acad. Sci. Paris*, II **312**, 143-150.
[2] Hervé, E. and Zaoui, A. (1994) Morphological n-layered cylindrical pattern-based micromechanical modelling, *Proc. IUTAM/ISIMM Symposium on Anisotropy, Inhomogeneity and Nonlinearity in Solid Mechanics*, Nottingham, UK (*in press*).
[3] Hervé, E., Stolz, C. and Zaoui, A. (1991) A propos de l'assemblage de sphères composites de Hashin, *C. R. Acad. Sci. Paris*, II **313**, 857-862.
[4] Hashin, Z.(1962) The elastic moduli of heterogeneous materials, *J. Appl. Mech*, **29**, 143-150.
[5] Hashin, Z. and Rosen, B.W. (1964) The elastic moduli of fiber-reinforced materials, *J. Appl. Mech.*, **31**, 223-232.
[6] Kröner, E. (1978) Self-consistent scheme and graded disorder in polycrystal elasticity, *J. Phys. F: Metal Phys.*, **8**, 2261-2267.
[7] Christensen, R.M. and Lo, K.H. (1979) Solutions for effective shear properties in three phase sphere and cylinder models, *J. Mech. Phys. Solids*, **27**, 315-330.
[8] Bornert, M. (1994) *Private communication*.
[9] Hervé, E. and Zaoui, A. (1990) Modelling the effective behaviour of nonlinear matrix / inclusion composites, *Eur. J. Mech. A / Solids*, **9**, 505-515.
[10] Bornert, M., Hervé, E., Stolz, C. and Zaoui, A. (1994) Self-consistent approaches and strain heterogeneities in two-phase elastoplastic materials, *Appl. Mech. Rev.*, **47**, S66-76.
[11] Hervé E., Dendievel, R. and Bonnet G. (1994) Steady state power-law creep in "inclusion-matrix" composite materials (*submitted*).
[12] Suquet, P. (1985) Elements of homogenization for inelastic solid mechanics, in E. Sanchez-Palencia and A. Zaoui (eds), *Homogenization Techniques for Composite Media*, Springer Verlag, Berlin, pp. 193-278.
[13] Rougier, Y., Stolz, C. and Zaoui, A. (1993) Représentation spectrale en viscoélasticité linéaire des matériaux hétérogènes, *C. R. Acad. Sci. Paris*, II **316**, 1517-1522.
[14] Rougier, Y., (1994) *Ph. D. Dissertation*, Ecole Polytechnique, Palaiseau (France).
[15] Weng, G.J. (1981) Self-consistent determination of time-dependent behaviour of metals, *J. Appl. Mech.*, **48**, 41-46.
[16] Nemat-Nasser, S. and Obata, M. (1986) Rate-dependent finite elastoplastic deformation of polycrystals, *Proc. Roy. Soc. London*, A **407**, 343-375.
[17] Harren, S.V. (1991) The finite deformation of rate-dependent polycrystals, *J. Mech. Phys. Solids*, **39**, 345-383.
[18] Hill, R. (1965) Continuum micro-mechanics of elastoplastic polycrystals, *J. Mech. Phys. Solids*, **13**, 89-101.
[19] Hutchinson, J.W. (1976) Bounds and self-consistent estimates for creep of polycrystalline materials, *Proc. Roy. Soc. London*, A **348**, 101-127.
[20] Laws, N. and McLaughlin, R (1978) Self-consistent estimates for the viscoelastic creep compliances of composite materials, *Proc. R. Soc. Lond.*, A **359**, 251-273.
[21] Rougier, Y., Stolz C. and Zaoui, A. (1994) Self-consistent modelling of elastic-viscoplastic polycrystals, *C. R. Acad. Sci. Paris*, II **318**, 145-151.
[22] Zaoui, A. and Raphanel, J.L. (1993) On the nature of the intergranular accommodation in the modelling of elastoviscoplastic behavior of polycrystalline aggregates, in C. Teodosiu, J.L. Raphanel and F. Sidoroff (eds), *Large plastic deformations, Fundamentals and applications to metal forming*, Balkema, Rotterdam, pp. 185-192.
[23] Navidi, P. and Rougier, Y. (1994) *Private communication*.

SUBJECT INDEX

LIST OF PARTICIPANTS

Professor E. Aifantis, Aristotle University of Thessaloniki, Polytechnic School, Laboratory of Mechanics & Materials, 54006 Thessaloniki, Greece

Dr. S.I. Andersen, Materials Department, Risø National Laboratory, 4000 Roskilde, Denmark

Adjunkt J. Andreasen, Institute of Mechanical Engineering, Aalborg University, Pontoppidanstræde 101, 9220 Aalborg, Denmark

Dr. M. Arminjon, Chargé de Recherche CNRS, Laboratoire Sols, Solides, Structures, B.P. 53x, F-38041 Grenoble, France

Civ.ing. M.S. Axelsen, Institute of Mechanical Engineering, Aalborg University, Pontoppidanstræde 101, 9220 Aalborg, Denmark

Docent M.P. Bendsøe, Department of Mathematics, Technical University of Denmark, Building 303, 2800 Lyngby, Denmark

Dr. A.R. Boccaccini, Institut für Gesteinshüttenkunde, Rheinish-Westfälische Technische Hochschule, Mauerstrasse 5, D-52056 Aachen, Germany

Dr. B. Bochenek, Institute of Mechanics and Machine Design, Cracow University of Technology, Warszawska 24, 31-155 Cracow, Poland

Professor J. Botsis, Department of Civil Engineering, Mechanics and Metallurgy, The University of Illinois at Chicago, 2095 Engineering Research Facility, Box 4348, Chicago, Illinois 60680, U.S.A.

Lektor R. Brincker, Institute of Building Technology and Structural Engineering, Aalborg University, Sohngaardsholmsvej 57, 9000 Aalborg, Denmark

Dr. P. Brøndsted, Materials Department, Risø National Laboratory, 4000 Roskilde, Denmark.

Civ.ing. C. Burchardt, Institute of Mechanical Engineering, Aalborg University, Pontoppidanstræde 101, 9220 Aalborg, Denmark

Dr. H.J. Böehm, Institute of Light Weight Structures & Aerospace Engineering, Vienna University of Technology, Gusshausstrasse 27-29, A-1040 Vienna, Austria

Dr. G.P. Carman, UCLA, Mechanics, Aerospace & Nuclear Engineering, 405 Hilgard Avenue, 38-137, Eng. IV, Los Angeles, CA90024-1597, U.S.A.

Professor P. Ponte Castañeda, Department of MEAM, University of Pennsylvania, Philadelphia, PA 19104-6315, U.S.A.

Docent J. Christoffersen, Department of Solid Mechanics, Technical University of Denmark, Building 404, 2800 Lyngby, Denmark

Dr. A. Clarke, The University of Leeds, Physics Department, LS2 9JT, Leeds, United Kingdom

Professor N. Claussen, Advanced Ceramic Group, Technical University of Hamburg-Harburg, P.O. Box 901052, D-2100 Hamburg 90, Germany

Lektor K. Dalgaard-Jensen, Institute of Mechanical Engineering, Aalborg University, Pontoppidanstræde 101, 9220 Aalborg, Denmark

Professor G.J. Dvorak, Department of Civil Engineering, Renselaer Polytechnic Institute, Troy, NY 12180-3590, U.S.A.

Dr. N.A. Fleck, Cambridge University, Engineering Department, Trumpington Street, Cambridge CB2 1PZ, United Kingdom

Professor K. Friedrich, Institute of Composite Materials, University of Kaiserslautern, Erwin Schroedinger Strasse, Bldg. 57, D-6750 Kaiserslautern, Germany

Civ.ing. P. From, Institute of Mechanical Engineering, Aalborg University, Pontoppidanstræde 101, 9220 Aalborg, Denmark

Professor P. Gudmundson, Department of Solid Mechanics, Royal Institute of Technology, S-10044 Stockholm, Sweden

Professor Z. Hashin, Department of Solid Mechanics, Materials and Structures, Tel-Aviv University, Tel-Aviv 69978, Israel

Professor K. Herrmann, Laboratory of Applied Mechanics, University of Paderborn, Pohlweg 47-49, 33098 Paderborn, Germany

Dr. M. Hori, Department of Civil Engineering, Tokyo University, 7-3-1 Hongo, Bunkyo-ku, Tokyo 113, Japan

Dr. M. Jørgensen, National Institute of Agricultural Engineering, Bygholm, 8700 Horsens, Denmark

Professor B. Karihaloo, School of Civil and Mining Engineering, The University of Sydney, N.S.W. 2006, Australia

Ing.doc. A. Kildegaard, Institute of Mechanical Engineering, Aalborg University, Pontoppidanstræde 101, 9220 Aalborg, Denmark

Dr. P. Kim, École Polytechnique Fedérale de Lausanne, Laboratoire de Technologie des Composites, DMX-H, CH-1015 Lausanne, Switzerland

Professor I. Kimpara, Department of Ship and Ocean Engineering, University of Tokyo, 7-3-1 Hongo, Bunkyo-ku, Tokyo 113, Japan

Civ.ing. E.S. Knudsen, Institute of Mechanical Engineering, University of Aalborg, Pontoppidanstræde 101, 9220 Aalborg, Denmark

Civ.ing. K.B. Kristensen, Institute of Mechanical Engineering, Aalborg University, Pontoppidanstræde 101, 9220 Aalborg, Denmark

Dr. D. Kujawski, Department of Mechanical Engineering, University of Alberta, Edmonton T6G 268, Canada

Professor R.M. McMeeking, Materials Department and Mechanical Engineering, College of Engineering, University of California, Santa Barbara, C.A. 93106, U.S.A.

Adjunkt B. Mikkelsen, Institute of Mechanical Engineering, Aalborg University, Pontoppidanstræde 101, 9220 Aalborg, Denmark

Professor G.W. Milton, Courant Institute of Mathematical Sciences, New York University, New York, N.Y. 10012, U.S.A.

Dr. H. Moulinec, Laboratoire de Méchanique et d'Acoustique, 31 Chemin Joseph Aiguier, F-13402 Marseille Cedex 20, France

Lektor O.Ø. Mouritsen, Institute of Mechanical Engineering, Aalborg University, Pontoppidanstræde 101, 9220 Aalborg, Denmark

Professor Z. Mróz, Institute of Fundamental Technical Research, Polish Academy of Science, Swietokrzyska 21, 00-049 Warsaw, Poland

Professor S. Murakami, Department of Mechanical Engineering, Nagoya University, Nagoya 464-01, Japan

Professor O.B. Naimark, Institute of Mechanics of Continuous Media, Russian Academy of Science, 1 Acad., Korolev Street, 614061 Perm, Russia

Professor A. Needleman, Division of Engineering, Brown University, Providence, RI 02912, U.S.A.

Professor S. Nemat-Nasser, Department of Applied Mechanics and Engineering Science, University of California, San Diego, La Jolla, CA 90293, U.S.A.

Professor F. Niordson, Geelsvej 19, 2840 Holte, Denmark

Professor N. Olhoff, Institute of Mechanical Engineering, Aalborg University, Pontoppidanstræde 101, 9220 Aalborg, Denmark

Lektor J.J. Pedersen, Institute of Mechanical Engineering, Aalborg University, Pontoppidanstræde 101, 9220 Aalborg, Denmark

Docent P. Pedersen, Department of Solid Mechanics, Technical University of Denmark, Building 404, 2800 Lyngby, Denmark

Dr. T. Petersen, Danish Technological Institute, Department of Polymer Technology, Gregersensvej, 2620 Taastrup, Denmark

Dr. K. Pietrzak, Wólczyńska 133, 01-919 Warsaw, Poland

Dr. V. Petrova, Research Institute of Mathematics, Voronezh State University, 1, University SQ, Voronezh 394693, Russia

Professor M.R. Piggott, Department of Chemical Engineering and Applied Chemistry, University of Toronto, Toronto, Ontario M5S 1A4, Canada

Professor A. Plumtree, Department of Mechanical Engineering, University of Waterloo, Waterloo, Ontario N2L 3G1, Canada

Docent R. Pyrz, Institute of Mechanical Engineering, Aalborg University, Pontoppidanstræde 101, 9220 Aalborg, Denmark

Professor J.N. Reddy, Department of Mechanical Engineering, Texas A & M University, College Station, TX 77843-3123, U.S.A.

Professor K.L. Reifsnider, Department of Engineering Science and Mechanics, VPI & SU Blacksburg, VA24061-0219, U.S.A.

Civ.ing. O. Sigmund, Department of Solid Mechanics, Technical University of Denmark, Building 404, 2800 Lyngby, Denmark

Professor B. Storåkers, Department of Solid Mechanics, Royal Institute of Technology, S-10044 Stockholm, Sweden

Professor P. Suquet, Laboratoire de Mécanique et d'Acoustique, 31 Chemin Joseph Aiguier, F-13402 Marseille Cedex 20, France

Professor S. Suresh, M.I.T., Department of Materials Science & Engineering, Room 13-5056, Cambridge, Massachusetts 02139-4307, U.S.A.

Dr. B.F. Sørensen, Materials Department, Risø National Laboratory, 4000 Roskilde, Denmark

Professor V. Tamužs, Institute of Polymer Mechanics, Latvian Academy of Science, 23 Aizkraukles Street, Riga 22 6606, Latvia

Professor M.F. Thorpe, Department of Physics and Astronomy and Center for Fundamental Material Research, Michigan State University, East Lansing, Michigan 48824-1116, U.S.A.

Dr. S. Toll, Laboratoire de Technologie des Composites et Polymères, EPFL, CH-1015 Lausanne, Switzerland

Professor S. Torquato, Department of Civil Engineering and the Princeton Mathematical Institute, E-307 Engineering Quadrangle, Princeton University, Princeton, NJ 08544, U.S.A.

Professor V. Tvergaard, Department of Solid Mechanics, Technical University of Denmark, Building 404, 2800 Lyngby, Denmark

Adjunkt J.P. Ulfkjær, Institute of Building Technology & Structural Engineering, Aalborg University, Sohngaardsholmsvej 57, 9000 Aalborg, Denmark

Dr. J. Varna, Department of Material & Production Engineering, Luleå University of Technology, S-95187 Luleå, Sweden

Professor J. Weitsman, University of Tennessee, Department of Engineering Science & Mechanics, 307 Perkins Hall, 37996-2030 Knowville, U.S.A.

Dr. E.A. Werner, Institute of Metal Physics, Montan University Leoben, Jahnstrasse 12, A-8700 Leoben, Austria

Civ.ing. K.H. Winter, Institute of Mechanical Engineering, Aalborg University, Pontoppidanstræde 101, 9220 Aalborg, Denmark

Professor Cz. Woźniak, Institute of Fundamental Technological Research, Swietokrzyska 21, 00-049 Warsaw, Poland

Dr. N. Yu, College of Engineering, Science and Mechanics, University of Tennessee, 310 Perkins Hall, Knoxville, Tennessee 37996-2030, U.S.A.

Professor A. Zaoui, Laboratoire de Mécanique des Solides, Ecóles Polytechnique, 91128 Palaiseau Cedex, France

442

Mr. D. Zhao, Department of Civil Engineering, Mechanics and Metallurgy, The University of Illinois of Chicago, 2095 Engineering Research Facility, Box 4348, Chicago, Illinois 60680, U.S.A.

Mechanics

SOLID MECHANICS AND ITS APPLICATIONS

Series Editor: G.M.L. Gladwell

Aims and Scope of the Series

The fundamental questions arising in mechanics are: *Why?*, *How?*, and *How much?* The aim of this series is to provide lucid accounts written by authoritative researchers giving vision and insight in answering these questions on the subject of mechanics as it relates to solids. The scope of the series covers the entire spectrum of solid mechanics. Thus it includes the foundation of mechanics; variational formulations; computational mechanics; statics, kinematics and dynamics of rigid and elastic bodies; vibrations of solids and structures; dynamical systems and chaos; the theories of elasticity, plasticity and viscoelasticity; composite materials; rods, beams, shells and membranes; structural control and stability; soils, rocks and geomechanics; fracture; tribology; experimental mechanics; biomechanics and machine design.

Kluwer Academic Publishers – Dordrecht / Boston / London

Mechanics

SOLID MECHANICS AND ITS APPLICATIONS
Series Editor: G.M.L. Gladwell

Kluwer Academic Publishers – Dordrecht / Boston / London

Mechanics

FLUID MECHANICS AND ITS APPLICATIONS
Series Editor: R. Moreau

Aims and Scope of the Series

The purpose of this series is to focus on subjects in which fluid mechanics plays a fundamental role. As well as the more traditional applications of aeronautics, hydraulics, heat and mass transfer etc., books will be published dealing with topics which are currently in a state of rapid development, such as turbulence, suspensions and multiphase fluids, super and hypersonic flows and numerical modelling techniques. It is a widely held view that it is the interdisciplinary subjects that will receive intense scientific attention, bringing them to the forefront of technological advancement. Fluids have the ability to transport matter and its properties as well as transmit force, therefore fluid mechanics is a subject that is particularly open to cross fertilisation with other sciences and disciplines of engineering. The subject of fluid mechanics will be highly relevant in domains such as chemical, metallurgical, biological and ecological engineering. This series is particularly open to such new multidisciplinary domains.

Kluwer Academic Publishers – Dordrecht / Boston / London

Mechanics

Kluwer Academic Publishers – Dordrecht / Boston / London

Mechanics

From 1990, books on the subject of *mechanics* will be published under two series:
FLUID MECHANICS AND ITS APPLICATIONS
  Series Editor: R.J. Moreau
SOLID MECHANICS AND ITS APPLICATIONS
  Series Editor: G.M.L. Gladwell

Prior to 1990, the books listed below were published in the respective series indicated below.

MECHANICS: DYNAMICAL SYSTEMS
Editors: L. Meirovitch and G.Æ. Oravas

MECHANICS OF STRUCTURAL SYSTEMS
Editors: J.S. Przemieniecki and G.Æ. Oravas

Mechanics

MECHANICS OF ELASTIC AND INELASTIC SOLIDS
Editors: S. Nemat-Nasser and G.Æ. Oravas

MECHANICS OF SURFACE STRUCTURES
Editors: W.A. Nash and G.Æ. Oravas